S0-BLK-849

Knowledge Visualization and Visual Literacy in Science Education

Anna Ursyn
University of Northern Colorado, USA

A volume in the Advances in Educational Technologies and Instructional Design (AETID) Book Series

Published in the United States of America by
Information Science Reference (an imprint of IGI Global)
701 E. Chocolate Avenue
Hershey PA, USA 17033
Tel: 717-533-8845
Fax: 717-533-8661
E-mail: cust@igi-global.com
Web site: http://www.igi-global.com

Library of Congress Cataloging-in-Publication Data

Names: Ursyn, Anna, 1955- editor.
Title: Knowledge visualization and visual literacy in science education / Anna Ursyn, editor.
Description: Hershey PA : Information Science Reference, [2016] | Includes bibliographical references and index.
Identifiers: LCCN 2016010959| ISBN 9781522504801 (hardcover) | ISBN 9781522504818 (ebook)
Subjects: LCSH: Science--Study and teaching--Graphic methods. | Visual learning.
Classification: LCC Q181 .K6135 | DDC 507.1--dc23 LC record available at https://lccn.loc.gov/2016010959

This book is published in the IGI Global book series Advances in Educational Technologies and Instructional Design (AE-TID) (ISSN: 2326-8905; eISSN: 2326-8913)

British Cataloguing in Publication Data
A Cataloguing in Publication record for this book is available from the British Library.

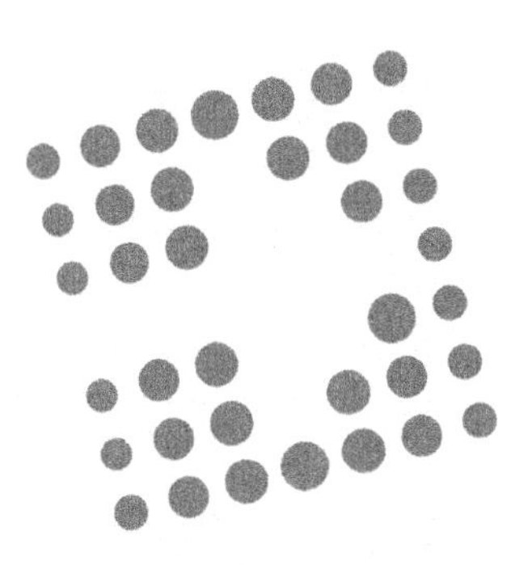

Advances in Educational Technologies and Instructional Design (AETID) Book Series

Lawrence A. Tomei
Robert Morris University, USA

ISSN: 2326-8905
EISSN: 2326-8913

Mission

Education has undergone, and continues to undergo, immense changes in the way it is enacted and distributed to both child and adult learners. From distance education, Massive-Open-Online-Courses (MOOCs), and electronic tablets in the classroom, technology is now an integral part of the educational experience and is also affecting the way educators communicate information to students.

The **Advances in Educational Technologies & Instructional Design (AETID) Book Series** is a resource where researchers, students, administrators, and educators alike can find the most updated research and theories regarding technology's integration within education and its effect on teaching as a practice.

Coverage

- Educational Telecommunications
- Classroom Response Systems
- Virtual School Environments
- Adaptive Learning
- Curriculum Development
- Instructional Design Models
- Bring-Your-Own-Device
- K-12 Educational Technologies
- Collaboration Tools
- Game-Based Learning

IGI Global is currently accepting manuscripts for publication within this series. To submit a proposal for a volume in this series, please contact our Acquisition Editors at Acquisitions@igi-global.com or visit: http://www.igi-global.com/publish/.

The Advances in Educational Technologies and Instructional Design (AETID) Book Series (ISSN 2326-8905) is published by IGI Global, 701 E. Chocolate Avenue, Hershey, PA 17033-1240, USA, www.igi-global.com. This series is composed of titles available for purchase individually; each title is edited to be contextually exclusive from any other title within the series. For pricing and ordering information please visit http://www.igi-global.com/book-series/advances-educational-technologies-instructional-design/73678. Postmaster: Send all address changes to above address.

Titles in this Series

For a list of additional titles in this series, please visit: www.igi-global.com

Wearable Technology and Mobile Innovations for Next-Generation Education
Janet Holland (Emporia State University, USA)
Information Science Reference • copyright 2016 • 364pp • H/C (ISBN: 9781522500698) • US $195.00 (our price)

Creating Teacher Immediacy in Online Learning Environments
Steven D'Agustino (Fordham University, USA)
Information Science Reference • copyright 2016 • 356pp • H/C (ISBN: 9781466699953) • US $185.00 (our price)

Revolutionizing Education through Web-Based Instruction
Mahesh Raisinghani (Texas Woman's University, USA)
Information Science Reference • copyright 2016 • 391pp • H/C (ISBN: 9781466699328) • US $185.00 (our price)

Emerging Tools and Applications of Virtual Reality in Education
Dong Hwa Choi (Park University, USA) Amber Dailey-Hebert (Park University, USA) and Judi Simmons Estes (Park University, USA)
Information Science Reference • copyright 2016 • 360pp • H/C (ISBN: 9781466698376) • US $180.00 (our price)

Handbook of Research on Cloud-Based STEM Education for Improved Learning Outcomes
Lee Chao (University of Houston - Victoria, USA)
Information Science Reference • copyright 2016 • 481pp • H/C (ISBN: 9781466699243) • US $300.00 (our price)

User-Centered Design Strategies for Massive Open Online Courses (MOOCs)
Ricardo Mendoza-Gonzalez (Instituto Tecnologico de Aguascalientes, Mexico)
Information Science Reference • copyright 2016 • 323pp • H/C (ISBN: 9781466697430) • US $175.00 (our price)

Handbook of Research on Estimation and Control Techniques in E-Learning Systems
Vardan Mkrttchian (HHH University, Australia) Alexander Bershadsky (Penza State University, Russia) Alexander Bozhday (Penza State University, Russia) Mikhail Kataev (Tomsk State University of Control System and Radio Electronics, Russia & Yurga Institute of Technology (Branch) of National Research Tomsk Polytechnic University, Russia) and Sergey Kataev (Tomsk State Pedagogical University, Russia)
Information Science Reference • copyright 2016 • 679pp • H/C (ISBN: 9781466694897) • US $235.00 (our price)

Handbook of Research on Learning Outcomes and Opportunities in the Digital Age
Victor C.X. Wang (Florida Atlantic University, USA)
Information Science Reference • copyright 2016 • 851pp • H/C (ISBN: 9781466695771) • US $425.00 (our price)

www.igi-global.com

701 E. Chocolate Ave., Hershey, PA 17033
Order online at www.igi-global.com or call 717-533-8845 x100
To place a standing order for titles released in this series, contact: cust@igi-global.com
Mon-Fri 8:00 am - 5:00 pm (est) or fax 24 hours a day 717-533-8661

Editorial Advisory Board

Table of Contents

Detailed Table of Contents

Section 1
Cognition and Visual Literacy

This section discusses essential notions pertaining to cognitive and visual thinking. As cognitive mental activities support us in acquiring knowledge with the use of our senses, thoughts, and exposure to nature, authors of this section's chapters inquire into the sensory and cognitive functions that seem to be necessary in understanding and learning processes.

This chapter brings about concepts about implementing a framework for teaching and learning across the disciplines by introducing topics and activities pertaining to science, computing, and graphic arts as a unified cognitive and visual learning experience. First, theoretical framework is presented, to support designing integrative projects for cognitive learning. This part provides basic information about brain and mind, feelings and emotions, cognitive thinking, intelligence and cognitive styles, along with some basics about technologies used for studying cognitive activity. Then follows short introduction to ways to communicate knowledge, visual thinking, visual literacy, and knowledge visualization concepts and methods. Next, concerns about science education draw attention to a need of including into curriculum new developments in science and information about currently emerging disciplines. The goal is to enhance technological literacy of students, activate their interest, motivation, abstract thinking, and elicit a wish to achieve their aims.

This chapter examines research from psychology of perception and cognition as well as select developments in the visual arts that inspired the design of the split-brain user interface developed for the interactive documentary Anita und Clarence in der Hölle: An Opera for Split-Brains in Modular Parts (Garvey, 2002). This experimental interface aims at 'enhanced' interaction while creating a new aesthetic experience. This emergent aesthetic might also be described as induced artificial cognitive dissonance and recalls select innovations in the rise of modernism notably the experiments of the Surrealists. The split-brain interface project offers a model for further investigations of human perception, neural processing and cognition through experimentation with the basic principles of stereo and binocular vision. It is conceivable that such an interface could be a design strategy for augmented or virtual reality or even wearable computing. The chapter concludes with a short discussion of potential avenues for further experimentation and development.

Traditionally, the subject of "scientific visualization" focuses on the creation of novel or innovative graphical representations: essentially, new types of images to perceive. A truly complete approach to scientific visualization should include not only the perceived object, but also the abilities of the perceiver. Human "visual common sense" is a product of evolution, suited to the survival of the species; but it has severe and recurring limitations for the purposes of scientific understanding and education. People cannot readily understand phenomena that are too fast, slow, or complex for their visual systems to take in; they cannot see wavelengths outside visual spectrum; they have difficulty understanding three-dimensional (or, even worse, four-dimensional) objects. This chapter explores a variety of ideas and design themes for approaching scientific visualization by enhancing the powers of human vision.

The fourth dimension is a complex concept that deals with abstract reasoning, our sense of perception, and our imagination. Mathematics posits that a four-dimensional space is a geometric space with four dimensions. For many the fourth dimension is the element of time added to the three parameters of length, height and depth. How does a geometer incorporates time in the description of a structure, or a visual artist integrates time in a two dimensional flat surface image, when they both rely on well-defined principles that are a tangible descriptive of our reality? This chapter gives a brief overview of the different schools of thought in the Humanities and in Science, offers a possible definition of this elusive element needed to anchor the fourth dimension in our larger abstract reasoning consensus, and focuses on the specific of Mathematics and visual imaging to illustrate the particular benefit of collaborating on a simple, usable descriptive to create a sound outcome.

This section describes instances of knowledge visualization in selected areas: biology, mathematics, digital media, and music. Authors present their conceptions of visualization with the use of implements used in their domains.

Foreword

In meeting Anna Ursyn the first time, I was struck by the fire of curiosity burning in her eyes, wide open to a universe of inquiry and suspense most don't even know exists. Her persistent sense of underlying connection – a kind of faith in the orderly mystery of the universe – and her relentless effort to seek out and grapple with new information is contagious, which continues to inspire new generations of artists, scientists, and generally philosophical millennia alike. That's the mark of a charismatic teacher!

It seems to me our first discussion embraced the grand paradigm of the last century, that light can be identified as both a point and a wave, existing in more than one place at a time, depending on the expectations of the Observer. Activating that concept as a visual metaphor in the connectivity of knowledge itself is as paradoxical as understanding the Nature of Life, Time, and Meaning altogether. And that is exactly what artists and scientists continue to do. This book, Anna Ursyn's fifth in a series of exciting research compendiums, explores the intriguingly diverse connections between knowledge, language, and visualization. The lambent flame of education is an energizing yet quixotic force, demanding update and renewal at every turn. Thanks to a shift in current curricular research, renewed educational insight will be afforded not by S.T.E.M., but by S.T.E.A.M., the integration of Science, Technology, Art, and Mathematics. Clearly, this work has already begun.

Anna Ursyn's book provides a glittering summary of the key elements in teaching the nature of cognition and visual literacy, while at the same time its very structure illuminates overarching paradoxes in generality vs. specificity, connection vs. disconnection — what Dennis Summers so artfully addresses as the GAPS and SEAMS in grappling with both congruent and dissonant visual information. His erudite treatise on this subject is a resourceful reiteration of the polemics of the 20th century art history of collage, used as a method for sharing analogous information, as well as "creating realties that might not otherwise exist." His discussion serves as an evolving model for the language of visual inquiry and synthesis, in a Digital Age where ingenious tools and enterprising systems are becoming ever more powerful in conveying scientific ideas.

In contrast, Hervé Lehning uses mathematical diagrams and SANGAKU to elicit a parallel conclusion, "That popularization at the levels of teaching and research, visualization is not only a simple illustration. It can help the understanding of some concepts, but can also bring new ideas." The geometric figures and calculations Hervé Lehning chose for his Chapter were eloquent, although within the overall context of this book, the urge to read them in 3D and 5D rather than 2D would not be surprising. Each of the authors in this book explores ideas about knowledge, synthetic relevance, logic, and imagination from within a series of unique esthetic perspectives directed towards an augmentation of visual learning enhancement.

Greg Garvey's surrealistic audio-acoustic art installation… which the reader is not privileged to witness or experience… questions the nature of "noise" vs. congruent meaning in art, gender discrimination, and politics. The artwork itself is primarily left to the imagination, yet provokes the cogent platform for a fascinating, in depth, bio-technical discussion of right and left brain hemisphere inter-functionality. Amazing! And profound! Greg is nothing short of an agent provocateur, not alone in this artful role when assessing and assimilating the diversity of info-authorship represented in this sometimes infuriatingly contrary yet inspired book.

Curious in a book on information visualization, Maura Flannery chose to forego using illustrations. In her aquatic case study, Victorian women are lauded for creating illustrational masterpieces of fungus and seaweed, drawings that reveal truth and beauty as a consequence of pristine line, shape, form and design – elementary esthetic descriptors that serve to separate them from the muck and ooze in which the plants themselves exist. The art collections served as a precedent for establishing scientifically comprehensive taxonomic archives, and at the same time exemplified a sotto voce feminist commentary on the leisure expectations of privileged 19th century women who chose to work as pioneering visionary biologists rather than dabblers in the arts. Wouldn't they be surprised to know that a century later, the infamous Southern California taxonomist John Ljubenkov knew for a fact that, "In a healthy inter-tidal zone, more than 20,000 animals live in every cubic foot of sea mud… so many, that scientists are forced to assign numbers rather than names to new species!" Thanks to macro lenses, Bio-Chemistry has not yet preempted old school biological research.

I cannot pretend to summarize the oft times overwhelmingly fascinating ideas packed into the many provocative Chapters of this book, much less the myriad associations, puzzles, and lesson plans inspired by it. Reading and rereading it is a bit like digesting one of Rudyard Kipling's *Just So Stories*, wherein "You might discover, my dearest and most beloved reader, that the spots on the leopard are better studied if you peek behind that rock in the picture." Contrarian that he was, Kipling himself might ask Mohammad Majid al-Rifaie of Goldsmiths, University of London, UK, or Md Fahimul Islam of Queens College CUNY, NY, USA, "How can one reduce the turbulent Tango to a fascinating, abstract map of footwork in motion, without the pulse of improvisational eroticism, press of flesh, and sudden sweat of sweet temptation?" What will you yourself sense while looking at the Josh Solomon's fascinating dance diagrams? Can the Reader appreciate the stunning mathematical genius and logistical self control required to dance passionately without reading more about the history of Tango itself? So much for *The Butterfly That Stamped*! Not to mention the curious Reader's need to pair such disparately engaging concepts with perhaps a complete investigation and analysis of Mondrian's *Broadway Boogie Woogie*! Or fast forward to more existential questions concerning pattern and movement, perhaps in reference to killer whales and penguins in Antarctica caught in a state of global warming? Further, how might the movement of these creatures be documented IF diagramed as 2D, 3D, 4D or 5D informatics models strung together as mathematical constructs? And, what impact might their extinction have on the Web of Life? Will jellyfish someday rule the Earth? And how would Stephen Hawking's *Theory of Everything* impact that assessment? Say what!

Readers should be advised to have an open laptop anchored nearby while reading this book in whole or part – for grounding – as much to 'see' what IS and IS NOT there! As Theodor Wyeld would agree, framing the unfamiliar with the familiar reduces cognitive overload, while enhancing the joy of learning. But what is not present and not discussed can be as inspiring as that which is! In reading Anna's awesome first Chapter, I found myself continually distracted by questions like "What else?" and "Is there more?" Quick Wikipedia and GOOGLE searches punctuated my enjoyment of the book. Anna's aggregate research is mind bending – the more so for those less up to date. Humble enough, even the 'restless genius', futurist Ray Kurzweiler cannot keep pace with the exponential rate of invention today. How can we? What inventions and leaps of faith will have been made by the time Readers finish the Introduction, much less this book in its entirety? What essential expertise might be added? Near-Infrared Spectroscopy (NIRS)? Event Related Optical Signals (EROS)? KEOPS beyond Kepler? Alien vs. inalienable existential rights?

Educators might well be advised to require Readers (of any age) to prepare Computer Bookmark/ Browsing Histories to share with one another while reading this book — not only to achieve consonance in agreed meaning associated with Key Words, Vocabulary, Expositions, and Conclusions, but to more fully investigate and appreciate tools, concepts, strategies, conditions, applications, and other intriguing data presented for discussion! Information cannot exist without interpretation. Knowledge cannot exist without perception. Creating an educational framework that respects existing language and expertise, yet at the same time inspires mastery, creative growth, and innovation across disciplines is a worthy enterprise, as amply evidenced by this book – and can be fun.

Please look forward to reading the empty pages between the Chapters as much as the Chapters themselves, resting points affording reflection on what has just been said, or what might happen next, a grand opportunity to exercise humility in assessing expression, congruency, and contradiction. The self-avowed Minimalist might find respite in attributes afforded by the emptiness of interleaving pages – flatness, surface, form, shape, scale, mass, texture, reflectivity, absorbency, or other aspects of 'it-ness' devoid of metaphor or suggestion. The Taoist, however, might consider the mystery of emptiness, a space for qi (chi), a cosmological term, which is formless, but bestows life. Mysticism associated with the sacred geometry of numbers and proportions is ancient, reflected in Cabalistic as well as Sanskrit, Classical Chinese, and Greek texts. For centuries, artists were trained in the esoteric mysteries of existence, the mystical and metaphysical, hand in spirit with mathematics, geometry, composition, and design.

Despite our desire to assign particularity in meaning based on materiality, there is the immateriality of meaning itself, its ephemeral nature within the elusive matrix of time, sequence, and duration. The paradox of the moment compels us to wonder about the nature of existence and experience, its measurable reality, its consequence, its place in the continuum. Changes in consciousness provoked by new connections afforded by readers and authors in aggregate are inevitable. Appropriate to those who create something out of nothing, we need to recognize how to recognize and teach these principals. Naivety is one thing, stupidity another. Knowing the difference is fundamental to Perception, Inception, and Deception. The Identity, Ethos, Ethics, and Mind of a new generation depend on it.

Jennifer Grey
California State University, Long Beach, USA

Jennifer Grey *is an Art Professor, awarded Emerita status after working at California State University Long Beach since 1975, a premiere urban campus in the USA. Over time, her career focused on Drawing and Painting in any media, from works of art on rock, paper, canvas, public buildings, intertidal zones, desert salt flats, computer screens, time-based projections, and 5D immersive C.A.V.E. painting – an old school artist bridging curricular reform in studio and electronic art. She is a professional artist, freelance writer, curator, producer, and Native American flute musician working under a variety of more or less secret avatars and pseudonyms in Real Life (RL) and in Virtual Reality (VR), as needed to maintain a certain degree of contemplative privacy. Since 1999, however, JEN ZEN® became the principal identity exhibiting internationally, including a series of* *groundbreaking exhibitions with ACM SIGGRAPH and D'Art Conferences; a one-person show at The Museum of Modern Art in Dalian, China; and, a less auspicious yet more experimental solo show on Museum Island in Second Life. She has received many art grants, awards and public commissions including the National Endowment for the Arts Award to Individual Artists, the Fulbright Teacher Exchange, and Creative Services Grants, California State Legislative Grants, and the Professional Artists Fellowship sponsored by Public Corporation for the Arts of the City of Long Beach. Education: 1969-71 James Scholar, Graphic Design, University of Illinois, Urbana, USA 1971-73 BFA Summa Cum Laude, Bradley University, Peoria, Illinois, USA 1973-75 MFA, Hoffberger School of Painting, Maryland Institute College of Art, Baltimore, USA.*

Preface

The book entitled "Knowledge Visualization and Visual Literacy in Science Education" is a collection of essays written by science related professionals, specialists in selected areas, and digital artists inspiring themselves with science. The book comprises visual and verbal information introducing the work done by individual authors – contributors to this book. Chapters pertain to several scientific fields, digital art, computer graphics, and new media. Co-authors discuss the possible ways that visuals, visualization, simulation, and interactive knowledge presentation can help understand and share the content of scientific thought, research, and practice. Works created by the co-authors, students, and professional artists illustrate their views and concepts.

CHALLENGES AND POSSIBLE SOLUTIONS

The reason for writing this edited book was to create a first-hand, innovative handbook about presenting in visual way the selected science domains. This book comprises visual and verbal information pertaining to particular scientific fields, digital art, programming, computer graphics, and new media. Concepts and processes related to science have been introduced for teachers and learners through graphical interpretation and explanation of science-based occurrences, processes, products, and facts. Physics has been examined at the macro-, micro-, and nanoscale level, for example by telling about the nanotechnology related tools such as biophotonics. Mathematics can be seen through the natural world (Fathauer, 2016); math proofs, works, and animations can be presented in a visual and tactile way without words and numbers (Gurke, 2016; Li, 2016). In this book, knowledge about the events and processes is integrated with visual, literary, and musical achievements related to the theme, facts about human existence on our planet, and the actual global issues.

Concepts and processes are presented as learning projects based on these materials. Information about several fields of science and teaching strategies with visual approach draws from recent developments in the cognitive approach to teaching and learning. A visual approach to learning became necessary because science, education, and media communication unify pictorial and written data. Information is often displayed with the use of knowledge visualization techniques, when computers transform data into information, and visualization converts information into picture form.

The book may serve as a resource book for cognitive learning with the use of visualization, programming, computer graphics, new media, and digital art disciplines. Materials for learning can be used as a resource for art history, scientific visualization, and technology-related education. Topics and projects presenting life and nature related disciplines in terms of concept visualization draw from the biology-inspired events existing in real life, but familiarize the readers with more complex disciplines and their applications and specialty pursuits.

A rationale for writing on integrative approach to the book's themes comes from a great need for the inclusion into instruction the familiar concepts and applications used everyday by students, like computers, communication media, TV, multimedia, online protocols, and games. Every medium can be put in service of teaching as the container for stories or narratives, with plots and characters, which allows packaging information into various forms. It seems reasonable to assume that merging verbal and visual ways with communication media makes the central part in scientific visualization and simulation, virtual reality environments, web based environments, web graphics, game design, visualizations of big sets of data, semantic web, data mining, and many other tasks and areas of interest.

Teaching the Science: Art Context with Integrative Approach

The book comprises chapters contributed by cooperating specialists who share their knowledge and insight about communication and learning with the use of relevant graphics, music, and text. Their chapters may serve as an in-depth, timely reference source; they are targeted towards the interests of instructors, fellow researchers, and students, for course adoption, and for university libraries as reference materials.

Effective education involves technology-based communication between teachers and students comprising the verbal, auditory, and visual modes. The multi sensory approach became needed because of the importance of knowledge visualization in teaching and learning, the existing multimodal input from online resources, and the multimedia content of interactive textbooks and online educational materials involving text, audio, still images, animation, and video footage.

Traditional schooling had been focused on developing memory skills. At present, a necessity of coping with large amount of information creates a need to expand abilities of higher order thinking, visualization, and understanding of abstract concepts. This model of learning supports learning by visualization of concepts and shifts the college and high school students' attention toward imaging processes and products involved in an event under study. It thus stimulates thinking about links between concepts rather than rote-based learning. That means this book develops the functional and conceptual ways of learning, instead of mechanical repetition of concepts to be learned by rote. Research results suggested that students working in a group that visualized concepts scored better at the geology final tests and the lab tests (Ursyn, 1997). In addition, drawings may reveal what are the students' preconceptions.

Concerns about Science Education

Current disciplines and the fields of study are necessarily based on collaboration of specialists within different disciplines (such as in the domain of archeology) and of people with different types of talents (e.g., in film industry or big entertainment companies such as Walt Disney Company). Hence, division of specific subject areas such as physics, chemistry, or biology, which is practiced in the high school curricula, is not supportive of the scientific concepts' understanding, nor it is encouraging the learners to further their studies in the more specific but integrative areas such as biochemistry, geophysics, or astronomy.

Many disciplines (such as medicine where a basic background knowledge of chemistry, physics, biochemistry, and physiology is needed) remain almost totally unknown and unexplored for high school students who look for orientation about further studies or a job that would fit their talents and interests. This is so because school curricula do not include these topics into the subject matter under study.

Teaching science with an integrative approach and making learning topics closer to future professional tasks may augment students' motivation and interest in learning. Developments in new materials result in developments in ubiquitous computing and wearable apps, with huge educational implications. Teaching with the use of current applications and devices may arise motivation and support students' wish for achievements. Leonard Sax (2009) discusses factors driving the growing epidemic of unmotivated boys and underachieving young men. This author considers five factors: video games, teaching methods, prescription drugs (especially medications for ADHD – attention deficit/hyperactivity disorder), environmental toxins (which he regards as the endocrine disruptors), and devaluation of masculinity being the major agents that cause the decline in numbers of motivated learners and future achievers.

Along with studies on recent advances in the selected fields of science, the Handbook contains learning projects for the readers, both the learners and their instructors, especially teachers from the school districts that adhere to the STEM or STEAM programs. Learning projects offered in this book would comply with the actual trends in education described as the STEM (science, technology, engineering, and mathematics) and the STEAM (science, technology, engineering, arts, and mathematics) programs in education. The book provides background information related to selected projects, and then offers integrative learning assignments designed in the spirit of the STEAM education.

Many educators have been working toward transforming STEM into STEAM as a framework for teaching across the disciplines. A growing number of schools move from STEM to STEAM (STEM to STEAM, 2014; Gonzalez & Kuenzi, 2012). The STEM fields of study are becoming the STEAM fields in education (science, technology, engineering, arts, and mathematics) (Maeda, 2012). This trend results from the presence of technological literacy in small children who can now use software teaching them science, mathematics, programming, and storytelling. Not only we should integrate scientific, technical, and artistic topics, but also construct, envision, and explain through visuals and interactivity (Honey, Pearson, & Schweingruber, 2014). Involving programming and solving problems through knowledge visualization should enhance programs for Studio Art. Students should learn, create, and play together globally through games, VR, apps, social networking, and meaningful communication, K through PhD (DeVry University, 2014). The innovative ideas and timely research will create beneficial contribution to the research community.

Developing imagination may happen due to digital animation and filming techniques that make that the impossible actions of Superman or Batman look real. However, fantasy and imagination presented in digitally created stories rarely surpass accomplishments of the pre-computer works such as Gulliver's Travels (1726) by Jonathan Swift, Frankenstein (1818) by Mary Shelley, Alice's Adventures in Wonderland (1865) by Lewis Carroll, The Lord of the Rings series (1954-55) by John Ronald Reuen Tolkien, and fictional characters such as Count Dracula, Nosferatu, Dr Jekyll and Mr Hyde (1886, by Robert Louis Stevenson), Dorian Grey (1891, by Oscar Wilde) or The Little Prince (1943, by Antoine de Saint-Exupéry). It is not surprising that authors of comic books, games, animations, and movie scenarios remake these works in many ways, with the results that overshadow many other attempts.

ORGANIZATION OF THE BOOK

The book is organized in four sections that are introduced shortly below.

Section 1, entitled *Cognition and Visual Literacy* comprises four chapters. In Chapter 1, *Teaching and Learning Science as a Visual Experience* Anna Ursyn discusses teaching and learning across the disciplines by introducing topics and activities pertaining to science, computing, and graphic arts as a unified cognitive and visual learning experience. Theoretical framework is presented, to support designing integrative projects for cognitive learning, and then the ways to communicate knowledge, visual thinking, visual literacy, and knowledge visualization. Concerns about science education draw attention to a need of including into curriculum new developments in science and information about currently emerging disciplines. Greg P. Garvey examines in Chapter 2, *Exploring Perception, Cognition, and Neural Pathways of Stereo Vision and the Split Brain Human Computer Interface* research from psychology of perception, cognition, and the visual arts, and describes the experimental the split-brain user interface. Chapter 3, *Better Visualization Through Better Vision* by Michael Eisenberg discusses the developing of animated simulations, interactive interfaces to large information spaces, or embedding aural cues within diagrams as the way to improve scientific visualization. Jean Constant offers in Chapter 4, *Visual Approach to the 4th Dimension in Mathematics, Computing and Art* an overview of the different schools of thought in the humanities and science about the concept of 4th dimension, especially in mathematics and visual imaging.

Section 2, *Visual Communication and Knowledge Visualization* describes instances of knowledge visualization in selected areas: biology, mathematics, digital media, and music. Authors present their conceptions of visualization with the use of implements used in their domains. Maura Flannery describes in Chapter 5, *Visualization in Biology: An Aquatic Case Study* algae – photosynthetic organisms that live in aquatic environments, as a good subject to explore the intersection of art and science with students, and discusses observation, comparison, and drawing as key tools in learning about the living world. In Chapter 6, *Visualisation and Communication in Mathematics*, Hervé Lehning offers the ways of communication in mathematics using the aesthetic drawings to set a problem, to understand it, to find a method to tackle it and to illustrate it. In Chapter 7, *Understanding Collage Strategy as a Learning Method and Its Use in Digital Media* Dennis Summers analyzes and defines collage in some of its many forms including artists' books, cinematic film, and digital media; he reviews the use of digital software and the pedagogical implications of collage. Robert C. Ehle examines in Chapter 8, *How we Hear and Experience Music* the occurrences and events associated with composing, playing, or listening to music, virtual music, and then recounts an experiment on the nature of pitch and psychoacoustics of resultant tones. Then he discusses the prenatal origins of musical emotion as the case for fetal imprinting.

Section 3, Computing and Programming delves into selected methods aimed at assisting the learners in acquiring computing and programming skills with the use of video tutorials and metaphorical visualization. In Chapter 9, *Using Video Tutorials to Learn Maya 3D for Creative Outcomes: a Case Study in Increasing Student Satisfaction by Reducing Cognitive Load*, Theodor Wyeld describes the transition from traditional front-of-class software demonstration of Autodesk's Maya 3D to the introduction of video tutorials, and finds that the introduction of video tutorials for learning Maya 3D reduced frustration and freed up time for more creative pursuits. Anna Ursyn, Mohammad Majid al-Rifaie, & Md. Fahimul Islam offer in Chapter 10, *Metaphors for Dance and Programming: Rules, Restrictions, and Conditions for Learning and Visual Outcomes* an overview of programming with ready to follow codes, and visual explanation on how to code using dance as a metaphor.

Section 4, *Educational Applications and Cognitive Learning* provides theoretical and practical materials supporting teaching and learning science. In Chapter 11, *Optimizing Students' Information Processing in Science Learning: A Knowledge Visualization Approach*, Robert Zheng & Yiqing Wang discuss a theoretical framework for designing effective visual learning in science education, and identify the variables in visual learning and the instructional strategies that aim at the improvement of visual performance. Chapter 12, *Integrative Visual Projects for Cognitive Learning* by Anna Ursyn comprises integrative studies on selected processes, events, and related technologies associated with several science categories, and offers learning projects designed around the themes derived from the nature and science in terms of concept visualization, including selected subjects pertaining to physics, chemistry, biology, geography, and biology-inspired computing and modeling. Donna Farland Smith, & Kevin Finson V*isualizing Scientists & Engineers: Uncovering How Children View Science Related Careers and it's Importance to Science Identity* discuss in Chapter 13 multiple tools for analyzing children's representations of scientists and engineers.

REFERENCES

DeVry University. (2014). Web Game Programming Degree Specialization. Retrieved April 3, 2014 from http://www.devry.edu/degree-programs/college-media-arts-technology/web-game-programming-about.html

Fathauer, R. (2016). Artist statement in Fathauer, R., & Selikoff, N. (Eds.), *2016 Joint Mathematics Meetings Exhibition of Mathematical Art*. Tesselation Publishing.

Gonzalez, H. B., & Kuenzi, J. J. (2012). *Science, Technology, Engineering, and Mathematics (STEM) Education: A Primer.* Congressional Research Service. Retrieved July 5, 2014, from http://fas.org/sgp/crs/misc/R42642.pdf

Gurkewitz, R., & Arnstein, B. (2016). Artist statement in Fathauer, R., & Selikoff, N. (Editors*), 2016 Joint Mathematics Meetings Exhibition of Mathematical Art*. Tesselation Publishing.

Honey, M., Pearson, G., & Schweingruber, H. (Eds.). (2014). *STEM Integration in K-12 Education: Status, Prospects, and an Agenda for Research. National Academy of Engineering and National Research Council*. Washington, DC: The National Academies Press.

Li, H.-L. (2016). Artist statement in Fathauer, R., & Selikoff, N. (Eds.), *2016 Joint Mathematics Meetings Exhibition of Mathematical Art.* Tesselation Publishing.

Maeda, J. (2012). STEM to STEAM: Art in K-12 is Key to Building a Strong Economy. *Edutopia*. Retrieved July 5, 2014, from http://www.edutopia.org/blog/stem-to-steam-strengthens-economy-john-maeda

Sax, L. (2009). *Boys adrift: the five factors driving the growing epidemic of unmotivated boys and underachieving young men*. Basic Books.

STEM to STEAM. (2014). Retrieved July 5, 2014, from http://stemtosteam.org/

Ursyn, A. (1997). Computer Art Graphics Integration of Art and Science. *Learning and Instruction. The Journal of the European Association for Research on Learning and Instruction*, *7*(1), 65–87. doi:10.1016/S0959-4752(96)00011-4

Acknowledgment

The editor of this book wishes to thank many individuals for their input and help with this book:

First, I would like to thank the members of the IGI-Global publishing team: Erika Carter, Vice President, Editorial and Lindsay Johnston, Acquisitions Editor for inviting me to work on the book; Editorial Director, for setting up the project; Jan Travers, Agreement Facilitator for legal help; Janine Haughton, Developmental Editor–Books for working on Editorial Content, all and all the team members for their kind help and cheerful, personal assistance with this project.

I would like to thank Art Professor Jennifer Grey, California State University, Long Beach for writing a Foreword and all the Co-authors who contributed to this book by writing chapters and providing images illustrating their projects, for their cooperation, in order as they appear in the book: Gregory P. Garvey, Michael Eisenberg, Jean Constant, Maura C. Flannery, Hervé Lehning, Dennis Summers, Robert C. Ehle, Theodor Wyeld, Mohammad Majid al-Rifaie, Md Fahimul Islam, Robert Zheng, Yiqing Wang, Donna Farland-Smith, and Kevin D. Finson. Also, many thanks go to my students who contributed to this book by providing images illustrating their projects.

Many thanks go to the Reviewers who diligently provided supportive suggestions and critiques, and thus helped to make each part of the book better: Mohammad Majid al-Rifaie, *Goldsmiths, University of London, UK*; Jean Constant, *Hermay.org*, Switzerland, *Santa Fe, NM, USA;* Donna Farland-Smith, *Ohio State University, USA*; Maura C. Flannery, *St. John's University, NY;* Hervé Lehning, *AC-HL, France;* Terry Scott, University of Northern Colorado, USA; Dennis Summers, *Strategic Technologies for Art, Globe and Environment, USA;* Ying Tan, University of Oregon, Theodor Wyeld, *Flinders University, Australia;* and Robert Zheng, *University of Utah, USA.*

I wish also to thank the members of the book's Advisory Board for their help with working on this book: Mohammad Majid al-Rifaie, *Goldsmiths, University of London, UK*; Donna Farland-Smith, *Ohio State University, USA*; Maura C. Flannery, *St. John's University, NY*; Hervé Lehning, *AC-HL, France*; Dennis Summers, *Strategic Technologies for Art, Globe and Environment, USA*; Theodor Wyeld, *Flinders University, Australia.*

Also I want to express my gratitude to my family and friends.

Section 1
Cognition and Visual Literacy

This section discusses essential notions pertaining to cognitive and visual thinking. As cognitive mental activities support us in acquiring knowledge with the use of our senses, thoughts, and exposure to nature, authors of this section's chapters inquire into the sensory and cognitive functions that seem to be necessary in understanding and learning processes.

Chapter 1
Teaching and Learning Science as a Visual Experience

Anna Ursyn
University of Northern Colorado, USA

ABSTRACT

This chapter brings about concepts about implementing a framework for teaching and learning across the disciplines by introducing topics and activities pertaining to science, computing, and graphic arts as a unified cognitive and visual learning experience. First, theoretical framework is presented, to support designing integrative projects for cognitive learning. This part provides basic information about brain and mind, feelings and emotions, cognitive thinking, intelligence and cognitive styles, along with some basics about technologies used for studying cognitive activity. Then follows short introduction to ways to communicate knowledge, visual thinking, visual literacy, and knowledge visualization concepts and methods. Next, concerns about science education draw attention to a need of including into curriculum new developments in science and information about currently emerging disciplines. The goal is to enhance technological literacy of students, activate their interest, motivation, abstract thinking, and elicit a wish to achieve their aims.

INTRODUCTION

The ways of cognitive thinking and learning are changing along with the advancements in instructional technologies, but the notions of creativity, talent, problem solving, aesthetics, and beauty retain their value. Education is often perceived less valued now, when ready solutions are available online, countless students are learning using YouTube, and we may observe the lowering interest in studying at the universities. However, due to the access to information people untrained for example in musicology, computer science, or mechanical engineering may work and produce collectively. Materials and learning projects encourage the use of senses and enhance a feeling of trust in one's own abilities while creating visual solutions.

Advances in technologies provide science education with tools for applying knowledge visualization and enhancing visual literacy. The developments in new materials, the resulting developments in

DOI: 10.4018/978-1-5225-0480-1.ch001

ubiquitous computing, wearable apps, and the use of new materials in architecture and design, all offer captivating educational implications and possibilities. Projects combining science, technology, and art offer a response to the economical and social demands for the arts-based development training, which became used in corporations. The art theorist and perceptual psychologist Rudolf Arnheim posed that some of the objectives attributed to art are means of making visual thinking possible (Arnheim, 1969/2004, p. 254). Visual approach to learning facilitates the comprehending of the core concepts in programming for art, web, and everyday applications. There is a trend to include the arts in business supporting teambuilding, communication, and leadership. Tools for enhancing visual literacy and thus supporting learning about science may comprise toys, games, puzzle, apps, models (often involving 3D printing), animations, simulations (automatic pilot, control tower), and augmented reality environments, among other solutions.

THE FRAMEWORK

Brain, Mind, and Cognition

The process of learning depends on our attempts made by a conscious and curious mind, which has nonphysical nature hinging on the functioning and physiology of a brain. The brain makes a mind using neurons, which are the specialized cells, the basic brain structures for their activity. Neurons, mostly located in the central nervous system send signals to other neurons, muscles, different cells, and the outside world. Signals going outside and toward the brain are transmitted and transformed through synapses, the specialized areas at the ends of the neuronal extensions called axons. There are more than 100 billions of individual nerve cells (Kandel & Schwartz, 2012) and trillions of the synaptic contacts in the brain (Damasio, 2012). Large neural networks act as neural circuits consisting of smaller circuits. Patterns of the large networks' activity change briefly according to signals going from the body, the external world, and from the patterns of other circuits in the brain. Our environmental space may mean different things for different people (Figure 1). It can be pragmatic (related to where we live), perceptual (showing what we experience), existential (introducing social and cultural issues), cognitive, (based on thinking), and logical (abstract) space.

As a neuroscientist and psychologist David Marr put it, the mind has access to systems of internal representations: "mental states are characterized by asserting what the internal representations currently specify, and mental processes by how such internal representations are obtained and how they interact" (Marr, 1982/2010, p. 6). The brain forms maps, which are concrete or abstract images (visual, auditory, visceral, tactile, etc.) that represent patterns resulting from objects and events. Cognitive mapping means mental transformations, to arrange information in everyday spatial environment. This is not a map but a metaphor, a process rather than product, by making routes, network and metrical descriptions of relative positions. They may become subject of conscious experience (Cavanna & Nanni, 2014). Imagery is the image-making function of the mind. Maps can reflect the physiological condition of the body tissues and organs (interoceptive maps that provide feelings such as pain, hunger, temperature, itch, or visceral sensation; Craig, 2003), state of the organism's skeletal muscles, tendons, and joints (proprioceptive maps) and the environment external to the organism detected by senses (exteroceptive maps). As for external signals, our perception relies on imagery, which makes that sensations (sounds, shapes, colors and motions) convey meaning. For example, the rustle of leafs means danger for a small animal (Brody,

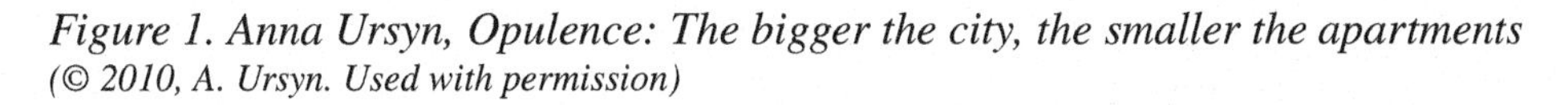

Figure 1. Anna Ursyn, Opulence: The bigger the city, the smaller the apartments
(© 2010, A. Ursyn. Used with permission)

1987). There is also haptic perception recognizing objects by the sense of touch, along with primitive relationships that build topological space, such as proximity and separation, order and enclosure (or surrounding), and continuity.

Spatial cognition is a developmental process; it depends on the child's cognitive level. Spatial skills and abilities mean competence in spatial visualization or orientation. Causes for variation in spatial skills can be age-related, sex-related, genetic, hormonal, neurological, and induced by the environmental influences. Sociocultural influences act in favor of males, e.g. by the choice of toys and plays, Lego, games, and blocks. It is possible to improve one's spatial abilities, especially if somebody is at formal operational level of cognitive development. Several years ago the U.S. Employment Service listed over 80 occupations with spatial ability being important for the achievement in a career. Practicing freehand drawing may improve spatial skills. However, somebody's high spatial visualization and orientation do not insure drawing ability and does not lower the "drawing barrier" between one's mental representation and what is actually drawn (Arnheim, 1969/2004).

Feelings and Emotions

Antonio Damasio (2012) described changes in intensity and frequency of firing in small neuron circuits. He pointed up that some circuits are synchronously activated to meet some conditions of network connectivity. According to Damasio, "the size and complexity of neural networks scale up to produce cognition and feeling," for example, from a single muscle cell to the whole muscle. The brain thus monitors the interior and exterior environment, produces neural foundations of our minds, and makes them conscious (Damasio & Carvalho, 2013, p. 151). While maps respond to internal and external events, mind combines the actual and recalled images; they may be abstract such as those active in composing music or computing. Mindfulness – focus of one's emotions, thoughts and sensations occurring in the present moment depends on improvements of attentional functions and cognitive flexibility, and can be trained by meditation practices (Moore & Malinowski, 2009). Spontaneous cognition, for example involuntary musical imagery providing the everyday experience of having music in one's head is recruiting brain networks involved in perception, emotions, memory, and spontaneous thoughts (Farrugia, Jakubowski, Cusack, & Stewart, 2015).

Antonio Damasio's (1999) research about the neurophysiology of mind and behavior was focused on understanding of brain processes such as memory, language, emotions, and decision-making. His studies on the relations of cognition and emotions were based on imaging techniques, as attitudes and behaviors became more understandable with the advent of these techniques. Damasio compared human behavior to horse driving: we have to control horses when each of them tries to take our carriage in different directions. He wrote about the neural basis for the emotions and the central role of emotions in cognition and decisions, "Pain and pleasure are the levers the organism requires for instinctual and acquired strategies to operate efficiently. In all probability they were also the levers that controlled the development of social decision-making strategies. … Feeling can serve as barometers of life management" (Damasio, 2012, p. 60). To become conscious, the brain needs subjectivity, the feeling that pervades the images we experience subjectively (Damasio, 2012, p. 10). According to the author, research results show that while neocortex performs reasoning and analysis, it is wired up through the old biological brain (Damasio, 1999). Thus, as stressed by Ralph Lengler, "emotion leads to action while reason leads to conclusion; emotions and feelings will always be formed precognitively and preattentively before any information processing takes place" (Lengler, 2007, p. 384). Emotional component in the arts can be seen in several media: in abstract expressionistic paintings, sculptures, dramatic performances, stage choreography, acting, improvisation, and other media allowing individual or group expression. Finding entertainment in social networking builds a need for connectivity and addiction to social media. Online relationships can be just as real as those conducted offline. Those who tweet many hours a day, when suddenly deprived of an access to the Internet, get a feeling as if they were without their friends and family.

Cognitive Thinking

Consciousness is defined in reference books as the quality or state of being aware, especially of something within oneself: being characterized by sensation, emotion, volition, and thought. This is the upper level of mental life of which the person is aware as contrasted with unconscious processes (Merriam-Webster, 2015). The central nervous system has several main parts: the spinal cord, the brain stem with the medulla oblongata, the pons, and the midbrain, the cerebellum, the diencephalon, and the cerebrum with two cerebral hemispheres (Kandel, Schwartz, Jessell, Siegelbaum, & Hudspeth, 2012). However, consciousness in the brain cannot be referred to a single brain structure. It is compounded of many activities, in a similar way as in a team sport or a philharmonic orchestra. Consciousness depends on the presence of a feeling of knowing one's self, in addition to basic mind processes. Humans have both a core self and an autobiographical self. According to Damasio (2012), animals that may have both kinds of self are wolfs, apes, marine mammals, elephants, cats, and dogs. Mind's consciousness results first from activity of the brain stem, which gathers information about the internal state of the body and provides a basic level of feelings. Then, other parts of the brain, including cerebral cortex, build further features of cognition. The self-memory system, which comprises the working self, autobiographical memory, and episodic memory, takes place in the remembering-imaging system. According to Conway & Loveday (2015) all memories are to some degree false and the main role of memories lies in generating personal meanings.

Cognitive thinking involves activities of the brain aimed to seek out information, select potentially useful information from the total, organize it into a memory store, and retrieve information from memory for use in decision making. This may involve perception, memory, comprehension, decision-making, problem solving, and reasoning; they all may occur at the individual or the group level. Information is compared in the brain to something previously learned, stored in memory, and organized in a cognitive structure. Information is transformed into codes to be transferred to long-term memory: an auditory or semantic code with verbal language rules; a visual or iconic code. Rudolf Arnheim wrote in 1969,

By "cognitive' I mean all mental operations involved in the receiving, storing and processing of information: sensory perception, memory, thinking, learning ... therefore I must extend the terms "cognitive" and "cognition" to include perception ... The cognitive operations called thinking are not the privilege of mental processes above and beyond perception but the essential ingredients of perception itself. I am referring to such operations as active exploration, selection, grasping of essentials, simplification, abstraction, analysis an synthesis, completion, correction, comparison, problem solving, as well as combining separating, putting in context. ... These are the manner in which the minds of both man and animal treat cognitive material at any level (Arnheim, 1969/2004, p. 13).

It has been already widely accepted that cognitive thinking involves both short-term and long-term memory. The working (short-term) memory is a temporary scratch pad for conscious thoughts and feelings. Short-term memory contains a sensory storage; it holds pictures, smells, and sounds. Information about objects is depicted and manipulated in a visual buffer or a working memory – a mental space for manipulating, scanning, and inspecting visual images; we may redraw maps from memory. Visual buffer has limited resolution; it fades if not refreshed, and may draw from the long-term memory. It can be rotated and scaled at will. The attention window selects a region within the visual buffer for a detailed further processing. The size of the window in the visual buffer can be altered and it can be shifted. People can scan visual mental images, even when their eyes are closed. For example, you may want to close your eyes and imagine an indoor or outdoor place where you liked to play when you were ten year old. Then, you may scan to portions of this place that initially were 'off screen' and then draw what you have just 'seen' as a series of sketches like a short storyline for animation.

Semir Zeki (1993, 1998, 2001, 2009, 2011) explored the domain of neuroaesthetics, a study of the neural basis of artistic creativity and achievement, while examining brain activity associated with the elementary perceptual process: perception of images. He arrived at conclusion, "artists are, in a sense, neurologists who unknowingly study the brain with techniques unique to them." The visual brain functions in search for dependable qualities to obtain knowledge about the external world (Zeki, 1999). Our survival in the world may thus strongly depend on the accuracy and completeness of mental models used by our mind to represent real life. Cognitive thinking plays the decisive role in communication through art (but there is no Museum of Cognitive Art yet). Computer art images are helpful in knowledge comprehension. When drawing computer-generated graphics, one can notice topological relations – qualitative characteristics of spatial arrangement, properties, and configurations. Margaret Boden (2016), a researcher in artificial intelligence, psychology, philosophy, cognitive, and computer science stressed that new art was associated with positive fascination for technology; many times digital artwork production involves electrical engineering and/or electronic technology.

Rudolf Arnheim (1990) posed that the sensory system is a primary resource in cognitive life. According to this author, when we think about what we have just been reading and feel pity for fictional characters, we absorb through our senses a printed sequence of letters, and then convert them into vivid mental experiences and potent emotions. Imaging process, such as inspecting an object or memorizing something lights up specific areas of the brain. The long-term memory system allows for making comparisons. New links are created between neurons and old links are strengthened in this process, augmenting neural networks. Old memory must be used in order to comprehend a word and support he language organization: information flows from visual (reading) and auditory (listening speech) reception to areas into the left temporal lobe for comprehension and then on to frontal areas for speech production. Imagery plays an indirect role in making associations. Association is a mental connection between thoughts, feelings, ideas, or sensations. We can remember (or even imagine) a feeling, emotion, or sensation, which is linked to a person, object, or idea. Imaging techniques provide pictures of sites where the working memory is in action. For example, it is possible to record how the short-term and the long-term memory are involved in recognizing an image of a familiar face.

According to Arnheim (1988), gravity makes the space asymmetrical, not in a geometrical but in dynamical sense, because an upward movement requires energy, whereas downward movement can be done by removing any support that keeps an object from falling. We perceive this asymmetry by two senses, with kinesthesia (awareness of the tension in the muscles and joints of the body) and vision.

Intelligence and Cognitive Styles

Cognitive thinking can be only indirectly linked to the concept of intelligence, which involves studies of the behavior of intelligent organisms or intelligent programs and their ability to perform intellectual tasks. People are behaving intelligently when they choose courses of action that are relevant to their goals, reply coherently and appropriately to questions that are put to them, solve problems, and create or design something useful, or beautiful, or novel.

Individual peak performance may depend on the differences in abilities that have been described for example, by Joy Paul Guilford (1950, 1959, 1967, 1968), Robert Sternberg (2007, 2011), and Howard Gardner (1993, 1997). Gardner defines intelligence as the capacity to solve problems or make things that are valued in a culture (at least one cultural setting or community).

The multiple intelligence theory (Gardner, 1983/2011, 1993/2006) is a system of classifying human abilities and a study about encouraging learning in ways that respect individual interests and strengths. Human cognitive competence can be described in terms of abilities, talents, and mental skills, which we call intelligences.

Howard Gardner described distinct forms of intelligence. His list of intelligences include

- **Linguistic Intelligence:** The ability to use language to express meaning, tell a story, react to stories, learn new vocabulary or languages. Poets exhibit this ability in its fullest form.
- **Logical/Mathematical Intelligence:** Prized in schools and especially in school examinations
- **Spatial Intelligence:** The ability to form a mental model of a spatial world and be able to carry out that model, and find one's way around a new structure.
- **Musical Intelligence:** Capacity to create and perceive musical patterns
- **Bodily Kinesthetic Intelligence:** The ability to use the body or parts of the body (hands, feet, etc.) to solve problems or to fabricate products, as in playing a ballgame, dancing, or making objects with the hands. Dancers, athletes, surgeons and crafts people all exhibit highly developed bodily-kinesthetic intelligence.
- **Interpersonal Intelligence:** Oriented toward the understanding other people: what motivates them, how they work, how to work cooperatively and effectively with them. Successful marketing and sales people, politicians, teachers, clinicians, and religious leaders are all likely to be individuals with high degree of interpersonal intelligence.
- **Intrapersonal Intelligence:** The ability turned inward to form an accurate model of oneself, understand things about oneself: how one is similar to or different from others, and how to soothe oneself when sad and use that model to operate effectively in life.
- **Naturalistic Intelligence:** The apprehension of the natural world as epitomized by skilled hunters or botanists. The ability to recognize species of plants or animals in one's environment.

Gardner (1997) carried also the investigation on extraordinary people, and described four forms of extraordinary individuals:

- A master who gains complete mastery over one or more domains of accomplishment,
- A maker who devotes energy to the creation of a new domain,
- An introspector who explores his/her inner life, and
- An influencer who is often a leader influencing other individuals.

Cognitive styles may represent the individual modes of remembering, thinking, decision-making, and problem solving. Individual cognitive styles, which usually act in an unconscious manner, refer to the ways in which one receive, analyze, process, store, retrieve, and transmit information. The Myers-Briggs Type Indicator or MBTI (Scholl, 2001) is often used as the measure of cognitive style. It assesses cognitive style on four dimensions:

- Extraversion (E) versus Introversion (I).
- Sensing (S) versus Intuition (N).
- Thinking (T) versus Feeling (F).
- Judging (J) versus Perceiving (P).

This instrument measures one's cognitive preferences but not the cognitive skills. Its accuracy depends on one's honesty, often honesty with oneself, and the frame of reference that affects the score, including factors such as work or family and social environment (Scholl, 2001).

Figure 2 shows a work of Computer Graphics student at the University of Northern Colorado, Emrey Winter. The work entitled 'Bird's Eye View' was created in a 3D program Maya with an emphasis placed on textures. After selecting a point on a map students were drawing an orthographical depiction of that place, as if they'd ride a bike, or a car. Then they changed the view to the bird's eye view. They focused on the relations of particular events to the whole theme, and the ways of how previous circumstances could affect the next occurrences. This exercise resulted in this almost abstract image. The artist focused on the use of real texture in a somewhat stylized context. The texture of the grass and a road provide a peaceful juxtaposition of nature and the man made structure. Bird's eye view may be useful for constructing a timeline for some events, or for a city planner who decides on an architectural design of an urban complex along with its surrounding green environment. This kind of view may be also helpful for producers in defining their ideas and visions for a movie.

Technology Used for Studying Cognitive Activity

Researchers use the imaging techniques to study the regions of the brain that are involved with cognition.

Imaging techniques provide pictures of sites where the working memory is in action. Methods for researching imagery may include physiological recordings (e.g., cerebral blood flow, electroencephalography) and clinical neuropsychology (e.g., split-brain patients).

The nature of consciousness is examined by using current technologies, simulations, computer models of brain functions, virtual reality and robotic systems, neuroaesthetics based on studies on human brain, artificial intelligence applied to the study of behavioral and learning processes, among other approaches. Researchers are studying interactions of cognitive systems with their environment or visualization abilities in humans and animals that support cognitive learning by enhancing perception.

Figure 2. Emrey Winter, Bird's Eye View
(© 2014, E. Winter. Used with permission)

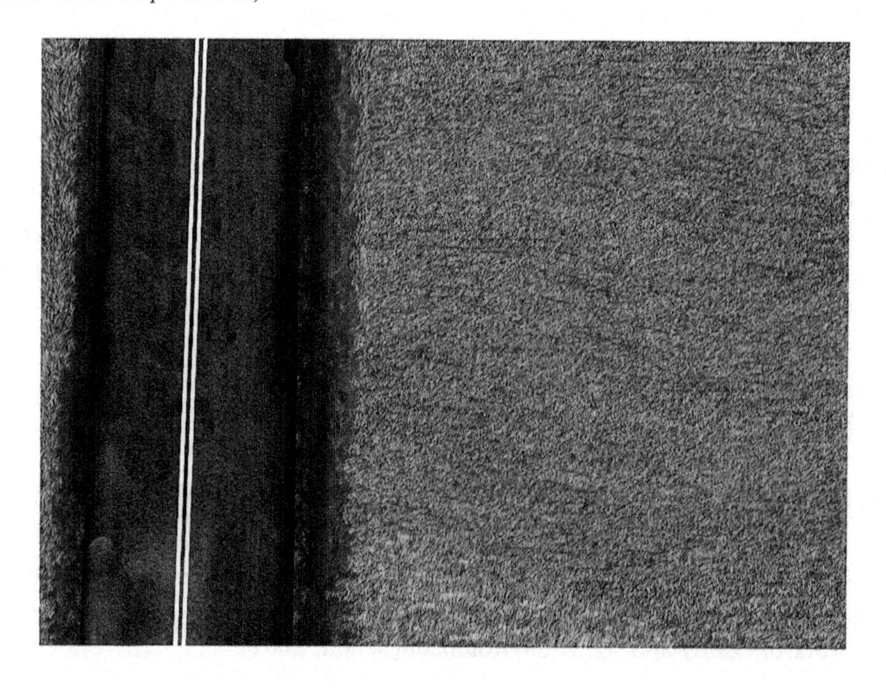

The 2014 Nobel Prize in Physiology or Medicine was awarded with one half to John O'Keefe and the other half jointly to May-Britt Moser and Edvard I. Moser for their study on grid cells. Grid cells provide a context-independent metric representation of local environment (Giocomo, Moser, & Moser, 2011). The firing of the grid cells forms a positioning system in the brain telling about the animal's speed and running direction at each location in the environment (Moser, Moser, & Roudi, 2014). Computational models of grid cells present how the neurons fire within a tessellating pattern, and how firing properties depend not primary on sensory signals but rather on internal self-organizing principles.

Studies of electrical activity of the brain include recordings of the responses evoked by the electrical or other stimuli, often involving single cells and their event-related potentials, which allows examining links between brain functions in specific regions of cognitive activities. Computers produce three-dimensional pictures of the brain structures. Current methods such as atomic force microscopy or photonics allow recording visual objects, events, and signals in a micro and nano scale. Several angiography techniques (arteriograms and venograms) visualize the blood vessels' interior. With computing and analyzing electromagnetic changes in energy of the subatomic particles contained in the brain researches get metabolic imaging and functional patterns.

Functional neuroimaging techniques show what's going inside the brain. They provide pictures of sites where the working memory is in action. Brain imaging techniques, including physiological recordings (e.g., cerebral blood flow, electroencephalography) and clinical neuropsychology (e.g., split-brain patients), serve also for researching cognitive imagery. Several techniques make possible mapping mental activity of the brain. Neuroimaging techniques allow structural and functional imaging to conduct research on neurological, cognitive, and psychological processes, and also develop interfaces between a brain and a computer. As listed in textbooks (e.g., Sternberg, 2011, Sternberg & Kaufman, 2011) and Wikipedia, types of functional neuroimaging include several techniques.

- Computed tomography (CT) and computed axial tomography (CAT) scanning use x-ray beams according to a computer program; CT scan technique provides 3D representations telling how tissues absorb radioisotope energy.
- Diffuse optical imaging (DOI) or diffuse optical tomography (DOT) and high-density diffuse optical tomography. HD-DOT are techniques that use near infrared light and rely on the absorption spectrum of hemoglobin depending on its oxygenation status (Eggebrecht, White, Chen, Zhan, Snyder, Dehlgani, & Culver, 2012).
- Magnetic resonance imaging (MRI) uses magnetic fields and radio waves to quickly construct a 2D or 3D image of the brain structure on a computer, and its changes over time, without use of x-rays or radioactive tracers.
- Functional magnetic resonance imaging (fMRI) (Logothetis, Pauls, Augath, Trinath, & Oeltermann, 2001) shows images of changing blood flow in the brain. Cognitive activity evoked by various stimuli results in changes in the amount of blood flow in different regions of the brain. Changes in blood flow are associated with perception, thought, and action related to different tasks: it reflects reasoning, the processing of emotions, conflict resolution, making moral judgments, or feeling reward and pleasure after making a proper conclusion.
- Positron emission tomography (PET) is the functional brain imaging method that tracks radioactively labeled chemicals (such as glucose with radioactive atoms) in the blood flow. Glucose is metabolized in proportion to the brain activity, so radioactivity is concentrated in the most active areas and provides multicolored 3D images of brain areas in action.

- Single photon emission computed tomography (SPECT) uses gamma ray radioisotopes and a gamma camera to construct 2D or 3D images of active brain regions. The brain rapidly takes up injected radioactive tracer, reflecting cerebral blood flow at the time of injection.
- Electroencephalography (EEG) is an early functional technique that detects a summary of the electrical activity of all cortex dendrites located under electrodes placed on a scalp; a recorder encephalogram can be compared with a normal activity of cortex.
- Magnetoencephalography (MEG) directly measures the magnetic fields produced by electrical activity in the brain using extremely sensitive superconducting quantum interference devices (SQUIDs). Neural activity measured by MEG is less distorted by surrounding tissue (particularly the skull and scalp) compared to the electric fields measured by EEG.

WAYS OF COMMUNICATING KNOWLEDGE

Teaching and learning may become more effective when we look after many possible ways to convey information. Theories and factual applications of the present-day approaches to visual, beyond the verbal education may include the use of communication media, education through gaming, or the use of augmented and virtual reality in specialized instruction. Online and classroom instruction may apply several means of communication.

- Visual communication and its implementation in teaching may refer to applying signs, symbols, analogies, icons, visual plus verbal imaginative metaphors, and visualizations, among other signals. Images become messages, and the distinction is made between a signal and its referent. By thinking about our imagery, we can see relation of signals, symbols, and signs to their referents. Signs may take a form of objects, actions, or events. They may contain a message, indicate some occurrences, or contain information. Signs usually act according to the rules of visual semiotics, but also they are often used in the robotic vision integrated with sensors, and computer vision applications (Chen, 2011). Visual communication techniques often include small multiple drawings that represent the sets of data with miniature pictures, to reveal repetition, change, pattern, and facilitate comparisons (Tufte, 1990). Since their introduction, cartoon faces evolved into a tool for presenting data as empathic facial expressions (Loizides, 2012; Loizides & Slater, 2001, 2002). Figures 3 and 4 show two works providing visual communication. The work "Signs" by Andrew Hudges conveys a message rather than information. "The Ripper" by Travis Brandl informs and provides a warning about possible threat caused by a dark figure under the streetlight.
- Verbal communication may involve semantics, abstract ideas and messages, and also storytelling, using a story as a container (a particular setting appropriate for telling the same story for another medium, such as an interactive novel (where a user is given a power to dictate how the story would unfold). Human cognition (and memory) have separate but interconnected verbal and imaginal systems. Nonverbal communication includes gesture, face expression often used by software designers for creating glyphs, body language, and pitch, among other means of expression. Some hold that humans, and maybe some animals are experts (without training) of picking up emotional expression in faces (Loizides, 2012).
- Communication through numbers is older than communication using numerals. Numeral activities involve more than one sense and several brain areas. In children counting happens before

Figure 3. Andrew Hudgen, Signs
(© 2014, A. Hudgen. Used with permission)

naming things by speaking, reading, or seeing written names. Without counting, we know when a mosquito bit us 3 times, we know that a whistle sounded two times, or recognize a SOS alert conveyed, for example in a Morse code. The same message can be received by various senses: signaled visually with flags or acoustically with sounds. Howard Gardner considered the perception of numerical patterns to be the core of logical mathematical intelligence (Gardner, 1983/2011; 1993/2006). Rudolph Arnheim assumed that "visual perception lays the groundwork of concept formation" (Arnheim, 1969/2004, p. 294), so he proposed a concept of perceptual sensitivity, which he described as the ability to see a visual order of shapes as images of patterned forces that underlie our existence (Arnheim, 1974/1983). This notion may correspond well to the way computer scientists talk about the codes in terms of patterns. Animals, children, and some indigenous peoples in South America, and Australia think about the number quantities without first learning and mastering verbal counting. The Mayans counted with fingers and toes using a system based on groups of twenty units (Maya Mathematical System, 2015), in contrast with the Arabic decimal system used today almost everywhere. Maybe, the covering of toes with shoes might limit a possibility to use this counting tool and thus influence the developing of a new counting system based on tens.

- Sound and musical sounds, speech and sound effects go together in a sound studio for a broadcast production, film, or video. Visual music combines sensory modalities such as acoustical, visual, oral, gestural.
- Temperature and thermal information provided by infrared sensors allow, for example seeing from a helicopter whether an object is alive, dead, or inanimate.

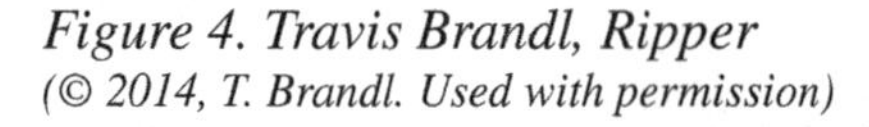

Figure 4. Travis Brandl, Ripper
(© 2014, T. Brandl. Used with permission)

- Several other ways to communicate may involve faculties usually described as senses, which include many internal and external receptors answering to changes, for example of temperature, acceleration, or velocity of the body. We may feel and communicate our kinesthetic sense, a sense of motion, proprioception, a feel of direction, responsiveness to pheromones, and sensitivity to pain, among other possibilities. We may sense someone's feelings or mood through the tone of their voice, body language, even from the look in their eyes; it may happen also in one's communication with animals.

Biometric measurements, which provide information on human sensory response, are currently used to improve security in health care, financial services, transportation security, military and government institutions (Nash, 2015). For example, biometric devices may include hand-held laser scanners and readers detecting skin temperature; software authenticating whether an iris is live or fake; scanners using infrared light to authorize users and passengers by identifying their personal patterns in finger veins; software for analyzing voiceprints from phone calls; and even video or radar analyzers that recognize individual gait.

Computer languages allow communicating with a machine using programs as instructions. For example, HTML (HyperText Markup Language) allows displaying information on the Internet, e.g., on the websites or in web browsers, while Processing, an open source programming language and integrated development environment, is not only a programming language but also a source for creating an online community. Environment for developments supported by haptic actions and data or info collection are possible with the Arduino boards.

A Computer Graphics student at the University of Northern Colorado, Andrew Hudgen responded to the abstract character behind signs with his imperative-based composition (Figure 3), while another student, Travis Brandl responded to the notion of signs by creating an imaginative, story based warning (Figure 4).

Visual Thinking, Visual Literacy

Visual thinking involves generation and manipulation of images coming both from imagery and from abstract systems. Visual processing goes beyond the language and often cannot be translated into verbal, linear manner. Imagery is the essence of thinking because we generate and manipulate images when we think. When we produce graphic images, we apply graphic language to visual thinking, attain synthesis, use perceptual and mental imagery, rely on intuition, and work at various levels of consciousness (such as dreaming) outside the realm of language thinking. Thus mental images help us to compare objects. Knowledge is important in this process.

Imagery and perception are different processes. For example, when we see a square (perception), we know it is a square because we compare it to a mental image of a square previously experienced (imagery). As stated by Arnheim (1969/2004), perceptual sensitivity, the ability to see a visual order of shapes as patterned forces that underlie our existence, helps the most gifted minds with intuitive wisdom to avoid troubles with the formalistic thought operations due to their brilliant cross-circuits. Cognitive thinking with the use of making flexible, generalized concepts enables students to think abstractly and draw conclusions (The Farlex Medical Dictionary, 2015). According to Arnheim (1988), abstractions are necessary to view art from a formal perspective, such as principles of design. Robert Sternberg (2007) discussed the interrelationships among intelligence, creativity, and wisdom in adapting to, shaping, and selecting environments. Interaction with environment involves aspects relative to creativity such as knowledge, styles of thinking, personality, and motivation, with balance of these personal attributes necessary to do creative work.

Swiss developmental psychologists Jean Piaget and Bärbel Inhelder (Piaget and Inhelder, 1971; Inhelder & Piaget, 1958) described structural schema developed in the mind to recognize and process visual patterns. Knowledge is a process of becoming "resulting from a construction of reality through the activities of the subject" (Inhelder, 1977, p. 339). According to Jean Piaget (1970), "...the aim of intellectual training is to form the intelligence rather than to stock the memory, and to produce intellectual explorers rather than mere erudition." The logical thinking is primarily non-linguistic. The roots of logic are in actions and not in words. Piaget points out that "... many educators limit themselves to showing the objects without having the children manipulate them, or, still worse, simply present audio-visual representations of objects (pictures, films, and so on) in the erroneous belief that the mere fact of perceiving the objects and their transformations will be equivalent to direct action of the learner in experience" (In a Foreword to Schwebel & Raph, 1973, pp. ix-x).

Visual imagery serves for generation, inspection, recoding, maintenance, and transformation of images. Generation is necessary because we do not have images all of the time. Images come and go, through short-term memory representations. One can "mentally draw" in one's imagery, producing images or patterns never actually seen. Then, we must have a way to make an inspection and interpret the patterns of images, 'zoom in' on isolated parts of them, or scan across them. Recoding means that we can encode the patterns of images into memory, remember new combinations of patterns or imaging new patterns. Maintenance serves for maintaining images, which require effort to remember them. The more perceptual units that are included in an image, the more difficult it is to maintain. Transformation lies at the heart of the use of imagery in reasoning. For example, we can rotate patterns in images, which we recall in three dimensions, so that we can 'see' new portions as they come into view. We also can imagine objects growing or shrinking.

The role and meaning of visual literacy and visual thinking evolved from early concepts and uses applied by Babylonian, Japanese, Chinese, Arab, Greek, Roman, and other cultures to the contemporary educational trends in instructional technologies applied to alleviating developmental problems. Students versed and talented in their preferred fields may display visual, verbal, literary, mathematical, or musical literacy. Visual literacy may be thus developed in modes selected by learners: within the semiotic, semantic, and computing-based frames of reference. The use of metaphors, iconic concepts and objects, cloud and tree visual presentations, and other design models for visual thinking and learning may advance visual literacy and thus support learning. The role of poetry evolves in a contemporary visual literature taking form of the texts in musicals and songs. Songs in movies often explain the action, or summarize the message.

Visual Aspects of Mathematics, Computing, Programming, and Learning with Software

Visual aspects of mathematics and computing allow presenting theories and their proofs as two dimensional and three dimensional constructs. Placing a great emphasis on developing visual literacy in students may be beneficial in fulfilling this task. For example, students may create mathematically programmed sculptures, or the fractal based art works, sceneries, and backdrops. Some versed in programming would be coding music- and dance-related shapes and forms, thus creating music visualization. Current means of delivering knowledge, for example, with the use of 3D printing technologies (Mercuri & Meredith, 2014), augmented reality (Bredl, Groß, Hünniger, & Fleischer, 2012), and open source printers just arrived to the school environment (Irwin, Opplinger, Pearce, & Anzalone, 2015); Schelly, Anzalone, Wijnen & Pearce, 2015). In 3D printing, based on the rapid prototyping process, additive processes are used, with successive layers of thermoplastic material laid down according to a computer program (Grujović, Radović, Kanjevac, Borota, Grujović, & Divac, 2011; Ravikumar, Khan, Mohanty, Sageer, & Aigali, 2015). Thus for example, the Voronoi tower can be created. Two- and three-dimensional Voronoi diagrams (Voronoi tessellations) are now commonly applied in architectural concepts, and other science and technology fields. Using the 3D printing technology and dinosaur fossils from the American Museum of Natural History paleontology collections, a group of high school students learned to think like paleontologists while producing models of dinosaurs as part of the innovative program "Capturing Dinosaurs: Reconstructing Extinct Species Through Digital Fabrication" (American Museum, 2013).

Scientists and artists see a purpose in applying visual way of presentation while working on particular scientific branches. For example, mathematicians, anthropologists, artists, designers, and architects conduct computer analysis of facades, friezes, and some architectural details. Researchers in many fields of natural sciences, medicine, pharmacology, biology, geology, or chemistry examine and visualize symmetry in natural and human-made structures. Many artists have created masterpieces this way.

Artists used to transform patterns and repetitions to apply the unity or symmetry in their compositions (for example, by examining a Fibonacci sequence, prime numbers and magic squares, a golden section, or tessellation techniques). Mathematicians, computing scientists, and artists used to apply visual metaphors as a cognitive tool to visualize the world's structure and our knowledge. For example, hierarchical structures are predominantly analyzed with the use of a tree metaphor. Manuel Lima called the tree figure the most ubiquitous and long-lasting visual metaphor, "through which we can observe the evolution of human consciousness, ideology, culture, and society" (Lima, 2014, p. 42). Figure 5 presents "Standards for Novelty" by Anna Ursyn.

Figure 5. Anna Ursyn, Standards for Novelty: New tools change our rules of defining novelty and allow us to feel comfortable wherever we go
(© 2010, A. Ursyn. Used with permission)

Visual literacy supports the quality design and product aesthetics. Definitions of the design aesthetics often put the form and content together. For Edward Tufte (1983, 1990), good design has two key elements found in simplicity of design and complexity of data. Frøkjær, Hertzum, & Hornbæk (2000) connected usability with effectiveness – the accuracy and completeness with which users achieve certain goals; efficiency – the relation between effectiveness and the resources expanded in achieving them; and satisfaction – the user's comfort with and positive attitudes towards the use of the system.

KNOWLEDGE VISUALIZATION CONCEPTS AND METHODS

Advances in several technologies enrich science education with knowledge visualization. Possibly, visualization is the best way of learning, teaching, or sharing the data, information, and knowledge because it amplifies cognition, outperforms text-based sources and increases our ability to think and communicate. Knowledge visualization ability should be taught and trained since kindergarten.

At present, visualization means using the computer, which transforms data into information, and then visualization converts information into picture forms and creates graphic images and symbols to convey and express meaning; this lets us comprehend data and make decisions. Software and programming solutions are used in the visualization industry. Visualization is frequently applied in numerous disciplines such as the cultural, historical, and architectural research. Examples may include a study on the Napoleon's campaign (Tufte, 1990, 1997) or a study on Giotto's frescoes (Wyeld, 2015). Time-based, 3D, and augmented reality applications serve for the military, intelligence, aviation and air traffic control, medical education, and other purposes.

Visualization means the communication of information with graphical representations. Knowledge visualization means the use of visual representations to transfer insights and provide new knowledge in the process of communicating between at least two persons. The most important domains in visualization are: data-, information-, and knowledge visualization. They change numerical data (which may be 1D, linear, 2D, or 3D) into graphs, clouds, tree visualizations (Chen, 2010, Shneiderman, 2014), and metaphorical visualization designs (Lima, 2014). Visual metaphors make a basic structure in visualization because they describe relations among data, organize information in a meaningful way, and combine creative imagery with the analytic rationality of conceptual diagrams. A metaphor indicates one thing as representing another, difficult one, thus making mental models and comparisons.

Data visualization enables us to go from the abstract numbers in a computer program (ones and zeros) to visual interpretation of data. Infographics refers to tools and techniques involved in graphical representation of data, information, and knowledge, mostly in journalism, art, and storytelling.

Information Visualization, often characterized as representation plus interaction, means the use of computer-supported, interactive visual representations of abstract data to amplify cognition (Bederson & Shneiderman, 2003) and derive new insights. According to the classic Ben Shneiderman's information visualization seeking mantra (Shneiderman, 1996), visualization should support seven high level tasks: overview, zoom, filter, detail-on-demand, relate, history, and extract information; these tasks were later elaborated in prescriptive way by Craft and Cairns (2005, 2008). Creating visualizations includes transforming raw data into structured data as data tables, converting data for calculations of their attributes, and visual mapping of important structures into the abstract visual structures. Users transform these structures on a screen changing their shape, color, size, or location. To this aim, designing IV may involve importing data, combining visual representations with textual labels, data mining large volumes of data, and collaborating with others.

Knowledge Visualization (KV) uses visual representation to transfer insights to create, integrate, and apply knowledge, rather than data, between individuals; it concentrates on the recipients, other types of knowledge, and on the process of communicating different visual formats (Burkhard, Meier, Smis, Allemang, & Honish, 2005). Techniques are focused on users, explanation, and presentation of knowledge, so they enhance cognitive processes and reduce cognitive load for working memory. A mind map drawn with a pen on paper is an example of knowledge visualization generated without a computer. Knowledge visualization could contribute design- and user-specific representations, e.g., a map, metro, aquarium, solar system, or flower metaphor for users with limited visual literacy (Kienreich, in Bertschi et al., 2011).

Scientific Visualization was established in 1985 at the National Science Foundation panel and deals with physically-based data defined, selected, transformed, and represented according to space coordinates, such as geographic data or computer tomography data of a body for medical use.

Information Aesthetics forms a cross-disciplinary link between IV and visualization art. Visualization is not only making the unseen visible – it is building a meaningful net of associations and connotations.

Figure 6 shows a work of Computer Graphics student at the University of Northern Colorado, Arturo Bugarin Correra. He visualizes key points in the insect–human biological relations resulting from the life cycle of a fly.

Figure 6. Arturo Bugarin Correra, The Fly and Us: Animal Cycle
(© 2015, A. Bugarin Correra. Used with permission)

Visualization tools derive their form from many domains. One may discuss them in terms of cognitive instruments such as utilization of a metaphor or the use of layers, details, and complexity in a website. One may examine and use scanners, microscopes, and cameras as instruments for visualization while recording and measuring in real time applications, micro agents, or bots. In terms of computing one may apply computer graphics tools such as Photoshop, and also the computer networking related concepts such as cloud computing.

Data visualization applies tools that have been listed as techniques for spatial and geospatial data, imaging multivariate data, visualization of trees, graphs, and networks, text representations, and interaction techniques (Ward, Grinstein, & Keim, 2010). Clustering technique (clusters are subsets of observations) is used in data mining for statistical analysis, pattern recognition, and bioinformatics. Concept Mapping represents the structure of information visually and builds knowledge models useful for strategic planning, product development, market analysis, decision-making, and measurement development.

Knowledge maps as a subset of knowledge visualization show changes, interrelationships, help to design strategies and build assessment. Network and web-search result visualization became the important carrier of information. Visual search engines use information visualization, data mining, and semantic web. Visual Analytics provides automated analysis of large amounts of dynamic information and the closed loop approach with visualization and interaction utilizing multi-touch surfaces; combines computational and visual methods using visualization, data mining, and statistics; uses abstract visual metaphors, mathematical deduction, and human intuitive interaction. Visual analytics serves for studying the entire genome of an organism at many abstraction levels: cells, organisms, and ecosystems, in formats and scales such as molecules, gene networks, and signaling networks. It also serves medicine, environmental research, and national security. Tag cloud visualization is a text-based visual representation of a set of tags.

CONCERNS ABOUT SCIENCE EDUCATION

Current disciplines and the fields of study are necessarily based on collaboration of specialists within different disciplines (such as in the domain of archeology) and of people with different types of talents (e.g., in film industry or big entertainment companies such as Walt Disney Company). Creating an environment for combining art and computing related skills makes collaborative efforts easier. Arts-based development training became widely used in corporations to support teambuilding, communication, and leadership. There is a growing development of various digital visual productions including (and often combining) video, immersive virtual reality, the Web, wireless technology, performance, large-scale urban art installations, and interactive exhibitions.

Hence, division of specific subject areas such as physics, chemistry, or biology, which is practiced in the high school curricula, is not supportive of the scientific concepts' understanding, nor it is encouraging the learners to further their studies in the more specific but integrative areas such as biochemistry, geophysics, or astronomy. Arnheim observed, "our educational system, including our intelligence tests, is known to discriminate not only against underprivileged and handicapped but equally against the most gifted. … To what extend go our schools and universities serve to weed out and retard the most imaginative minds? (Arnheim, 1969/2004, p. 207). Leonard Sax (2009) alerted educators to the growing epidemic of unmotivated boys and underachieving young men, and discussed factors driving this problem. Teaching with the use of current applications and devices may arise motivation and support students' wish for achievements. After graduation from high school, many disciplines (such as medicine where a basic background knowledge of chemistry, physics, biochemistry, and physiology is needed) remain almost totally unknown and unexplored for students who look for a direction of further study or a job that would fit their talents and interests, because school curricula do not include these topics into the subject matter under study. However, introduction of the new content and its novel ways of presentation has been limited by a so-called 80–20 principle adopted by many textbook producers, who tend to allot eighty percent of the text to information based on the existing general standards and high frequency of occurrences, with the remaining twenty percent of the textbook available to presentation of a new material.

Along with studies on recent advances in the selected fields of science, further parts of this book contain learning projects for the readers, both the learners and their instructors, which would comply with the actual trends in education described as the STEM (science, technology, engineering, and mathematics) and the STEAM (science, technology, engineering, arts, and mathematics) programs in education. Combining science, technology, and art offers a response to the actual trends in education described as the STEAM (science, technology, engineering, art, and mathematics). Many educators have been working toward transforming STEM into STEAM as a framework for teaching across the disciplines (STEM to STEAM, 2014; Gonzalez & Kuenzi, 2012; Maeda, 2012). This trend results from the presence of technological literacy in small children who can now use software teaching them science, mathematics, programming, and storytelling. It is also a response to the economical and social demands for the arts based development training, which became used in corporations; there is a trend to include the arts in business supporting teambuilding, communication, and leadership. Not only we should integrate scientific, technical, and artistic topics, but also construct, envision, and explain through visuals and interactivity (Honey, Pearson, & Schweingruber, 2014). Involving programming and solving problems through knowledge visualization should enhance programs for Studio Art. Students should learn, create, and play together globally through educational games, VR, teaching oriented apps, social networking, and meaningful communication, K through PhD (DeVry University, 2014). This chapter discusses the existing and coming tendencies and solutions in educational and instructional solutions, technologies, and practices. Tools aimed at enhancing visual literacy and thus supporting learning about science may include toys, games, puzzles, apps, models (often involving 3D printing), animations, simulations (automatic pilot, control tower), and augmented reality environments, among other solutions. Perhaps there is a need for focus on games, apps, and ways to interact through social networking that would bring more learning opportunities, actively converting exchange of ideas into somehow curiosity-based learning realm.

Developing imagination may happen due to digital animation and filming techniques that make that impossible actions of Superman or Batman look real. However, fantasy and imagination presented in digitally created stories rarely surpass accomplishments of the pre-computer works such as Gulliver's Travels (1726) by Jonathan Swift, Frankenstein (1818) by Mary Shelley, Alice's Adventures in Wonderland (1865) by Lewis Carroll, The Lord of the Rings series (1954-55) by John Ronald Reuen Tolkien, and fictional characters such as Count Dracula, Nosferatu, Dr Jekyll and Mr. Hyde (1886, by Robert Louis Stevenson), Dorian Grey (1891, by Oscar Wilde) or The Little Prince (1943, by Antoine de Saint-Exupéry). It is not surprising that authors of comic books, games, animations, and movie scenarios remake these works in many ways, with the results that overshadow many other attempts.

CONCLUSION AND DISCUSSION

The focus of this chapter is on contributing on visual ways of learning with inspiration coming from sciences. This way the learner grasps information in less controlled atmosphere, while inspiration can motivate them to develop one's own, personal approach to creating visuals that are unique, have explanatory power, and make the viewer think, digest, and control own thoughts as a result of appreciating those graphical representations. The experience gathered in these case studies confirms that students developed better capabilities of grasping information, and grew up as artists.

Combining science, technology, and art in curriculum provides students with better preparation for future careers. The idea of including into instruction familiar concepts and applications used everyday by students, such as computers, communication media, TV, multimedia, online protocols, and games, draws from current developments in the cognitive approach to teaching and learning, and offers a response to the actual trends in education. Part of integrative concepts in education should perhaps be based on collaborative efforts; therefore an individual performs less and less sole individual tasks in the workspace. There is more of teamwork in a workspace, with discussions, brainstorming, and talent-based assignments of task and duties. Thus, the annual Autodesk's training for educators at the Siggraph 2015 conference was entitled "Don't Be an Asshole," to stress the value of sharing and supporting a group. They teamed up with some experts from the Vancouver School of Film, to introduce Computer Graphics professors to the notion of a collaborative workflow as an outcome of student's training.

New tools, solution, and shortcuts allow co-workers to learn faster by sharing, playing, exchanging ideas, and supporting fellow co-workers. With current state of tools and technological solutions, new ideas, which are not reality-based, can be depicted in a convincing way faster, safer, easier, and friendlier.

This also affects our daily routines through access to ever growing solutions. For example, a panel at the 2015 SIGGRAPH conference was discussing 3-D printed foldable furniture for students who often change apartments. Some view a future travel as a luggage-free event, as hotels will have an access to costume wireframes, which can be easily rendered with selected textures and then printed inside of a hotel room as 3-D objects.

New inventions and their adaptations require the abstract and visual thinking, and should be combined with special types of instant connotations and imagination that would allow a student quickly grasp the idea behind the software, equipment, computer language, and some interactions between them. Creating, interactions, gaming, all types of social networking, and internet channels might support learning and develop special cognitive skills, which are so different for every classmate. When combined with collaborative effort they may produce important and novel approaches to the self-promoted solutions. Creativity has been recognized by many as a currency of the 21st century. While creativity can be taught, it can also be nurtured. The adaptability to a new workspace can be developed by contributing to non-judgmental atmosphere, positive and supportive class critiques where students look at the progress, highlights, and heights of artistic and aesthetics solutions of their classmates, and finding ways to nurture and support their future growth through pleasant discussions and supportive comments. Further research is needed, to assess introduction of computing and programming into K-12 curriculum, beginning from kindergarten, so the task becomes natural in students' later careers.

REFERENCES

American Museum of Natural History. (2013). *Students Use 3D Printing to Reconstruct Dinosaurs.* Retrieved August 8, 2015, from https://www.youtube.com/watch?v=_KBxG1_WO8k

Arnheim, R. (1969/2004). *Visual Thinking*. University of California Press.

Arnheim, R. (1974/1983). *Art and visual perception: A psychology of the creative eye* (2nd ed.). University of California Press.

Arnheim, R. (1988). *The Power of the Center, A Study of Composition in the Visual Arts*. Univ. of California Press.

Arnheim, R. (1990). *Thoughts on art education* (Occasional Paper Series Vol 2). Oxford University Press.

Bederson, B., & Shneiderman, B. (2003). *The Craft of Information Visualization: Readings and Reflections*. San Francisco, CA: Morgan Kaufmann Publishers.

Bertschi, S., Bresciani, S., Crawford, T., Goebel, R., Kienreich, W., & Lindner, M. et al.. (2011). What is Knowledge Visualization? Perspectives on an Emerging Discipline.*Proceedings of the Information Visualisation 15th International Conference* (pp. 329-336), London. doi:10.1109/IV.2011.58

Boden, M. (2016). Skills and the Appreciation of Computer Art. In *Computational Creativity, Measurement and Evaluation, Connection Science*. Taylor & Francis. doi:10.1080/09540091.2015.1130023

Bredl, K., Groß, A., Hünniger, J., & Fleischer, J. (2012). The Avatar as a Knowledge Worker? How Immersive 3D Virtual Environments may Foster Knowledge Acquisition. *Electronic Journal of Knowledge Management*, *10*(1), 15–25.

Broudy, H. S. (1987). *The Role of Imagery in Learning.* Occasional Paper 1, The Getty Center for Education in the Arts.

Burkhard, R., Meier, M., Smis, J. M., Allemang, J., & Honish, J. (2005). Beyond Excel and PowerPoint:Knowledge Maps for the Transfer and Creation of Knowledge in Organizations. In *International Conference on Information Visualisation* (pp. 403-408). London, UK: IEEE Computer Society Press.

Cavanna, A. E., & Nanni, A. (2014). *Consciousness: Theories in Neuroscience and Philosophy of Mind.* Springer. doi:10.1007/978-3-662-44088-9

Chen, C. (2010). *Information Visualization: Beyond the Horizon* (2nd ed.). Springer.

Chen, C. H. (2011). *Emerging topics in computer vision and its applications*. World Scientific Publishing Company.

Conway, M. A., & Loveday, C. (2015). Remembering, imagining, false memories & personal meanings. *Consciousness and Cognition*, *33*, 574–581. doi:10.1016/j.concog.2014.12.002 PMID:25592676

Craft, B., & Cairns, P. (2005). Beyond guidelines: What can we learn from the visual informationseeking mantra? In *Proceedings of the 9th International Conference on Information Visualization* (pp. 110-118). IEEE. doi:10.1109/IV.2005.28

Craft, B., & Cairns, P. (2008). Directions for methodological research in information visualization. In *Proceedings of the 12th International Conference on Information Visualisation*. IEEE. doi:10.1109/IV.2008.88

Craig, A. D. (2003). Interoception: The sense of the physiological condition of the body. *Current Opinion in Neurobiology*, *13*(4), 500–505. doi:10.1016/S0959-4388(03)00090-4 PMID:12965300

Damasio, A. (2012). *Self Comes to Mind: Constructing the Conscious Brain*. Vintage.

Damasio, A., & Carvalho, G. B. (2013). The nature of feelings: Evolutionary and neurobiological origins. *Nature Reviews. Neuroscience*, *14*(2), 143–152. doi:10.1038/nrn3403 PMID:23329161

Damasio, A. R. (1999). *The feeling of what happens: Body and emotion in the making of consciousness*. New York: Harcourt Brace.

DeVry University. (2014). *Web Game Programming Degree Specialization*. Retrieved April 3, 2014 from http://www.devry.edu/degree-programs/college-media-arts-technology/web-game-programming-about.html

Eggebrecht, A. T., White, B. R., Chen, C., Zhan, Y., Snyder, A. Z., Dehlgani, H., & Culver, J. P. (2012). A quantitative spatial comparison of high-density diffuse optical tomography and fMRI cortical mapping. *NeuroImage*, 2012. PMID:22330315

Farrugia, N., Jakubowski, K., Cusack, R., & Stewart, L. (2015). Tunes stuck in your brain: The frequency and affective evaluation of involuntary musical imagery correlate with cortical structure. *Consciousness and Cognition*, *35*, 66–77. doi:10.1016/j.concog.2015.04.020 PMID:25978461

Frøkjær, E., Hertzum, M., & Hornbæk, K. (2000). Measuring Usability: Are Effectiveness, Efficiency, and Satisfaction Really Correlated? In *CHI 2000 Conference Proceedings*, (pp. 345-352). ACM Press.

Gardner, H. (1983/2011). *Frames of mind: The theory of multiple intelligences* (3rd ed.). Basic Books.

Gardner, H. (1993/2006). *Multiple intelligences: New horizons in theory and practice*. Basic Books.

Gardner, H. (1993). *Art, Mind, and Brain: A Cognitive Approach to Creativity*. New York: Basic Books, A Division of Harper Collins Publishers.

Gardner, H. (1997). *Extraordinary minds: Portraits of exceptional individuals and an examination of our extraordinariness*. Basic Books, Harper Collins Publishers.

Giocomo, L. M., Moser, M.-B., & Moser, E. I. (2011). Computational models of grid cells. *Neuron*, *71*(4), 589–603. doi:10.1016/j.neuron.2011.07.023 PMID:21867877

Gonzalez, H. B., & Kuenzi, J. J. (2012). *Science, Technology, Engineering, and Mathematics (STEM) Education: A Primer*. Congressional Research Service. Retrieved July 5, 2014, from http://fas.org/sgp/crs/misc/R42642.pdf

Grujović, N., Radović, M., Kanjevac, V., Borota, J., Grujović, G., & Divac, D. (2011). 3D printing technology in education environment. *34th International Conference on Production Engineering* (pp. 29-30). Academic Press.

Guilford, J. P. (1950). Creativity. *The American Psychologist*, *5*(9), 444–454. doi:10.1037/h0063487 PMID:14771441

Guilford, J. P. (1959). Traits of creativity. In *Creativity and its cultivation*. New York: Harper and Row.

Guilford, J. P. (1967). *The nature of human intelligence*. McGraw-Hill.

Guilford, J. P. (1968). *Intelligence, creativity and their educational implications*. San Diego, CA: Robert Knapp, Publ.

Honey, M., Pearson, G., & Schweingruber, H. (Eds.). (2014). *National Academy of Engineering and National Research Council. In STEM Integration in K-12 Education: Status, Prospects, and an Agenda for Research*. Washington, DC: The National Academies Press.

Inhelder, B. (1977). Genetic epistemology and developmental psychology. *Annals of the New York Academy of Sciences*, *291*(1), 332–341. doi:10.1111/j.1749-6632.1977.tb53084.x

Inhelder, B., & Piaget, J. (1958). *The Growth of Logical Thinking from Childhood to Adolescence*. London: Routledge and Kegan Paul. doi:10.1037/10034-000

Irwin, J. L., Opplinger, D. E., Pearce, J. M., & Anzalone, G. (2015). Evaluation of RepRap 3D Printer Workshops in K-12 STEM (Program/Curriculum Evaluation).*122nd ASEE Annual Conference & Exposition*. doi:10.18260/p.24033

Kandel, E. R., Schwartz, J. H., Jessell, T. M., Siegelbaum, S. A., & Hudspeth, A. J. (2012). *Principles of Neural Science* (5th ed.). McGraw-Hill Education / Medical.

Lengler, R. (2007). How to induce the beholder to persuade himself: Learning from advertising research for information visualization. In *Proceedings of the 11th International Conference on Information Visualization*, (pp. 382-392). IEEE Computer Society Press. doi:10.1109/IV.2007.66

Lima, M. (2014). *The book of trees: Visualizing the branches of knowledge*. New York: Princeton Architectural Press.

Logothetis, N. K., Pauls, J., Augath, M., Trinath, T., & Oeltermann, A. (2001). Neurophysiological investigation of the basis of the fMRI signal. *Nature*, *412*(6843), 150–157. doi:10.1038/35084005 PMID:11449264

Loizides, A. (2012). *Andreas Loizides research home page*. Retrieved August 1, 2015, from http://www0.cs.ucl.ac.uk/staff/a.loizides/research.html

Loizides, A., & Slater, M. (2001). The empathic visualisation algorithm (EVA): Chernoff faces revisited. *Technical Sketch SIGGRAPH*, *2001*, 179.

Loizides, A., & Slater, M. (2002). The empathic visualisation algorithm (EVA) - An automatic mapping from abstract data to naturalistic visual structure.*Proceedings of 6th International Conference on Information Visualisation* (pp. 705-712). Los Alamitos, CA: IEEE. doi:10.1109/IV.2002.1028852

Maeda, J. (2012). *STEM to STEAM: Art in K-12 is Key to Building a Strong Economy. Edutopia*. Retrieved July 5, 2014, from http://www.edutopia.org/blog/stem-to-steam-strengthens-economy-john-maeda

Marr, D. (1982/2010). *Vision: A Computational Investigation into the Human Representation and Processing of Visual Information*. MIT Press.

Maya Mathematical System . (2015). Yucatan's Maya calendar. Retrieved August 10, 2015, from http://www.mayacalendar.com/f-mayamath.html

Mercuri, R., & Meredith, K. (2014). An educational venture into 3D Printing.*Integrated STEM Education Conference (ISEC)*. IEEE.

Merriam-Webster Dictionary and Thesaurus. (2015). *An Encyclopedia Britannica Company*. Retrieved August 8, 2015, from http://www.merriam-webster.com/

Moore, A., & Malinowski, P. (2009). Meditation, mindfulness and cognitive flexibility. *Consciousness and Cognition*, *18*(1), 176–186. doi:10.1016/j.concog.2008.12.008 PMID:19181542

Moser, E. I., Moser, M.-B., & Roudi, Y. (2014). Network mechanisms of grid cells. *Phil. Trans. R. Soc. B*, *369*(1635), 20120511. doi:10.1098/rstb.2012.0511 PMID:24366126

Nash, K. S. (2015, October 14). The Next Security Frontier: The Human Body. *The Wall Street Journal*.

Piaget, J. (1970). *Science of education and the psychology of the child*. New York: Orion Press.

Piaget, J., & Inhelder, B. (1971). *Mental Imagery in the Child. A study of the development of imaginal representation* (P. A. Chilton, Trans.). New York: Basic Books Inc.

Ravikumar, R., & Mohsin Khan, I. (2015). Design & Development of a 3D Printer. *Proceedings of 12th IRF International Conference*. Academic Press.

Sax, L. (2009). *Boys adrift: the five factors driving the growing epidemic of unmotivated boys and underachieving young men*. Basic Books.

Schelly, C., Anzalone, G., Wijnen, B., & Pearce, J. M. (2015). Open-Source 3-D Printing Technologies for Education: Bringing Additive Manufacturing to the Classroom. *Journal of Visual Languages and Computing*, *28*, 226–237. doi:10.1016/j.jvlc.2015.01.004

Scholl, R. W. (2001). *Cognitive style and the Meyers-Briggs type inventory*. The University of Rhode Island, The Charles T. Schmidt, Jr. Labor Research Center. Retrieved August 9, 2015, from http://www.uri.edu/research/lrc/scholl/webnotes/Dispositions_Cognitive-Style.htm

Schwebel, M., & Raph, J. (Eds.). (1973). Piaget in the Classroom. New York: Basic Books, Inc.

Shneiderman, B. (1996). The Eyes Have It: A Task by Data Type Taxonomy for Information Visualizations. In *Proceedings of the IEEE Symposium on Visual Languages*. Washington, DC: IEEE Computer Society Press. doi:10.1109/VL.1996.545307

Shneiderman, B. (2014). *Treemaps for space constrained visualization of hierarchies*. Retrieved August 10, 2015, from: http://www.cs.umd.edu/hcil/treemap-history/

STEM to STEAM. (2014). Retrieved September 9, 2015, from http://stemtosteam.org/

Sternberg, R. (2011). *Cognitive Psychology* (6th ed.). Wadsworth Publishing.

Sternberg, R. J. (2007). *Wisdom, intelligence, and creativity synthesized.* New York: Cambridge University Press.

Sternberg, R. J., & Kaufman, S. B. (Eds.). (2011). *The Cambridge Handbook of Intelligence (Cambridge Handbooks in Psychology).* Cambridge University Press. doi:10.1017/CBO9780511977244

The Farlex Medical Dictionary. (2015). Retrieved August 20, 2015, from http://medical-dictionary.thefreedictionary.com/abstract+thinking

Tufte, E. R. (1990). *Envisioning information* (2nd ed.). Graphics Press.

Tufte, E. R. (1997). *Visual explanations: Images and quantities, evidence and narrative.* Graphics Press.

Ward, M., Grinstein, G. G., & Keim, D. (2010). *Interactive data visualization: foundations, techniques, and applications.* Natick, MA: A K Peters Ltd.

Wyeld, T. G. (2015). Re-Visualising Giotto's 14th-Century Assisi Fresco "Exorcism of the Demons at Arezzo". In *Handbook of Research on Maximizing Cognitive Learning through Knowledge Visualization* (pp. 374-396). Hershey, PA: IGI Global.

Zeki, S. (1993). Vision of the brain. Wiley-Blackwell.

Zeki, S. (1998). Art and the Brain. *Journal of Conscious Studies: Controversies in Science and the Humanities, 6*(6/7), 76–96.

Zeki, S. (2001). Artistic creativity and the brain. *Science, 293*(5527), 51–52. doi:10.1126/science.1062331 PMID:11441167

Zeki, S. (2009). *Splendors and miseries of the brain: Love, creativity, and the quest for human happiness.* Wiley-Blackwell.

Zeki, S. (2011). *The Mona Lisa in 30 seconds.* Blog. Retrieved December 12, 2011, from http://profzeki.blogspot.com/

ADDITIONAL READING

Arnheim, R. (1974). *Art and Visual Perception.* Berkeley: University of California Press.

Klanten, R., Bourquin, N., Ehmann, S., van Heerden, F., & Tissot, T. (Eds.). (2008). *Data Flow: Visualising Information in Graphic Design.* Die Gestalten Verlag.

Lima, M. (2011). *Visual complexity: Mapping patterns of information.* New York: Princeton Architectural Press.

Lima, M. (2014). *The Book of Trees: Visualizing the Branches of Knowledge.* New York: Princeton Architectural Press.

Yau, N. (2011). *Visualize this: The flowing data guide to design visualization and statistics.* Wiley.

Yau, N. (2013). *Data Points: Visualization That Means Something.* Wiley.

KEY TERMS AND DEFINITIONS

Cognitive Development: Jean Piaget and Bärbel Inhelder described the child's cognitive developmental levels as sensorimotor, intuitive, concrete operational, and formal operational problem solving (Piaget, J. & Inhelder, B., 1971, Mental Imagery in the Child. A study of development of imaginal representation. New York: Basic Books Inc., Publishers).

Cognitive Load: The amount of information and its necessary processing placed on the working memory of a learner (the part of short-term memory involved in conscious perceptual and linguistic processing).

Data: Factual information, especially organized for analysis, reasoning or making decisions.

Graphic: Image represented by a graph or relating to graphics. Graphic display is often generated by a computer.

Haptic: Relates to the sense of touch; the senses of touch and proprioception enable the perception and manipulation of objects.

Icon: Represents a thing or refers to something by resembling or imitating it; thus a picture, a photograph, a mathematical expression, or an old-style telephone may be regarded as iconic objects. The iconic object has some qualities common with things it represents, by looking, sounding, feeling, tasting, or smelling alike.

Information: Knowledge derived from study, experience, or instruction, a collection of facts or data.

Pattern: A regular order existing in nature or in a humanmade design. In nature patterns can be seen as symmetries (e.g., snowflakes), and structures having fractal dimension such as spirals, meanders, or surface waves. In computer science, design patterns serve in creating computer programs. In the arts, pattern is an artistic or decorative design made of recurring lines or any repeated elements. We can see patterns everywhere in nature, mathematics, art, architecture, and design. A pattern makes a basis of ornaments, which are specific for different cultures. Owen Jones (1856) made a huge collection of ornaments typical for different countries. He wrote an amazing monographic book entitled "The Grammar of Ornament."

Semiotics: The study about the meaningful use of signs, symbols, codes, and conventions that allow communication. The name 'semiotics' is derived from the Greek word 'semeion' which means "sign". "Meaning" is always the result of social conventions, even when we think that something is natural or characteristic, and we use signs for those meanings. Therefore, culture and art is a series of sign systems. Semioticians analyze such sign systems in various cultures; linguists study language as a system of signs, and some even examine film as a system of signs. The semiotic content of visual design is important for non-verbal communication applied to practice, especially for visualizing knowledge.

Sign: A conventional shape or form telling about facts, ideas, or information. It is a distinct thing that signifies another thing. Natural signs signify events caused by nature, while conventional signs may signal art, social interactions, fashion, food, interactions with technology, machines, and practically everything else.

Signage: Signs, icons, and symbols are collectively called signage. Icons and symbols help compress information in a visual way.

Spatial Cognition: Mental reflection and reconstruction of space in thought, with a distinction between the perception (what is seen, when we process sensory information) and cognition (what is assimilated by a person, based on person's cognitive structures).

Spatial Visualization: An ability to mentally rotate, twist, or invert pictorially presented objects.

Spatial Orientation: The understanding of the array of objects, utilizing body orientation of the observer.

Symbol: Represents an abstract concept, not just a thing, and is comparable to an abstract word. Highly abstracted drawings that show no realistic graphic representation become symbols. Examples of symbols present in our life may include an electric diagram, which uses abstract symbols for a light bulb, wire, connector, resistor, and switch; an apple for a teacher or a bitten apple for a Macintosh computer; a map – typical abstract graphic device; a 'slippery when wet' sign. Symbols don't resemble things they represent but refer to something by convention. We must learn the relationship between symbols and what they represent, such as letters, numbers, words, codes, traffic lights, and national flags.

Topological Space: A notion of a mathematical space used in topology as a branch of mathematics, which means a set of points within their neighborhoods as defined by related axioms. These notions may refer to concepts such as continuity, connectedness, and convergence.

Variable: A quantity capable of assuming any set of values, or a symbol representing such a quantity. Variables represent characteristic traits, which take on different amounts or numbers under changing conditions.

Chapter 2
Exploring Perception, Cognition, and Neural Pathways of Stereo Vision and the Split-Brain Human Computer Interface

Gregory P. Garvey
Quinnipiac University, USA

ABSTRACT

This chapter examines research from psychology of perception and cognition as well as select developments in the visual arts that inspired the design of the split-brain user interface developed for the interactive documentary Anita und Clarence in der Hölle: An Opera for Split-Brains in Modular Parts (Garvey, 2002). This experimental interface aims at 'enhanced' interaction while creating a new aesthetic experience. This emergent aesthetic might also be described as induced artificial cognitive dissonance and recalls select innovations in the rise of modernism notably the experiments of the Surrealists. The split-brain interface project offers a model for further investigations of human perception, neural processing and cognition through experimentation with the basic principles of stereo and binocular vision. It is conceivable that such an interface could be a design strategy for augmented or virtual reality or even wearable computing. The chapter concludes with a short discussion of potential avenues for further experimentation and development.

INTRODUCTION

Form is henceforth divorced from matter. In fact, matter as a visible object is of no great use any longer, except as the mould on which form is shaped. Give us a few negatives of a thing worth seeing, taken from different points of view, and that is all we want of it. (Oliver Wendell Holmes, 1859)

DOI: 10.4018/978-1-5225-0480-1.ch002

The split-brain user interface was first developed for the interactive installation *Anita und Clarence in der Hölle* (trans. "Anita and Clarence in Hell"): *An Opera for Split-Brains in Modular Parts*. This project used documentary video (C-SPAN 1991) from the 1991 Senate Judiciary Committee hearings on the nomination of Judge Clarence Thomas to be Associate Justice of the United States Supreme Court (United States Congress Senate Committee on the Judiciary 1993). During these televised hearings lawyer Anita Hill came forward to testify under oath and accused nominee Clarence Thomas of sexual harassment when she worked under him at the United States Equal Opportunity Employment Commission in Washington, D.C. Thomas vehemently denied these charges as "a high tech lynching." On October 15, 1991 the full Senate voted 52 to 48 to confirm Thomas as a Justice on the United States Supreme Court.

The split-brain interface presents the testimony of Anita Hill simultaneously with that of Clarence Thomas before the Senate Judiciary Committee. The principle behind a split-brain interface is to deliver two independent video and audio streams separately to each eye and ear so that the content of those streams is experienced or perceived independently and simultaneously by each hemisphere of the brain.

These two diametrically opposed versions of the same events delivered in parallel to each hemisphere of the brain induce a kind of artificial cognitive dissonance. The right and left hemisphere of the viewer hears a different version of the events. The 'whole' brain is left to sort out the truth.

The title of the original installation (Garvey 2002), "refers literally to the ordeal of the two main participants. It is a story with tragic characters, a clear conflict and a well-defined beginning, middle and end. In the tradition of grand opera, it is an epic narrative that emerged from the realm of the personal and private to play out on the national televised stage. Against a backdrop of gender, race and class, this drama pits the political forces of the right and left, and a man and a woman, against one another in a lurid spectacle of 'she said, he said.'" This work was originally developed as a co-production at the Banff New Media Institute in Alberta, Canada (Banff Centre 1999) during a residency and was later part of the exhibition *Sleuthing the Mind* in the fall of 2014 at Pratt Manhattan Gallery in New York City curated by Ellen K. Levy (2014).

The original prototype developed at the Banff Centre required the viewer to wear the Virtual Research Systems Head Mounted Display (1998–2000) to view the digital video. The 2014 installation at the Pratt Manhattan Gallery used the ScreenScope (Stereo Aids, n.d.) handheld mirror stereoscope for viewing the digital video displayed side-by-side on a large flat screen display (Figure 1 shows the Screenscope and monitor at the Pratt Manhattan Gallery). Normally a stereoscope creates the perception of depth by projecting two slightly different images to each eye. The brain fuses these two images into a single coherent percept (perception) of three dimensions.

When the two eyes receive independent images this is technically known "dichoptic" stimulation (Blake 2005). The split-brain or dichoptic interface uses the ScreenScope viewer so the left eye sees only video of Anita Hill and simultaneously the right eye only sees the video of Clarence. Therefore this video installation included in the Sleuthing the Mind exhibit was titled *The Split-Brain (Dichoptic) Interface: Thomas v. Hill* (1999/2014). Through headphones the left ear hears only the testimony of Anita Hill while the right ear hears that of Clarence Thomas (Figures 2 and 3 show a user viewing the installation).

BACKGROUND

The split-brain interface uses a technique similar to the Divided Visual Field technique (Banich 2003), which exploits the lateralization of the visual system. The visual pathways as shown in Figure 4 exit

Figure 1. The ScreenScope Stereoscopic Viewer for the split-brain (dichoptic) interface: Thomas v. Hill (1999/2014) at Pratt Manhattan Gallery in New York City curated by Ellen K. Levy
(© 2015, Gregory P. Garvey. Used with permission)

Figure 2. The split-brain (dichoptic) interface: Thomas v. Hill (1999/2014) at Pratt Manhattan Gallery in New York City curated by Ellen K. Levy
(© 2015, Gregory P. Garvey. Used with permission)

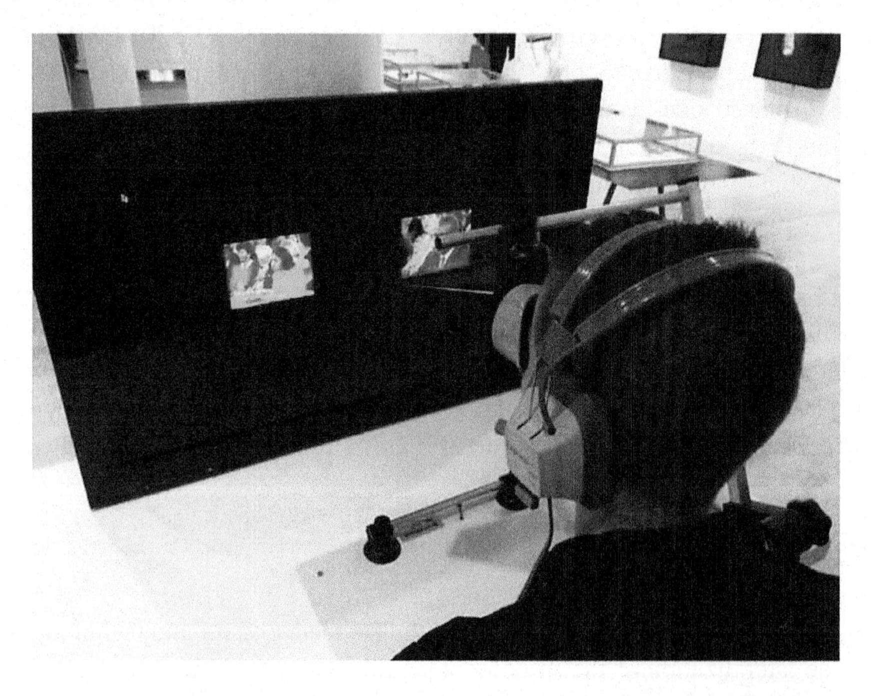

Figure 3. The split-brain (dichoptic) interface: Thomas v. Hill (1999/2014) at Pratt Manhattan Gallery in New York City
(© 2015, Gregory P. Garvey. Used with permission)

each eye and continue to the same side (temporal) hemisphere and also split at the optic chiasma and cross to the opposite side hemisphere. In the scientific literature the terms lateralization, hemispheric specialization or localization are used interchangeably. When a visual stimulus is presented in the left visual field (LVF) it is initially processed by the right cerebral hemisphere. Similarly, when a visual stimulus is presented in the right visual field (RVF) it is initially processed by the left cerebral hemisphere.

Fibers from nasal (medial) and temporal (lateral) sides of the retina of each eye exit at the back of the eyeball. The temporal fibers continue to the Lateral Geniculate body (nucleus) on the same side hemisphere. The fibers from the nasal retina cross the Optic Chiasma and continue to the Pulvinar nuclei making up 40% of the thalamus. Fibers continue to the cortex of the Occipital Lobes (Figure 4 shows the visual pathways).

With normal stereovision, what is seen in the left visual field is viewed by the nasal retina of the left eye and the temporal retina of the right eye. Conversely, what is seen in the right visual field is viewed by the nasal retina of the right eye and the temporal retina of the left eye (Figure 5 shows these relationships).

The lens of the eye functions as a biconvex lens so the image on the retina is inverted and left-right reversed (Figure 6 shows how the image is reversed and inverted).

The split-brain interface presents digital video on only one half of each visual field. The digital videos registered to the lateral (temporal) edge to the center of the right and left visual fields and in principle, are seen only by the nasal retinas. Conversely the digital videos registered to the medial (nasal) edge to the center of the right and left visual fields are seen only by the temporal retinas.

When properly holding the ScreenScope, the viewer looks straight ahead and fixates both eyes on a point or target that is effectively 'in the middle.' The viewer "sees" each speaker separately with the right and left eye. Therefore Clarence Thomas seen only by the right eye is perceived by left brain, while simultaneously Anita Hill seen only by the left eye is perceived by the right brain. While binocular and

Figure 4. The visual pathways based on Gray's Anatomy depiction of the optic nerves & nuclei, optic chiasma and the optic lobes in a human brain. Plate 722
(Adopted from Gray (1918)© Public Domain)

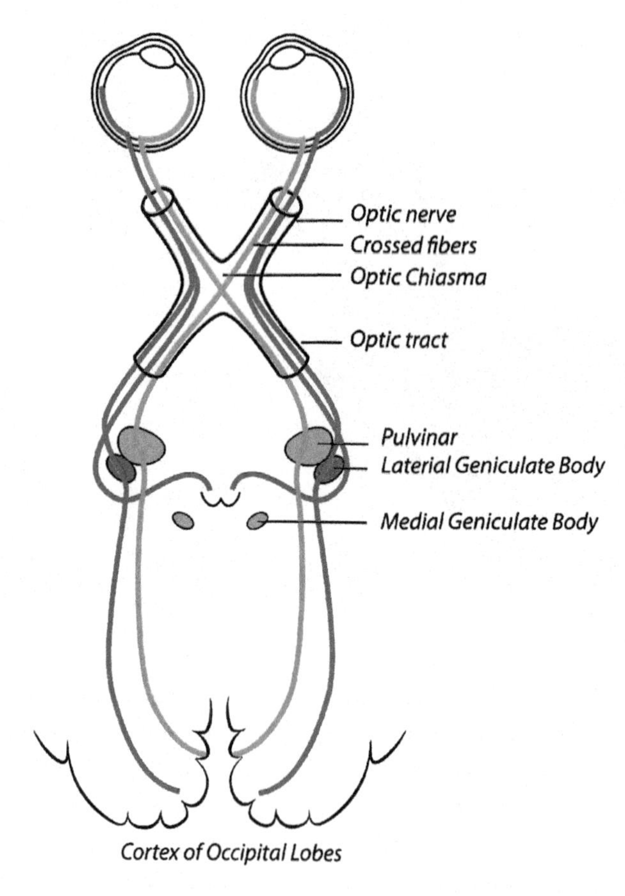

Figure 5. Each eye has a biconvex lens so the image viewed in the left visual field is viewed by the nasal retina of the left eye and also by the temporal retina of the right eye. The image is inverted, left-right reversed
(Adopted from Byrne (1997–Present) © 2015, Gregory P. Garvey. Used with permission)

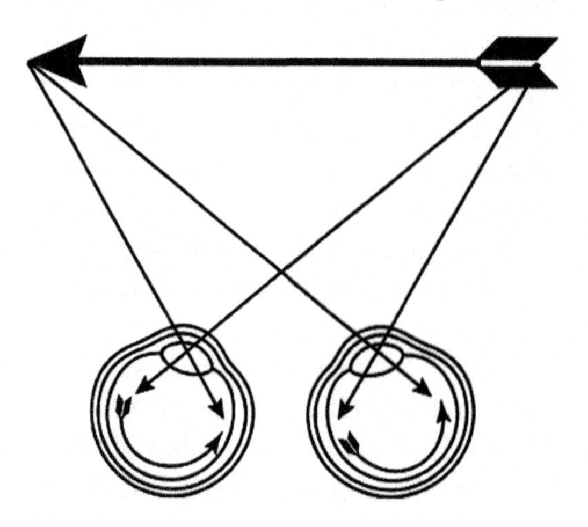

Figure 6. The lens of the eye functions as a biconvex lens so the image on the retina is inverted and left-right reversed
(Adopted from Uzwiak (n.d.) © 2015, Gregory P. Garvey. Used with permission)

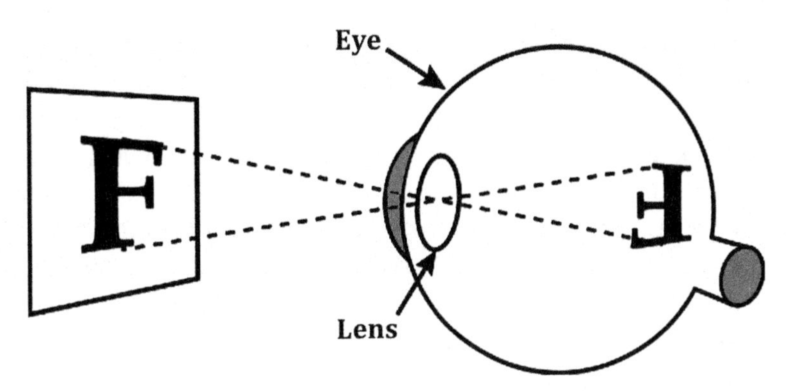

retinal rivalry, hemispheric dominance and suppression are factors (discussed below) most viewers can perceive a fusion of both separate inputs where the heads and faces merge into a composite perceived image.

Split-Brain Research

The split-brain interface was inspired by accounts of split-brain research, which led to important discoveries regarding specialization of brain functions in the left and right hemispheres in both animals and humans (Gazzaniga 2005). Neuropsychologist Roger Sperry (1968), who won the Nobel Prize in 1981, is generally credited with launching split-brain research conducting the first split-brain studies with Joseph E. Bogen at California Institute of Technology (Caltech). Sperry's collaborators also included his graduate student at the time, Michael S. Gazzaniga who separately made remarkable findings based on studies performed with split-brain patients (Gazzaniga, Bogen, & Sperry, 1962). These patients had undergone a surgical procedure that severs the bundle of connections between the two hemispheres known as corpus callosum. Sperry (1966) observed of these patients: "Everything we have seen indicates that the surgery has left these people with two separate minds, that is, two separate spheres of consciousness. What is experienced in the right hemisphere seems to lie entirely outside the realm of awareness of the left hemisphere. This mental division has been demonstrated in regard to perception, cognition, volition, learning and memory."

This procedure was first performed by Van Wagenen and Herren, in 1939 in order to limit the spread of seizures from one hemisphere to the other on a group of 26 people in Rochester, New York with intractable cases of epilepsy (Van Wagenen and Herren 1940). With a callosotomy the hippocampal commissure along with the corpus callosum are removed. A more radical procedure known as a commissurotomy, sections the corpus callosum, hippocampal commissure and the anterior commissure (Hirstein 2005). These procedures are considered a treatment of last resort and have been largely discontinued with the advent of more powerful drugs that could control epileptic seizures. Roughly thirty years after the initial experiments (Gazanniga 2005), Vogel and Bogel performed a complete commissurotomy (c) on a World War II veteran plagued with severe seizures due to a war time head injury (Bogen & Vogel,

1962). Gazanniga (1982, 2005) was able to study this patient known as "WJ" for the next five years. Later researchers were able to locate select others who had undergone the same procedure and study them.

These studies were designed to take advantage of the visual pathways where a visual stimulus presented in the left visual field is seen by the right hemisphere and a visual stimulus presented in the right visual field is seen by the left hemisphere. A typical experimental procedure involved having a right handed split-brain subject to sit in front of a screen and focus on a central fixation point (Kitterle, Christman, & Hellige, 1990; Rapaczynski & Ehrlichman, 1979; Van Kleek, 1989; Weismann & Banich, 1999, 2000).

If an image is flashed for 100–200 milliseconds in the left visual field it is perceived only by the right hemisphere. The brevity of the stimulus does not leave time for the subject to shift his or her eyes away from the central fixation point (Hirstein 2005). The left hemisphere, which has the capacity for language has not seen the visual stimulus and cannot say what it was. Miller (Wolman 2012) notes in regard to these experiments that "the right hemisphere is experiencing its own aspect of the world that it can no longer express, except through gestures and control of the left hand".

Other experiments used goggles or lenses to maintain image separation to guarantee lateralization of the visual stimulus. Bourne (2006) recommends using for stimulus presentation either eye-tracking equipment or electrooculography. Sperry (1968) used a procedure that permitted the split-brain subject to use one hand to feel unseen objects. An image is flashed for 100 milliseconds (to prevent the eyes from moving and engaging the other visual field).

While Sperry died in 1994 Gazzaniga continued his split-brain tests with these patients. Using a similar procedure Gazzaniga reports another experiment where the right brain sees a wintery scene of a snow man in front of a house covered in snow and the left brain sees a chicken claw. Eight cards with images are placed before the subject. Four of the images are related to the snow scenes and the other four images are related to the image of the chicken claw. Both the left and right hemisphere can see all eight cards. When told to select cards the split-brain subject chose with his right hand (left hemisphere) an image of chicken. With his left hand (right hemisphere) the subject chose an image of a shovel. Why asked why he made these choices the split brain subject said: "Oh, that's simple. The chicken claw goes with the chicken, and you need a shovel to clean out the chicken shed." Gazzaniga (1998) explains that because the split-brain subject's left brain did not know what the right brain saw (the snowy scene) the left brain rationalizes or confabulates the choice of a shovel as necessary to "clean out the chicken shed."

From numerous experiments he concluded that for right handed split-brain subjects the left hemisphere of the brain is dominant in language related skills such as writing, speaking, mathematical calculation and reading. Gazzaniga (2005) further argues that the right hemisphere is specialized for visuo-spatial processing. "Studies with split-brain patients have revealed right hemisphere superiority for various tasks involving such components as part–whole relations (Nebes 1972), spatial relationships (Nebers 1973), apparent motion detection (Forster, Corballis, & Corballis, 2000), mental rotation (Corballis & Sergent, 1988), spatial matching (Corballis, Funnell & Gazzaniga, 1999) and mirror image discrimination (Funnell, Corballis, & Gazzaniga, 1999)". However these capabilities may be distributed differently with left-handed or ambidextrous patients (Geschwind & Crabtree 2003). In a majority of left-handed split-brain subjects speech functions are in both hemispheres. Few left-handers have speech localized in the right hemisphere (McCarthy & Warrington, 1990).

The lateralization of skills in the two hemispheres is thought to be a result of the evolutionary history of bilaterally symmetric animals inherited from a common ancestor that appear over 570 million years ago. Left right symmetry apparent in humans in utero confers obvious survival benefits: having symmetric limbs make locomotion quick and energy efficient. As a basic body plan this is left right symmetric led to two eyes, ears, antennae and two hemispheres emerges in the neocortex – the outer layers of the brain. However there are asymmetries such as the heart or the appendix and differences between the hemispheres. Damage to Broca's area in the left hemisphere impairs speech but damage to the corresponding area in the right hemisphere does not. The right half of the facial fusiform area on the underside of the brain is critical for recognizing faces. This can be tested by viewing a face with the left eye (hence the right hemisphere) and subjects do a better job at recognizing faces than if they look only with their right eye. It is speculated that the optic nerve crosses to the opposite side hemisphere so the appropriate hemisphere controls the side of the body that correspondences to what is seen in retina.

Issues, Controversies, Problems

Since the original pioneering work by Sperry, Gazzaniga and others, split-brain research has continued at leading institutions primarily in the United States and in Italy. These studies have largely confirmed the same results (Gazzaniga 2005): "Severing the entire callosum blocks the interhemispheric transfer of perceptual, sensory, motor, gnostic and other forms of information in a dramatic way, allowing us to gain insights into hemispheric differences as well as the mechanisms through which the two hemispheres interact." Questions have been raised on whether some connections persisted at the cortical level (the cerebral cortex) in split-brain patients or that there were still subcortical (a part of the brain below the cerebral cortex) connections that remain untouched by surgical procedures. Lambert (1991) has shown there is evidence for interhemispheric integration persisting after surgeries. Franz, Waldie, & Smith, (2000) designed experiments that required split-brain subject to perform bimanual skills and concluded that such skills in split-brain subjects are still coordinated at the subcortical level. Other research (Corballis 1994) indicates that the two hemispheres share processing resources.

In the normal, 'intact' brain, the two hemispheres work together. Bourne (2006) describes four experimental methods used to measure interhemispheric cooperation using the Divided Field methodology. These include the Poffenberger paradigm (Poffenberger, 1912), still in today (e.g., used by Sperry and Gazzaniga as described above), where visual targets are presented to both the left and right visual field and the subject responds with either the opposite or same side hand as the stimulus: "the crossed uncrossed difference can be used as an estimate of the amount of time it takes for information to be transferred from one hemisphere to another." This is known as interhemispheric transfer time or IHTT. A second method is the redundant target paradigm (Todd, 1912) where the reaction times of presenting just one stimulus to one eye (one hemisphere) is compared with presenting two copies of the same stimulus to both eyes (and hemispheres). A faster reaction in the later instance indicates likely interhemispheric cooperation. Similarly a third method (Dimond and Beaumont 1972) compares the time to process pairs of stimuli presented unilaterally versus pairs presented bilaterally. A fourth method (Banich & Shenker, 1994) involves presenting three stimuli using alphabetic (Weissman & Banich, 2000), geometric (Weissman & Banich, 1999) and face examples (Compton, 2002). Bourne (2006) describes that stimuli are presented: "two in the top half of the display, one in each visual field, and the third in the bottom half of the display. The participant's task is to decide whether the bottom stimulus matches either of the stimuli presented above it."

The validity of these four experimental methods have been tested with both normal and split-brain subjects. For example in bilateral trials split-brain patients have no advantage (Mohr, Pulvermuller, Rayman, & Zaidel, 1994). In addition to work using functional Magnetic Resonance Imaging (fMRI) new brain imaging techniques combined with behavioural methodologies such as ITHH (Colvin, Funnell, Hahn, & Gazzaniga, 2005) promise further advances in understanding functional connectivity and recruitment through the corpus callosum between the two hemispheres in healthy individuals (Gazzaniga 2005). Gazzaniga (2008) describes the advantage of interhemispheric cooperation as follows: "The right hemisphere maintains an accurate record of events, leaving the left hemisphere free to elaborate and make inferences about the material presented. In an intact brain, the two systems complement each other, allowing elaborative processing without sacrificing veracity." Much has been made of the differences in capacities between the left and right brain hemispheres especially in the popular press spawning a mini industry of self-help and having a huge influence on educational theory.

Nobuyuki Kayahara's Spinning Dancer illusion (see Other Online Resources in the Appendix) has been promoted as a test for left or right brain dominance. However it is simply an example of an ambiguous or bi-stable image (Parker-Pope 2008). Several websites offer 'scientifically validated' self-administered tests to determine "which side of your brain do you wake up on in the morning? Find out with this test!" (psychtests n.d.). After taking the results are reported on a simple one dimensional scale from 0% to 100%. If a score is closer to 0% the personality is "more characteristically left-brained." A score closer to 100% reflects a right brain personality. A score in the middle represents a balance between the left and right brain "a level of perfect harmony – rather than trying to dominant each other, they work together to create a unique and well-balanced "you." Your spontaneous, impulsive, and free-flowing right brain creates an exciting and adventurous world, while your left brain helps you make sense of it and keep track of everything." The website includes the following disclaimer that "This report is intended for personal growth purposes only." However such psychological assessments are utilized by human resource professionals, therapists, athletic coaches and consultants and can lead to real impacts in hiring and retention.

In a paper that was greatly influential outside of neuroscience, Joseph E. Bogen (1975) noted that, "the brain is double, in the sense that each cerebral hemisphere is capable of functioning independently, each in manner different from the other." His argument is essentially that the research of the time suggested that a dichotomous theory of intelligence could be reflected, as Arthur Jensen suggests in a more diversified "curricula, instructional methods and educational goals and values that will make it possible to for children ranging over a wider spectrum of abilities and proclivities genuinely benefit from their years in school" (Jensen 1972). Many readers and later authors seized upon Bogan's discussion of "two kinds of intelligence" or parallel "ways of knowing" and his linkage to hemispheric specialization. Bogan (1975) quotes cognitive psychologist Neisser: "Historically, psychology has long recognized the existence of two different forms of mental organization. The distinction has been given many names: 'rational' vs. 'intuitive,' 'constrained' vs. 'creative,' 'logical' vs. 'prelogical,'...".

Neisser himself makes a distinction between "sequential processing" vs. "multiple processing" which involves simultaneous or independent thinking (Bogan, 1975). Bogan compiled a list of authors associated with dichotomous pairs of terms that represented two "types of intelligence" or "cognitive styles." In the 1980s a series of academic publications addressed the potential impacts of split-brain theory on education and the arts (e.g. Brooks, 1980; LeCompte & Rush, 1981; Henry, 1981; Hopkins, 1984; McLuhan, 1979). Murr and Williams, observe that spatial reasoning, symbolic processing and pictorial interpretation associated with the right brain are ignored in education. They call for "whole brain" learning where both hearing and seeing a story forces "sensory connections" to both the left and right hemispheres. The authors conclude with this celebratory invocation of the personal computer that should "become the facilitator of whole-brain learning, the essence of the corpus callosum, bridling the left and right hemispheres and connecting the book and the video cassette tape, the newspaper and television, visual statics and dynamics, science and the liberal arts, the West and the East, creating a world convergence toward common perceptions and singular realities" (Murr and Williams, 1988).

In many ways Betty Edwards' *Drawing with the Right Side of the Brain* epitomizes this faith in the under utilized powers of the right brain. Edwards introduces practical techniques for achieving accurate representations from observational drawing and its success can be measured in the number of editions published since the first edition. Much of the popularized versions of left-brained or right-brained tests originate in the demonstrated lateralization of brain function. This research has shown that the left hemisphere controls features of language and logical while the right hemisphere is specialized for processing visual comprehension and spatial information. However, research using neuroimaging techniques (Nielsen et al. 2013) challenges these earlier results: "*It has been conjectured that individuals may be left-brain dominant or right-brain dominant based on personality and cognitive style, but neuroimaging data has not provided clear evidence whether such phenotypic differences in the strength of left-dominant or right-dominant networks exist.*" The exaggerations and over generalization of pop-psychology about right-brain/left-brain theories have largely been debunked (Rogers 2013). Later research has shown that the brain is not nearly as dichotomous as once thought. For example, recent research has shown that abilities in subjects such as math are strongest when both halves of the brain work together. Today, neuroscientists know that the two sides of the brain collaborate to perform a broad variety of tasks and that the two hemispheres communicate through the corpus callosum.

The exhibition of the Split-Brain interface does not use the rigorous controls of a laboratory experiment. In the gallery setting users sit and lean slightly forward to peer through ScreenScope which is attached to an armature. There can still be a great deal of variation in how each individual sees the display. While the ideal is to achieve fusion of the two separate images of Clarence Thomas and Anita Hills this is not always possible for all viewers. As noted earlier binocular rivalry can lead to the suppression of one of the 'incongruent' or dissimilar images–and the viewer sees only one view. If the viewer has uncorrected amblyopia (lazy eye), which occurs in up to 5% of the population (Webber & Wood, 2005) it is likely the individual is viewing with one eye. It is generally thought the cause resides in the brain where one eye is 'turned off' to compensate for double vision (Levi 2013). This can also occur with strabismus (misaligned eyes or heterotropia) often referred colloquially as being cross–eyed or wall-eyed. Other factors can inhibit has severe refractive problems (anisotropia) colloguially known as extreme near or far sightedness or astigmatism. Should any of these conditions be present the viewer may be unable to see equally each image. The ability to view straight ahead with both through the ScreenScope with both eyes replaces a central fixation point or a chin and forehead rest used experimentally to stabilize the viewer's gaze. If the viewer is unable to fixate straight ahead then it is unlikely that an image is seen only by the opposite side cerebral hemisphere. Rather the eye fixates in such a way that the image falls on both the temporal (lateral) and nasal (medial) retina.

The Interpreter Mechanism

Yet with right handed split-brain patients Gazzaniga theorized that the left hemisphere served as a kind of "interpreter mechanism" with a tendency to fabricate or confabulate explanations. Gazzaniga describes an experiment, which displayed two stories separately to each hemisphere of a split-brain subject. While this individual's left hemisphere was dominant verbally, the right hemisphere had some verbal capabilities. "Two stories are presented, one to each half-brain. The left hemisphere quickly reports its story, followed by the right hemisphere offering its story in bits and pieces. After the left brain hears these semantic items, it combines both stories into yet a new one" (Gazzaniga 1985). This experiment in particular led to the conception of a split-brain interface.

The source story would normally be read from left to right as follows:

Story 1*: Mary Ann, May Come, Visit Into, The Town, Ship Today.*

The source story is decomposed into two stories made by selecting every other word from the original resulting in Story 2 and 3.

Story 2: *Ann Come Into Town Today*
Story 3: *Mary May Visit The Ship*

Using a specially constructed screen following the principles described above, word pairs are displayed so that the right and left hemisphere of the split-brain subject each see a single word. For example:

Left Side of Screen Right Side of Screen
Right Hemisphere (Story 3) Left Hemisphere (Story 2)
Mary Ann

Word pairs from Story 2 and 3 continued to be shown one word at a time as follows:

Left Side of Screen Right Side of Screen/
Right Hemisphere (Story 3) Left Hemisphere (Story 2)
May Come
Visit Into
The Town
Ship Today

Gazzaniga reports that when the split-brain subject was asked to recall the story, the subject responded "Ann come into town today." This is the story seen by the verbally dominant left hemisphere. The subject was then prompted to respond if that was the whole story. According to Gazzanniga the subject's right hemisphere "blurted out" an invented or confabulated story based on what it saw: "on a ship . . . to visit . . . to visit Ma!'" The subject was asked again to repeat the entire story again. Now that the left hemisphere has heard what the right hemisphere said the subject synthesized the two source stories into a new story: "Ann came into town today to visit Ma on the boat!" (Gazzaniga, 1985).

Gazzaniga speculates that this "interpreter mechanism" of the left hemisphere has a "creative narrative talent" which "is constantly looking for order and reason, even when there is none – which leads it continually to make mistakes. It tends to over generalize, frequently constructing a potential past as opposed to a true one." Gazzaniga suggests that the left hemisphere is prone to invent an explanation based on partial information it already knows. By contrast the right hemisphere is thought to be literal and truthful.

These intriguing results led to consideration of what if two related stories were displayed separately to the left and right hemispheres of a "normal subject" with an intact corpus callosum? How would two conflicting stories heard or read separately by each hemisphere affect interpretation and understanding? A split-brain interface was designed to test this conjecture–to independently target each hemisphere of a normal brain by delivering video and audio directly to the separate visual and auditory pathways.

Binocular Vision, Depth Perception and Retinal Disparity

With binocular vision (Latin bini for double combined with oculus for eye) a creature uses two eyes together creating overlapping field of views, which allows for the perception of depth. In contrast with monocular vision a creature may have two eyes, but each eye sees a different image. Fish, many birds, reptiles and lizards have monocular vision where the two eyes are located on the sides of the head (lateralized) and see two separate fields of view. The eyes of the chameleon are controlled by a rotating turret like, anatomical structure. This allows each eye to see independently. The chameleon can also move the eyes to converge the fields of view to focus the eyes together with overlapping visual fields. Some birds of prey are able to converge their eyes forward to create overlapping fields of view and thereby more accurate perception of depth.

Depth perception can be further classified into monocular (single eye) and binocular (two eyes) cues. Examples of monocular depth include: motion parallax when moving objects nearby quickly pass but distant object move slowly or even appear stationary; single point, two point and three point perspective–one point perspective is seen with the recession of a railroad track from a vantage point in the middle between the two rails. Three point perspective can be seen with tall buildings; aerial or atmospheric perspective occurs as objects get further and further in the distance they lose sharpness and the color becomes muted losing saturation (purity of the hue) and contrast (difference between black and white) due to thickness of the atmosphere; overlap or interposition–when one object overlaps another or blocks the view it is perceived to be closer; texture gradient discrimination–textures close by appear sharp and distinct whereas textures in the distant appear blurry and lack detail.

In human binocular vision two eyes are front facing and the visual fields overlap (Figure 7 shows the overlap of the two visual fields forming the binocular visual field). Typically in an adult with normal binocular vision the two eyes are, on average, 6.5 centimeters apart. Due to the location of the eyes, the retina of each eye sees near objects from a slightly different angle. This difference between the two retinal images is called retinal disparity. Ocular convergence occurs when coordinated movements of each eye look at the same object (Herring 1868). When the object is closer the degree of turning the eyes inward is greater. The region of binocular overlap is 120 degrees wide by 135 degrees in height (Wandell, n.d.).The brain fuses the two similar images from each eye into the perception of a single composite visual image known as binocular fusion. This single mental image is sometimes called the Cyclopean eye because the fused image is perceived to be between the two eyes (Julesz, 1971).

This area of overlap provides visual cues for the perception of depth, which is known as stereopsis. Depth perception makes it possible to view the world in three dimensions. Stereopsis can be readily demonstrated by holding a finger in front of the nose while focusing on a point in the far distance. Focusing on the finger causes convergence where both eyes turn inward toward each other. By closing and opening one eye after the other it is possible to observe the difference in angle of view of each eye and the resulting retinal disparity as the left and right eye see different sides of the finger.

Stereopsis and retinal disparity were described by Sir Charles Wheatstone (1838) in his seminal monograph published by the Royal Society, Wheatstone initially suggests these principles can be deduced

Figure 7. Binocular visual field
(Adopted from Figure 7 in Aprile, Ferrarin, Padua, Di Sipio, Simbolotti, Petroni, Tredici & Dickmann (2014) © 2015, Gregory P. Garvey. Used with permission).

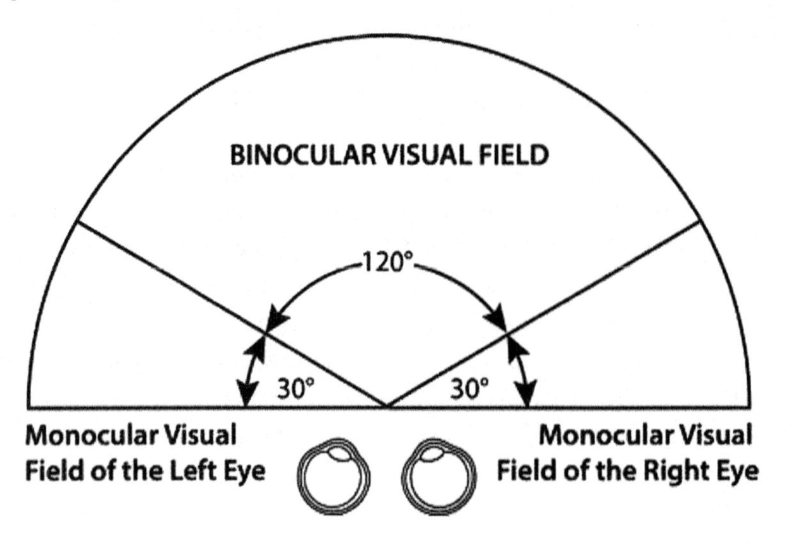

by simple perspective: "The appearances, which are by this simple experiment rendered so obvious, may be easily inferred from the established laws of perspective; for the same object in relief is, when viewed by a different eye, seen from two points of sight at a distance from each other equal to the line joining the two eyes." He illustrates this with the following figure, which shows a 'wireframe' of a cube as if seen from the left (a) and right (b) eyes (Figure 8 shows the cube as stereo image pair):

Without the aid of a stereoscope, the reader can fuse the wireframe images of the cube by using the cross-eyed viewing method (Simanek n.d.). This is accomplished by viewing the two cubes at a distance of approximately 6 to 7 inches and then gradually cross the eyes so the two images of the cubes begin to overlap. When three images appear focus on the center image and try to bring it into sharp focus and it will appear to be a 3 dimensional image. Normally the cross-eyed viewing method requires that the image to be seen by the right eye is on the left and the image to be seen by the left eye is on the right. This is particularly true with photographic pairs. However the cross-eyed method works with the above

Figure 8. A recreation of original figure 13 from Wheatstone illustrating Wireframe Cube Image Pair.
(Adapted from Wheatstone (1838). Public Domain)

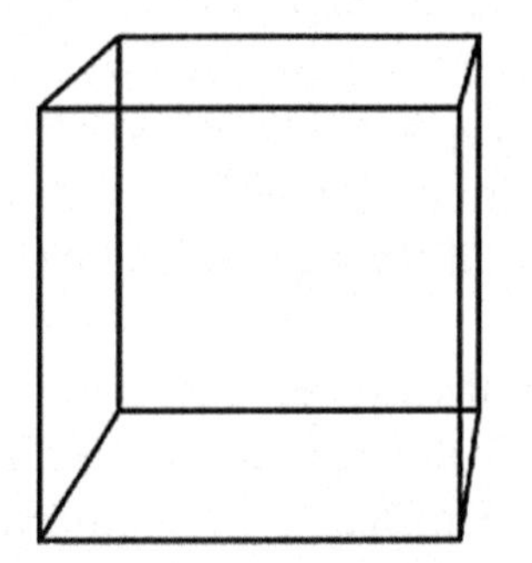

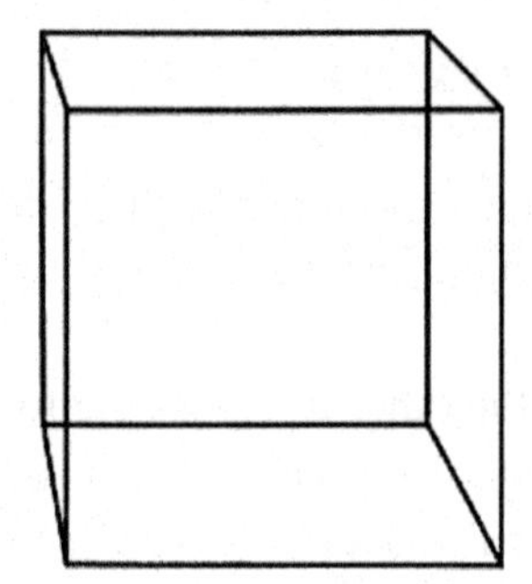

images because the cube is rendered as a wireframe and all 6 sides are visible. The illusion of 3D is possible because of corresponding or congruent points that fall on the retina (Figure 9 shows how these points fall on the retina).

In his original monograph Wheatstone (1838) describes retinal disparity as: "the projection of two obviously dissimilar pictures on the two retinæ when a single object is viewed, while the optic axes converge, must therefore be regarded as a new fact in the theory of vision." He restates this principle as follows and then poses a question which led to the design of the stereoscope: "It being thus established that the mind perceives an object of three dimensions by means of the two dissimilar pictures projected by it on the two retinæ, the following question occurs: What would be the visual effect of simultaneously presenting to each eye, instead of the object itself, its projection on a plane surface as it appears to that eye?" Wheatstone answered this question with his design of the first stereoscope (Figure 10 shows the original design of the Wheatstone Stereoscope). Wheatstone's design is seen from the front, which depicts the functional representation of how the stereoscope works.

Instead of the Object Itself

Writing about the stereoscope twenty-one years after Wheatstone's invention, Oliver Wendell Holmes (1859) inspired by Wheatstone's conjecture ("instead of the object itself, its projection on a plane surface as it appears to that eye?") designed a low cost stereoscope that quickly outsold more expensive European designs. In 1861 Holmes introduced a low cost hand held stereoscope (Figure 11 shows an

Figure 9. Corresponding points on retina
(Adopted from Encyclopedia Britannica. (1994) © 2015, Gregory P. Garvey. Used with permission)

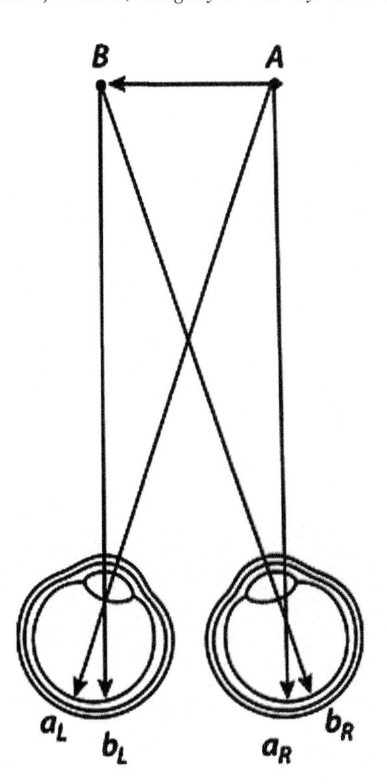

Figure 10. The Wheatstone Stereoscope
Wheatstone (1838). (© Public Domain).

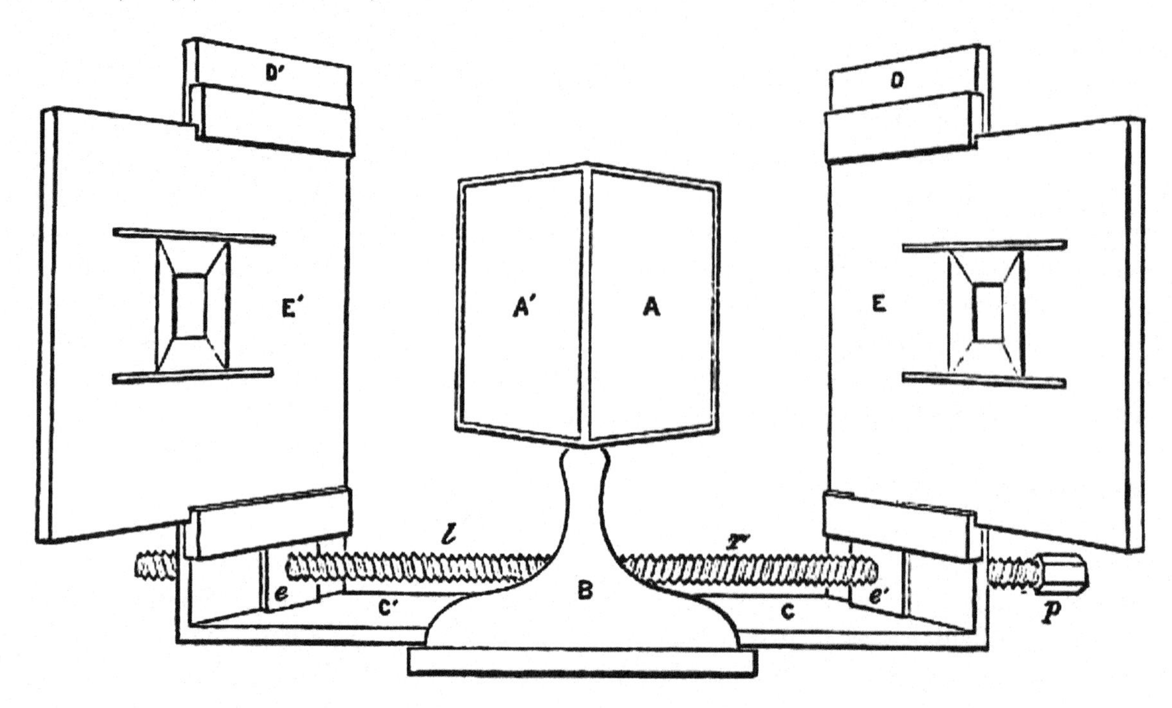

Figure 11. The Holmes stereoscope, with the inventions and improvements added by Joseph L. Bates, circa 1869)
(© Public Domain, Unknown, circa 1869)

advertisement for Holmes' stereoscope). He deliberately chose not to file a patent on his design – a kind of 'open source" invention which arguably was the first Virtual Reality viewer!

It soon became one of the most popular designs of its time. Commenting on its success Holmes wrote: "There was not any wholly new principle involved in its construction, but, it proved so much more convenient than any hand-instrument in use, that it gradually drove them all out of the field, in great measure, at least so far as the Boston market was concerned."

Wendell proclaimed a new epoch "in the history of human progress" where "Form is henceforth divorced from matter. In fact, matter as a visible object is of no great use any longer, except as the mould on which form is shaped. Give us a few negatives of a thing worth seeing, taken from different points of view, and that is all we want of it." What is remarkable about this statement is that Holmes could be describing Virtual Reality. The 'mould' is the polygonal model. The virtual camera can simulate 'different points of view' and all we need is the motion tracking head mounted display.

The Stereoscopes of the nineteenth century were further enabled by the still new technology of photography. Holmes recounts the necessary inventions of his time that led him to design an affordable hand held stereoscope. He acknowledges Wheatstone's invention and then mentions the development of the Daguerrotype in 1837 which "has fixed the most fleeting of our illusions, that which the apostle and the philosopher and the poet have alike used as the type of instability and unreality." Next is the light fast and permanent photograph affixed to a paper substrate first invented by William Henry Fox Talbot (Chisholm 1911) which "has completed the triumph, by making a sheet of paper reflect images like a mirror and hold them as a picture." Holmes celebrates the invention of photography as "This triumph of human ingenuity is the most audacious, remote, improbable, incredible" but he laments "It has become such an everyday matter with us, that we forget its miraculous nature, as we forget that of the sun itself, to which we owe the creations of our new art." Holmes goes one to describe the principles of binocular vision and stereopsis: "These exceptions illustrate the every-day truth, that, when we are in right condition, our two eyes see two somewhat different pictures, which our perception combines to form one picture, representing objects in all their dimensions, and not merely as surfaces." Two photographs of the same subject taken from slightly different angles (determined by the disparity of interpupillary distance) create the illusion of depth.

Holmes next describes stereo-photography: "A first picture of an object is taken, then the instrument is moved a couple of inches or a little more, the distance between the human eyes, and a second picture is taken. Better than this, two pictures are taken at once in a double camera." He mentions two ways to view a 'stereograph'. One is by squinting but this method as with the cross eyed method described above is difficult and fatiguing. The other method is to cut a convex lens in two, grind down one side to be flat and to join the two resulting lens together. Declaring this a 'squinting magnifier' it affords the ability to "with its right half we see the right picture on the slide, and with its left half the left picture, it squints them both inward so that they run together and form a single picture." David Brewster is credited with the introduction of lenses. Brewster's stereoscope from 1849 (Figure 12 shows Brewster's stereoscope) is a lenticular design (having lenses) permitted a more compact housing (Brewster 1856).

Holmes declares the stereograph of Ann Hathaway's Cottage in Shottery, England (the birthplace of Shakespeare's wife) to be "the most perfect, perhaps, of all the paper stereographs we have seen." (Figure 13 shows the Hathaway Cottage Stereograph). A copy of this stereograph is in the collection of the Boston Public Library and is reproduced here under a creative commons license (Keystone View Company1879–1930).

Figure 12. The Brewster Lenticular Stereoscope circa 1849 (Public Domain).

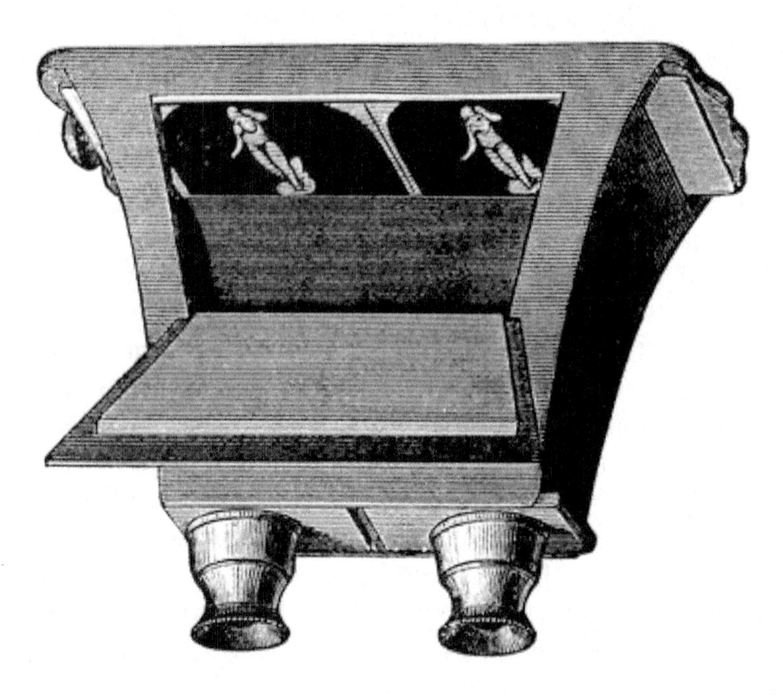

Figure 13. The Ann Hathaway Cottage, Shottery, England (Creative Commons BY-NC-ND).

He repeats his argument that "Matter in large masses must always be fixed and dear; form is cheap and transportable. We have got the fruit of creation now, and need not trouble ourselves with the core. Every conceivable object of Nature and Art will soon scale off its surface for us." Holmes envisions a future having billions of pictures requiring vast libraries "where all men can find the special forms they particularly desire to see as artists, or as scholars, or as mechanics, or in any other capacity. Already a workman has been travelling about the country with stereographic views of furniture, showing his employer's patterns in this way, and taking orders for them. This is a mere hint of what is coming before long." Holmes' predictions were in part dead on. Eventually mail order catalogs like Sears, Roebuck and Co. featured photography to promote hundreds of different goods including furniture and entire homes. However stereoscopes were still too cumbersome to be of practical use to salesmen. The Holmes stereoscope remained in the home and parlor and was relegated to the status of a curiosity and diversion. Interestingly one hundred twenty years later Jaron Lanier, Virtual Reality pioneer and proselytizer did predict that one day you could design your kitchen in Virtual Reality while inhabiting the body of a octopus (Huffman n.d.).

Binocular Rivalry

When discordant views are seen separately by the left and right eyes there are alternating moments of dominance and suppression. Wheatstone (1838) accurately described what happens when each eye views a dissimilar image having corresponding and non-corresponding, discordant or incongruent elements. (Figure 14 shows a recreation of Wheatstone's figure 25 depicting the letter "S" and "A") "each [the letter "S" and "A"] presented at the same time to a different eye, the common border will remain constant, while the letter within it will change alternately from that which would be perceived by the right eye alone to that which would be perceived by the left eye alone. At the moment of change the letter which has just been seen breaks into fragments, while fragments of the letter which is about to appear mingle with them, and are immediately after replaced by the entire letter."

Wade and Tgo (2013) credits Porta (1593) with first describing what happens when each eye sees two different images. Porta used a partition so each eye sees a different page of a different book: "To separate the two eyes, let us place a book before the right eye and read it; then someone shows another book to the left eye, it is impossible to read it or even see the pages, unless for a short moment of time the power of seeing is taken from the right eye and borrowed by the left." (Porta, 1593, pp. 142–143) Blake (2001) notes that Porta observed an essential feature of binocular rivalry where the vision in his right eye was dominant and his left eye suppressed. Later Dutour (1760) described an alternation between the perceived color when each eye viewed a different color (O'Shea 1999), Charles Wheatstone did a more systematic study using outline drawings. Randolf Blake (2001) points out that Wheatstone correctly observed "key aspects of rivalry, including the complete suppression of one of two discordant stimuli, the alternations in dominance between the eyes, the spatial fragmentation of the two images

Figure 14. A recreation of Wheatstone's original figure 25 illustrating dissimilar image pair
(Adopted from Wheatstone (1838) Public Domain).

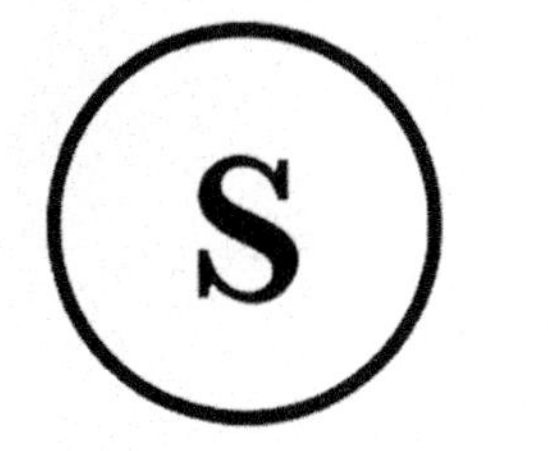

during times of transition, and the dependence of predominance on the physical characteristics of the rival stimuli." When different images are presented simultaneously and independently to the right and left eye "observers are presented with an ambiguous stimulus which supports two distinct interpretations, perception alternates between these interpretations in a random manner." This alternation in perceptual dominance is part of a more general phenomenon, known as perceptual bi-stability (Blake 2001), which can also be described as perceptual rivalry.

In addition to Wheatstone's original work Blake recounts other important work on perceptual instability including binocular rivalry (Levelt, 1968; Blake, 1989; Logothetis, 1998; Blake, 2001; Tong, 2001), ambiguous motion displays (Wallach, 1935, 1996; Hupé and Rubin, 2003), ambiguous depth perception (Necker, 1832; Rubin, 1921, 2001), and random dot kinematograms (Julesz, 1971). Bitstable images are another example of perceptual instability. The Schroeder Stairs, Necker Cube, and the Vase/Faces (each shown by Figure 15) are well-known examples that can be perceived with monocular viewing (Breese, 1899; Campbell and Howell, 1972). (For *The Young Lady Versus Old Lady Optical Illusion* see Other Online Resources in the Appendix). For some researchers these bistable images are hypothesized to result from underlying common mechanisms (Andrews and Purves, 1997). O'Shea (2004) suggests that the processing of binocular rivalry occurs at a lower level: "Processing of the other studied bistable phenomena must involve neurons sensitive to the orientation, colour and location of stimuli on the retina. Such neurons are located at low levels of the visual system." The alternation between figure and ground (Rubin, N. 2001) and the underlying neuronal mechanisms also likely play a role.

Elsewhere researchers (Meenes, 1930) also describe subjects reporting "patchwork" rivalry similar to Wheatstone's account "fragmentation of the two images" in place an alternation between the dominance of one eye's view versus the other eye's view. Longer viewing times of so-called rival (discordant) targets or images lead to an increase in reporting the perception of this "patchwork" rivalry (Hollins & Hudnell, 1980).

Measurable features of binocular rivalry include that of predominance of the total view time for a rival pattern, the stimulus strength as compared with the rival pattern. Stimulus variables also include contour density, spatial frequency, motion, and pattern contrast (Levelt, 1965). Blake (2001) notes that "a 'stronger' rival target (e.g., one that is higher contrast than the other) enjoys enhanced predominance, defined as the total percentage of time that a given stimulus is visible during an extended viewing period."

An important finding is the Eye Exchange Experiment (see Online Demonstrations in the Appendix). Normally we are not aware of which eye is dominant–we simply see the stimulus pattern regardless of which eye is doing "the seeing." This experiment allows a subject to view two competing patterns and

Figure 15. Common bi-stable images: Schroeder Stairs; Necker Cube; Face/Vase (Ⓟ Public Domain).

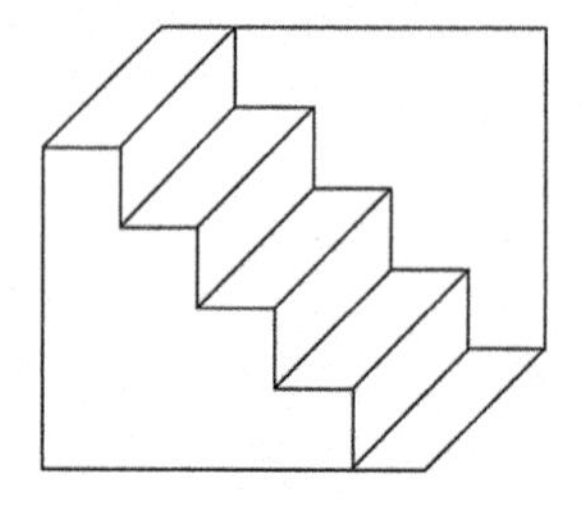
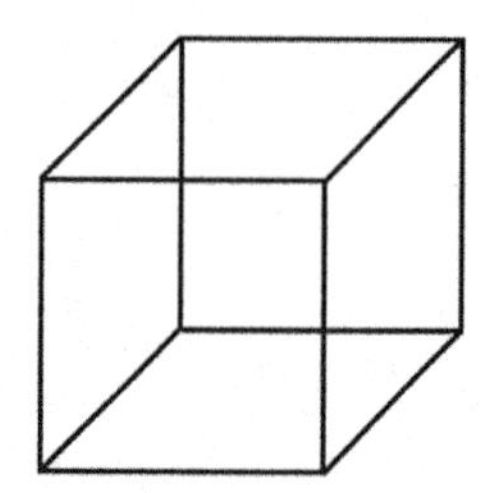
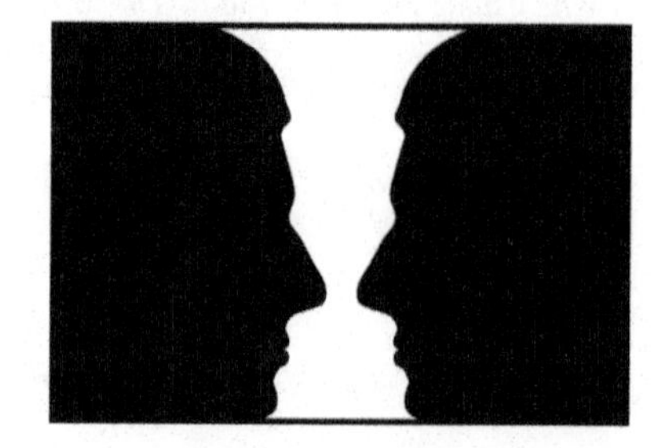

one is dominant. However the subject can press a button, which causes the two rival patterns to switch from one eye to the other. Under controlled conditions Blake (1979) drew the following conclusion: "the dominant pattern abruptly becomes invisible and the previously suppressed pattern becomes dominant, implying that it was the region of an eye that was dominant, not a particular stimulus." This outcome holds for meaningful objects such house or face patterns (Blake, Westendorf, & Overton 1979; Walker & Powell, 1979).

The Cheshire Cat

The phenomenon of binocular rivalry was considered to be something of a "laboratory curiosity" but has been of greater interest in neuroscience for not only understanding perception but as a way to study the neural basis of visual awareness at the level of individual neurons (Logothetis, 1998; Engel et al, 1999) and studies using brain imaging (Tong et al, 1998; Polansky et al, 2000). Blake and Logothetis (2002) argue that "Instead, multiple neural operations are implicated in rivalry, including: registration of incompatible visual messages arising from the two eyes; promotion of dominance of one coherent percept; suppression of incoherent image elements; and alternations in dominance over time."

To further the understanding of the links between visual awareness and consciousness Crick and Koch (1992, 2003) underscored the importance of the need for new theories and research into the neural basis of attention and short term memory supported by studies at the molecular, neurobiological level in conjunction with clinical imaging studies. They speculate that binocular rivalry may involve sets of neurons firing that correspond directly to visual awareness. The Cheshire Cat effect is vivid demonstration of binocular rivalry, attention and the relationship to awareness. The viewer holds a mirror up to the nose so the field of vision is divided. The mirror reflects a blank wall on the right seen only by the right eye. The left eye sees only a cat seated on stool. If the viewer waves his/her right hand in the visual field of the right eye (which sees the blank wall reflected in the mirror) and then into the visual field of the left eye (which sees the cat), "The result is that the cat may disappear. Or if the viewer was attentive to a specific feature before the hand was waved, those parts —— the eyes or even a mocking smile —— may remain (Crick and Koch 1992)." The phenomenon of the lingering facial feature, led to this effect being called the Cheshire Cat Effect referring to the depiction of Cheshire Cat's smile in Lewis Carroll's (the pseudonym of Charles Lutwigde Dodgson) Alice's Adventures in Wonderland (1865). Important research continues (Blake et al. 2014) in how binocular rivalry offers a powerful framework for the study of the neural correlates of conscious and visual perception.

A comprehensive list of research papers on binocular rivalry compiled by is provided in the Appendix in the Other Online Resources section.

The Auditory Pathways and the Split-Brain Interface

The split-brain interface requires the use of stereo headphones to properly experience the work. The proper orientation of the headphones should be where the left and right stereo channels match what is seen in the left and right visual fields. Specifically for *The Split-Brain (Dichoptic) Interface: Thomas v. Hill* Anita Hill is seen by the left eye and heard by the left ear. Similarly Clarence Thomas is seen by the right eye and heard with the right ear. However where the ScreenScope viewer exploits the visual pathways to deliver the digital video signal to the opposite side hemisphere, the auditory pathways while initially lateralized (starting with sounds heard in each ear) project to both sides the brain in a complex of neural connections still poorly understood.

The ear has three major anatomical parts: the outer, middle and inner ear. The sense organs of the inner govern both hearing, posture equilibrium, head and eye movements (Hawkins n.d.) When sound waves reach the ear the outer ear or pinna reflects and attenuates the sound. These changes contain information about direction. The auditory canal is approximately 2 centimeter in length serves to amplify the sound. As sound pressure waves impinge on the ear drum or the tympanic membrane the wave energy is converted to mechanical energy in the air filled middle ear. The mechanical energy is conducted by ossicles–the malleus (hammer), the incus (anvil) and stapes (the stirrup), which articulates (connects) to the oval window. Working together the ossicles mechanically convert the lower pressure waves to higher pressure waves required for the fluid filled inner ear. The cochlea in the inner converts the mechanical wave information to nerve impulses (Figure 16 shows the major anatomical features of hearing). Within the cochlea the organ of Corti has hair cells that directly transform the liquid wave forms into nerve impulses. It is beyond the scope of this chapter to describe further anatomical and bio chemical details of the inner ear or to cover the vestibular system that governs balance.

Nerve fibers from the auditory nerve innervate (connect to) these hair cells, continue and join the vestibular nerve forming the vestibule cochlear nerve also know as the eighth cranial nerve. The auditory pathways from each ear are bilateral meaning they go to both sides of the brain. The cochlear nerve connects to cochlear nuclei, which in turn sends auditory information to olivary nuclei on both sides of the brain stem. The medial olivary nuclei respond to arrival time differences and the lateral olivary nuclei

Figure 16. The anatomy of the human ear
(Adopted from Chittka and Brockmann (2005)© Public Domain)

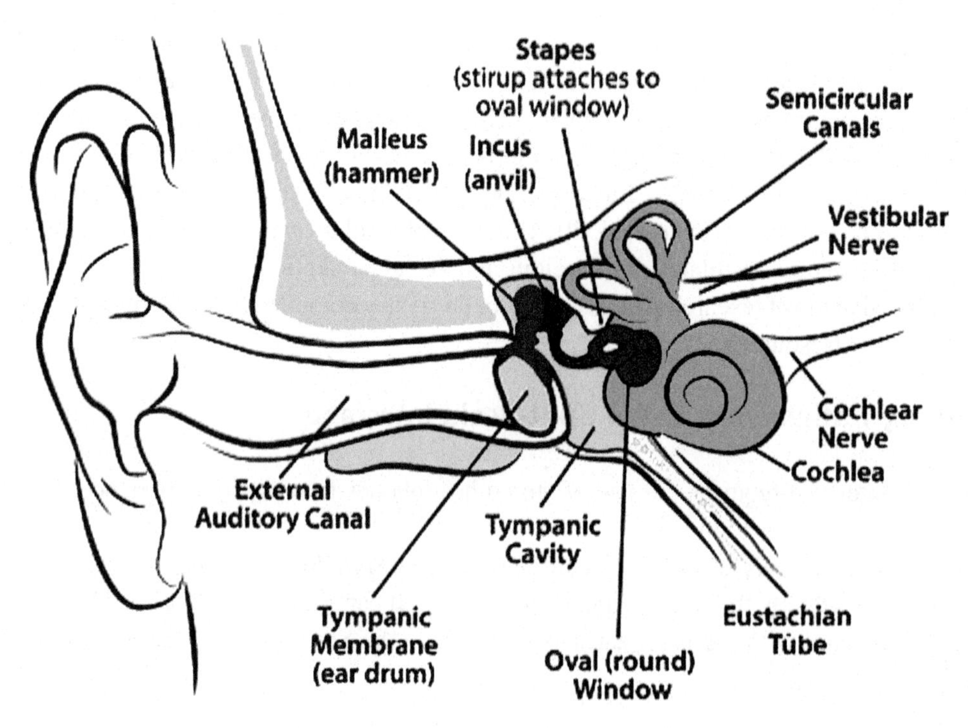

Figure 17. The auditory pathways and the primary auditory cortex
(Public Domain, Recreation under CC BY-SA 3.0 and GFDL)

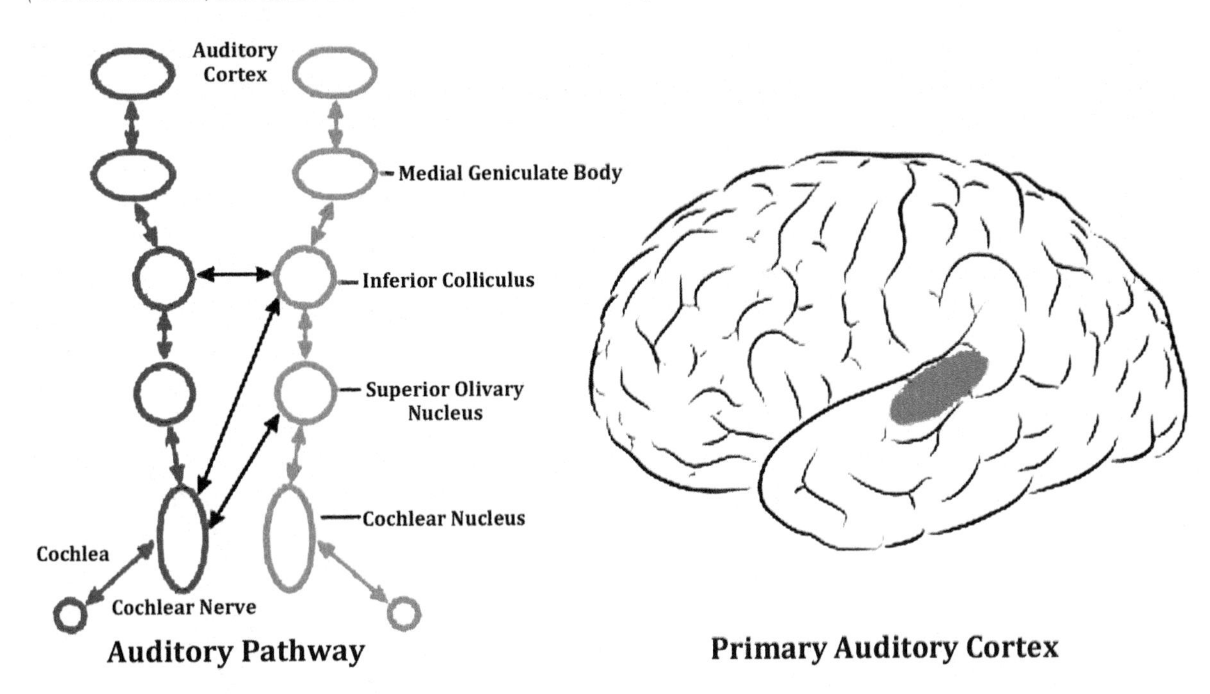

respond to amplitude differences. The cochlear nuclei also send information up to the mid brain to the inferior colliculus in both sides of the brain. Pathways continue from inferior colliculus on each side to the medial geniculate nucleus and then to the primary cortex (Gray n.d.) located in the temporal lobe (Figure 17 shows the auditory pathways and the location of the primary auditory cortex). Because the complexity of projections to both sides of the brain the auditory pathways have been difficult to study.

As previously noted (Garvey, 2002), hearing simultaneously the opening statements of the two protagonists (Thomas and Hill) is a kind of multi-tasking polyphony (literally many voices). Testimony is given slowly by actors in this drama. Each word of their sworn testimony is carefully articulated. Rather than hearing each word at the same moment, the spoken words alternate. For many viewers (listeners) it is possible to follow both. The viewer/listener multi-tasks by shifting attention and focus rapidly back and forth from one speaker to the other.

Another factor at play may be what is called selective auditory attention or selective listening. This phenomenon was first described by cognitive scientist Colin Cherry (1953) became known as the cocktail problem where one is able to follow one conversation against a background of other noisy conversations. Cherry's experimental method played two different messages to the left and right ear. Subjects were instructed to pay attention to just one message in one ear while ignoring the other message in the unattended ear. This is known as shadowing.

Dichotic listening tests were introduced by Broadbent (1954) and further developed by Kimura (1961a, 1961b) and are used to test different models of attention. With the dichotic method two different auditory stimuli are presented to each ear and the subject is tested on the ability to accurately identify these sounds. Dichotic listening tests offer another experimental method for the study of laterality or hemispheric differentiation. Broadbent concluded from his early work (Broadbent, 1954), that subjects attended to a message one ear at a time. Kimura (1961a, 1961b) used a dichotic auditory technique where competing verbal stimuli are presented to the two ears simultaneously. Kimura describes a "right-ear effect." Also known as "right-ear advantage" it is thought this is due to left hemispheric dominance for language processing and the crossing of auditory pathways from the ear to the brain.

Moray (1959) presented subjects' own names to their unattended ear and they responded with recognition. Treisman (1964) contributed to an attenuation model of attention based on work that showed that subjects understood words heard by the unattended ear if these words were somehow related to the message that the subject was attending to. Perhaps this is the better explanation for ability of most listeners to follow and grasp the testimony of both Hill and Thomas.

An interesting historical footnote to a discussion of the neural pathways of hearing is the work of Julian Jaynes (1976, 2000). Jaynes proposed an intriguing theory of the rise of the conscious mind. In Jaynes account prior to 1000 BCE humans were preconscious or possessed what Jaynes called the bicameral mind. Auditory hallucinations in the right hemisphere were interpreted as the voices of the gods. Commands heard in the right hemisphere were communicated via the corpus callosum to the left hemisphere and immediately put into action "to hear is to obey." The development of the technology of writing contributed to the "breakdown" this "bicameralism" of the left and right hemispheres. This in turn led to the development of consciousness, as we understand it today. While criticized as being an untestable theory Jaynes' account offers compelling ideas that are fertile ground for artistic explorations.

The Arts and Ideas

While the concept of a "split-brain interface" is directly inspired by experiments with split-brain patients there are many other discoveries and developments in the arts as well as in neuroscience and psychology. Binocular disparity and dissimilar visual inputs have long be observed and commented on. In 1664 Rèné Descartes published *Traité de l'homme* (Treatise on Man). According to Howard and Rogers (1995) Descartes based his account of vision on Kepler's ideas of image formation but retained "Galen's notion of animal spirits and ventricular projection." An illustration (Figure 18 shows this image) from this treatise accurately depicts binocular vision, depicting how the optic nerve of each eye projects back to the same side of the brain so that fibers originating in the retina project to the lining of the ipsilateral cerebal ventricle. According to Descartes "particles" from each point on the retina are then mapped to the pineal gland (the purportive center of the soul) to form a single image. In Descartes account the pineal gland, in turn manipulates "fluids" to control muscles.

A century and half before Wheatstone designed the first stereoscopic Sébastian Le Clerc accurately described retinal disparity and took Descartes to task. In 1679 Le Clerc wrote the following criticism of Descartes' theories of vision:

Monsieur Descartes having considered that according to his principles external objects should make an impression on both eyes, and that the soul neverless had only one perception believed that the images of the same object found in the two eyes are reunited in the brain, but if this great genius had reflected a little more on the demonstrations which he gave in his Treatise on Man, he would have recognized that the images in the two eyes although produced by the same object, are different, and because of these differences their reunion is impossible. (Le Clerc, 1679, 44-46).

Figure 18. Illustration showing binocular vision from Traité de l'homme by Rèné Descartes (1644) (Public Domain).

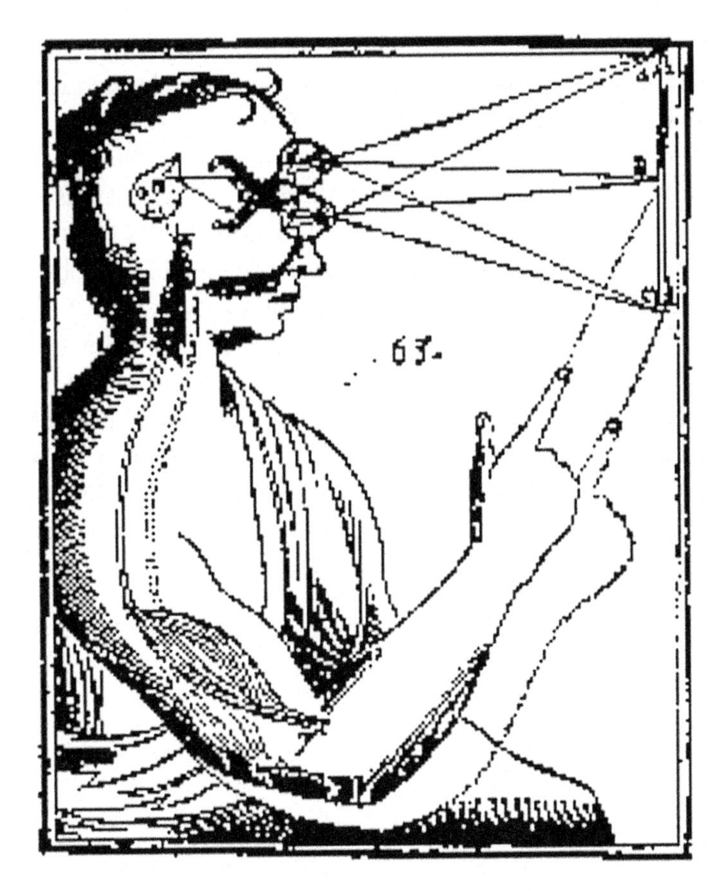

It turns out that Descartes was right about the projection of fibers from the retina of the eye into the brain where a unitary image is formed. However his claim that this took place in the pineal gland was wrong.

It was noted above that Dutour (1760) first described an alternation between the perceived color when each eye viewed a different color (O'Shea 1999). The 19th century French scientist Michel Eugène Chevreul (1860) noticed similar properties of colors and termed this the principle of simultaneous contrast. Chevreul developed dyes and in describing the perceptual effects of juxtapositions of pure patches of color. Chevreul's work inspired in part, the experiments of the French Impressionists (e.g. Monet) with broken brushwork. At a certain distance individual brushstrokes of pure color would perceptually fuse and be perceived as a third color. A patch of yellow next to blue is perceived as green. Subsequently the post-impressionists (notably George Seurat) used small dots or points of color to created vibrant effects. This became known pointillism or divisionism. Such techniques suggest that the artifacts of the artwork are perceived as a subjective experience seen only in the mind's eye. Moreover these techniques opened the door to the exploration of viewing simultaneously dissimilar elements in images.

An important development in parallel was the shift in the subject matter of art: paintings of gods, goddesses, heroes, kings, generals and aristocrats gave way to the depiction of the working class, the common people, the everyday and the pedestrian. As a genre till life painting allowed artists the luxury of exploring colors, textures and simulations, which set the stage for abstraction. In the late Renaissance Giuseppe Arcimboldo (1527–1650) did a series of paintings on the subject of the four seasons. Winter, spring, summer and fall depicted faces in profile made up of plants and vegetables commonly available at that time of year. Contemporary American artist Philip Haas (Figure 19 shows Haas's outdoor sculpture) created scale models of these paintings in fiberglass as monumental sculptures.

The Spanish surrealist Salvador Dalí was fascinated by such work. Many of Dalí's paintings incorporate bi-stable images. For Dalî these images opened the door to a kind of reasoning madness essential to his paranoiac-critical method. The surrealists explored "abnormal" mental states pursuing techniques such as automatic writing in an attempt to unlock the unconscious (inspired by the writings of Freud) to stimulate creativity and challenge conventional notions of reality.

Dalí's pursuit of "optical insecurity" led to an interest in stereoscopic images. Dalí painted pairs of photorealistic stereoscopic images, which were to be viewed by means of two angled mirrors placed at 45 degrees, similar to the first stereoscopes invented by David Brewster and Charles Wheatstone. One painting "Fire in the Borgo; Is Athens Buring?" presents the viewer with copies of two paintings by Raphael: "The School of Athens" and "Fire in the Borgo." (see Other Online Resources in the Appendix). While each painting depicts similar architectural features (e.g. an arch) each painting is dissimilar. Dalí

Figure 19. Philip Haas. The Four Seasons (After Arcimboldo), 2010–2011. Painted and pigmented fiberglass
(Photograph © 2015, Gregory P. Garvey. Used with Permission).

unified the paintings by superimposing a spiral of multi-colored squares on each of the two paintings. These squares are offset on each painting to create a vivid stereoscopic three-dimensional effect. Today the reader can easily experiment with similar setup by placing two mirrors at 90 degrees to one another and setting up either printed images or even display monitors to be reflected by the mirrors (see figure 10). To view the effect position one self so you see the reflected images from each mirror as indicated in Wheatstone's original depiction.

The French surrealist writer, Raymond Roussel (1935, 1995) describes how he used "a tiny pendant opera-glass, of which each tube, two millimeters across and made to go tight against the eye, contains a photograph on glass: one of Cairo bazaars, the other of a quay at Luxor". He found inspiration for his writing in such unusual perceptual effects. Viewing two dissimilar images in this fashion is a direct historical precursor to the Split-Brain Interface.

It is part of a tradition of experimentation for visual artists to explore different mental, psychological and perceptual effects as part of the artistic process. The reader can easily experiment with such aesthetic perceptual effects in which the viewer has to consciously struggle to merge two dissimilar images. The reader could create a similar viewing experience by the use of two mobile phones and a piece of card-board. Hold the cardboard in front of ones nose so that each eye sees only the image or video on the phone immediately in front of the eye. The reader can use Google Cardboard (https://www.google.com/get/cardboard/) along with a compatible Android phone to explore related optical effects that deliver dissimilar images to each eye.

Today artists, computer graphic modelers, animators, game designers and virtual world builders use these techniques in combination to create compelling immersive and engaging 3D experiences. For the computer graphics artist and game designer software and game engines render 3D views where many of these effects are by product of the rendering engine. Yet many digital games explore alternative worlds consisting of two and half dimensional spaces for side-scrolling platformer games. These games restrict movement along a 2D plane yet have limited depth that provides the illusion of dynamic movement through motion parallax.

In conventional drawing, painting and animation a student learns how to create each effect separately and then in combination. In basic drawing typically one point perspective is introduced first, followed by two point and three point techniques. So called 'scientific perspective' based on projective geometry can then be combined with other visual depth cues such as relative object size, overlap, motion parallax and atmospheric perspective to create a vivid visual experience of depth and space. While it can be tedious to analyze individual objects it heightens observational skills. In combination with learning about art history students learn that there are alternative systems of visual representation such as the canon of representation of ancient Egypt, the paraline, isometric or non-converging elevated viewpoints of seen in Chinese and Japanese screen paintings or of course the ground breaking experiments of the cubists in combining multiple angles of viewpoints in a single 2D framed painting. The fragmentation of viewpoints into individual facets or planes became an aesthetic end in itself.

Inspiration

Wheatstone's original monograph (1838) includes an illustration of a stereoscopic image pair depicting a single line. (Figure 20 shows Wheatstone's original figure 23). He describes this illustration as follows: "Present, in the stereoscope, to the right eye a vertical line, and to the left eye a line inclined some degrees from the perpendicular (Figure 23); the observer will then perceive, as formerly explained, a line, the extremities of which appear at different distances before the eyes."

This simple illustration from Wheatstone's monograph inspired the design of an interactive installation entitled *The Wheatstone Stereoscopic Random Line Pair Generator with Fitness Function for Non-objective Art* (Garvey 2011). Wheatstone's stereoscopic pair shown above by Figure 20 inspired the initial random pair generation (Figure 21 shows that initial seed pair initial pair). Viewers in the gallery wear stereoscopic glasses in order to see the stereoscopic effect. The motion of each viewer will trigger the projection of randomly positioned stereoscopic pairs of lines projected on each wall of the gallery forming a dynamic, evolving, emergent, stochastic, 'non-objective' composition. The goal is the creation of 3D generated stereoscopic wire-frame pairs both as stills and animations. The brain attempts to fuse the two inputs into one coherent but virtual image seen only in the "mind's eye."

The Generative Composition

In the actual installation motion by the viewer wearing a stereoscopic headset will trigger the generation of randomly positioned pairs of lines that are displayed within demarcated rectangular areas. These designated rectangular areas are like picture frames in which an abstract or non-objective composition emergences. In the original proposal, audio is created by using line length to calculate audible frequencies. Z-depth will be mapped to generate loudness or softness. The 2D Fitness Function is a simple evolutionary algorithm described as follows: "Overlapping lines will be discarded but the position of these lines will be used to generate a new pair that is constrained to a best fit without overlap. Surviving pairs will be added to the database. The attributes (position, length) of 'parents' are inherited by next generation with random perturbations (Garvey, 2011)." As viewers move through the gallery new stereoscopic pairs are generated. If the newly generated line pairs overlap (Figure 22 shows a sample overlapping stereoscopic pair), those overlapping lines are discarded. The elimination of overlapping lines is done to preserve the illusion of 3D depth.

New lines are generated that inherit properties of the initial pair. If these lines are a best fit and don't overlap they 'survive' into the next generation (Figure 23 shows non-overlapping pair).

As the system responds to the viewer's movement a complex image emerges (Figure 24 shows an sample emergent image after 10 generations).

Figure 20. Wheatstone's original figure 23 demonstrating stereoscopy
(Adopted from Wheatstone (1838), © Public Domain).

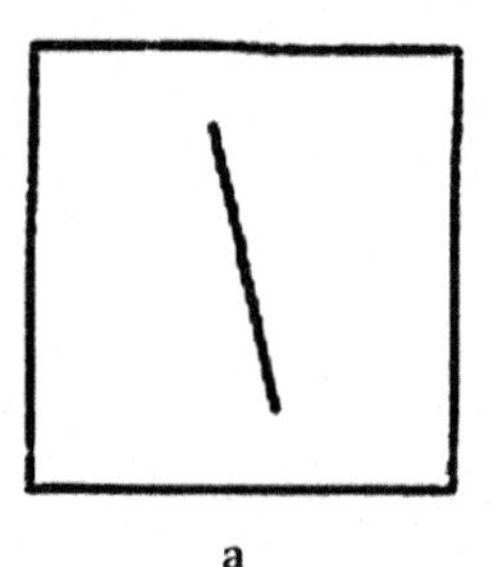

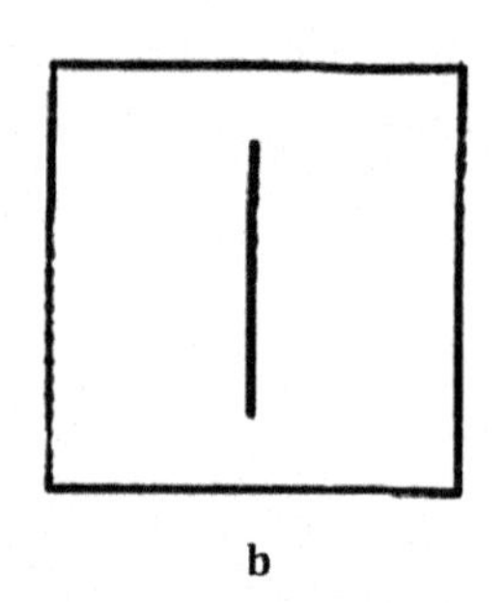

Figure 21. The Wheatstone Seed based on figure 19.
(© 2011, Gregory P. Garvey. Used with permission).

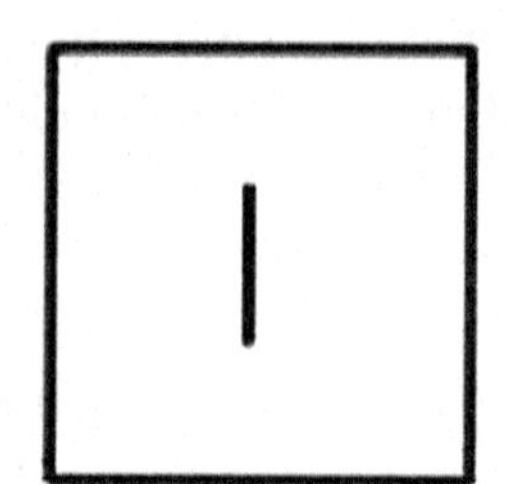

Figure 22. Random pair generation
(© 2011, Gregory P. Garvey. Used with permission).

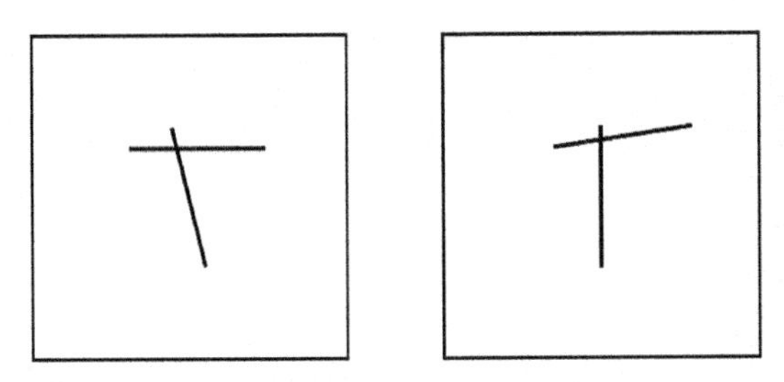

Figure 23. After discarding overlapping lines, replacement is with non-overlapping best fit lines
(© 2011, Gregory P. Garvey. Used with permission).

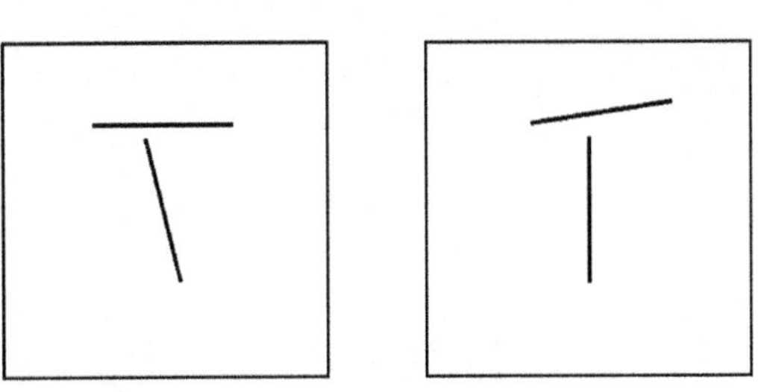

The displayed images continue to add stereo image pairs building complexity for multiple generations until the viewer leaves the gallery space (Figure 25 shows 50 generations).

The original proposal for the installation of *The Wheatstone Stereoscopic Random Line Pair Generator with Fitness Function for Non-objective Art* (Garvey 2011) planned for multiple displays updating continuously triggered by viewers' movements in the gallery space. (Figure 26 is a computer graphics visualization of the proposed installation).

Figure 24. After 10 generations of random stereoscopic pairs
(© 2011, Gregory P. Garvey. Used with permission).

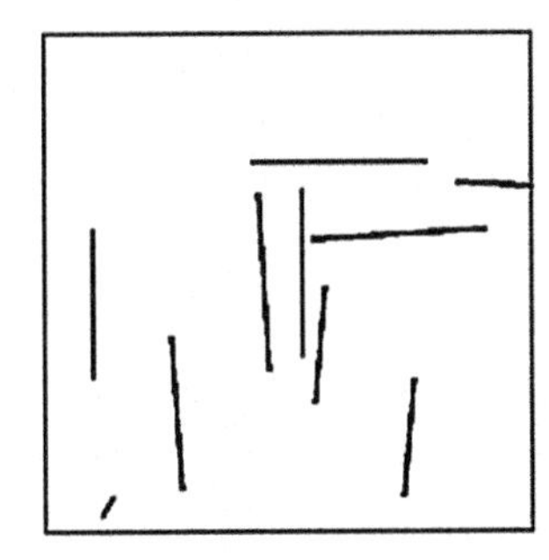

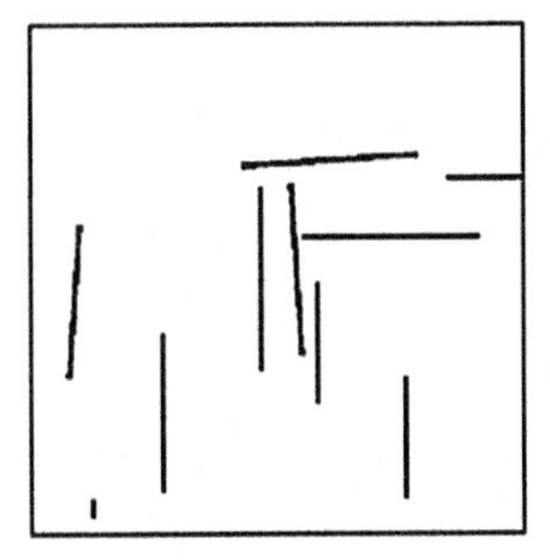

Figure 25. After 50 generations of random stereoscopic pairs
(© 2011, Gregory P. Garvey. Used with permission).

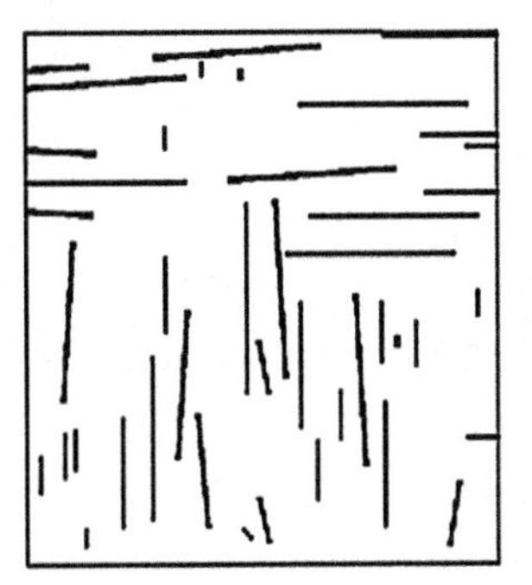

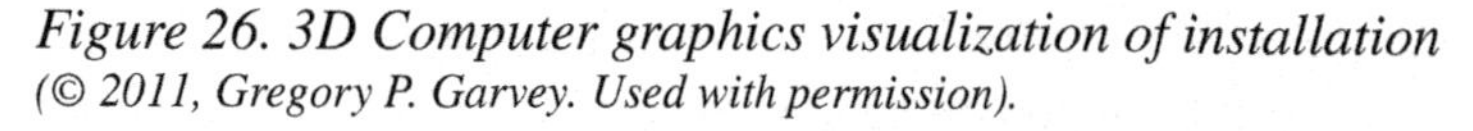

Figure 26. 3D Computer graphics visualization of installation
(© 2011, Gregory P. Garvey. Used with permission).

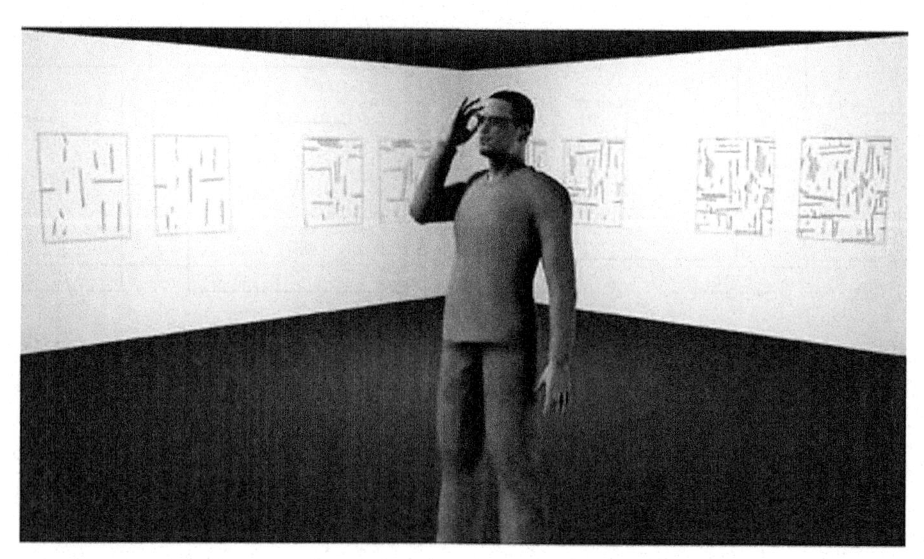

Homage to the Square

Joseph Albers (1975) explored the "interaction of color" first as a series of paintings and then in book form. This publication and the documentation of the painting series were hugely influential especially for artists studying color theory. Albers built upon the discoveries of Chevreul (discussed above). By reducing or abstracting to a simple depiction of a flat square within a square Albers systematically explored hundreds of color relationships: analogous, complementary and value contrasts. Using Albers Homage to The Square series as a point of departure a 2D computer generated animation [Garvey 2005] was created entitled "Homage to Square: Suprematist Composition." This series of work is given a name that also references the early 20th century Suprematist paintings of the Russian constructivist Kasimir Malevich. Malevich sought to reduce art to pure sensation by the use of pure form. This series explores color relationships much like the work of Albers while keeping a strict restriction of the simple form of the square.

The installation shows an animation of two similarly size squares presented side by side. In each square another smaller square is inscribed. Colors pairs continuously vary from warm to neutral to cool (Figures 27, 28, and 29 show sample inputs of different warm, cool and neutral values).

When viewed the ScreenScope Viewer most people can fuse the two dissimilar inputs into a coherent overlay of one square on top of the other (Figure 30 shows a simulation of what the viewer's perception of the overlapping images).

SOLUTIONS AND RECOMMENDATIONS

Recent developments in Virtual Reality Head mounted displays offer accessible technology to explore stereovision and possibly effect of binocular rivalry. With the introduction of low cost headsets like the Oculus Rift, virtual reality is now an accessible and robust technology for playing 3D games and even

Figure 27. Warm pairs
(© 2005, Gregory P. Garvey. Used with permission).

Figure 28. Cool pairs
(© 2005, Gregory P. Garvey. Used with permission).

Figure 29. Neutral pairs
(© 2005, Gregory P. Garvey. Used with permission).

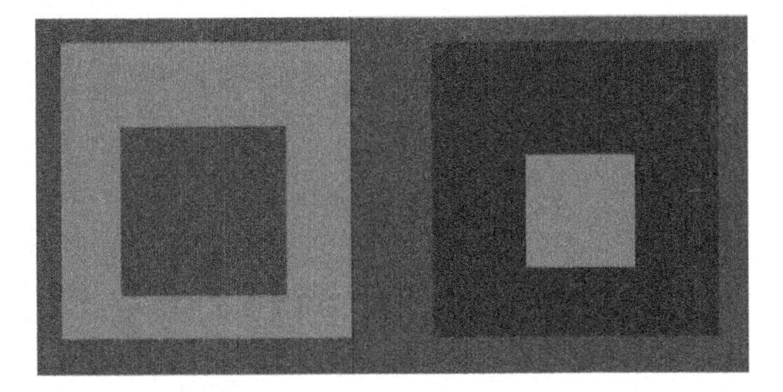

viewing movies. The Oculus Rift combines stereo-optics and head-tracking to offer a fully immersive 3 dimensional viewing experience. In 2014 Facebook paid 2 billion dollars to acquire Oculus spurring other manufacturers to rush to market competing headsets. Other leading systems include Sony's Project Morpheus, HTC Vive (Steam/Valve), Samsung's Gear VR, Microsoft HoloLens and Zeiss VR One. The FOVE VR includes eye tracking and simulated depth-of-field. The Avegant Glyph is unique

Figure 30. Simulation of the merger into one image of two dissimilar inputs for Homage to the Square
(© 2005, Gregory P. Garvey. Used with permission).

in its lightweight minimalist design using an array of micro-mirrors to reflect images directly to the lens onto the retina of the eye. Google's Cardboard offers an inexpensive do-it-yourself solution that is compatible with Android Phones.

These modern day systems not only have the virtue of affordability, but have largely solved latency issues that triggered motion sickness and headaches after prolonged use that plagued the VR systems of the late eighties and early nineties. All of these systems seek to reproduce a 3 dimensional viewing experience by using principles of stereopsis.

FUTURE RESEARCH DIRECTIONS

Visual illusions offer a rich resource for further exploration and for controlled experiments in perception. For example the Müller-Lyer or Zöllner illusions (Figure 31 shows the Müller-Lyer illusion. Figure 32 shows the Zöllner illusion) could be used as a basis for a new fitness function for an interactive genera-

Figure 31. Recreation of Müller-Lyer Illusion
(Ⓟ Public Domain).

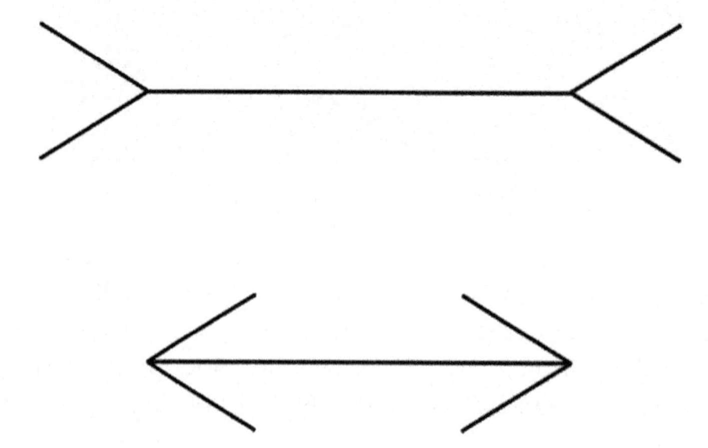

Figure 32. Recreation of Zöllner Illusion
(© Public Domain)

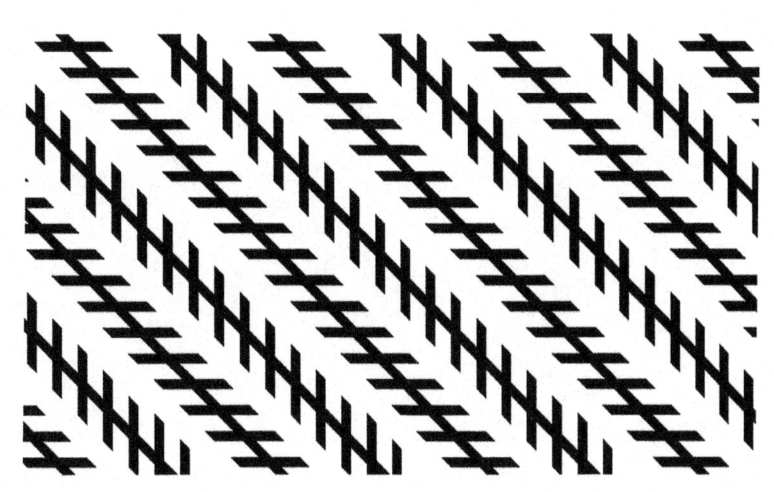

tive work. At the end of this chapter a list of online resources for optical illusions is provided. Blind spot, random dot stereograms and spreading waves of dominance demonstrations are rich in possibilities for further artistic experimentation. The simple demonstration of retinal disparity using the finger held close to the center of the face described above is another area underexplored by artists. The Cheshire Cat Effect also discussed, suggests intriguing possibilities for additional experiments that explore binocular rivalry combining dissimilar inputs with motion erasure effects.

Game engines such as UNITY or UNREAL offer the prospect of real time generation of 3D stereoscopic images that create an immersive environment when viewed with low cost headsets like the Oculus Rift. Motion in this case is generated by head movement. Point of view (POV) is a natural way to determine what is rendered and visible since game engine rendering automatically discards hidden line and occluded polygons. The artistic works describe above that employ dissimilar inputs using recorded digital video, simple line 'drawings' or flat fields of color point the way towards the expanded possibilities of 3D generated stereoscopic pairs of incongruent images. How will the brain attempt to fuse such 3D dissimilar inputs into the singular perception of the 'cyclopean' eye? Other possible ways to monitor user interaction includes low cost eye tracking for using gaze and focus of attention as triggers for algorithmic generation of image pairs. The S2 Eye Tracker along with Eyeworks software is a powerful and reliable system. Other systems include the EYE-TRAC Head Mounted Display and the Mobile Eye from Applied Science Laboratories. (see Other Online Resources in the Appendix). While not technically not an eye or motion tracking affordance Google Glass is a fairly reliable technology that enables augmented reality approaches.

CONCLUSION

There are striking examples of Roman mosaic floors patterns, which depict a kind of faux 3 dimensional space. In particular a pattern of cubes are depicted where three shades of tesserae are used to create a repeating 3-D box pattern made from polygonal lozenge shapes (Figure 33 shows a copy of one such pattern from the Piazza della Vittoria in Palermo, Italy).

Figure 33. First century 3D box pattern mosaac from the Terme di Diocleziano, Palermo Italy (Adopted from Ace of the Fungal Kingdom (2006) (© Creative Commons Noncommercial license. http://creativecommons.org/licenses/by-nc-sa/2.5/ BY-NC-SA 2.5).

Such patterns have long engaged and delighted the eye. The mind's eye is literally captivated by the illusion of a 3D shape that emerges from a pattern on a flat form. This is a kind of play that visual artists have long explored. It is pursued also in a spirit of investigation and exploration leading to the creation and discovery of striking visual illusions. Exploiting the human visual system opens doors to new experiences that can result from leveraging the power of 3D computer graphics, binocular vision and stereoscopy, coupled to new 3D viewing devices supported by an understanding the visual pathways. Today we can open the doors of perception through a determined study and understanding of the "eye" and "brain."

REFERENCES

10 cool optical illusions. (n.d.). Retrieved September 17, 2015 from http://psychology.about.com/od/sensationandperception/tp/cool-optical-illusions.htm

Ace of the Fungal Kingdom. (2006). *Terme di Diocleziano.* Retrieved December 10, 2015, from http://www.tiedyedfreaks.org/ace/diocleziano/DSCN5314crop.jpg

Aids, S. (n.d.). *ScreenScope - Mirror Stereoscope.* Retrieved August 3, 2015, from http://www.stereo-aids.com.au

Albers, J. (1975). *Interaction of Color* (Revised Edition). New Haven, CT: Yale University Press.

Andrew, T. J., & Purves, D. (1997). Similarities in normal and binocular rivalrous viewing. *Proceedings of the National Academy of Sciences of the United States of America, 94*(18), 9905–9908. doi:10.1073/pnas.94.18.9905 PMID:9275224

Animation, P. (n.d.). Retrieved September 17, 2015 from http://www.psy.vanderbilt.edu/faculty/blake/rivalry/BR.html

Aprile, I., Ferrarin, M., Padua, L., Di Sipio, E., Simbolotti, C., Petroni, S., & Dickmann, A. et al. (2014, July). Walking strategies in subjects with congenital or early onset strabismus. *Frontiers in Human Neuroscience*, *8*, 484. doi:10.3389/fnhum.2014.00484 PMID:25071514

Banff Centre New Media Institute. (1999). BNMI Co-Production Archives 'G'. *The Split-Brain Human Computer User Interface*. Retrieved July 10, 2015, from http://www.banffcentre.ca/bnmi/coproduction/archives/s.asp#thesplit

Banich, M. T. (2003). The divided visual field technique in laterality and interhemispheric integration. In K. Hughdahl (Ed.), *Experimental Methods in Neuropsychology* (pp. 47–63). New York: Kluwer. doi:10.1007/978-1-4615-1163-2_3

Banich, M. T., & Shenker, J. I. (1994). Investigations of interhemispheric processing: Methodological considerations. *Neuropsychology*, *8*(2), 263–277. doi:10.1037/0894-4105.8.2.263

Binocular rivalry bibliography. (n.d.). Retrieved October 2, 2015, from https://sites.google.com/site/oshearobertp/publications/binocular-rivalry-bibliography

Blake, R. (1989). A neural theory of binocular rivalry. *Psychological Review*, *96*(1), 145–167. doi:10.1037/0033-295X.96.1.145 PMID:2648445

Blake, R. (2001). A Primer on Binocular Rivalry, Including Current Controversies. *Brain and Mind*, *2*(1), 5–38. doi:10.1023/A:1017925416289

Blake, R. (2005). Landmarks in the History of Binocular Vision. In D. Alais & R. Blake (Eds.), *Binocular Rivalry* (pp. 1–27). Cambridge, MA: MIT Press.

Blake, R., Brascamp, J., & Heeger, D. J. (2014). Can binocular rivalry reveal neural correlates of consciousness? *Philosophical Transactions of the Royal Society of London. Series B, Biological Sciences*, *369*(1641). doi:10.1098/rstb.2013.0211 PMID:24639582

Blake, R., & Logothetis, N. K. (2002, January). Visual Competition. *Nature Reviews. Neuroscience*, *1*(1), 13–21. doi:10.1038/nrn701 PMID:11823801

Blake, R., Westendorf, D., & Overton, R. (1979). What is suppressed during binocular rivalry? *Perception*, *9*(2), 223–231. doi:10.1068/p090223 PMID:7375329

Blind Spot. (n.d.). Retrieved September 17, 2015 from http://www.exploratorium.edu/snacks/blind_spot/index.html

Bogen, J. E. (1975, Spring). Some Educational Aspects of Hemispheric Specialization. *UCLA Educator*, *17*(2), 24–32.

Bogen, J. E., & Vogel, P. J. (1962). Cerebral commissurotomy in man. *Bulletin of the Los Angeles Neurological Society*, *27*, 169–172.

Bourne, V. J. (2006). The divided visual field paradigm: Methodological considerations. *Laterality*, *11*(4), 373–393. doi:10.1080/13576500600633982 PMID:16754238

Breese, B. B. (1899). On inhibition. *Psychological Monographs*, *3*(1), 1–65. doi:10.1037/h0092990

Brewster, D. (1856). *The Stereoscope; its History, Theory, and Construction, with its Application to the fine and useful Arts and to Education: With fifty wood Engravings*. London: John Murray.

Broadbent, D. E. (1954). The role of auditory localization in attention and memory span. *Journal of Experimental Psychology*, *44*, 51–55. doi:10.1037/h0056491 PMID:13152294

Brooks, R. (1980, January). Hemispheric Differences in Memory: Implications for Education. *The Clearing House: A Journal of Educational Strategies, Issues and Ideas*, *53*(5), 248–250. doi:10.1080/00098655.1980.9959221

Byrne, J. H. (Ed.). (1997–Present). *Chapter 14. Visual Processing: Eye and Retina: Figure 14.7*. Neuroscience Online. Department of Neurobiology and Anatomy. The University of Texas Medical School at Houston. Retrieved November 21, 2015, from http://neuroscience.uth.tmc.edu/s2/chapter14.html

C-SPAN. (1991). *Thomas Confirmation Hearings*. Retrieved September 6, 2015, from http://www.c-span.org/search/?searchtype=Videos&sort=Newest&seriesid[]=24

Campbell, F. W., & Howell, E. R. (1972). Monocular alternation; a method for the investigation of pattern vision. *The Journal of Physiology*, *225*, 19–21. PMID:5074381

Carroll, L. (1865). *Alice's Adventures in Wonderland*. London: MacMillan and Company.

Cherry, C. (1953). Some experiments on the recognition of speech, with one and with two ears. *The Journal of the Acoustical Society of America*, *25*(5), 975–979. doi:10.1121/1.1907229

Cheshire Cat Experiment. (n.d.). Retrieved September 17, 2015 from http://www.exploratorium.edu/snacks/cheshire_cat/

Chittka, L., & Brockmann, A. (2005). Perception Space–The Final Frontier. *PLoS Biology*, *3*(4), e137. doi:10.1371/journal.pbio.0030137 PMID:15819608

Colvin, M. K., Funnell, M. G., Hahn, B., & Gazzaniga, M. S. (2005). Identifying functional channels in the corpus callosum: Correlating interhemispheric transfer time with white matter organization. *Journal of Cognitive Neuroscience*, *139*(suppl. 5), 2409–2419.

Compton, R. J. (2002). Interhemispheric interaction facilitates face processing. *Neuropsychologia*, *40*(13), 2409–2419. doi:10.1016/S0028-3932(02)00078-7 PMID:12417469

Contrast Influences Predominance. (n.d.). Retrieved September 17, 2015 from http://www.psy.vanderbilt.edu/faculty/blake/rivalry/BR.html

Corballis, M. C. (1994). Split decisions: Problems in the interpretation of results from commissurotomized subjects. *Behavioural Brain Research*, *64*(1-2), 163–172. doi:10.1016/0166-4328(94)90128-7 PMID:7840883

Corballis, M. C., & Sergent, J. (1988). Imagery in a commissurotomized patient. *Neuropsychologia*, *26*(1), 13–26. doi:10.1016/0028-3932(88)90027-9 PMID:3362338

Corballis, P. M., Funnell, M. G., & Gazzaniga, M. S. (1999). A dissociation between spatial and identity matching in callosotomy patients. *Neuroreport*, *10*(10), 2183–2187. doi:10.1097/00001756-199907130-00033 PMID:10424695

Crick, F. C., & Koch, C. (1992, September). The Problem of Consciousness. *Scientific American, 267*(3), 152–159. doi:10.1038/scientificamerican0992-152 PMID:1502517

Crick, F. C. & Koch. (2003 February). A Framework for Consciousness. *Nature Neuroscience, 6*(2), 119-126. Retrieved November 2, 2015, from http://www.klab.caltech.edu/koch/crick-koch-03.pdf

Descartes, R. (1644). *Traité de l'homme*. Paris: Angot.

Dimond, S., & Beaumont, G. (1972). Processing in perceptual integration between the cerebral hemispheres. *British Journal of Psychology, 63*(4), 509–514. doi:10.1111/j.2044-8295.1972.tb01300.x PMID:4661080

Dutour, É. F. (1760). Discussion d'une question d'optique. Mémoires de Mathématique et de Physique Présentés par Divers Savants. *L'Académie des Sciences, 3*, 514-530.

Edwards, B. (2012). *Drawing on the Right Side of the Brain: The Definitive* (4th ed.). New York: Tarcher/Penguin.

Encyclopaedia Britannica. (1994). Depth Perception: Correspondence of Points. *Encyclopaedia Britannica*. Retrieved December 20, 2015, from http://www.britannica.com/topic/depth-perception

Engel, A. A. K., Fries, P., Konig, P., Brecht, M., & Singer, W. (1999). Temporal binding, binocular rivalry and consciousness. *Consciousness and Cognition, 8*(2), 128–151. doi:10.1006/ccog.1999.0389 PMID:10447995

Exchange, E. (n.d.). Retrieved September 17, 2015 from http://www.psy.vanderbilt.edu/faculty/blake/rivalry/BR.html

Experiments. (n.d.). Retrieved September 17, 2015 from http://www.richardgregory.org/experiments/

Forster, B. A., Corballis, P. M., & Corballis, M. C. (2000). Effect of luminance on successiveness discrimination in the absence of the corpus callosum. *Neuropsychologia, 38*(4), 441–450. doi:10.1016/S0028-3932(99)00087-1 PMID:10683394

Framing Game. (n.d.). Retrieved December 26, 2015 from http://www.vision3d.com/frame.html

Franz, E. A., Waldie, K. E., & Smith, M. J. (2000). The effect of callosotomy on novel versus familiar manual actions: A neural dissociation between controlled and automatic processes. *Psychological Science, 11*(1), 82–85. doi:10.1111/1467-9280.00220 PMID:11228850

Funnell, M. G., Corballis, P. M., & Gazzaniga, M. S. (1999). A deficit in perceptual matching in the left hemisphere of a callosotomy patient. *Neuropsychologia, 38*, 441–450. PMID:10509836

Garvey, G. P. (2002). The Split Brain Human Computer User Interface. *Leonardo, 35*(3), 319–325. doi:10.1162/002409402760105352

Garvey, G. P. (2005). *Homage to Square: Suprematist Composition*. Flash Animation. Artist Collection.

Garvey, G. P. (2011). The Wheatstone Stereoscopic Random Line Pair Generator with Fitness Function for Non-objective Art. *Proceedings of the 8th Conference on Creativity & Cognition*. Atlanta, GA: The High Museum of Art. doi:10.1145/2069618.2069728

Gazzaniga, M. S. (1982). Split brain research: A personal history. *Cornell University Alumni Quarterly.*, *45*, 2–12.

Gazzaniga, M. S. (1985). *The Social Brain. Discovering the Networks of the Mind*. New York, NY: Basic Books, Inc.

Gazzaniga, M. S. (1998, June). The Split Brain Revisited. *Scientific American*. Retrieved July 20, 2015, from http://www.scientificamerican.com/article/the-split-brain-revisited

Gazzaniga, M. S. (2005). Forty-five years of split-brain research and still going strong. *Nature Reviews. Neuroscience*, *6*(8), 653–659. doi:10.1038/nrn1723 PMID:16062172

Gazzaniga, M. S. (2008). Spheres of Influence. *Scientific American*, *19*(3), 32–39. Retrieved from http://issuu.com/samuelsantos52/docs/scientific_american_mind_vol_19__3_ PMID:18783046

Gazzaniga, M. S., Bogen, J. E., & Sperry, R. W. (1962). Some functional effects of sectioning the cerebral commissures in man. *Proceedings of the National Academy of Sciences of the United States of America*, *48*(10), 1765–1769. doi:10.1073/pnas.48.10.1765 PMID:13946939

Geschwind, D. H., & Crabtree, E. (2003). Handedness and Cerebral Laterality. In H. Whitaker (Ed.), *Concise Encyclopedia of Brain and Language* (pp. 221–230). Oxford, UK: Elsevier Science.

Gray, H. (1918). *Anatomy of the Human Body*. Philadelphia, PA: Lea and Febiger. Retrieved December 2, 2015, from https://commons.wikimedia.org/wiki/File:Gray722.png

Gray, L. (n.d.). Auditory System: Pathways and Reflexes. In *Neuroscience Online, the Open-Access Neuroscience Electronic Textbook*. The University of Texas Health Science Center at Houston (UTHealth). Retrieved September 6, 2015, from http://neuroscience.uth.tmc.edu/

Hawkins, J. E. (n.d.). The Human Ear Anatomy. *Encyclopeadia Britannica*. Retrieved September 6, 2015, from http://www.britannica.com/science/ear

Henry, L. H. (1981). The Economic Benefits of the Arts: A Neuropsychological Comment. *Journal of Cultural Economics*, *5*(1), 52–60. doi:10.1007/BF00189206

Herring, E. (1868). The Theory of Binocular Vision. Plenum.

Hirstein, W. (2005). *Brain Fiction: Self-deception and the Riddle of Confabulation*. Cambridge, MA: MIT Press.

Hollins, M., & Hudnell, K. (1980). Adaptation of the binocular rivalry mechanism. *Investigative Ophthalmology & Visual Science*, *19*, 1117–1120. PMID:7410003

Holmes, O. W. (1859, June). The Stereoscope and the Stereograph. Boston: *The Atlantic Monthly*. Retrieved September 12, 2015, from http://www.theatlantic.com/magazine/archive/1859/06/the-stereoscope-and-the-stereograph/303361/

Hopkins, R. L. (1984, Summer). Education and the Right Hemisphere of the Brain: Thinking in Patterns for a Computer World. *Journal of Thought*, *19*(2), 104–114.

Howard, I. P., & Rogers, B. J. (1995). *Binocular Vision and Stereopsis*. Oxford, UK: Oxford University Press.

Huffman, K. R. (n.d.). *The Electronic Arts Community Goes Online: A Personal View*. Retrieved September 30, 2015, from http://www.lehman.cuny.edu/vpadvance/artgallery/gallery/talkback/issue3/centerpieces/huffman.html

Hupé, J., & Rubin, N. (2003, March). The Dynamics of Bi-stable Alternation in Ambiguous Motion Displays: A Fresh Look at Plaids. *Vision Research*, *43*(5), 531–548. doi:10.1016/S0042-6989(02)00593-X PMID:12594999

Illusions, V. (n.d.). Retrieved November 2, 2015, from http://faculty.washington.edu/chudler/chvision.html

Jaynes, J. (1976, 2000). The Origin of Consciousness in the Breakdown of the Bicameral Mind. New York: Houghton Mifflin/Mariner Books.

Jensen, A. R. (1972, October12). Genetics and Education. A second look. *New Scientist*, 96–99.

Julesz, B. (1971). *Foundations of Cyclopean Perception*. Cambridge, MA: MIT Press.

Keystone View Company. (1879-1930). *Keystone View Company*. Retrieved September 18, 2015, from http://ark.digitalcommonwealth.org/ark:/50959/sq87dd56f

Kimura, D. (1961a). Cerebral dominance and the perception of verbal stimuli. *Canadian Journal of Psychology*, *15*(3), 166–171. doi:10.1037/h0083219

Kimura, D. (1961b). Some effects of temporal-lobe damage on auditory perception. *Canadian Journal of Psychology*, *15*(3), 156–165. doi:10.1037/h0083218 PMID:13756014

Kitterle, F. L., Christman, S., & Hellige, J. B. (1990). Hemispheric differences are found in the identification, but not the detection, of low versus high spatial frequencies. *Perception & Psychophysics*, *48*(4), 297–306. doi:10.3758/BF03206680 PMID:2243753

Lambert, A. J. (1991). Interhemispheric interaction in the splitbrain. *Neuropsychologia*, *29*(10), 941–948. doi:10.1016/0028-3932(91)90058-G PMID:1762673

Le Clerc, S. (1679). Discours Touchant de Point de Veue, dans lequel il es prouvé que les chose qu'on voit distinctement, ne sont veues que d'un oeil. In Destined for Distinguished Oblivion: The Scientific Vision of William Charles Wells (1757-1817). New York: Springer-Science+Business Media.

LeCompte, N., & Rush, J. C. (1981). The Jack Sprat Syndrome: Can Split-Brain Theory Improve Education by Including the Arts? *The Journal of Education*, *163*(4), 335–343.

Levelt, W. (1965). *On Binocular Rivalry*. Soesterberg, The Netherlands: Institute for Perception RVO-TNO.

Levi, D. (2013). Linking assumptions in amblyopia. *Visual Neuroscience*, *30*(5-6), 277–287. doi:10.1017/S0952523813000023 PMID:23879956

Levy, E. K. (2014). Sleuthing the Mind: Curator's Introduction. *Leonardo*, *47*(5), 427–428. doi:10.1162/LEON_a_00864

Logothetis, N. K. (1998). Single units and conscious vision. *Philosophical Transactions of the Royal Society of London. Series B, Biological Sciences*, *353*(1377), 1801–1818. doi:10.1098/rstb.1998.0333 PMID:9854253

Magic Eye Random Dot Stereograms. (n.d.). Retrieved September 17, 2015 from http://www.magiceye.com/

McCarthy, R. A., & Warrington, E. K. (1990). *Cognitive Neuropsychology: A Clinical Introduction*. New York: Academic Press.

McLuhan, M. H. (1979). Figure and Grounds in Linguistic Criticism. Interpretation of Narrative by Mario J. Valdés, Owen J. Miller. *A Review of General Semantics, 36*(3), 289-294.

Meenes, M. (1930). A phenomenological description of retinal rivalry. *The American Journal of Psychology*, *42*(2), 260–269. doi:10.2307/1415275

Mohr, B., Pulvermuller, F., Rayman, J., & Zaidel, E. (1994). Interhemispheric cooperation during lexical processing is mediated by the corpus-callosum: Evidence from the split-brain. *Neuroscience Letters*, *181*(1-2), 17–21. doi:10.1016/0304-3940(94)90550-9 PMID:7898762

Moray, N. (1959). Attention in dichotic listening: Affective cues and the influence of instructions. *The Quarterly Journal of Experimental Psychology*, *11*(1), 56–60. doi:10.1080/17470215908416289

Murr, L. E., & Williams, J. B. (1988). Half-Brained Ideas about Education: Thinking and Learning with Both the Left and Right Brain in a Visual Culture. *Leonardo*, *21*(4), 413–419. doi:10.2307/1578704

Nebes, R. (1972). Superiority of the minor hemisphere in commissurotomized man on a test of figural unification. *Brain*, *95*(3), 633–638. doi:10.1093/brain/95.3.633 PMID:4655286

Nebes, R. (1973). Perception of spatial relationships by the right and left hemispheres of a commissurotomized man. *Neuropsychologia*, *7*, 333–349. PMID:4792179

Necker, L. A. (1832). Observations on some remarkable optical phaenomena seen in Switzerland; and on an optical phaenomenon which occurs on viewing a figure of a crystal or geometrical solid. *London and Edinburgh Philosophical Magazine and Journal of Science*, *1*(5), 329–337.

Neisser, U. (1966). Cognitive Psychology. New York: Appleton.

Nielsen, J. A., Zielinski, B. A., Ferguson, M. A., Lainhart, J. E., & Anderson, J. S. (2013). An evaluation of the left-brain vs. right brain hypothesis with resting state functional connectivity magnetic resonance imaging. *PLOS One*. Retrieved September 17, 2015, from http://www.plosone.org/article/info%3Adoi%2F10.1371%2Fjournal.pone.0071275

Nobuyuki Kayahara's Spinning Dancer Illusion. (n.d.). Retrieved September 19, 2015, from http://www.procreo.jp/labo/silhouette.swf

O'Shea, R. P. (1999). *Translation of Dutour (1760)*. Retrieved October 1, 2015, from https://sites.google.com/site/oshearobertp/publications/translations/dutour-1760

O'Shea, R. P. (2004, June). Psychophysics: Catching the Old Codger's Eye. *Current Biology*, *14*(12), R478–R479. doi:10.1016/j.cub.2004.06.014 PMID:15203021

Optical Illusion Pictures. (n.d.). Retrieved September 17, 2015 from http://brainden.com/optical-illusions.htm

Optical Illusions & Visual Phenomena 123 of them. (n.d.). Retrieved September 17, 2015 from http://michaelbach.de/ot/

Parker-Pope, T. (2008). *The Truth About the Spinning Dancer.* Retrieved October 1, 2015, from http://well.blogs.nytimes.com/2008/04/28/the-truth-about-the-spinning-dancer/?_r=0

Poffenberger, A. (1912). Reaction time to retinal stimulation with special reference to the time lost in conduction through nervous centers. *Archives de Psychologie*, *23*, 1–73.

Polansky, A., Blake, R., Braun, J., & Heeger, D. (2000). Neuronal activity in human primary visual cortex correlates with perception during binocular rivalry. *Nature Neuroscience*, *3*(11), 1153–1159. doi:10.1038/80676 PMID:11036274

Porta, J. B. (1593). *De Refractione. Optices Parte. Libri Novem.* Carlinum and Pacem, Naples Retrieved October 1, 2015, from http://testyourself.psychtests.com/testid/3178

Psychology Concepts. (n.d.). Retrieved October 1, 2015, from http://www.psychologyconcepts.com/depth-perception/

Rapaczynski, W., & Ehrlichman, H. (1979). Opposite visual hemifield superiorities in face recognition as a function of cognitive style. *Neuropsychologia*, *17*(6), 645–652. doi:10.1016/0028-3932(79)90039-3 PMID:522978

Rivalry, F. (n.d.). Retrieved September 17, 2015 from http://www.psy.vanderbilt.edu/faculty/blake/rivalry/BR.html

Rogers, M. (2013). *Researchers debunk myth of "right brain" and "left-brain" personality traits*. University of Utah, Office of Public Affairs. Retrieved September 17, 2015, from http://www.plosone.org/article/info%3Adoi%2F10.1371%2Fjournal.pone.0071275

Roussel, R. (1995). *How I Wrote Certain of My Books* (T. Winkfield, Trans.). Boston, MA: Exact Change. (Original work published 1935)

Rubin, E. (2001). Readings in perception. In S. Yantis (Ed.), *Visual perception: Essentials readings*. Philadelphia: Psychology Press. (Original work published 1921)

Rubin, N. (2001). Figure and ground in the brain. *Nature Neuroscience*, *4*(9), 857–858. doi:10.1038/nn0901-857 PMID:11528408

Severed Corpus Callosum. (2008, June 25). Scientific American Frontiers. Retrieved September 17, 2015, from https://www.youtube.com/watch?v=82tlVcq6E7A

Simanek, D. (n.d.). *How to View 3D Without Glasses*. Retrieved September 28, 2015, from http://www.lhup.edu/~dsimanek/3d/view3d.htm

Smith, E. E. (2015, July 27). One Head, Two Brains. *The Atlantic*. Retrieved August 24, 2015, from http://www.theatlantic.com/health/archive/2015/07/split-brain-research-sperry-gazzaniga/399290/

Sperry, R. (1966). Brain bisection and the neurology of consciousness. In J. C. Eccles (Ed.), Brain and conscious experience, (pp. 298-313). New York: Springer-Verlag.

Sperry, R. (1968). Hemisphere Deconnection and Unity in Conscious Awareness. *The American Psychologist*, *23*(10), 723–733. doi:10.1037/h0026839 PMID:5682831

Spreading Waves of Dominance. (n.d.). Retrieved September 17, 2015 from http://www.psy.vanderbilt.edu/faculty/blake/rivalry/BR.html

The S2 Eye Tracker. Eyeworks software. (n.d.). Retrieved September 25, 2015, from (http://www.mirametrix.com/products/?gclid=CN3Jl8Xz5sgCFZWRHwodaBEBdA)

The blind spot. (n.d.). Retrieved September 17, 2015 from http://faculty.washington.edu/chudler/chvision.html

The Young Lady Versus Old Lady Optical Illusion. (n.d.). Retrieved September 22, 2015, from http://brainden.com/face-illusions.htm

Todd, J. W. (1912). *Reaction to multiple stimuli*. New York: The Science Press. doi:10.1037/13053-000

Tong, F., Nakayama, K., Vaughan, J. T., & Kanwisher, N. (1998). Binocular rivalry and visual awareness in human extrastriate cortex. *Neuron*, *21*(4), 753–759. doi:10.1016/S0896-6273(00)80592-9 PMID:9808462

Treisman, A. (1964). Verbal cues, language and meaning in selective attention. *The American Journal of Psychology*, *77*(2), 206–209. doi:10.2307/1420127 PMID:14141474

Unitary vs. Piecemeal Rivalry. (n.d.). Retrieved September 17, 2015 from http://www.psy.vanderbilt.edu/faculty/blake/rivalry/BR.html

United States. Congress. Senate. Committee on the Judiciary. (1993). Nomination of Judge Clarence Thomas to be Associate Justice of the Supreme Court of the United States: hearings before the Committee on the Judiciary, United States Senate, first session ... Washington: U.S. G.P.O. For sale by the U.S. G.P.O., Supt. of Docs. Unknown. *The Holmes stereoscope, with the inventions and improvements added by Joseph L. Bates*. Center for the History of Medicine: OnView. Retrieved October 4, 2015, from http://collections.countway.harvard.edu/onview/items/show/6277

Uzwiak, A. (n.d.). *Vision*. Retrieved December 2, 2015, from http://www.rci.rutgers.edu/~uzwiak/AnatPhys/Sensory_Systems.html

Van Kleek, M. H. (1989). Hemispheric differences in global versus local processing of hierarchical visual stimuli by normal subjects: New data and a meta-analysis of previous studies. *Neuropsychologia*, *27*(9), 1165–1178. doi:10.1016/0028-3932(89)90099-7 PMID:2812299

Van Wagenen, W. P., & Herren, R. Y. (1940). Surgical division of commissural pathways in the corpus callosum relation to spread of an epileptic attack. *Archives of Neurology and Psychiatry*, *44*(4), 740–759. doi:10.1001/archneurpsyc.1940.02280100042004

Virtual Research Systems. (1998–2000). *Company Profile: Virtual Research Systems, Inc*. Retrieved September 4, 2015, from http://www.virtualresearch.com/company.html

Wade, N. J. (2000). *A Natural History of Vision*. Cambridge, MA: MIT Press.

Wade, N. J., & Tgo, T. T. (2013). Early views on binocular rivalry. In S. M. Miller (Ed.), *The constitution of visual consciousness: Lessons from binocular rivalry, Edition: Advances in Consciousness Research* (Vol. 90, pp. 77–108). Philadelphia: John Benjamins Publishing Company. doi:10.1075/aicr.90.04wad

Walker, P., & Powell, D. J. (1979). The sensitivity of binocular rivalry to changes in the nondominant stimulus. *Vision Research, 19*(3), 247–249. doi:10.1016/0042-6989(79)90169-X PMID:442549

Wallach, H. (1935). Uber visuell wahrgenommene Bewegungsrichtung. *Perception, 25*, 1319–1368.

Wandell, B. A. (n.d.). *Useful Numbers in Vision Science*. Retrieved September 4, 2015, from http://web.stanford.edu/group/vista/cgi-bin/wandell/useful-numbers-in-vision-science/

Webber, A. L., & Wood, J. (2005). Amblyopia: Prevalence, Natural History, Functional Effects and Treatment. *Clinical & Experimental Optometry, 88*(6), 365–375. doi:10.1111/j.1444-0938.2005.tb05102.x PMID:16329744

Weissmann, D. H., & Banich, M. T. (1999). Global local inference modulated by communication between the hemispheres. *Journal of Experimental Psychology. General, 128*(3), 283–308. doi:10.1037/0096-3445.128.3.283 PMID:10513397

Weissmann, D. H., & Banich, M. T. (2000). The cerebral hemispheres cooperate to perform complex but not simple tasks. *Neuropsychology, 14*(1), 41–59. doi:10.1037/0894-4105.14.1.41 PMID:10674797

Wheatstone, C. (1838). Contributions to the physiology of vision – Part the first. On some remarkable, and hitherto unobserved, phenomena of binocular vision. *Philosophical Transactions of the Royal Society of London. Series B, Biological Sciences, 128*, 371–394. Retrieved from https://www.stereoscopy.com/library/wheatstone-paper1838.html

Wolman, D. (2012, March14). The split brain: A tale of two halves. *Nature, 483*(7389), 260–263. doi:10.1038/483260a PMID:22422242

Zimmer, C. (2009, April 15). The Big Similarities & Quirky Differences Between Our Left and Right Brains. *Discover*. Retrieved October 1, 2015, from http://discovermagazine.com/2009/may/15-big-similarities-and-quirky-differences-between-our-left-and-right-brains

ADDITIONAL READING

Alasis, D., & Blake, R. (2005). *Binocular Rivalry*. Cambridge, MA: MIT Press.

American Psychological Association. *Retrieved* September 2, *2015*, from http://apa.org/

Annett, M. (2002). *Handedness and Brain Asymmetry: The Right Shift Theory*. New York: Psychology Press.

Banich, M. T. (1998). The missing link: The role of interhemispheric interaction in attentional processing. *Brain and Cognition, 36*(2), 128–157. doi:10.1006/brcg.1997.0950 PMID:9520311

Benson, D. F., & Zaidel, E. (Eds.). (1985). *The dual brain: Hemispheric specialization in humans*. New York: Guilford Press.

Blake, R., & Sekuler, R. (2006). *Perception* (5th ed.). New York: McGraw Hill.

Bogen, J. E. (1985). The dual brain: some historical and methodological aspects. In D. F. Benson & E. Zaidel (Eds.), *The Dual Brain*. New York: Guilford Press.

Breese, B. B. (1909). Binocular rivalry. *Psychological Review*, *16*(6), 410–415. doi:10.1037/h0075805

Brown, L. (1993). *The New shorter Oxford English Dictionary on Historical Principles*. Oxford, England: Clarendon.

Concepts, P. http://www.psychologyconcepts.com/*Retrieved September 8, 2015* Psychology Concepts. Retrieved September 6, 2015, from http://www.psychologyconcepts.com/depth-perception/

Consmelli, D., & Thompson, E. (2007). Mountains and valleys: Binocular rivalry and the flow of experience. Subjectivity and the Body. *Consciousness and Cognition*, *16*(3), 623–641. doi:10.1016/j.concog.2007.06.013 PMID:17804257

Crick, F. (1996). Visual perception: Rivalry and consciousness. *Nature*, *379*(6565), 485–486. doi:10.1038/379485a0 PMID:8596623

Department of Neurobiology and Anatomy. Neuroscience Online. Retrieved October 30, *2015*, from http://neuroscience.uth.tmc.edu/index.htm

Deutsch, D. (1974, September). An auditory illusion. *Nature*, *25*(5473), 307–309. doi:10.1038/251307a0 PMID:4427654

DIY Hacks and How Tos (n.d.). *3D Stereoscopic Photography*. Retrieved September 28, 2015, from http://www.instructables.com/id/3D-Stereoscopic-Photography/step3/How-to-View-Cross-eyed-3D-Images/

Doherty, P. (1995). *The Cheshire Cat and Other Interactive Experiments in Perception*. New York: Wiley.

Duane's Clinical Ophthalmology. (2004, October 25). New York: Lippincott Williams & Wilkins.

Eye, M. (n.d.) Retrieved September 20, 2015 from http://www.magiceye.com/

Gardner, H. (1978). What We Know (and Don't Know) about the Two Halves of the Brain. *Harvard Magazine*, *80*, 24–27.

Gazzaniga, M. S. (2005). Forty-five years of split-brain research and still going strong. *Nature Reviews. Neuroscience*, *6*(8), 653–659. doi:10.1038/nrn1723 PMID:16062172

Greenstein, B., & Greenstein, A. (2000). *Color Atlas of Neuroscience: Neuroanatomy and Neurophysiology*. New York: Thieme.

Gregory, R. L. (1974). *Concepts and Mechanisms of Perception*. London: Duckworth.

Gregory, R. L. (1992). Evolution of the Eye and Visual System. J. R. Cronly-Dillon & R. L. Gregory (Eds.). Vol. 2 of Vision and Visual Dysfuction. London: Macmillan.

Gregory, R. L. (1995). *The Artful Eye* (R. L. Gregory, J. Harris, P. Heard, & D. Rose, Eds.). Oxford: Oxford University Press.

Gregory, R. L. (1998). *Eye and Brain: The Psychology of Seeing* (5th ed.). Princeton: Princeton University Press.

Gregory, R. L. (1978). Illusions and Hallucinations. E. C. Carterette, & M. P. Freidman (eds.) Handbook of Perception 9 (Chapter 9).

Gregory, R. L., & Gombrich, E. (Eds.). (1973). *Illusion in Nature and Art*. London: Duckworth.

Helige, J. B. (2001). *Hemispheric Asymmetry: What's Right and What's Left*. Cambridge, MA: Harvard University Press.

Hugdahl, K., & Davidson, R. J. (Eds.). (2003). *The Asymmetrical Brain*. Cambridge: MIT Press.

Human, B. F. T. Online Psychology Laboratory. Retrieved September 10, 2015, from http://opl.apa.org/contributions/EC/BrainFly.htm

Ivry, R. B., & Robertson, L. C. (1998). *The Two Sides of Perception*. Cambridge: MIT Press.

Joseph, R. (1990). *Neuropsychology, Neurospyschiatry, and Behavioral Neurology*. New York: Plenum Press. doi:10.1007/978-1-4757-5969-3

Kellogg, R. T. (2016). *Fundamentals of Cognitive Psychology*. Thousand Oaks, CA: Sage Publishing.

Kosslyn, S. M. (1987). Seeing and imagining in the cerebral hemispheres: A computational approach. *Psychological Review*, *94*(2), 148–175. doi:10.1037/0033-295X.94.2.148 PMID:3575583

Levy, J., Trevarthen, C., & Sperry, R. A. (1972). Perception of bilateral chimeric figures following hemispheric deconnexion. *Brain*, *95*(1), 61–78. doi:10.1093/brain/95.1.61 PMID:5023091

Lipton, L. (1982). *Foundations of the Stereo-Scopic Cinema, A Study in Depth. NewYork*. Van Nostrand Reinhold.

Marek, P. (Ed.). (n.d.) Online Psychology Laboratory. Retrieved September 8, 2015, from http://opl.apa.org/Main.aspx

Miller, M. B., Sinnot-Armstrong, W., Young, L., King, D., Paggi, A., Fabri, M., & Gazzaniga, M. S. et al. (2010). Abnormal Moral Reasoning in complete and partial callosotomy patients. *Neuropsychologia*, *48*(7), 2215–2220. doi:10.1016/j.neuropsychologia.2010.02.021 PMID:20188113

Monaghan, P., & Pollmann, S. (2003). Division of labor between the hemispheres for complex but not simple tasks: An implemented connectionist model. *Journal of Experimental Psychology. General*, *132*(3), 379–399. doi:10.1037/0096-3445.132.3.379 PMID:13678374

Motz, B. A., James, K. H., & Busey, T. A. (2012). The Lateralizer: A tool for students to explore the divided brain. *Advances in Physiology Education*, *36*(3), 220–225. doi:10.1152/advan.00060.2012 PMID:22952261

Ornstein, R. (1977). *The Psychology of Consciousness*. New York: Harcourt.

Sensation and Perception. Online Psychology Laboratory. *Retrieved* September 8, *2015*, from http://opl.apa.org/Resources.aspx#Sensation

Sherwood, L. (2004). *Human Physiology: From Cells to Systems* (7th ed., pp. 197–208). Belmont, CA: Brooks/Cole Cengage Learning.

The Society for the Teaching of Psychology. Division 2 of the American Psychological Association. *Retrieved* September 10, *2015*, from http://teachpsych.org/page-1588384

Unknown, "The Holmes stereoscope, with the inventions and improvements added by Joseph L. Bates, " *Center for the History of Medicine: OnView*, Retrieved November 29, 2015, http://collections.countway.harvard.edu/onview/items/show/6277

Valyus, N. A. (1962). *Stereoscopy*. New York: The Focal Press.

Wade, N. J. (1996). Descriptions of visual phenomena from Aristotle to Wheatstone. *Perception*, *25*(10), 1137–1175. doi:10.1068/p251137 PMID:9027920

Wagemans, J. (Ed.). (2015). *The Oxford Handbook of Perceptual Organization*. Oxford, UK: Oxford University Press. doi:10.1093/oxfordhb/9780199686858.001.0001

Wandell, B. A. (1995). *Foundations of Vision*. Sunderland, MA: Sinauer Associates.

Whitaker, H. A. (2010). *Concise Encyclopedia of Brain and Language*. Oxford, UK: Elsevier Ltd.

Wikipedia, (n.d.). The Auditory System. Retrieved September 5, 2015, from https://en.wikipedia.org/wiki/Auditory_system

Yeh, Y. Y., & Silverstein, L. D. (1990). Limits of Fusion and Depth Judgment in Stereoscopic Color Displays. *Human Factors*, *32*, 45–60.

Zaidel, E., & Iacoboni, M. (Eds.). (2003). *The Parallel Brain: The Cognitive Neuroscience of the Corpus Callosum*. Cambridge: MIT Press.

KEY TERMS AND DEFINITIONS

Binocular Vision: Is in which creatures having two use them together. The word binocular comes from two roots, *bini* for double, and *oculus* for eye. https://en.wikipedia.org/wiki/Binocular_vision.

Corpus Callosotomy: Corpus callosotomy is a palliative surgical procedure for the treatment of seizures. As the corpus callosum is critical to the interhemispheric spread of epileptic activity, the procedure seeks to eliminate this pathway. The corpus callosum is usually severed in order to stop epileptic seizures. Once the corpus callosum is cut, the brain has much more difficulty sending messages between the hemispheres. Although the corpus callosum is the largest white matter tract connecting the hemispheres, some limited interhemispheric communication is still possible via the anterior commissure and posterior commissure. When tested in particular situations, it is obvious that information transfer between the hemispheres is reduced.

Cerebral Cortex: Is the cerebrum's (brain) outer layer of neural tissue in humans and other mammals. It is divided into two cortices, along the sagittal plane: the left and right cerebral hemispheres divided by the medial longitudinal fissure. The cerebral cortex plays a key role in memory, attention, perception, awareness, thought, language, and consciousness. The human cerebral cortex is 2 to 4 millimetres (0.079 to 0.157 in) thick. https://en.wikipedia.org/wiki/Cerebral_cortex#Areas.

Cerebral Hemispheres: The vertebrate cerebrum (brain) is formed by two cerebral hemispheres that are separated by a groove, the medial longitudinal fissure. The brain can thus be described as being divided into left and right cerebral hemispheres. Each of these hemispheres has an outer layer of grey matter, the cerebral cortex, that is supported by an inner layer of white matter. In eutherian (placental see: https://en.wikipedia.org/wiki/Eutheria) mammals, the hemispheres are linked by the corpus callosum, a very large bundle of nerve fibers. Smaller commissures, including the anterior commissure, the posterior commissure and the hippocampal commissure also join the hemispheres and these are also present in other vertebrates. These commissures transfer information between the two hemispheres to coordinate localized functions. https://en.wikipedia.org/wiki/Cerebral_hemisphere.

Commissures or Commissural Fibers: The fibers or transverse fibers are coherent white-matter structures that connect the two hemispheres of the brain. https://en.wikipedia.org/wiki/Commissural_fiber.

Commissurotomy: In, as a treatment for severe epilepsy, the corpus callosum, or the area of the brain that connects the two hemispheres, would be completely bisected. By eliminating the connection between the two hemispheres of a patient's brain, electrical communication would be cut off greatly diminishing the amount and severity of the epileptic seizures. For some, seizures would be completely eliminated. https://en.wikipedia.org/wiki/Commissurotomy.

Corpus Callosum: (/ˈkɔrpəs kəˈloʊsəm/; Latin for "tough body"), also known as the callosal commissure, is a wide, flat bundle of neural fibers about 10 cm long beneath the cortex in the eutherian brain at the longitudinal fissure. It connects the left and right cerebral hemispheres and facilitates interhemispheric communication. It is the largest white matter structure in the brain, consisting of 200–250 million contralateral axonal projections. https://en.wikipedia.org/wiki/Corpus_callosum.

Cortical: 1. of, relating to, or consisting of cortex. 2. involving or resulting from the action or condition of the cerebral cortex. http://www.merriam-webster.com/dictionary/subcortical.

Cyclopean Image: Is a single mental image of a scene created by the brain by combining two images received from the two eyes. Cyclopean image is named after the mythical Cyclops with a single eye. Literally it refers to the way stereo sighted viewers perceive the centre of their fused visual field as lying between the two physical eyes, as if seen by a cyclopean eye. Alternative terms for cyclopean eye include third central imaginary eye and binoculus. https://en.wikipedia.org/wiki/Cyclopean_image.

Dichoptic: (From the Greek words δίχα *dicha*, meaning "in two," and ὀπτικός *optikos*, "relating to sight") is viewing a separate and independent field by each eye. In dichoptic presentation, stimulus A is presented to the left eye and then stimulus B is presented to the right eye. https://en.wikipedia.org/wiki/Dichoptic_presentation.

Electrooculography: Is a technique for measuring the corneo-retinal standing potential that exists between the front and the back of the human eye. The resulting signal is called the electrooculogram. Primary applications are in and in recording.

Monocular Vision: Is in which both are used separately. By using the eyes in this way, as opposed by, the is increased, while is limited. The eyes of an animal with monocular vision are usually positioned on opposite sides of the animal's head, giving it the ability to see two objects at once. The word monocular comes from the root, *mono* for one, and the root, *oculus* for eye. https://en.wikipedia.org/wiki/Monocular_vision.

Split-Brain: Split-brain is a lay term to describe the result when the corpus callosum connecting the two hemispheres of the brain is severed to some degree. It is an association of symptoms produced by disruption of or interference with the connection between the hemispheres of the brain. The surgical operation to produce this condition results from transection of the corpus callosum, and is usually a last resort to treat refractory epilepsy. https://en.wikipedia.org/wiki/Split-brain.

Stereopsis: (From the Greek στερεο- *stereo-* meaning "solid", and ὄψις *opsis*, "appearance, sight") is a term that is most often used to refer to the perception of depth and 3-dimensional structure obtained on the basis of visual information deriving from two eyes by individuals with normally developed binocular vision. https://en.wikipedia.org/wiki/Stereopsis.

Subcortical: Of, relating to, involving, or being a part of the brain below the cerebral cortex. http://www.merriam-webster.com/dictionary/subcortical.

APPENDIX

Online Demonstrations

10 cool optical illusions. Retrieved September 17, 2015 from http://psychology.about.com/od/sensation-andperception/tp/cool-optical-illusions.htm

The blind spot. Retrieved September 17, 2015 from http://faculty.washington.edu/chudler/chvision.html

Blind Spot. Retrieved September 17, 2015 from http://www.exploratorium.edu/snacks/blind_spot/index.html

Cheshire Cat Experiment. Retrieved September 17, 2015 from http://www.exploratorium.edu/snacks/cheshire_cat/

Contrast Influences Predominance. Retrieved September 17, 2015 from http://www.psy.vanderbilt.edu/faculty/blake/rivalry/BR.html

Experiments. Retrieved September 17, 2015 from http://www.richardgregory.org/experiments/

Eye Exchange. Retrieved September 17, 2015 from http://www.psy.vanderbilt.edu/faculty/blake/rivalry/BR.html

(The) Framing Game. Retrieved December 26, 2015 from http://www.vision3d.com/frame.html

Fusion Rivalry. Retrieved September 17, 2015 from http://www.psy.vanderbilt.edu/faculty/blake/rivalry/BR.html

Magic Eye Random Dot Stereograms. Retrieved September 17, 2015 from http://www.magiceye.com/

Optical Illusion Pictures. Retrieved September 17, 2015 from http://brainden.com/optical-illusions.htm

Optical Illusions & Visual Phenomena 123 of them. Retrieved September 17, 2015 from http://michaelbach.de/ot/

Probe Animation. Retrieved September 17, 2015 from http://www.psy.vanderbilt.edu/faculty/blake/rivalry/BR.html

Spreading Waves of Dominance. Retrieved September 17, 2015 from http://www.psy.vanderbilt.edu/faculty/blake/rivalry/BR.html

Stereo Pictures for Cross-Eyed Viewing. Retrieved September 17, 2015 from http://www.lhup.edu/~dsimanek/3d/stereo/3dgallery.htm

Unitary vs. Piecemeal Rivalry. Retrieved September 17, 2015 from http://www.psy.vanderbilt.edu/faculty/blake/rivalry/BR.html

Visual Illusions. Retrieved November 2, 2015, from http://faculty.washington.edu/chudler/chvision.html

Other Online Resources

Binocular rivalry bibliography. Retrieved October 2, 2015, from https://sites.google.com/site/oshearobertp/publications/binocular-rivalry-bibliography

Fire in the Borgo; Is Athens Buring? Retrieved September 10, 2015, fromhttp://www.wikiart.org/en/salvador-dali/athens-is-burning-the-school-of-athens-and-the-fire-in-the-borgo-1980

Nobuyuki Kayahara's Spinning Dancer Illusion. Retrieved September 19, 2015, from http://www.procreo.jp/labo/silhouette.swf

The S2 Eye Tracker, Eyeworks software. Retrieved September 25, 2015, from (http://www.mirametrix.com/products/?gclid=CN3Jl8Xz5sgCFZWRHwodaBEBdA).

EYE-TRAC Head Mounted Display, the Mobile Eye from Applied Science Laboratories. Retrieved September 26, 2015, from (http://www.asleyetracking.com/Site/).

Severed Corpus Callosum (2008, June 25), Scientific American Frontiers.Retrieved September 17, 2015, from https://www.youtube.com/watch?v=82tlVcq6E7A

The Young Lady Versus Old Lady Optical Illusion. Retrieved September 22, 2015, from http://brainden.com/face-illusions.htm

Chapter 3
Better Visualization through Better Vision

Michael Eisenberg
University of Colorado – Boulder, USA

ABSTRACT

Traditionally, the subject of "scientific visualization" focuses on the creation of novel or innovative graphical representations: essentially, new types of images to perceive. A truly complete approach to scientific visualization should include not only the perceived object, but also the abilities of the perceiver. Human "visual common sense" is a product of evolution, suited to the survival of the species; but it has severe and recurring limitations for the purposes of scientific understanding and education. People cannot readily understand phenomena that are too fast, slow, or complex for their visual systems to take in; they cannot see wavelengths outside visual spectrum; they have difficulty understanding three-dimensional (or, even worse, four-dimensional) objects. This chapter explores a variety of ideas and design themes for approaching scientific visualization by enhancing the powers of human vision.

INTRODUCTION: CHANGING HOW WE SEE

When the subject of "scientific visualization" comes up in the context of educational research, the usual assumption is that we are talking about creating better or more informative graphics. Perhaps the way to improve scientific visualization is through developing animated simulations; or interactive interfaces to large information spaces; or embedding aural cues within diagrams.

All of these approaches are promising, and thoroughly deserve to be pursued; but at the same time, a truly complete view of visualization research needs to look beyond an exclusive focus on the perceived object. Visualization, after all, requires both an object and a perceiver. We might thus seek to explore new types of visualization both by creating novel objects-to-perceive and by creating new tools and agents of perception. That is, by remaking the nature and equipment of vision itself, it might well be possible to achieve an expanded understanding of scientific ideas and the world in which we are embedded.

The purpose of this (frankly speculative) chapter is to suggest and enumerate a variety of potential avenues for creating visual technologies to enhance scientific education and understanding. The central

DOI: 10.4018/978-1-5225-0480-1.ch003

theme of these examples is to think about the human visual system as the biological core of what could be an expandable visual apparatus, combining both biological and non-biological elements. The process of evolution has provided us with visual powers tuned to the survival challenges that have historically faced our species; but at the same time, those powers constitute a sort of "visual common sense" that is not always, and not necessarily, suited to understanding ideas that challenge our inherited intuitions.

An Example: High Speed Vision

An introductory example might help to illustrate the motivation behind this chapter. In Fischbein (2002), there is a discussion of diSessa's notion of "phenomenological primitives" (or *p-prims*) that act, essentially, as intuitive "building block" scenarios for understanding the behavior of physical objects. One provocative example involves explaining the reason that some objects (say, a ping pong ball) will bounce when dropped onto a hardwood floor, whereas others (say, a ball of wet clay) will not. When diSessa asked one student to explain the difference,

She could not think by herself of springiness, and the interviewer suggested the compression of a spring. The subject had a clear intuitive understanding of the behavior of a compressed spring but nonetheless could not accept that the same explanation holds for the ball and generally, for every kind of piece of matter (for instance a ball made of steel). Her justification was that many objects are rigid and then they cannot be squished (deformed). For that subject rigidity and "squishiness" were p-prims, that is to say properties which may be understood intuitively by themselves and which, in turn, may explain other phenomena (Fischbein, p. 169).

For this student, the reason that a steel ball bounces is not identified with the deformation of the ball, but precisely because (in her view) the ball does *not* deform at all. And the ability for a ball to bounce is somehow–though not entirely clearly–associated with the "rigidity" (and thus, "bounciness"?) of the substance itself.

Why would someone have this intuition–that the act of bouncing involves no deformation? The most natural hypothesis is simply that we cannot see the deformation take place: it occurs too rapidly. My own intuitions about this scenario were fundamentally altered long ago when I saw a high-speed photograph (taken by Harold Edgerton) of a man hitting a softball with a bat: the photograph clearly shows what the naked eye cannot see, namely, a surprisingly deformed softball poised to "spring" off the bat (which is also visibly slightly bent in contact). Our visual limitations prevent us from seeing the deformation directly, without the aid of high-speed photography; and as a result our "common sense" notions of how the act of bouncing takes place are impaired.

Indeed, one of the very earliest anecdotes in "scientific visualization" similarly highlights the surprising difficulty of unaided vision of rapid motion. In 1872, Leland Stanford (later the founder of Stanford University) commissioned the photographer Eadweard Muybridge to photograph a horse in motion, with a particular interest in determining whether all four of the horse's hooves were above the ground at any point in its stride. Muybridge's classic series of photographs showed definitively that the answer was positive for a galloping horse; and the result was a triumph for the extension of human intuition via technology (Cf. Braun, p. 68ff).

To return, then, to the theme with which this chapter opened: a standard "scientific visualization" approach to the difficulty of high-speed perception would be to increase our access to photographs such as those of Edgerton and Muybridge. This, inarguably, would be a positive step. At the same time, we might wonder about whether it might be possible to develop tools that augment our own vision so that (in appropriate situations) we could actually view events directly through a "stop-motion" interface. Conceivably, a consciously and controllably expanded visual system of this sort might serve, over time, to change the parameters of "common sense" itself. This is one instance of how educational visual enhancement might be achieved, and it suggests more broadly the motivating idea behind the discussion that follows.

A MAP OF THE TERRITORY

The remainder of this chapter explores a variety of plausible ways in which, through the development of technological tools, we might pursue the notion of enhancing vision for the purpose of scientific visualization and (more broadly) education in general. In the following sections, we discuss possible techniques for seeing in new ways: seeing motion, seeing beyond the visible electromagnetic spectrum, seeing pre-interpreted phenomena through the aid of intelligent sensors, and so forth. Most of these techniques involve ways of seeing the natural world; but the final suggestion involves ways of creating novel types of "artificial" graphics, designed in tandem with extra-ocular visual apparatus. In the final section of the paper, we take a step back and return to the topic of what constitutes "visual common sense"; and acknowledge a variety of cautionary notes in the pursuit of enhanced vision.

Seeing Moving Objects

The introductory example that we used above – "seeing motion" through photography – is perhaps a good jumping-off point for a more extended discussion. Some phenomena (e.g., the deformation of a bouncing ball) occur too swiftly to be seen by the human eye; others (the movement of a clock's hands) occur too slowly; and there are still others (arguably including the eddying movement of fluids) that, in a sense, leave an impression of movement but are especially difficult for people to interpret in real time. From the evolutionary standpoint, human vision is (at least for survival purposes) as good as it needs to be; but from the educational standpoint, our limited sensitivity has intellectual consequences. Balls apparently bounce without deforming; continents and land masses don't move (unless the movement is catastrophic in human terms); a bird's wings seem to move up and down, and a running horse's legs may (for all we know) never entirely lose contact with the ground. Much of science education, in consequence, is devoted to convincing us of the fallibility of our day-to-day visual interpretation of the world.

How might we approach these issues through visual apparatus? In 1945, Vannevar Bush (1945) in his famous article "As We May Think" speculated on the rapid advances in mobile photography, and envisioned small head-mounted cameras with which people could take photographs of day-to-day phenomena for later recording and interpretation. To a significant extent, this scenario has progressively been realized through the ubiquitous presence of cell phone cameras (and, in the case of head-mounted devices, through artifacts such Google Glass). We might nonetheless imagine an extension of these (increasingly day-to-day) artifacts by designing versions that could be used specifically for scientific education or understanding.

It might, for instance, be possible to create a head-mounted camera that automatically notes the presence of fast motion and takes the occasional stop-action photograph in case the phenomenon being viewed is of interest. The wearer of this camera, then, might witness a variety of phenomena during the course of the day; and on returning home, the camera might alert her to the occurrence of potentially interesting dynamic phenomena that she might have seen (but were occurring too rapidly to see in detail). A student who sees a baseball player hitting a ball (or, for that matter, a student lucky enough to see a running horse) might be able to answer questions that just two centuries ago were unapproachable.

What about *slow* phenomena? Could we devise photographic interfaces to help us (e.g.) see grass growing, or the movement of the moon across the night sky? This might be a trickier sort of design problem: after all, when nature photographers deal with these subjects, they generally position a single camera in one position for a long time to take successive images. Conceivably, we might imagine devising computational interfaces for cameras that could "recognize" user-selected sites for successive images over time: for instance, a student might take a picture of a tree (in more-or-less the same position) on successive days over the course of a month, and the software might create a dynamic image of the tree changing using these images as input. Or one might imagine taking a relatively brief video of a wall clock (not the digital sort!) and the software might use the very slight motion detected during a brief time as the basis for an extrapolated succession of images showing how the movement could be continued over time.

Still another possibility might be to create "seeing devices" that are tuned to specific sorts of motions–complex motions that illuminate scientific phenomena of various sorts. Much as a frog's eyes are tuned toward the motion of small darting objects in front of them (like a potentially tasty fly), we might create visual devices that are specifically tuned to see (e.g.) rotary motion, or oscillations, or the complex eddies of flowing fluids or smoke. In effect, the goal here would be to create extra "special-purpose eyes" that are on the lookout for particular types of motions, and that are better at noticing these sorts of motions than we are.

Before leaving this particular topic, it is worth pausing to highlight two aspects of the ideas that we have just explored–aspects that will reoccur in the remainder of this chapter. First, there are many ways that we could imagine interfaces for the devices of the previous paragraphs. A user might position a camera with a strap on her forehead (that's what Vannevar Bush seemed to envision in his 1945 paper), or on a cap; or the user might carry several such devices oriented in different directions. The operation of the device could be imagined along a spectrum of user control: for instance, the user might switch on a "rapid-motion-detecting device" only under certain conditions, or for selected periods of time. Second, and perhaps more interesting, the design of most (if not all) such artifacts involves a combination of hardware and (potentially challenging) software innovation. A day-to-day camera interface that can "look for" oscillations in the world is a non-trivial task; we might begin with a device that sees only a small subset of the oscillations out there in the world, and improve the intelligence of the device over time. (The reader might, in fact, note that this is not entirely unlike the likely evolutionary development of the eye itself, from a simple light detector to the much more complex organ present in higher vertebrates.) Crucially, then, we should not view these proposed devices as merely exercises in the creation (say) of specific lenses, but rather as potentially complex combinations of accessible hardware and maximally rapid (if occasionally flawed) visual pattern recognition–again, rather like the biological eye.

Seeing Beyond the Visible Spectrum

One of the most obvious limitations of human vision is our electromagnetic imprisonment within the visual spectrum. We are generally unable to see the near infrared or ultraviolet versions of the world; nor can we see (e.g.) radio waves or X-rays. Indeed, so obvious are these limitations that it is not particularly eccentric to imagine devices to overcome them. The technology for (e.g.) creating wearable infrared goggles is by now well established for such tasks as night vision; while for the most part ultraviolet light is something that people want filtered out (through protective eyewear) rather than noticed.

In any event, rather than think of these unseen wavelengths as approached directly through eyewear, we might instead create interfaces rather like those envisioned for motion in the previous section. That is, a camera interface might not be worn over the eyes, but rather in clothing; and the goal of the camera might not be to transfer images directly to the eyes, but rather to "notice" specific phenomena in infrared or ultraviolet wavelengths over the course of a day. Such an interface might enable students to go back over the events of the day and re-interpret scenes from the point of view of a viewer sensitive to wavelengths outside the range of human vision.

Seeing with Intelligent Sensors

As already noted, the types of visual apparatus that we are imagining in this essay involve the interwoven design of hardware and software. The previous examples focused on the resolution and frequency limitations of human vision; but there are many more boundaries that could be tested. "Sensors" of various sorts could be tuned, algorithmically, to look for particular objects or phenomena in the world. To begin with a now-familiar example: current-day photographers routinely own cameras that (in fractions of a second) can detect the presence of human faces in the field of the camera's view, and tune the camera's resolution accordingly. The (surprisingly effective) software behind sensors of this type looks for visual patterns and features indicative of a human face, and in this sense loosely mimics the evolutionarily-derived attention that human beings devote to identifying the presence of faces.

Beginning with this as an existing proof-of-concept, one could imagine equipping students with a variety of "intelligent", software-intensive sensors, ranging from the plausible to the futuristic. If a sensor can detect a human face, perhaps one can be devised to make a rapid Sherlock-Holmes-like inventory of unusual features in people (a distinctive article of clothing or jewelry). Likewise, if a sensor can detect a human face, it seems reasonable that we might create sensors for the presence (e.g.) of birds sitting on tree limbs, or butterflies on leaves, or animal prints in the ground, or particularly interesting cloud formations, or constellations in the night sky. Perhaps, moving towards more abstract subject matter, we could create sensors that look for classical geometric forms, such as parabolic or elliptical outlines. In short, then, we might imagine intelligent sensors as ever-alert aids to noticing certain features or phenomena in the external world.

There is no reason to stop with these particular examples; it might be instructive, for research purposes, to imagine still other types of sensors. Might it be possible to create a sensor to distinguish (with reasonable accuracy) between inanimate and living entities? Or a sensor to detect Gaussian distributions (e.g., of the sizes of many similar objects in one's field of view)? Or a sensor to look for abstract machines of various sorts (levers, gears, linkages)? Again, such examples might prove difficult to engineer–or it might only be possible to engineer relatively error-prone versions–but the challenge itself would be instructional for educational researchers.

Seeing In and Beyond Three Dimensions

One of the recurring issues in science education is the difficulty of visualizing three-dimensional objects and events. Molecular structures, crystal lattices, the interior of the Earth, the structure of the human brain, and myriads of other scientific topics require, for their understanding, an ability to envision, and often mentally manipulate, spatial forms. Often, such difficulties only emerge upon reflection: the classic misconceptions that many people have about the cause of Earth's seasons, or the phases of the moon, derive from the necessity of three-dimensional thinking for understanding.

In some respects, this difficulty should be puzzling from the evolutionary standpoint. We *live*, after all, in a three-dimensional (spatial) world: why should we find such difficulty visualizing three-dimensional objects? The question is profound, and there is no reason to expect a resolution here. Nonetheless, at least two points should be noted as likely elements of a complete answer: first, that the initial stages of the human visual system are (at least when monocular vision is considered) two-dimensional, with light rays projected onto the retinal surface; thus, the effortful recovery of three-dimensional from two-dimensional structure is arguably the central task of human vision (Stone, 2012). The tools employed in this task include (among others) binocular vision, head and eye movements, multisensory integration (especially between vision and touch), and heuristics derived from background knowledge (both conscious and unconscious). Even with all these techniques at our disposal, humans are still capable of systematic errors in interpreting three-dimensional scenes, such as the famous "Ames Room" illusion (Behrens, 1997).

A second point to be considered in understanding the difficulty of three-dimensional visualization, at least for the purposes of science education, is that we (as students) are often called upon to visualize objects that in fact we have never actually seen or touched: protein molecules, star clusters and galaxies, the interior of a cell. Indeed, the creation of graphical or tangible models to aid this sort of visualization has historically been the main business of educational visualization in the past.

What would it mean to imagine visual aids to three-dimensional thinking? Here, we are faced with potentially more tricky engineering (and human interface) issues than have been considered up until now in this chapter. We might experiment with certain types of visual apparatus: for example, glasses that allow the wearer to see the world as if through a pair of eyes placed somewhat further apart than human eyes generally are, and thus to see sharper binocular disparity than we are capable of. (In the animal world, hammerhead sharks seem to have evolved a head shape with widely spaced eyes for this purpose.) Or we might investigate human response to still stranger or more exotic visual apparatus: perhaps lenses that attempt to integrate multiple (more than two) eyes, or cylindrical eyes that can take in a 360-degree field of view, or a visual apparatus that is designed to surround an object and "see" it from multiple angles simultaneously. It might well be that some of these potential visual devices would be unhelpful (or even disconcerting) but others might provide unorthodox intuitions into 3D structure–this is an empirical question, and one that might also require caution (in the form, say, of animal testing) in the early stages of implementation.

Another approach to these issues – one with perhaps a somewhat greater chance of success – might be to integrate relatively straightforward visual apparatus with motor control and tactile sensation. One might design visual apparatus whose binocular disparity can be controlled by hand or finger movements (so that the apparent distance between the two eyes is completely and smoothly within the control of the user), and simultaneously signals the disparity through, e.g., a variable pressure applied to the user's fingertips or forearm.

Having broached the topic of aids to three-dimensional vision, it is perhaps only a mild leap to consider devices whose goal is to explore the possibility of "seeing" (or experiencing) four-dimensional mathematical objects. The notion of understanding unseen dimensions (beyond the familiar three of our spatial world) is one that has long tantalized mathematicians and geometers; and as Rucker (2014) notes in a marvelous short book on the subject, there have been a few (very few) individuals who have in fact claimed to be able to visualize four-dimensional objects.

Again, the standard approach of the educational visualization community would be to attempt to design graphical effects that provide intuitions about higher dimensional objects (see, for instance, the lovely book by Banchoff, 1990). This, as noted before, is focusing on the problem of the *object perceived* as opposed to the *perceiver*. What would it mean, in contrast, to design apparatus to approach four-dimensional vision (or visualization)? As in the three-dimensional case, a possible approach might be to employ tactile sensation to register input along an unseen dimension, and to use some sort of motor control to move one's apparent gaze in that dimension. The overall apparatus might involve a headset to see a projection (in 3D) of a four-dimensional object like a tesseract; and then to use motor control to (say) rotate the object in the fourth dimension while sensing through pressure the diameter (in the fourth dimension) of the rotated object. The design of such a device would be a welcome research topic for a variety of fields–an interweaving of exploration in mathematics education, cognitive science, and human-interface design.

Creating Graphics to Be Visualized with Extra-Visual Apparatus

Throughout this chapter, we have drawn a distinction between designing graphical effects (the traditional approach to visualization) and designing innovative perceptual apparatus that expands or refines sensory perception. It should be noted that these two approaches are not by any means mutually exclusive: it might be profitable to create scientific or educational visualizations whose appreciation is enhanced (or made possible) through perceptual apparatus.

Students are already familiar with such artifacts in the form of "3D drawings" that, in combination with inexpensive tinted plastic lenses, allow viewers to see drawn figures at a variety of depths by exploiting binocular disparity. These may be seen as a proof-of-concept (if one were needed) that special-purpose graphical design may be accompanied by special-purpose visual apparatus. The examples of this chapter could suggest more complex or challenging phenomena that educators might explore. Conceivably, one might create visualizations on a computer screen whose interpretation radically changes in the presence of appropriate visual tools. A dynamic display (or video) might include phenomena that occur too fast to be visible to the naked eye, but that can be noticed with appropriately-tuned motion lenses. A mathematical simulation might be viewed through lenses that filter out (or highlight) phenomena at particular frequencies. A coded display could hide a "secret message" using techniques of steganography (Cf. Abelson, Ledeen, & Lewis, 2008) and the message might be decoded through the use of special-purpose software-augmented visual apparatus. A physical demonstration for a classroom or museum might be created so that (e.g.) special-purpose lenses could note phenomena at given speeds, frequencies, or accompanied by specific wavelengths of light.

FUTURE RESEARCH DIRECTIONS

The Prospect of Rethinking Common Sense

The various examples and possibilities discussed in this chapter are perhaps new to the field of educational visualization; but they are hardly unprecedented (or even very daring) in other respects. The notion of augmented visual apparatus is a recurring idea in science fiction – including movies like *Robocop* and *The Terminator*, and such television staples as *Star Trek* (with its VISOR device) and the old *Eight Million Dollar Man* series. Numerous fictional characters in these settings have been endowed with vision that surpasses the biological, including (among other things) sensitivity to wavelengths outside the standard visible range, night vision, or telescopic sight.

Moreover, there are aspects of enhanced vision that are often discussed in current technological writing, but that have (with some deliberation) not been included in this discussion. The alert reader may have noticed that there is little discussion here of "Web-connected" vision (e.g., the sorts of scenarios in which a student views a famous artwork or historical site and simultaneously reads explanatory material about what she is viewing). Nor, for that matter, have we discussed related topics such as endowing glasses with abilities such as face recognition or email readers. The reason for highlighting the particular topics that we have treated in this chapter is to focus on the issue of sensory limitation for the purpose of scientific "common sense" and understanding. It is one (quite useful) thing to see an animal running and to read automatically retrieved text about the animal and its biology; but it is another thing to see the animal differently–through different visual capacities–than before. Arguably, much of our difficulty in understanding scientific concepts stems not from a lack of immediate information, but from the paucity or limitations of our bodily endowment.

A harbinger of the growing interest in sensory extension can be found in the fascinating recent book by Platoni (2015), particularly in its final chapters. The author describes, for example, a provocative (if somewhat discomfiting) subculture of "body modders" who experiment with techniques for expanding their sensory apparatus. A typical (if that word can be applied) choice is to implant magnets in one's hands or fingertips; those who have done this (not, it seems to me, without some medical risk) report that they are able to detect magnetic fields through an "add-on" to their sense of touch.

The purpose of this chapter is not to advocate for such radical embedded changes to the body (at least not without a great deal of caution and prior experimentation); but the individuals encountered in Platoni's book are, in the main, motivated by a sense of curiosity about the world and a defensible sense of frustration about the putative biological boundaries of human understanding. It is that notion–the notion that we are embodied observers who have the capacity to augment our inborn apparatus–that is shared by this chapter. Why can't we see flowers in ultraviolet wavelengths, as bees do? Why can't we, in high resolution, the flicker of a snake's tongue, or the movement of a bird's wings? Why can't we see the polarization of light? Why can't we see (in graphical form) four dimensional objects? Such questions approach basic questions of cognitive science – the relationship between human evolutionary endowment and the limitations of mind–that tend to be suppressed by an exclusive focus on (say) improving the quality of graphical presentations or chart design.

At the same time, it is inevitable that exploration in this area will entail not only intellectual or cognitive questions (e.g., how to convey information about high-speed phenomena), but also social and ethical issues. Might it be possible for people to become so used to "enhanced" vision that they are, in a sense, dependent on it? (The "magnetically enhanced" individuals in Platoni's book seem to have become so accustomed to sensing magnetic fields that they would feel unduly deprived if that sense were now to be taken away). The apparatus envisioned in this chapter is, in all cases, temporary and removable, but one might imagine students becoming (in a broad sense) addicted to particular sensory extensions, with debatable consequences. (An alarming corollary to this scenario is one in which students become uncomfortable or even compromised when the additional sensory apparatus is removed or unavailable). And there are still other societal issues to consider: might we end up with a world in which the already-troublesome class-driven "technological divide" extends to sensory abilities as well? It is at least possible that students with augmented senses would be (or would be perceived as) privileged or advantaged; and that those without such opportunities for sensory extension would be seen as lagging behind (perhaps not unlike those current-day students without mobile phones).

A brief discussion such as this one can hardly do justice to these (very real) issues–but as the history of the cell phone and personal computer show, many of these issues, including those of class distinctions or brain alteration, are hardly unique to the world of visual augmentation. Indeed, these types of issues have arisen throughout the long history of technological change (arguably dating back to the invention of agriculture, numbers, and the written word). The consistency of these issues does not, of course, absolve us of considering them in the face of new technology; but we should treat these issues not as sudden arrivals in human affairs, but as the latest manifestation of an enduring theme in our species' history.

In any event–and to return to the subject of education–regardless of how we choose to navigate these difficult questions, we nonetheless cannot escape the central problem of scientific understanding, which remains, as always, the limitations of "common sense." According to common sense, the earth is flat and doesn't move; the air surrounding us is not made up of particles but is rather a continuous substance; animals come in fixed "kinds" and do not evolve; continents don't move; nothing can be both a wave and particle at the same time; and so forth. Science education is a continual struggle, or interplay, between the immediate images of our sensory experience and the (far more effortful and unfamiliar) models derived from experiments. To the extent, then, that we can narrow the gap between sensory experience and the world revealed by experiments, we can aspire toward a deeper, more transcendent, science education; and a greater appreciation of the larger world in which we happen to live.

REFERENCES

Abelson, H., Ledeen, K., & Lewis, H. (2008). *Blown to Bits*. Boston, MA: Pearson.

Banchoff, T. (1990). *Beyond the Third Dimension*. New York: Scientific American Library.

Behrens, R. (1997). Eyed awry: The ingenuity of Del Ames. *The North American Review*, *282*(2), 26–33.

Braun, M. (2010). *Eadweard Muybridge*. London: Reaktion Books.

Bush, V. (1945). As we may think. In *From Memex to Hypertext*. San Diego, CA: Academic Press.

Fischbein, E. (2002). *Intuition in Science and Mathematics*. New York: Kluwer.

Platoni, K. (2015). *We Have the Technology*. New York: Basic Books.

Rucker, R. (2014). *The Fourth Dimension: Toward a Geometry of Higher Reality*. Mineola, NY: Dover Books. (Reprint, 1984 edition.)

Stone, J. (2012). *Vision and Brain: How We Perceive the World*. Cambridge, MA: MIT Press.

ADDITIONAL READING

Agar, N. (2013). *Truly Human Enhancement*. Cambridge, MA: MIT Press.

Savulescu, J., & Bostrom, N. (Eds.). (2009). *Human Enhancement*. Oxford, UK: Oxford University Press.

KEY TERMS AND DEFINITIONS

Ames Room: A well-known optical illusion created by Adelbert Ames, Jr. At full-scale (some science museums boast such a display), the room is an irregular shape which, when viewed from the outside through a carefully placed window, presents the illusion of a standard rectangular space. Viewers looking through the Ames Room window will mistakenly interpret two identical objects as having vastly different sizes when those objects are placed at opposite corners of the room. (The objects are in fact at significantly distinct distances from the viewer, but they are mistakenly interpreted as objects with different sizes, and at the same distance from the viewer).

Sensory Augmentation/Extension: A style of design (particularly, in this context, for education) emphasizing the ways in which technology can be used to alter or augment the human sensory apparatus.

Stop-Action Photography: A general term referring to a variety of techniques for capturing high-speed motion in a photograph.

Tesseract: A four-dimensional hypercube: one can think of this as an extension of the cube into four dimensions. Perhaps the easiest way to visualize such a thing is to imagine a (two-dimensional) square as the result of joining two parallel lines; the (three-dimensional) cube is formed by joining two parallel squares; and the (four-dimensional) hypercube is formed by joining two parallel cubes.

Chapter 4
Science and Art:
A Concerted Knowledge Visualization Effort for the Understanding of the Fourth Dimension

Jean Constant
Hermay.org, USA

ABSTRACT

The fourth dimension is a complex concept that deals with abstract reasoning, our sense of perception, and our imagination. Mathematics posits that a four-dimensional space is a geometric space with four dimensions. For many the fourth dimension is the element of time added to the three parameters of length, height and depth. How does a geometer incorporates time in the description of a structure, or a visual artist integrates time in a two dimensional flat surface image, when they both rely on well-defined principles that are a tangible descriptive of our reality? This chapter gives a brief overview of the different schools of thought in the Humanities and in Science, offers a possible definition of this elusive element needed to anchor the fourth dimension in our larger abstract reasoning consensus, and focuses on the specific of Mathematics and visual imaging to illustrate the particular benefit of collaborating on a simple, usable descriptive to create a sound outcome.

INTRODUCTION

There is something I 'know,' which is that spatial dimensions beyond the Big 3 exist. I can even construct a tesseract or a hypercube out of cardboard. (David Foster Wallace, W. W. Norton & Company, 2004)

Scientists and philosophers have studied the concept of dimension since the beginning of time. Greek philosopher Aristotle and later a mathematician Ptolemy argued that no more than three spatial dimensions were possible. Inversely, Plato in the Allegory of the Cave tells us that concept and perception are two distinct things: "Relying on your physical senses alone is a sure way to make yourself effectively blind," said Socrates to Glaucon. (Asscher & Widger, 2008).

DOI: 10.4018/978-1-5225-0480-1.ch004

It was not until much later that philosophers and scientists looked again in the possibility of additional dimensions in our immediate environment. French physicist Jean-Baptiste le Rond d'Alembert introduced the notion that one should consider time as a fourth dimension time in the late 1700s as "an idea that may be contested but having some merit if only because of its novelty" (d' Alembert, 1751). According to a mathematician and writer E. T. Bell, a mathematician J. L. Lagrange (1811) said a few years later in *Mécanique Analytique* that the science of mechanics could be considered as the geometry of a space of four dimensions – three Cartesian coordinates with one time coordinate. Although many differ about the meaning and exactitude of this translation, most agree that the fourth element refers to the idea of time, as both a material and a spiritual dimension. This concept paved the way for later studies on a fourth dimension in physics and humanitie

Between 1826 and 1829 mathematicians Nicolai Lobachevsky and János Bolyai constructed a self-consistent system of geometry in hyperbolic space that permits an infinite number of straight lines to pass through a point parallel to a given line. Their theory for the first time in two thousand years questioned the fundamental principle of Euclidian geometry, which stated that, given any straight line and a point not on the line, there is only one other straight line passing through the point, which is parallel to that line (Casey, 2007). It led a German mathematician George Riemann to develop a new kind of differential geometry of space with any numbers or dimensions and curvature, and opened the door to the Einstein's general theory of relativity where the fourth dimension is associated to the idea of space-time (Overduin, 2007).

THE CONCEPT OF A FOURTH DIMENSION

The Philosophical Order

Philosophers and scientist have been well aware of the possibility of higher dimensions in our universe. However they always had difficulties to identify and demonstrate in a rational way where and how these new elements could fit in the larger construct of our physical reality. Scientists, in particular physicists have been studying time since antiquity to assign it an objective quantity of measurement (Figure 1).

Three-dimensional space is represented by three coordinate axes x, y, and z. Each one defines a unique parameter of height, width, and depth. Each axis is orthogonal to the other two. A 4-dimensional space then should have an extra coordinate axis, orthogonal to the other three. Aristotelian logic tells us that time does not have a beginning and it is impossible to define the first moment of time, since it would have to be identified from an earlier period of time – which is inconsistent with it being the first moment of time (Barnes, 1984).

German philosopher Immanuel Kant asserts that since we are dependent on our limited senses, all we can know is the way things appear as they are represented to us through our senses and cognition.

The world of 'things--in-themselves' is beyond the reach of our sensory--cognitive faculties and hence cannot be known to us. However he admits that the law, which determines that space "has the property of threefold dimension ... is arbitrary, and that God could have chosen another law [from which] an extension with other properties and dimensions would have arisen (Kemp Smith, 1933).

Figure 1. "Fleeting time". Medieval coin, circa 1500, Bargelo museum, Firenze, IT
(© 2014. Francesco Bini. Used with permission)

Literature and the Fourth Dimension

After the first findings generated by non-Euclidian geometry theories, time became an accepted component of this potential higher dimension. More than 1800 scientific and literary papers were published on the subject for various scientific, academic, and public audiences in the 19th century (Bork, 1964). Oscar Wilde (2001) in *The Canterville Ghost* published in 1887 used the fourth dimension as plot device to legitimate unexplainable events. H. G Wells (1995) in his book *The Time machine* devised a vehicle that could selectively travel time, and conquered the imagination of larger audiences.

In the late 1800s, English academic Edwin Abbott describes in a short story titled Flatland a universe where characters are geometric figures that explore various dimensions. The satirical essay did not attract much attention at the time but was rediscovered after Einstein's (1997) papers on general relativity were published. This became a clever way for a scientist and a non-scientist to visualize and explain time and space and how it could be articulated in a fourth dimension. Biochemistry professor and science fiction writer Isaac Asimov wrote in his introduction to the 1983 reprint of the novel by Edwin A. Abbott (1983), *Flatland: A Romance in Many Dimensions*, "It is the best introduction one can find into the manner of perceiving dimensions." More recently, as space became more and more a measure of the fourth dimension, a remarkable book by author Robert Heinlein (1940) derives from the larger understanding of hyperspace an entire plot around an unfolding tesseract, a term first used by a mathematician and author Charles Hinton in 1880, in a series of articles – speculation on the fourth dimension (Hinton, 1888)

Music and Time

For musicians, time is a way of measuring the rate at which things change. The universe of the fourth dimension and time as its main component had not been the object of many research or noticeable achievements. The popular belief that time is the fourth dimension has little value in the exploration of sound. It would require integrating back time in any forward going movement, which is challenging at most. Willard Van De Bogart (1970) writes in *Fourth Dimensional Theory for Electronic Music Composition*, "The requirement for such a task will necessitate thinking in terms of the fourth dimensional coordinate system. The four-dimensional system does not rely on time as a measurable quantity."

Engineers studying acoustics look into gravity or frequency as a possible element to add to populating the existing perceptual space. Gravity would decrease several times faster than it already does, because instead of an inverse square, it would decrease by an inverse cube. Computational biologist and geneticist Max Cooper's (2015) research is an example of new direction experimenting with new sound systems using omnidirectional speakers and special software to explore new means to deliver audio. Michio Kaku (1995), professor in Theoretical Physics and the Henry Semat chair at the City University of New York, states, "Matter is nothing but the harmonies created by this vibrating string. The laws of physics can be compared to the laws of harmony allowed on the string. The universe itself, composed of countless vibrating strings, would then be comparable to a symphony."

The Fourth Dimension in the Visual Arts

Architects, artists, and designers, like musicians, would have a difficult time rendering visually time or space-time: how do you incorporate an element of time on a two-dimensional flat surface?

Projection has been a known convention to render three-dimensional space in two-dimensional flat surface since antiquity. Vitruvius in "de Architectura" loosely approached the notion of linear perspective upon which Alberti and Brunelleschi drafted remarkable monuments during the Italian renaissance (Edgerton, 1975). Painters of the same era such as Leonardo da Vinci in one of his most famous painting "The Last Supper" demonstrated the full extent of the use of a one-point perspective on a flat surface (Figure 2).

The concept of the third dimension on a flat surface became such an integral part of pictorial representation that the first modernist abstract painters of the early 1900s such as Kazimir Malevich were almost unanimously rejected, so strong was the expectation of the public for a richer experience which an illusory perspective could provide. At the same time, following earlier Riemann theories, new discoveries in mathematics brought alternate hypotheses for the definition of the fourth dimension, adding a new element of space in the known three dimensional environment. Inventor Steven Richard Hollasch (1991) concluded in his science master thesis, "Mathematical operations used in the 3D rendering process extend effortlessly to four dimensions."

In Euclidean four-space, Pythagorean distance and angle can be easily extended from three-dimensional space relying on the same rationale that describes 3D objects on a flat surface. To represent the third dimension, geometers use a set of predefined lines x (length) and y (height) on an axis. Diagonal to the axis, a z line defines depth or perspective. The viewer then uses his or her imagination to simulate distances between points of reference.

Figure 2. Diagram. Leonardo Da Vinci, "The Last Supper"
(© 2015 Jean Constant. Used with permission).

In 1905 physicist Henri Poincaré argued that time is not a feature of reality to be discovered, but rather is something we've invented for our convenience. The fourth dimension is part of an abstract construct to map our environment, and is based on coordinates that can be described in real numbers (Murzi, 2015). French mathematician Maurice Princet is credited to have open the door, in a series of informal lectures and talks, to the understanding of Riemann and Poincaré theories to people like Picasso, Duchamp, or Metzinger. Albert Gleizes and Jean Metzinger (1912) discussed it at length in the book *Du cubisme.* The cubist perspective helped open the understanding of a material quantifiable fourth element that later Dali symbolically enshrined in the 1954 *Crucifixion (Corpus Hypercubus).* It depicts on a flat surface what the representation of a fourth dimension a hypercube or a tesseract could look like; it is now exhibited at the Metropolitan Museum of Art in New York City where it has drawn crowds ever since (Henderson, 1983).

Today, using of spatial representation to visualize the mathematical theory is becoming common both in art and architecture. The massive skyscraper *La Grande Arche* (Figure 3) was designed by architects Von Spreckelsen and Reitzelin to look like a hypercube projected on a 3 dimensional world (Reitzel, 1970). Located in an outskirt of Paris called La Defense, it is visited by more than 5 million tourists every year.

The computer aided media industry, the film industry in particular has been able to successfully use complex projective geometry technologies to induce viewers from a two dimensional space (the screen), and to make believe three dimensional worlds, for example in films like *Avatar* by James Cameron. More recently viewers are induced in the 4 dimensional spaces in Chris Nolan *Inception* where a multidimensional space, which folds back upon itself in space, is built like a Penrose stairways (Harshbarger, 2010), or in *Interstellar* where the gallery is a replica of a tesseract (Reitzel, 1970).

Figure 3. La Grande Arche
(© 2015. Jean Constant. Used with permission).

THE MATHEMATICAL ORDER

Time is not a physical dimension of space through which one could travel into the past or future. Kurt Friedrich Gödel (P. Schilpp, 1949)

Mathematical Definition

In mathematics, a four dimensional space is a geometric space composed of four dimensions. It can be created by extending the general rules of rules of vectors and coordinate to a space with four dimensions. In particular a vector with four elements (a 4-tuple) can be used to represent a position in four-dimensional space.

This space is a Euclidean space, so it has a metric and norm, and all directions are treated as the same: the additional dimension is indistinguishable from the other three. Mathematician Minkowski in 1907 attempted to demonstrate visually what the space-time dimension would look like by stacking up sequences of two-dimensional rectangles.

Amrit Sorli and David Fiscaletti (2012), physicists and founders of the Space Life Institute in Slovenia wrote an article for *Physic Essays* relating to quantifiable observation of time dilatation; which lead them to argue that while the concept of relativity was sound, the Minkowski space-time dimension has been at the origin of a serious misunderstanding regarding the identification of time as the fourth dimension: "By itself, *time* has only a mathematical value, and no primary physical existence."

Mathematical dimension is described in terms of lines. Two-dimensional lines are called polygons. In the third dimension a segment z is added to the x and y (length and height) polygons to create a polyhedron. There are five known convex polyhedra. A fourth projection on a polyhedron creates a new figure called a polytope. There are six convex 4-polytope analogues to the Platonic solids of the third dimension.

Mathematical Visualization

Mathematician and topologist Jeffrey Weeks developed a series of mathematical software for geometry students to help them visualize mathematical constructs from tori, curved spaces, symmetry, and other. One of the programs – a 4D Draw convincingly brings us closer to the perception of the fourth dimension on a two dimensional surface such as a computer monitor, a wall projection, or a piece of paper. The program 4D Draw is a vector–based program developed in part with a grant from the NSF - Division of Mathematical Sciences. The user can build 2D and 3D and 4D surfaces lining points and lines on a stage that can be tilted, turned upside down inverted and rotated 360°. Dr. Weeks has been a Mc. Arthur recipient and received the 2007 AMS- Conant prize [27] for his research on the Poincare dodecahedral space. Today he is lecturing worldwide to school and university audiences on the larger subject of mathematics, geometry, and the development of physical theories.

What starts as a deceptively simple positioning of points and line on a flat workspace (Figure 4) becomes more and more complex as vertices are added to the original shape. The program's algorithm monitors every new additional element introduced on the stage, accepts it, relines it if needed, or refuses if it is improperly placed. In addition points can be re-colored to help the viewer understand and differentiate their positioning and hierarchy in this virtual space. Following are examples of a transformation of a 2D space in a 4D space in 4D Draw (Figure 5).

Figure 4. A 4D Euclidean cube done in 4D Draw. The 4D Euclidean cube can be constructed by translating the central 3D-cube in an imagined fourth dimension
(© 2015. Jean Constant. Used with permission).

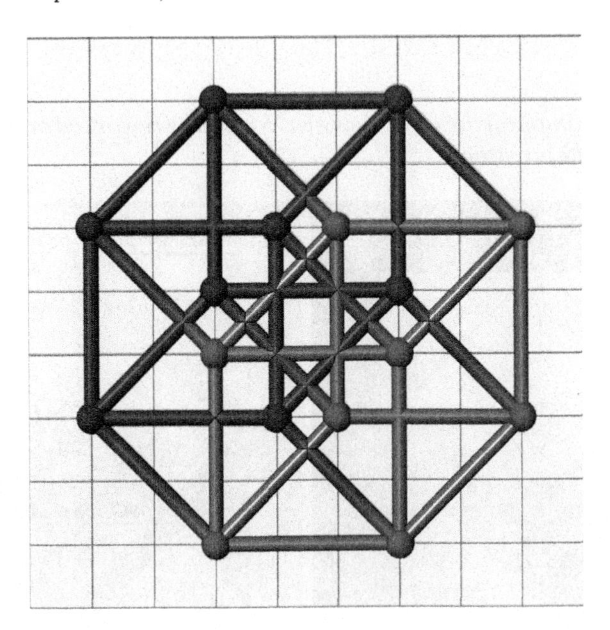

Figure 5. Left to right: from a 2D outline to a 3D model to a 4D projection
(© 2015. Jean Constant. Used with permission).

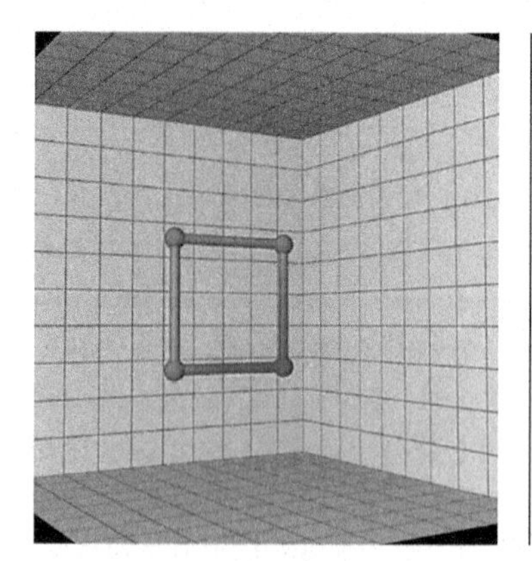
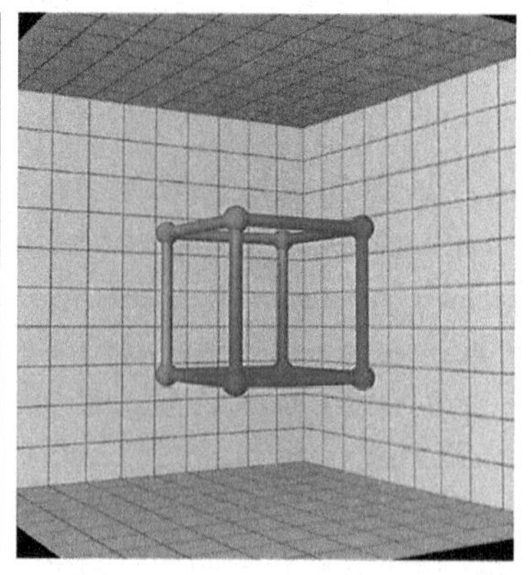
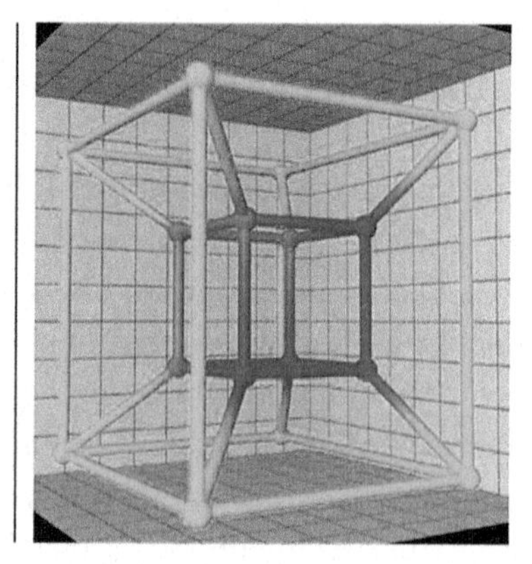

The 4D figure is also called a tesseract or a hypercube – It is composed of eight cubical cells that can only be perceived by rotation. Mathematicians, physicist, biologists, and economists use similar models for their research on the fourth dimensions. In virology it helps understand DNA sequences and virus mutation (Stewart, 2002).

Architects and visual artists have been well prepared to absorb the convention that makes possible the integration of a fourth, fifth, or sixth dimension in their work, having to deal everyday with the challenge of conveying on a flat surface a projection of a 3 dimensional space. Mathematics was the tool that helped 3D perspective to be accepted in the artistic two-dimensional world. Here again mathematics opens the possibility of investigating new dimension based on solid scientific data.

Here are a few examples of unconventional results one can obtain with the program. I extracted the original 4D Draw outlines and manipulated the images in various 2D and 3D modeling programs to extract and emphasize elements that seemed relevant to the final visual statement (Figure 6).

Figure 6. Jean Constant. Example of the front view of a hypercube reworked on various 2D graphic editors
(© 2015. Jean Constant. Used with permission).

The octaplex (Figure 7) composed of 96 triangular faces, 96 edges, and 24 vertices and has no equivalent in other dimension (Bourke, 1990). It is a good example of how a mathematic demonstration can inspire an artistic vision. This 2D representation is being studied today in a gemology lab to evaluate the possibility of converting it into a 3D object.

CONCLUSION

Perspective is the rein and rudder of painting. Leonardo da Vinci (CSIP, 2014)

Artists and architects are apt to adopt programs like 4D Draw or equivalent having been trained since antiquity to project 3D perspective on 2D drawing. From Vitruvius to Brunelleschi and Alberti in architecture, Piero della Francesca and Michelangelo in painting, the study of linear perspective brought us elements of depth in flat visualization, a concept widely accepted and understood by all today.

Computer aided visualizations such as ray tracers and CAD software reinforce further in the public consciousness today abstract projection. The emergence of new technologies like holography or virtual reality is bound to expand further and make more accessible the understanding of complex concepts of higher dimension. Multiple-lenses cameras brought us 3D movies, Oculus Rift, or Google VR; they are already attracting and challenging engineers and artists to outdo themselves and each other in this new environment.

Just like Möbius strip sculptures populate public space around the world, 3D printers explore and bring more and more new geometric forms. Theoretical physicist at the Fermi National Accelerator Laboratory Chris Quigg (2003) uses Zometool plastic vertices assembly system to explore and explain neutrino and quark.

Because of its simplicity program like 4D Draw is an excellent tool to begin this exciting journey and help new generations push further the frontier of science and knowledge based visualization.

Figure 7. Front view of a 24 cell-octaplex
(© 2015. Jean Constant. Used with permission).

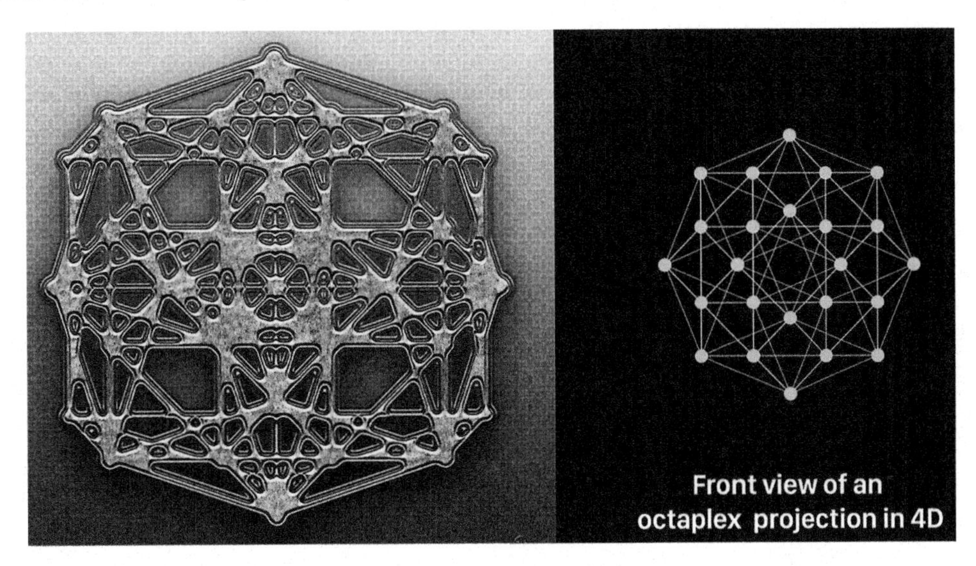

Mathematics is the language with which tangible discoveries are made. Visual art is the tool that will help us understand and communicate better these new and challenging concepts. Because they are grounded in physical reality, both mathematics and visual art are bound to benefit and grow from each other.

REFERENCES

Abbott, E. A. (1983). *Flatland: A Romance in Many Dimensions. Harper Perennial.*

D'Alembert, J. L. R. (1751). *Dictionnaire raisonné des sciences, des arts et métiers.* Briasson. Retrieved September 20, 2015, from http://www.lexilogos.com/encyclopedie_diderot_alembert.htm

AMS. (n.d.). *Conant prize.* Retrieved September 20, 2015, from http://www.ams.org/profession/prizes--awards/ams--prizes/conant--prize

Asscher, S., & Widger, D. (2008). *Plato - The Republic.* Project Gutenberg. Retrieved September 20, 2015, from http://www.gutenberg.org/files/1497/1497-h/1497-h.htm#2H_4_0004>

Barnes, J. (Ed.). (1984). *Aristotle, De Interpretatione. In The Complete Works of Aristotle.* Princeton University Press.

Bork, A. M. (1964). *The Fourth Dimension in Nineteenth-Century Physics.* The University of Chicago Press on behalf of The History of Science Society. Retrieved September 20, 2015, from http://www.jstor.org/stable/228574

Bourke, P. (1990). *Hyperspace User Manual.* Retrieved September 20, 2015, from http://paulbourke.net/geometry/hyperspace/

Casey, J. (2007). *The First Six Books of the Elements of Euclid.* Project Gutenberg. Retrieved September 20, 2015, from http://www.gutenberg.org/ebooks/21076

Cooper, M. (2015). *MaxCooper.* Retrieved September 20, 2015, from http://maxcooper.net

Edgerton, S. Y. (2008). The Renaissance Rediscovery of Linear Perspective. *ACLS Humanities.*

Einstein, A., Klein, M. J., Kox, A. J., Renn, J., & Schulman, R. (Eds.). (1997). The collected papers of Albert Einstein. Princeton University Press.

Gleizes, A. (1912). Du cubisme. Eugène Figuière, Éditeur. University of California Libraries.

Gödel, K. F. (1949). *A remark on the relationship between relativity theory and idealistic philosophy.* In P. Schilpp (Ed.), *Library of Living Philosophers* (Vol. 7, pp. 555–562). Open Court.

Harshbarger, E. (2010). Inception's Penrose Staircase. *Wired Magazine.* Retrieved September 30, 2015, from http://www.wired.com/2010/08/the-never-ending-stories-inceptions-penrose-staircase/

Heinlein, R. (1940). *And He Built A Crooked House.* Tor Books.

Henderson, L. D. (1983). *The Fourth Dimension and Non-Euclidean Geometry in Modern Art.* The MIT Press.

Hinton, C. H. (1888). *A New Era of Thought. Selected Writings of Charles H. Hinton.* Dover Publications.

Hollasch, S. R. (1991). *Four-Space Visualization of 4D Objects.* (Master of Science thesis). Arizona State University. Retrieved September 30, 2015, from http://steve.hollasch.net/thesis/

Kaku, M. (1995). *Hyperspace: A Scientific Odyssey Through Parallel Universes.* Anchor publisher.

Kemp Smith, N. (1933). *Kant, I. Critique of Pure Reason.* The Macmillan Press Ltd. Retrieved September 30, 2015, from https://archive.org/details/immanuelkantscri032379mbp

Lagrange, J.-L. (1811). *Mécanique Analytique.* Courcier.

Murzi, M. (2015). *Poincare, J. H.* Internet Encyclopedia of Philosophy. Retrieved September 20, 2015, from http://www.iep.utm.edu/poincare/

Overduin, J. (2007). *Einstein's Space-time.* Retrieved September 20, 2015, from https://einstein.stanford.edu/SPACETIME/spacetime2.html

Parker, M. (2014). *Things To Make and Do in the Fourth Dimension.* Farrar. *Straus and Giroux, LLC.*

Quigg, C. (2003). *Envisioning Particles and Interactions.* Retrieved September 20, 2015, from http://boudin.fnal.gov/~quigg/JGV/EnvPFintro.html

Reitzel, E. (1984). *Le Cube Ouvert, structures and foundations.* International Conference on Tall Buildings. Singapore. World Heritage Encyclopedia. Retrieved September 20, 2015, from http://self.gutenberg.org/articles/grande_arche

Da Vinci, L. (2014). The Notebooks of Leonardo Da Vinci. CreateSpace Independent Publishing Platform.

Sorli, A., & Fiscaletti, D. (2012). Special theory of relativity in a three--dimensional Euclidean space. *Physics Essays, 25*(1). Retrieved September 20, 2015, from http://physicsessays.org/browse-journal-2/category/28-issue-1-march-2012.html

Stewart, I. (2002). *The annotated Flatland. Basic Books.*

Van De Bogart, W. (1970). Retrieved September 20, 2015, from http://www.earthportals.com/Portal_Messenger/synthesizermusic.html

Wallace, D. F. (2004). *Everything and More: A Compact History of Infinity.* W. W. Norton & Company.

Wells, H. G. (1995). *The Time Machine.* Dover Publications.

Wilde, O. (2001). *The Canterville Ghost.* Dover Publications.

ADDITIONAL READING

Coxeter, H. S. M. (1973). *Regular Polytopes* (3rd ed.). Dover Publications.

Eckhart, L. (1969). *Four Dimensional Space.* Indiana UP.

Graziano, Aflalo TN. (2008). *Four-Dimensional Spatial Reasoning in Humans*. Journal of Experimental Psychology: Human Perception and Performance 34(5): 1066-1077. Retrieved September 20, 2015, from https://www.princeton.edu/~graziano/Aflalo_08.pdf

Heiserman, D. L. (1983). *Experiments in Four Dimensions*. Tab Books. Inc.

Manning, H. P. (1956). *Geometry of Four Dimensions*. Dover Publications.

Yourgrau, P. (2006). *A World Without Time: The Forgotten Legacy of Godel and Einstein*. Basic Books Publishers.

Penrose, R. (2005). *Road to Reality: A Complete Guide to the Laws of the Universe*. Alfred A. Knopf.

Rucker, R. (1977). *Geometry, Relativity and the Fourth Dimension*. Dover Publications.

KEY TERMS AND DEFINITIONS

Geometry: Is the study of figures in a space of a given number of dimensions and of a given type. (Wolfram MathWorld).

Hypercube: A hypercube's boundary consists of eight pieces, each of which is an ordinary cube. When projecting from 3D to 2D by ignoring one coordinate, a cube looks like a square. Similarly, when projecting from 4D to 3D by ignoring one coordinate, a hypercube looks like a cube. (Philip Spencer, University of Toronto Mathematics Network).

Mathematical Figure: A mathematical geometric figure is any set of points on a plane or in space. (Bruce Simmons, Mathwords.com).

Mobius Strip: A one-sided surface that is constructed from a rectangle by holding one end fixed, rotating the opposite end through 180 degrees, and joining it to the first end. (Merriam-Webster).

Perspective: In mathematics, perspective or linear perspective is a system for projecting the three-dimensional world onto a two-dimensional surface. It begins with a horizon line, which defines the farthest distance of the background and a central vanishing point. To this vanishing point, orthogonals may be drawn from the bottom of the picture plane, which defines the foreground of the space. The orthogonals, vanishing point, and horizon line establish the space. The points at which this line bisects the orthogonals establish the points at which horizontal lines, called transversals, may be placed. These lines represent the correct perspective regression of the square tiles into space. (Jim Elkins, What Is Perspective?").

Polytope: A regular polytope is a generalization of the platonic solids to an arbitrary dimension. The Swiss mathematician Ludwig Schläfli discovered the regular polytopes around 1852. The tesseract has 261 distinct nets. (EW Weisstein, Mathworld).

Projection: A projection is the transformation of points and lines in one plane onto another plane by connecting corresponding points on the two planes with parallel lines. The tesseract has 261 distinct nets. (EW Weisstein, Mathworld).

Space: Is a boundless three-dimensional extent in which objects and events occur and have relative position and direction. The geometrical understanding of space in non-Euclidean geometries and projective techniques forms part of mainstream mathematical research. A modern conception of space and spatial intuition has helped reshape pure and applied science, philosophy, and even the Visual Arts. (Vincenzo De Risi Max Plank Institute).

Tesseract: The tesseract is composed of 8 cubes with 3 to an edge. It has 16 vertices, 32 edges, 24 squares, and 8 cubes. It is one of the six regular polychora. The tesseract has 261 distinct nets. (EW Weisstein, Mathworld).

Time: Is the measurable period during which an action, process, or condition exists or continues. The non-spatial continuum is measured in terms of events which succeed one another from past through present to future.

Section 2

Visual Communication and Knowledge Visualization

This section describes instances of knowledge visualization in selected areas: biology, mathematics, digital media, and music. Authors present their conceptions of visualization with the use of implements used in their domains.

Chapter 5
Visualization in Biology:
An Aquatic Case Study

Maura C. Flannery
St. John's University, USA

ABSTRACT

This chapter deals with what are commonly called seaweed, but are more correctly termed algae, that is, photosynthetic organisms that live in aquatic environments. Algae are visually beautiful and therefore a good subject for a biology teacher looking to explore the intersection of art and science with students. These connections run deep into knowledge production because drawing is fundamental not only to communicating information about organisms but also to investigating their characteristics. Observation and comparison are key tools in learning about the living world, and drawing is essential to these processes. Also discussed here will be the importance of relating texts, photographs, specimens, and drawings in assisting students to learn about algae. In addition, online sources for these tools will be explored.

INTRODUCTION

A few years ago, I went on a tour of the Roger Williams Park Museum of Natural History in Providence RI. The curator showed us some of the treasures in its small but well-kept preserved plant collection, called a herbarium. She displayed not only sheets of pressed plants, the majority of specimens in such collections, but also albums, including several filled with seaweed collections. I have to admit that I had never seen anything like them, and they stayed in my mind when I left Providence. I soon discovered that these scrapbooks were popular souvenirs of seaside vacations in the 19th century and were manifestations of a more general interest in natural history. There were blank seaweed albums with decoratively bordered pages designed for amateur collectors, and there were also manuals on seaweed collecting, including one that ran to 270 pages (Hervey, 1881). The study of nature was considered a good thing to do both because it was a form of healthy outdoor entertainment, and more importantly, was a way to study God's creations (Keeney, 1992). Natural theology, the idea that learning about the living world was a means of learning about God, was an important element in 19th-century religion in both the United States and Britain.

DOI: 10.4018/978-1-5225-0480-1.ch005

COLLECTING SEAWEED OR ALGAE

Botany, the study of plants, was particularly popular among women because collecting plants was considered less strenuous than hunting animals or scaling the heights in search of rock specimens. Emily Dickinson made a herbarium in 1845 when she was a student at Amherst Academy and wrote to a friend: "Most all the girls are making one" (Dickinson, 2006). While Dickinson's specimens were all land plants, seaweed or algae collection was a rage that extended to the end of the 19th century in the United States after beginning in the late 18th century in Europe. The estate sale records for both Queen Charlotte and the Duchess of Portland list seaweed collections among their plant specimens, and Anne Christie (2011) has written about the seaweed calico prints created by the botanical illustrator and fabric designer William Kilburn in the 1790s. Christie sees these two types of interest as related and argues, as Emma Spary (2004) has, that a scholarly interest in a subject, such as seaweed, was often manifested in an interest in related design, as an outward manifestation of knowledge. This was particularly true for women who were less able to publically display their erudition.

One of the most spectacular displays of macroalgae in the 19th century was the work of Anna Atkins (1850) who produced hundreds of nature prints in the form of cyanotypes of seaweeds in her *Photographs of British Algae: Cyanotype Impressions* (1843-53). A cyanotype is a form of photograph created by placing a specimen on chemically treated paper that turns blue when exposed to light, usually by placing the preparation in bright sunlight. Each of Atkins's photographs is labeled with the species name, so the work is somewhat scientific, even if the date and location of collection are not given. Atkins's work is discussed in *Ocean Flowers*, a book on nature prints and botanical art, focusing on aquatic plants. In it, Carol Armstrong and Catherine de Zegher (2004) also describe the bound volumes of pressed seaweed specimens, *exsiccatae*, published by Mary Wyatt (1834). She was advised by Amelia Warren Griffths who was highly regarded for her botanical knowledge by such botanist as George Bentham and William Jackson Hooker, and sent many specimens to Robert Kaye Greville (1830) who later published the *Algae Britannicae*. The botanist William Henry Harvey (1849) suggests that readers of his *Manual of British Algae*, which was without illustrations, would do well to refer to Wyatt's *exsiccatae, Algae Danmonienses* (1834-1840) published in five volumes. He writes that the specimens were almost as good as illustrations, that they were beautifully dried and accurately named (Livingstone and Withers, 2011).

While Harvey had written about the land plants of South Africa, he came to specialize more and more in macroalgae, seaweed visible to the naked eye. In 1853, he set out on a collecting trip to Australia with the intention of devoting "special attention to Algae – making such a collection as never before seen in Europe. I fear you will think this a low, and mean, and slushy plan – and will be sending me to climb mountains or gather nobler plants. But I say, other collectors there are by score who look after such things – while no one minds poor Algae save a few scrap-picking folk and consequently we have little or no knowledge of the vegetation of the tropical seas" (Ducker, 1988). This says something of the inferior position algae held in the botanical world and Harvey's dedication to changing that image. He collected well. Though he had a position as director of the herbarium at Trinity College, Dublin, he was not a wealthy man, so he helped finance his expedition by sending out a prospectus that promised subscribers *exsiccatae*. He writes in the same letter quoted above that he planned to do at most 50 *exsiccatae*, each with from 200 to 600 different species.

Forty such collections are known today, amounting to over 20,000 specimens (Parshall & Millar, 2010). This is a massive output, and Harvey complained at times that a day's worth of collecting required three days of specimen preparation. Algae are particularly tricky to mount because they lack the rigid structure of land plants, which can be simply laid between two sheets of paper and then pressed between boards to dry and flatten them. Algae, instead, are floated in water; a piece of paper is put under the floating plant and then gently lifted. The sheet is then pressed and dried. No glue is required because the plant contains enough mucilaginous material on its surface to adhere it to the paper that is then pressed to keep it flat. It takes practice to have the often-filamentous algae end up beautifully spread on the sheet; it is a matter of aesthetics and craft as well as scientific knowledge of the organism.

Examination of a large number of sheets allows an observer to be able to distinguish between well done and poorly prepared specimens. Harvey's are definitely well mounted, and he kept precise records of the specimens. All have names clearly written on them, which is especially important for this collection, because as Harvey had claimed, the algae of Australia had not been well-studied, so over a third of the species he found were new to science, a very high proportion. The specimen upon which a new species is identified is particularly important in botany and becomes what is called the "holotype," the one referred to in naming the species. Harvey's own collection, which is kept at Trinity College, Dublin contains many holotypes. The other specimens of the same species collected at the same time are called isotypes and are also botanically important. Specimens from Harvey are found in collections around the world; I myself have seen some of them in England, Australia, and even a small town in rural Pennsylvania. Harvey not only was an avid collector, he published a great deal, and while his *Manual of British Algae* isn't illustrated, many of his other works are. He created his own drawings and even in some cases his own lithographs because he was wary of leaving it to artists to interpret plant form.

It might seem odd to open with the 19th century in a publication about knowledge visualization in the 21st, but there are several themes in this introduction that I plan to expand upon later especially the importance of images in biological inquiry and the relationship of botany to the larger culture. In the next section, I will introduce the portion of the living world that I want to focus upon here, namely photosynthetic organisms that live in water. Then I will discuss knowledge visualization and its significance both in developing our understanding of living things and in communicating it to others. I will go on to present different media used to accomplish these goals, and how 21st-century technology is changing how these goals are achieved.

AQUATIC ORGANISMS

Though I used the word "plant," the organisms discussed here are not all technically considered plants. Plants are organisms that produce their own food, namely sugar, through photosynthesis: using the energy in sunlight to convert carbon dioxide and water into the sugar, glucose, and oxygen, but there is a little more to it than that. From ancient times, living things were divided in into two groups: plants and animals, with the plants being characterized by green pigment and lack of movement. These two traits aren't absolute; there are fungi, such as mushrooms, that aren't green but seemed more plant-like than animal-like (Gibson, 2015). Then with the invention of a new visualization tool in the 17th century, the microscope, a whole new world opened up, and eventually a third kingdom was created to accommodate them, one we now call the Protista. Many one-celled photosynthetic organisms are classified as protists as are simple multicellular organisms including some of what we refer to as seaweed. This means that today, plants are considered only those photosynthetic organisms that are organized into structures such as leaves, though they may be tiny in the case of "simpler" plants such as liverworts and mosses; they also differ in reproductive methods. Some true plants live in aquatic environments, but they don't usually live fully submerged as most algae do.

Fungi turned out to be so genetically different that they were put in their own kingdom. However, since fungi, seaweeds, and photosynthetic protists had traditionally been studied by botanists, this continues to be true today. And there is one more kingdom to mention: the bacteria, which are the simplest and most primitive of organisms, and probably resemble the first life on earth. They do not have a separate structure, a nucleus, to enclose their DNA, as all the other organisms I've mentioned do. Some, such as the cyanobacteria, are photosynthetic, and there is good evidence that their photosynthetic mechanism is the basis for all photosynthesis on earth. Since this is a chapter on knowledge visualization, not plant science, I won't go into details on photosynthesis except to say that it's a complex process that involves an intricate array of molecules performing a series of chemical reactions. Working out the details took decades and involved a great deal of knowledge visualization at the molecular level with X-ray crystallography and electron microscopy.

Protists can be roughly divided into those that are visible to the naked eye and those that are microscopic, whether single-celled or small aggregations of cell. The microscopic algae make up a good portion of what is called plankton, the floating microscopic life in both salt and fresh water. While they are invisible to the naked eye, they are a powerful force in the web of life. They absorb about 50% of all the carbon dioxide produced by human activity such as the burning of fossil fuels. One-celled algae called diatoms are responsible for half of this absorption. They are encased in intricate glassy shells that make them among the most beautiful of algae. Other phytoplankton consist of webs of multicellular filaments, or small cell clumps. Phytoplankton are at the base of aquatic food webs and serve as food for countless heterotrophs, organisms that don't photosynthesize.

At the other end of the size spectrum are algae such as kelp that may be many meters long, yet they are relatively simple structurally, having no system for transporting nutrients within their tissues. This is why they are usually thin so they can absorb nutrients directly from their watery environment. It is no coincidence that more complex plants evolved on dry land where nutrients are not so easily accessible. The larger algae are organized into three classes – brown, red, and green – all have green photosynthetic pigments but in some these are masked by other pigments. Brown algae are primarily marine species and include rockweeds and kelps, including fucus. Most red algae also live in salt water, and some produce calcium armature that gives them a coral-like form. Finally, there are the green algae (Figure 1), the most diverse class, with over 17,000 species, both freshwater and marine (Anderson, 1992). Some are microscopic, such as the single-celled desmids that, like the diatoms, have been studied by amateurs because of their beauty. All three classes of macroscopic algae were subjects of interest in the 19th century and are to be found in seaweed albums, in fact, their different colors were at times exploited to create pictorial landscapes, with green algae used to represent leaves and red algae, flowers – definitely more works of art than science (Meier, 2014).

DRAWING AS KNOWING

The argument that I want to make here is one that has been very well presented in recent work by Omar Nasim (2013) and Barbara Wittmann (2013) namely that scientists use drawing not only for recording observations but for creating knowledge. The authors employ examples from two different centuries, the 19th and 21st respectively, and two different disciplines, astronomy and zoology, to make their case,

Figure 1. Herbarium specimen of Nitella montana collected on July 5, 2011 by K. G. Karol, J. D. Hall, and W. Perez at Show Low Lake, Arizona
(© The C. V. Starr Virtual Herbarium of New York Botanical Garden. Used by permission).

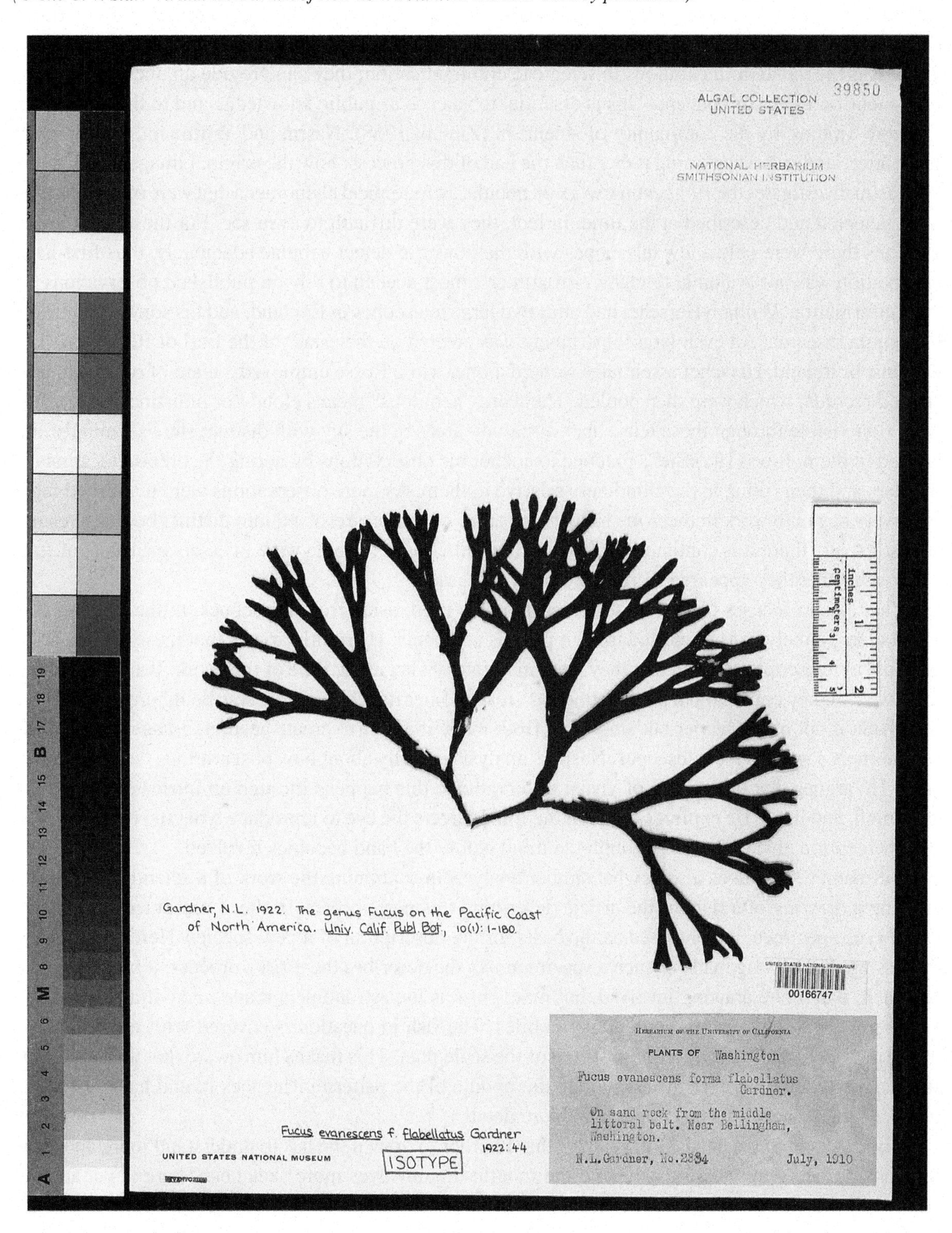

which I contend is also true in botany. This strengthens their argument: there is obviously something fundamental going on here. Both Nasim and Wittmann refer to the work of Bruno Latour (1990) who coined the term "immutable mobile" to describe a drawing or other illustration that could be printed multiple times without changing, something not really possible before the age of printing. Latour argues that this is the value of illustrations in scientific communication: they can provide an unchanging view of a scientific object or concept. This is essential to science as public knowledge and to the acceptance of a new finding by the community of scientists (Ziman, 1968). Nasim and Wittmann, however, are more interested in the beginning rather than the end of this process: how the original image comes to be.

Nasim investigates the 19th-century work on nebula, astronomical phenomena that were only beginning to be detected and described at the time. In fact, they were difficult to even see. For these early investigations there were only a few telescopes with the power to detect nebulae adequately, thus first-hand observation was not available to many astronomers; most needed to rely on published observations for their information. William Herschel had built two large telescopes in England, and his son used them for his nebula research. An even larger instrument was created on the estate of the Earl of Rosse, William Parsons, in Ireland. Herschel essentially worked alone, while Rosse employed a team of observers who created records, which were then pooled. The word "nebulous" means cloudy or indistinct, and nebula were first visible through these telescopes as cloudy areas in the sky with distinct stars seemingly embedded in them. It was Herschel's practice to anchor his observations by noting the precise locations of the stars and then filling in the cloudiness relative to them. As more observations were made, and other observers began to work in the same field, some areas of nebulae resolved into distinct bodies, presumably stars, and there was continuing debate as to whether or not nebula were in essence clusters of stars so far away that they appeared merely as cloudy masses.

What Nasim focuses on are the tools astronomers used, aside from telescopes, in the investigation of nebulae. Namely he is interested in their pencils and paper. He rightly argues that these are technologies just as telescopes are, and that they were in a state of flux at the time of this work. Pencils had been improved so they could hold a point better and create a finer line; also, paper was being produce in ways that made it not only cheaper but smoother. Both made it easier to create accurate images of what the astronomers saw with the telescopes. Nasim's analysis is really about how observations become knowledge. He argues that in the case of visual observations, this happens through an intricate interplay of eye, mind, and hand. He explores how can the mind directs the eye to reproduce what the eye perceives, with perception always filtered through the mind before the hand becomes involved.

Wittmann (2013) does a somewhat similar analysis in examining the work of a scientific illustrator creating a drawing of a fish for the article describing this new species. Earlier I presented the concept of the type specimen, the one used as the basis for the description of a new species. Here Wittmann is discussing the drawing made of such a specimen. As she describes the artist's process, it becomes clear that there is not one drawing involved, but many, just as the astronomers made many drawings before creating what would become immutable mobiles. The fish in question is covered with tiny scales, so the artist Nils Hoff first does a rough sketch of the scale plan. This makes him aware that the lateral line structure extending beneath the skin along the middle of the pattern influences it, and has to be taken into account, so he works out this area in more detail.

While sketching the head, Hoff realizes that in order to draw it, he has to understand more about the relationship between the nasal tube and the mouth; this involves more sketches. There is yet another sketch of all the spots on the scales; these are then overlaid on the final drawing. Wittmann's analysis makes it clear why scientists prefer to work with artists who have a scientific background. If Hoff had considered the mouth and nares as two unrelated orifices, he would have had difficulty realizing their connection, let alone drawing it. William Henry Harvey preferred to create his own illustrations because of this interplay of knowing and then rendering that knowledge visually. Such drawings are not just means of transmitting knowledge; they are actually means of creating it. Herschel's understanding of the structure of nebula evolved as he spent hours over many years drawing and redrawing these phenomena. The art historian James Elkins (2005) writes that every line he adds to a drawing not only changes the figure on the page but also the image of the object in his mind. In another work, Elkins (2003) argues that art historians should create art to have a deeper sense of what's involved. His comment on drawing suggests why.

Once the scientific drawing is completed, it can then become a public document as part of a presentation or publication; it can become science as public knowledge. There it may serve several functions. It is obviously used to communicate knowledge, but more than that it can serve as an argument for the validity of that knowledge. Drawings of similar species are often published together, in part to convince the reader that the species are indeed distinctly different and deserving of different names. Drawing has the advantage over photography in that the artist can tone-down details that aren't necessary to the argument – for example traits that are very similar in the two species – so that the structures that are important to the argument become more obvious. This is not a form of deception – nothing is added that is not there. It's just a matter of directing the eye and the mind's attention in a way the photograph cannot. The animal illustrator and zoologist Jonathan Kingdon (2011) notes that, since visual processing in humans works by detecting edges and emphasizing differences in brightness, black and white illustrations are more effective in transmitting information because they sync with the brain's visual processing mechanism.

For several years, I studied botanical illustration, having successfully completed enough courses at the New York Botanical Garden (NYBG) to earn a certificate. I should also note that I don't have the certificate because the portfolio of submitted was found wanting. Despite this, I am very glad that I've developed at least some skill in this area, and I continue to draw. At the very least, it helped me to understand Nasim's argument. Tools do matter, different pencils produce different effects, and fine detail requires a sharp point. This same detail is easier to achieve with watercolor than with color pencils, and detail is important: differences among species can often be measured in millimeters. The very need for images to accompany written descriptions arose from the difficulty of finding words to convey subtle structural differences (Ivins, 1953). But perhaps the most important lesson I've learned from drawing consistently is that drawing improves observation. Realistic drawing requires constant movement of the eyes between the subject and the paper, checking to see if the correspondence is accurate. Even before beginning a drawing, prolonged looking is necessary to get to know the subject. This is why I require my students to draw, not to make them artists, but to help them to realize just how much there is to see in the world around them.

Other Types of Visualizations

The similarities between Nasim's and Wittmann's analyses suggest that there is something fundamental about the process of converting observations into drawings whether or not an instrument is used. But other issues arise when the instrumentation is less direct. A traditional telescope or a light microscope merely amplify light signals in various ways before they hit the eye's retina. When an electron microscope is employed, the X-rays bouncing off the specimen are not visible but have to be made so through further processing. Learning about the structure of molecules is even more complex and entails complex manipulation of data with the aid of sophisticated mathematics and computer data processing. It's no wonder that images of proteins are so foreign to students: the molecules bear no resemblance to anything that's directly visible. Yet these representations are essential in helping students understand what is going on at the molecular level in a process such as photosynthesis.

As my students frequently ask: why is this important? Isn't it enough to know that carbon dioxide and water are converted by sunlight into sugar and oxygen? Well, no, I don't think it is, for several reasons. First, understanding the complexity of photosynthesis helps to explain why we aren't all doing it. Getting energy from the sun rather than scrounging around for food seems like an attractive idea in a world filled with malnutrition, yet turning people green is never mentioned as a viable solution to problem: photosynthesis requires many steps and therefore many genes and cell structures. Only by visualizing what that looks like can students appreciate why photosynthetic people are not on the horizon (Hoelzer, 2012). Another reason why it's important to develop an understanding of photosynthesis is so students can comprehend the relationship between environmental change and photosynthesis: why plants are so important to so many of the solutions to these problems, from creating biofuels to restoring forests to preventing the death of the oceans. All levels of biological organization from the ecological to the molecular are involved and none can be ignored.

But more than simulations of photosynthetic processes are needed. To deal with such global issues we need maps and other graphic organizers such as graphs and tables as well as the more sophisticated data visualizations that have become possible – and necessary – in the age of big data. I want to spend the remainder of this chapter examining ways to make information about the aquatic plant world, at all levels of organization, not only understandable to students but so compelling that they will want to delve ever deeper into this subject.

LEARNING ABOUT ALGAE

Ideally, the place to begin is with the organisms themselves. A trip to an ocean beach as the tide recedes would be the perfect way to begin. The advantage of this is not only access to fresh specimens, but to the ecosystem in which these organisms play a role. Perhaps more importantly, students would get a chance to handle this material, to experience the sliminess of kelp or the slipperiness of rocks covered with algal colonies. One of the main problems with most herbarium specimens is that they are bone-dry unless preserved in jars of alcohol. If they aren't kept dry, they are instant breeding grounds for fungi and bacteria that will destroy them in no time. However, we have the same problem today that botanists in the 16th century had when they were driven to create the first herbaria: they wanted to have access to plants at a distance in terms of time and place from the live organism. In other words, we don't live in a perfect world with everything at our fingertips. That's why drawings and photos and videos are so important in science education.

Without direct access to an ocean or lake, the next best thing is to have live specimens in the lab, something that is less common today than in the past because of budget constraints (where animals are involved there are often other issues that don't usually come into play here, though some (Abbott, 2008) have raised questions about the use of plants in science). This might leave just dried specimens, but less and less institutions even at the college and university levels have herbaria for teaching purposes (Funk, 2014), and if an herbarium is on-site it may only be available for research because frequent handling by classes of students could damage the fragile specimens. As these possibilities are closed, the focus moves to drawings and photographs in textbooks or on the web. The latter two choices are more and more becoming the same thing as textbooks become electronic and at the least, as textbook publishers produce more web resources. And then there are all the great videos of aquatic environments: floating through kelp forests, watching manatees chomp on seaweed, seeing the floating seaweed mats of the Sargasso Sea.

Recently I saw presentations on attempts to go beyond such documentaries. They were presented by two San Francisco institutions that created exhibits on wildlife in San Francisco Bay. The Exploratorium has a Bay Observatory Gallery (2015) that deals with the history, geography, and ecology of the Bay Area. Among the exhibits is an interactive table display that allows visitors to see the plankton growing in different parts of the Bay by moving a disc-shaped "lens" over the table. In other words, it enables viewers to see at the microscopic level while being reminded of the geographical context for these organisms.

At the California Academy of Sciences, the Morrison Planetarium has a show, Habitat Earth (2015), that takes visitors on a tour of the Bay from birds flying overhead, to otters swimming, to fish and reefs, and finally down to the microscopic level of plankton. Again the macroscopic and macroscopic are linked, and in this case, the connection is displayed graphically with a massive food web that appears on the screen showing how hundreds of life forms interact. The entire production was digitally produced with extremely life-like animations. The producers argued that it's nearly impossible to create videos of organisms doing precisely what a director wants them to do at precisely the right angle for a story line. There is also the appeal animation has for young audiences. While the presentation is very realistic, it is still obviously not photography and brings up interesting questions about the limits of realism.

Experiencing Algae

These exhibits are reminders of the sophisticated imagery that is commonplace today. There are electronic textbooks with elaborate simulations and videos integrated into them (Wilson, 2014). With such digital resources, what could possibly interest students about botanical illustrations and dried plant specimens? My answer is that these are different but equally important ways of looking at the living world, and moving among different types of visualization enhances learning. Anne Secord (2011) argues that amateur botanists were aided in learning about plant taxonomy through a variety of resources including living plants, written descriptions of plants and illustrated plant guides or flora, as well as collections of nature prints and *exsiccatae*. Amateurs, a broad population including people of both genders from a variety of social strata and jobs and with varying degrees of botanical expertise, used all these resources, as well as interaction with experts to improve their skills in identification. No one resource was sufficient, and there was no substitute for a trained eye accustomed to the same plant traits in dried and living material, in watercolors and in line drawings.

Today, we take this kind of visual literacy for granted as we focus on helping students understand complex digitally produced graphs and charts. What lies at the heart of plant identification is a form of knowing based on close observation of similarities and differences that can be useful in an endless number of contexts and can assist in understanding complex digital data displays. What botanists do when moving from one type of information about a plant to another is to compare: the size and placement of structures and the relationship of these structures to each other; there may also be key differences such as presence or absence of hairs on a leaf that are important. These are the same characteristics that Linnaeus, the founder of modern plant taxonomy, identified as important in the 18th century (Blunt, 1971).

This type of observation was at the heart of the pedagogy of Louis Agassiz, the founder of Harvard's Museum of Comparative Zoology and a 19th-century expert in fish biology. When a student asked to study with him, he would take one of his specimens out of its jar of alcohol, lay it on a dissecting tray, and ask the student to sit down, look at it closely, and come to him when finished. Since the fish was needed for future work, it wasn't to be dissected, just observed. After the initial viewing, students invariably would not be able to answer the three questions Agassiz asked, and they had to go back and observe some more. This cycle often lasted for days before the correct answers came. Years later, Lane Cooper (1945) interviewed many of these students, some of whom went on to careers in biology and some not, but they all pronounced this experience as life-changing: they never looked at the world in the same way again.

It is such involvement in observation that I think we should be giving our students today; the only difference is that we have a broader spectrum of visual experiences to offer. However, I do not think that there is any substitute for a live specimen. Even if the only algae available comes from a pet store, but so what, it provides students with an opportunity to see how flexible it is because it doesn't have to stand up against gravity the way land plants do, and they can feel its thinness, which relates to its simple structure. Finally they can touch its sliminess, still another difference with most land plants. One specimen can provide a host of experiences, observing with a variety of senses, and will make seaweed photographs, nature prints, and preserved specimens much more interesting. Written descriptions will also make more sense. Pointing out this interlocking of ways of knowing enhances students' sense of their own learning – all from plants that cost a few dollars at the pet store. If you have the opportunity to create pressed specimens with your students, so much the better, then they will understand that floating a seaweed onto a mounting sheet held under water might not be as easy as they had assumed. Then let them consider the work of May Wyatt who prepared thousands of specimens for her *exsiccatae*, and Anna Atkins who carefully laid out her algae on light-sensitive paper to create cyanotypes.

ONLINE RESOURCES

In the past, showing students the work of these women involved taking photographs or dragging books into class. Now large, clear images are available on the web, many through the Biodiversity Heritage Library (2015), one of the free tools accessible to any biology teacher. For those who are interested in using copyright-free images in class, there is a BHL Flickr site. These resources are helpful in linking the history of biology to the present day; the large number of illustrated books in this library indicates that attraction to beautiful images of organisms is hardly a product of the age of photography. The craving for pictures is fundamental to being human, as is asking questions about the visible world. While in the past images were relatively rare, and usually accessible only to the privileged classes, now they are available to most people.

This creates another problem: how to make sense of all this information. This is where curation comes into play. The current popularity of this term is related to its use in describing ways to deal with the ever-increasing amounts of information on the web. One answer is to have experts be the curators: let people who know what's out there pick and choose the best resources. If this is what a curator does, then teachers have been curating for generations. We select not only readings but classroom activities, approaches to assessment, etc. We have to follow standards and curricula and syllabi, but we still have room for creative curation. The problem is that the web so broadens the resources available to us that we need help. That's where large portals such as Merlot (2015) come in. It's a repository for teaching materials that have been peer-reviewed and rated, thus an initial screening for accuracy and efficacy has already been performed. Also, there's a good search function for locating items easily.

Still, just finding a variety of references isn't enough. There is more and more evidence that students learn best through stories. This can mean many things, but for me, it is a narrative that links information in a way that both makes sense and interests students. I find blog posts to be useful in this regard, particularly those created by natural history museums, botanical gardens, herbaria, and digital libraries of various types such as the Smithsonian Institution's Field Book Project (2015), where hundreds of notebooks created over the years by the Institution's scientists have been scanned, transcribed, and made available on the web. There are also several natural history organizations with interesting blogs. For example, the Natural Science Collections Association in Britain had a post on the algae collection of Margaret Gatty (1872), a knowledgeable 19th-century algologist who published a two-volume guide to British algae. There are thousands of specimens in this collection, approximately 500 of which are type specimens, some she received through her correspondence with the likes of William Henry Harvey. This post puts Gatty's work into perspective; the reader gets a sense of the times in which Gatty worked and why her algae are so valuable today (Gelsthorpe, 2013).

Besides using such posts as a way to lure students into the world of algae, another approach is to have students create their own posts. The assignment has to be carefully planned, not only must word length be stipulated, but also the minimum number of links to other materials and the number of images posted, if you are interested in visual learning. Again, narrative is the key. The post has to tell a story: about the discovery of a species, or its decline because of environmental change, or its link to historical events or to interesting people who have studied it. This isn't supposed to be just a fact sheet on a particular species; that isn't going to engage either the student or the reader. However, if a narrative can weave the information together, this makes all the difference. Martyn Kelly (2015) does this frequently in his blog posts on diatoms; he connects the importance of these planktonic organisms as indicators of water quality with information on environmental regulations in Britain. He may also describe the aquatic environments he visits in his work and include watercolors of diatoms and other photosynthetic plankton (Figure 2). Reading one of his posts is definitely more memorable than reading the same basic information in a text.

Since I am particularly interested in herbaria, I read a number of these blogs, and one of my favorites is called Herbology Manchester (2015), created by the herbarium staff at the Manchester Museum. This facility has just undergone a large-scale renovation, so many of the recent posts show various stages of the project. There have also been a series on particular plants. These include information on interesting characteristics of the species, its uses in medicine, cooking, and beyond, and photos of the living plant and of preserved specimens. This blog is a model for what students could put together from resources freely available on the web, since so many herbaria have specimens online: a particularly impressive is algae site is Seaweed Collections Online (2015) and another by the University and another is Phycological Research (2015) hosted by Jepson Herbarium of the University of California, Berkeley (2015).

Students could pair a specimen of a species with an illustration of the same species from BHL, and then use the Encyclopedia of Life (EOL) website to find information and photographs. EOL (2015) plans to eventually have a page for every species on earth and already has hundreds of thousands of the nearly two million named species. Some pages are much more developed than others, but still, this

Figure 2. "Round Loch of Glenhead" watercolor by Martyn Kelly. Filaments of the green alga Mougeotia and chains of the diatom Tabellaria quadriseptata, along with narrow cynaobacterial filaments and trapped particles, create a matrix within which the diatoms Frustulia and Navicula leptostriata are found (© Martyn Kelly. Used by permission).

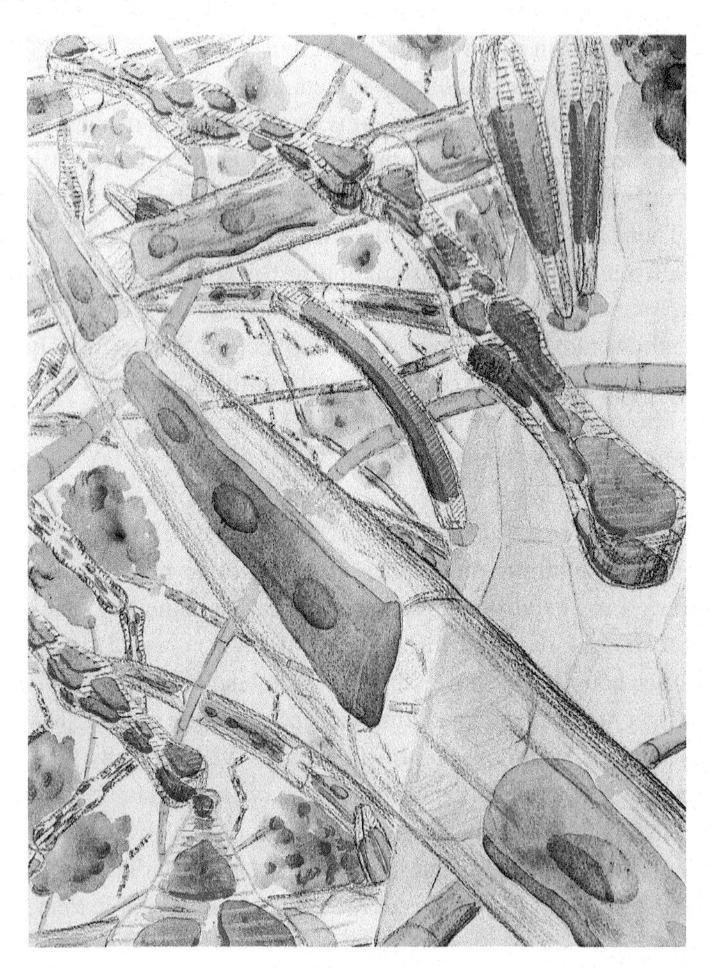

portal is useful for students because it combines text and visual information, including occurrence maps. Geographic data about a species is becoming more and more important as a way to track environmental change, not only from global warming but also from habitat degradation, which in many parts of the world is an equally ominous threat. The focus of this chapter doesn't allow for in-depth coverage of the topic of georeferencing specimens, or the tools such as Lifemapper (2015), which are available to assist biologists in their work. Google Maps and Google Earth are also widely used, but in every case, there is the caveat that all of these tools have a learning curve. Learning to read maps is not a trivial process and different maps require different visual skills, to say nothing of learning to use digital mapping tools (Wood & Fels, 2008). The essence of georeferencing is to provide the latitude and longitude coordinates for the site of collection, including an error radius because the precise spot where the plant was found is impossible to determine retroactively. For example, the error radius for a label giving the collection site as New York City will be much larger than for a collection within the borders of a small town.

CITIZEN SCIENCE

In the last few years, a new tool has become available to teachers in the form of citizen science projects. The field has burgeoned recently because scientists see it as a way to get free research help from amateurs. This is hardly a new concept; botanists like William Henry Harvey and Joseph Hooker had correspondence with networks of amateurs all over the world that provided them with specimens, seeds, and information on plants from remote areas. Though the term's use is relatively recent, citizen science has been going on for decades with events such as the annual Audubon bird count and bioblitzes that record all species found in a particular area at a designated time. Now a great deal of the assistance is given online. For example, citizen scientists helped with transcription in the Field Book Project, and they are transcribing herbarium specimen labels in many herbaria. This began several years ago in Britain with herbaria@home (2015), and now there is DigiVol (2015) in Australia, and a project called Zooniverse (2015), which is a portal for a number of citizen science projects in a variety of disciplines. There is also V Factor (2015) at the Natural History Museum, London that has several elements including one for imaging diatom slides. Some citizen science projects involve elements of gaming to keep participants involved. BHL hosts two games aimed at younger participants to aid in transcribing data from texts that have been imaged and are already available online (Duke, 2015). Transcription will make their content searchable and therefore more useful.

My own experience with citizen science involving online label transcription gave me some perspective on what a student could learn by doing this. As a NYBG volunteer for the Tri-Trophic Collection Network (2013), I was given a particular genus, *Arnica* in the daisy family, of which there were over 3000 specimens in the NYBG Steer Herbarium. All the specimens had been barcoded, photographed, and linked to skeleton digital records that just contained the specimen's name and its barcode number. I viewed a record on the screen next to the image, which I enlarged to read the label. Recent labels, those from the last 50 years, are usually typed and therefore easier to read; they also tend to contain more information, including geographic coordinates. However, there has been less plant collection done over the past 50 years, which means that many of those 3000 labels were handwritten, and some took time to decipher. I would average about 25 labels an hour, considered the norm. This meant that it took me a long time – almost a year – to finish the genus. However, that represented 3000 records that NYBG didn't have to pay someone to transcribe.

At first, I considered this work of limited importance to my understanding of herbaria, but eventually I realized that this was a great way to live, at least electronically, within a plant collection. I got a sense of the geographic range of this plant: primarily well above sea level in the Intermountain United States and Canada. Some collectors' names kept coming up again and again, and when I looked them up, I became acquainted with many of the giants of 19th and 20th-century botany of the West. I also developed a sense of what constituted a "good specimen:" carefully positioned on the page, not so full as to be bulky, and not so sparse as to look skimpy. In other words, my ability to discern subtle differences among sheets developed with more focused looking. I consider *Arnica* "my" genus. I am now learning to georeference, so I can eventually map all the specimens and see how geographic ranges may have varied with time. I am also planning to have my students contribute to such a project.

I would like to argue that citizen science can be a powerful force for bringing natural history back into the cultural limelight. The current rise in citizen science might be an indication of a new interest in natural history that could replicate that of the 19th century. I am not suggesting that people should be out collecting specimens as they did then, but rather they can use their iPhone cameras and send their data to iNaturalist (2015). They can also explore the wonderful illustrations of seaweeds that are available through BHL and its Flickr site. The BHL has a crowdsourcing project, The Art of Life (2012), to develop an infrastructure for data mining and crowdsourcing the identification and description of natural history illustrations from BHL. This combined with crowdsourcing of label transcription in the Macroalgal Digitization Project (2015) could interest a much broader audience in aquatic plant life. It's also a great way to develop visual literacy. It is impossible to work with these specimens and not learn about the aesthetics of plant placement, the subtle differences between the "styles" of different collectors, and the relationship between textual and visual information.

In the 19th century, many people studied nature as a way to approach God, the basic idea behind natural theology. Today, there are theologians such as Elizabeth Johnson (2014) who are now presenting arguments for preserving nature as necessary for our spiritual as well as physical wellbeing. There are also biologists like Ursula Goodenough (1998) who are making similar arguments. I am not going to take that tack, but instead argue from today's emphasis on civic engagement. Making careful observations of nature, digitizing information from natural history collections, and from BHL are ways to promote the public good. They are good things to do, and just as in the 19th century, what was good was also healthy and fun, the same is true today. These contributions make information available to scientists and the general public to support work on environmental issues. In addition, Lorna Hughes (2014) notes that the greatest benefit of such endeavors might not be the transcribed documents, but the transformation of people's experience of interaction with digital collections and of collaboration to produce new knowledge. Greater engagement with primary sources can create a democratization of research, something that was very much the case in 19th century natural history, when women on the Plains where sending specimens to Asa Gray in Boston and Australian pioneers were corresponding with Joseph Hooker in London.

ART AND ALGAE

In this chapter, I have in most cases remained on the scientific side of visualization and not ventured into the world of art. Admittedly, many consider Harvey's macroalgae illustrations and Anna Atkins's cyanotypes to be works of art. It was often the case that those who were drawn to seaweed by their beauty became so fascinated that they wanted to learn more about them. This was definitely true with Mary Philadelphia Merrifield, a 19th-century British artist and writer who produced several books on art and fashion, and then went on to publish *A Sketch of the Natural History of Brighton* (1860), where she lived. Not surprisingly, she became intrigued by algae. She studied them in such depth that she became on of the leading British algologists, publishing papers in such scientific journals as *Nature* and the *Journal of the Annals Botany.* Combining this level of knowledge with her artistic skill, Merrifield created hundreds of watercolors of individual species, all carefully labeled (Figure 3). They are found in the Cambridge University Herbarium. Merrifield's work belies the notion that women were simply attracted to the aesthetics of seaweeds or plants in general, just as the careful studies of fungi by Beatrice Potter do.

Diatoms are of both scientific and artistic interest to Martyn Kelly, an environmental scientist who does water quality studies in the British Isles. He has a blog called "Of Microscopes and Monsters" (2015) in which he discusses the phytoplankton he investigates and what they tell him about the health of aquatic ecosystems. Kelly began his studies not in biology but in art and manages to keep his hand in both arenas. He does beautiful watercolors of what he sees under the microscope. Some are definitely works of science, accurate renditions of diatoms, however he also creates works where he attempts to

Figure 3. Watercolor of seaweed by Mary Philadelphia Merrifield
(© Cambridge University Herbarium. Used by permission).

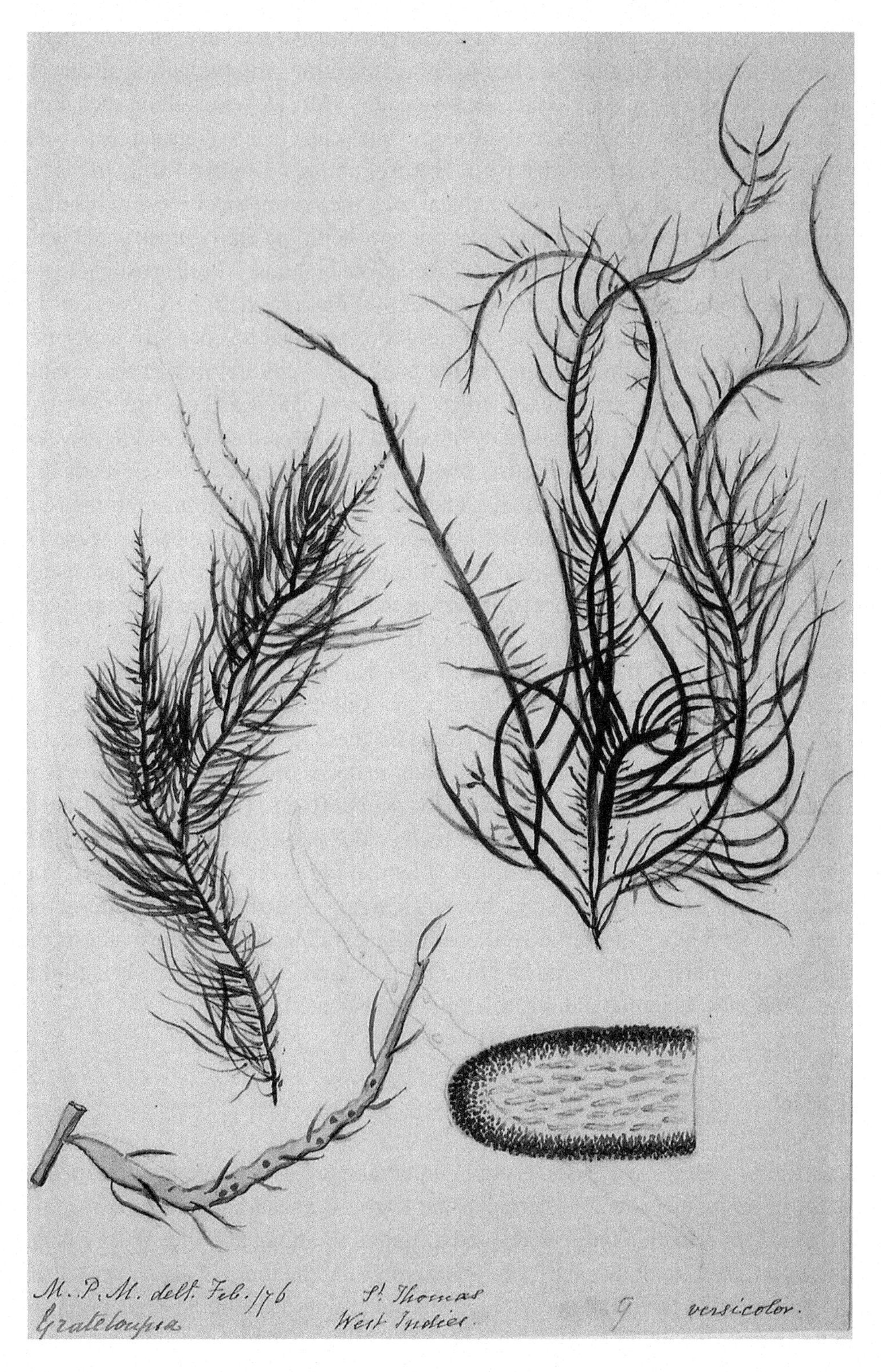

reproduce not what he sees through the microscope but what the organisms might look like in their natural habitat. These paintings are lovely pieces that definitely give the impression of looking into a watery world, while also accurately depicting the organisms that live on and around each other (Figure 2). Kelly argues that such visual representations are key in communicating with the public about what must be done to maintain rivers and streams in an ecologically healthy state. He writes about such representations as "guiding images" that make the aquatic microscopic landscape real to nonscientists (Kelly, 2012).

Diatoms were among the organisms that Ernst Haeckel portrayed so beautifully in his *Art Forms in Nature* (1904). He has been criticized for over emphasizing the symmetry of some of his drawings, such as those of siphonophores, but with diatoms he was merely recording the symmetry that is so evident in these organisms, or at least in their glassy shells. Haeckel is a biologist whose artistic talents, as well as his writing skill, made him a noted science popularizer. *Art Forms* is definitely a work of both science and art, though some would argue that because it is light on text that it is not serious science. Whether or not that is the case, it is nonetheless true that the book had a cultural impact that continues to this day. It was a major influence on Art Nouveau artists, architects, and designers. Henri Matisse was also intrigued by algae, and used their forms in many of the cutouts he created late in life (Buchberg, 2014).

There are a number of contemporary artists who are also using images of seaweeds in their work. Andrew McKeown is a British sculptor who has created over 30 large diatoms in iron-cast in a seaside park in Durham, making the microscopic definitely macroscopic (Kelly, 2015). Mark Dion (2010) is an American artist who creates interpretations of natural history collections. He made herbarium specimens with seaweeds and labeled them as being in the collection of Henry Perrine who was a 19th-century plant collector in Florida. Perrine's entire collection was destroyed in a fire. As in other work of this conceptual artist, Dion plays on the theme of what has been lost of the natural world of the past. He removes these specimens from nature still further by exhibiting not the sheets, but digital copies of them and leaves the place for the species name blank. So these are preserved specimens but of a very marginal type. Anselm Kiefer, who has used dried plant material in a number of his works, has created a vitrine containing six gold-plated clay objects along with a frond of dried seaweed. The clay pieces are parts of the body, including a heart, so the seaweed could be interpreted as a rib cage. The seaweed represents both life's fragility as well as the origin of human life in the sea (Biro, 2013). Finally, I have some of David Goodsell's watercolors of cell interiors hanging up in my dining room, yet these are also images that I use in my biology classes as well, including a simulation of photosynthesis that incorporates Goodsell's watercolors of photosynthetic proteins (Hoelzer, 2012). This is a beautiful marriage of art and science – visually beautiful and scientifically sophisticated.

CONCLUSION

Using such art works as examples, students could undertake projects where they create art with algae as a theme. This is yet another way of fostering visual learning, encouraging them to make works communicating not only information but feelings and attitudes about aquatic life. What I have attempted to present here is a repertoire of different ways of approaching the topic of algae, ways that will foster students' understanding of their biology while at the same time developing their visual literacy and appreciation for objectivity and its limits. These have ranged from drawing specimens to analyzing microscopic images to creating artworks. In each case there is an interplay of different types of knowing and learning and perceiving the world, all of which are important to a digital native of the 21st century.

FURTHER RESEARCH DIRECTIONS

Many of the areas I discussed in this chapter are in a state of rapid development. The work of Nasim and Wittmann are recent additions to the literature on knowledge visualization. I would hope that others would build on their work and use it as a basis for investigations of the use of drawing in various areas of science including botany. Citizen science is also in its infancy. References to it have exploded in the scientific literature in the past few years. This interest is spurred in part because it seems like an attractive means for scientists to get "free" labor in the form of interested volunteers. It also is a means for more public involvement in science, so they have a more positive attitude toward it – something that can only help in increasing pressure for public funding of scientific research. However, with the interest has come concern about the validity of the data that citizen scientists contribute and what can be done to insure data quality.

Also burgeoning now are online resources such as EOL, BHL, and databases for natural history collections. As more and more databases become available, investigations into how to make this information accessible in useable form to a diversity of audiences will become a focus for many in both science and education. The continuing efforts of institutions such as the California Academy of Sciences and the Exploratorium will be crucial to these efforts.

REFERENCES

Abbott, A. (2008). Swiss "dignity" law is threat to plant biology. *Nature*, *452*(7190), 919. doi:10.1038/452919a PMID:18441543

Anderson, R. A. (1992). Diversity of eukaryotic algae. *Biodiversity and Conservation*, *1*(4), 267–292. doi:10.1007/BF00693765

Armstrong, C., & de Zegher, C. (2004). *Ocean flowers: Impressions from nature*. Princeton, NJ: Princeton University Press.

Art of Life. (2012). Retrieved August 27, 2015, from http://biodivlib.wikispaces.com/Art+of+Life

Atkins, A. (1850). *Photographs of British Algae: Cyanotype impressions*. Academic Press.

Bay Observatory Gallery. (2015). *Observing Landscapes*. Retrieved August 27, 2015, from http://www.exploratorium.edu/visit/bay-observatory-gallery

Biodiversity Heritage Library. (2015). Retrieved August 27, 2015, from http://biodiversitylibrary.org/

Biro, M. (2013). *Anselm Kiefer*. New York, NY: Phaidon.

Blunt, W. (1971). *The compleat naturalist: A life of Linnaeus*. New York, NY: Viking.

Buchberg, K. (Ed.). (2014). *Henri Matisse: The cut-outs*. New York, NY: MOMA.

Christie, A. (2011). A taste for seaweed: William Kilburn's late eighteenth-century designs for printed cottons. *Journal of Design History*, *24*(4), 299–314. doi:10.1093/jdh/epr037

Cooper, L. (1945). *Louis Agassiz as a teacher; illustrative extracts on his method of instruction*. Ithaca, NY: Comstock.

Dickinson, E. (2006). *Emily Dickinson's herbarium*. Cambridge, MA: Harvard University Press.

DigiVol. (2015). Retrieved August 27, 2015, from http://volunteer.ala.org.au/

Dion, M. (2010). *Herbarium Perrine (Marine Algae)*. Retrieved December 16, 2010, from http://www.youtube.com/watch?v=F4voeIl-bXY

Ducker, S. C. (1988). *The contented botanist: Letters of W.H. Harvey about Australia and the Pacific*. Melbourne, Australia: Melbourne University Press.

Duke, G. (2015). *Smorball and Beanstalk: Games that aren't just fun to play but help science too*. Retrieved from http://blog.biodiversitylibrary.org/2015/08/smorball-and-beanstalk-games-that-arent.html

Elkins, J. (2003). *Visual studies: A skeptical introduction*. New York: Routledge.

Elkins, J. (2005). Letter to John Berger. In J. Savage (Ed.), *John Berger: Drawings of John Berger* (pp. 111–114). Cork, Ireland: Occasional Press.

Encyclopedia of Life. (2015). Retrieved August 27, 2015, from http://eol.org/

Field Book Project. (2015). Retrieved August 27, 2015, from http://www.mnh.si.edu/rc/fieldbooks/

Funk, V. (2014). The erosion of collections-based science: Alarming trend or coincidence? *Plant Press, 17*(4), 13-14.

Gatty, M. (1872). *British sea-weeds*. London, UK: Bella dn Daldy.

Gelsthorpe, D. (2013). *Margaret Gatty's algal herbarium in St Andrews*. Retrieved from https://naturalsciencecollections.wordpress.com/2013/08/27/margaret-gattys-algal-herbarium-in-st-andrews/

Gibson, S. (2015). *Animal, vegetable, mineral*. New York: Oxford University Press.

Goodenough, U. (1998). *The sacred depths of nature*. New York: Oxford University Press.

Greville, R. K. (1830). *Algae Britannicae*. Edinburgh, UK: Maclachlan and Stewart.

Habitat Earth. (2015). Retrieved August 27, 2015, from https://www.calacademy.org/exhibits/habitat-earth

Haeckel, E. (1904). Art forms in nature. New York: Dover.

Harvey, W. H. (1849). *A manual of British Algae*. London: van Voorst.

herbaria@home. (2015). Retrieved August 27, 2015, from http://herbariaunited.org/atHome/

Herbology Manchester. (2015). Retrieved August 27, 2015, from https://herbologymanchester.wordpress.com/

Hervey, A. B. (1881). *Sea mosses: A collector's guide and an introduction to the study of marine Algae*. Boston: S.E. Cassino. Retrieved from http://www.biodiversitylibrary.org/item/111603

Hoelzer, M. (2012). *SUN Chloroplast E-book*. Retrieved August 27, 2015, from http://www.markhoelzer.com/SUN-chlorophyllEbookWorking/index.html

Hughes, L. (2014). *Digital Collections as Research Infrastructure*. Retrieved June 24, 2014, from http://www.educause.edu/ero/article/digital-collections-research-infrastructure

iNaturalist. (2015). Retrieved August 27, 2015, from http://www.inaturalist.org/

Ivins, W. (1953). *Prints and visual communication*. Cambridge, MA: Harvard University Press.

Jepson Herbarium. (2015). Retrieved September 22, 2015, from http://ucjeps.berkeley.edu/jeps/

Johnson, E. A. (2014). *Ask the beasts: Darwin and the god of love*. New York, NY: Bloomsbury.

Keeney, E. (1992). *The botanizers: Amateur scientists in nineteenth-century America*. Chapel Hill, NC: University of North Carolina Press.

Kelly, M. (2012). The semiotics of slime: Visual representation of phytobenthos as an aid to understanding ecological status. *Freshwater Reviews*, *5*(2), 105–199. doi:10.1608/FRJ-5.2.511

Kelly, M. (2015). *Of Microscopes and Monsters*. Retrieved August 27, 2015, from https://microscopesandmonsters.wordpress.com/

Kingdon, J. (2011). In the eye of the beholder. In M. R. Canfield (Ed.), *Field notes on science and nature* (pp. 129–160). Washington, DC: Smithsonian Institution Press. doi:10.4159/harvard.9780674060845.c8

Latour, B. (1990). Drawing things together. In M. Lynch & S. Woolgar (Eds.), *Representation in scientific practice* (pp. 19–68). Cambridge, MA: MIT Press.

Lifemapper. (2015). Retrieved from http://lifemapper.org/

Macroalgal Herbarium Portal. (2015). Retrieved August 27, 2015, from http://macroalgae.org/portal/index.php

Meier, A. (2014). *From ocean to ornament, the Most Extraordinary Victorian Seaweed Scrapbook*. Retrieved November 11, 2014, from http://www.atlasobscura.com/articles/objects-of-intrigue-seaweed-scrapbook

Merlot. (2015). Retrieved August 27, 2015, from https://www.merlot.org/merlot/index.htm

Merrifield, M. P. (1860). *A sketch of the natural history of Brighton*. Brighton, UK: Pierce.

Nasim, O. W. (2013). *Observing by hand: Sketching the nebulae in the nineteenth century*. Chicago, IL: University of Chicago Press. doi:10.7208/chicago/9780226084404.001.0001

Parshall, N., & Millar, D. (2010). *The William Henry Harvey exsiccatae volumes*. Retrieved December 19, 2015, from http://www.aussiealgae.org/HarveyColl/manuscript.php

Phycological Research. (2015). Retrieved August 29, 2015, from http://ucjeps.berkeley.edu/CPD/algal_research.html

Seaweed Collections Online. (2015). Retrieved August 29, 2015, from http://seaweeds.myspecies.info/

Secord, A. (2011). Pressed into service: specimens, space and seeing in botanical practice. In D. Livingstone & C. Withers (Eds.), *Geographies of nineteenth-century science* (pp. 283–310). Chicago, IL: University of Chicago Press. doi:10.7208/chicago/9780226487298.003.0012

Spary, E. C. (2004). Scientific symmetries. *History of Science*, *42*(1), 1–46. doi:10.1177/007327530404200101

Tri-Trophic Collection Network Portal. (2013). Retrieved August 29, 2015, from https://www.google.com/search?q=tritrophic+portal&ie=utf-8&oe=utf-8

V Factor | Natural History Museum. (2015). Retrieved August 27, 2015, from http://www.nhm.ac.uk/about-us/jobs-volunteering-internships/volunteering-interns-information/v-factor/index.html

Wilson, E. O. (2014). Life on Earth. Palo Alto, CA: iBooks.

Wittmann, B. (2013). Outlining species: Drawing as a research technique in contemporary biology. *Science in Context*, *26*(2), 363–391. doi:10.1017/S0269889713000094

Wood, D., & Fels, J. (2008). *The natures of maps: Cartographic constructions of the natural world*. Chicago: University of Chicago Press.

Wyatt, M. (1834). *Algae Danmonienses*. Torquay: Cockrem.

Ziman, J. M. (1968). *Public knowledge*. Cambridge, UK: Cambridge University Press.

Zooniverse. (2015). Retrieved August 27, 2015, from https://www.zooniverse.org/#/

ADDITIONAL READING

Daston, L. (2008). On scientific observation. *Isis*, *99*(1), 97–100. doi:10.1086/587535

Daston, L., & Lunbeck, E. (Eds.). (2011). *Histories of scientific observation*. Chicago, IL: University of Chicago Press.

Drucker, J. (2014). *Graphesis: Visual forms of knowledge production*. Cambridge, MA: Harvard University Press.

Graham, J., Wilcox, L., & Graham, L. (2009). *Algae*. San Francisco, CA: Benjamin/Cummings.

Lynch, M. (2005). The production of scientific images: Vision and re-vision in the history, philosophy, and sociology of science. In *Visual cultures of science: Rethinking representational practices in knowledge building and science communication* (pp. 26–40). Hanover, NH: Dartmouth College Press.

Sousanis, N. (2015). *Unflattening*. Cambridge, MA: Harvard University Press.

KEY TERMS AND DEFINITIONS

Algae: A general term for aquatic organisms that have nucleated cells and also have the ability to photosynthesize.

Citizen Science: The designation for projects in which amateur scientists contribute to scientific research.

Desmids: Single-celled green algae found only in freshwater.

Diatoms: Single-celled algae enclosed in a two-part glassy shell.

Macroalgae: The biological term for seaweed, that is, aquatic organisms that photosynthesize and are visible to the naked eye.

Photosynthesis: The process by which carbon dioxide and water are converted into sugar and oxygen using the sun's energy.

Phytoplankton: Microscopic aquatic organisms that photosynthesize.

Chapter 6
Visualisation and Communication in Mathematics

Hervé Lehning
AC-HL, France

ABSTRACT

Communication in mathematics is necessary at several levels: popularisation, teaching and research. Even a small drawing is useful in every case: to set a problem, to understand it, to find a method to tackle it and to illustrate it. An appealing aesthetic can also draw the attention on the subject.

INTRODUCTION

In geometry, even the simplest proofs take an advantage of the light of a visualisation. For example, even if the concurrency of the three perpendicular bisectors of a triangle requires a small proof (the intersection of two of the lines belongs to the third), this proof is eased by a small drawing.

MISSING POINTS

In this first example, the visualisation is illustrating the proof rather than giving creative ideas. In similar cases, ideas come when we consider points *a priori* missing in the problem, but giving new lights on it. For example, to show that the three altitudes of a triangle are concurrent, we just need to construct another triangle in which these lines are the perpendicular bisectors (Figure 1).

A subtler problem, posed by Pierre de Fermat in 1636, changes when introducing new points:

In a given triangle ABC, locate a point whose distances from A, B and C have the smallest possible sum.

A simple idea is to bring back the question to the following result: “the shortest way between two points is the straight line”. Thus, given a point M, we look for a point D such that the sum AM + BM + CM is equal to the length of a broken line connecting C and D. The idea comes when constructing an

DOI: 10.4018/978-1-5225-0480-1.ch006

Figure 1. Hervé Lehning, from proof of the concurrency of the perpendicular bisectors to the concurrency of the altitudes
(© 2015, H. Lehning. Used with permission).

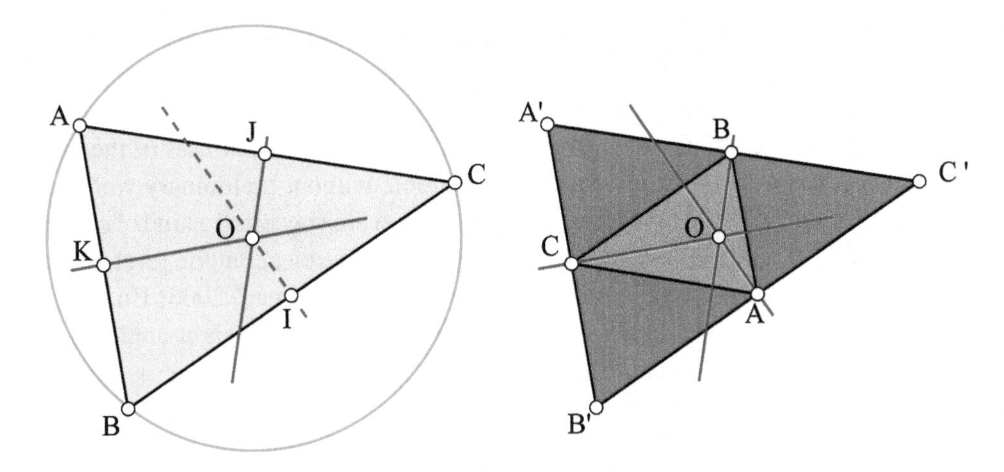

equilateral triangle ABD on side AB. This idea leads to some extra conditions (all angles of the triangle are less than 120° and B is the point where the angle is maximum) but remains rather general.

To answer the question, point M must be located on lines CD and AE (where E is the point similar to D construct on BC) thus at their intersection. The proof follows: if P is a point of the triangle ABC, the sum AP + BP + CP is equal to the length of the broken line CPQD, which is minimal if P is on line CD, and the same on AE (Figure 2).

Figure 2. Hervé Lehning, Fermat point
(© 2015, H. Lehning. Used with permission).

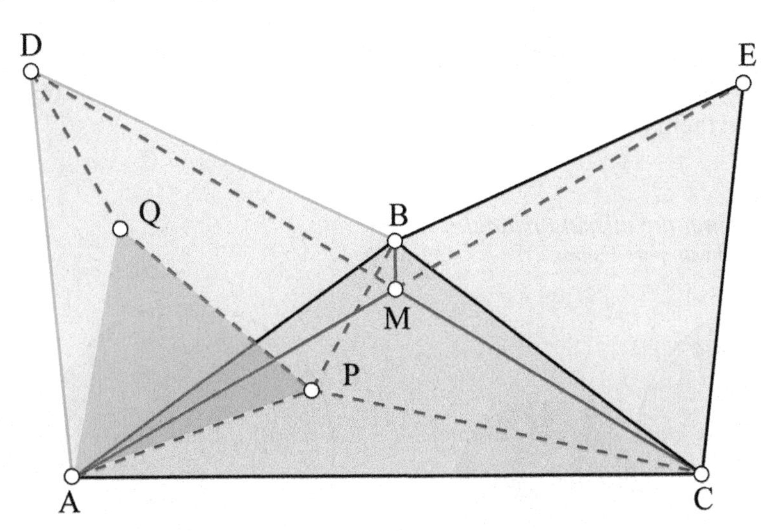

SANGAKU

If visualising a problem is particularly helpful in geometry to determine where this problem stands and to discover ideas leading to a solution, it also allows setting it in an enigmatic and aesthetic way. Let us see the difference between the approaches on a small exercise at high school level, written in a classical form:

Two circles externally tangent and a straight line tangent to both of them are given. Show that: $AB^2 = 4\ rR$ where A and B are the points of tangency on the line, R and r the radiuses of the two circles.

Visually presented, the exercise is immediately intelligible without preliminary work and far more exciting too. Thus, everyone has a direct access to the question, which becomes a kind of enigma, exciting to tackle. It leads us to the Japanese *Sangaku*, which are little masterpieces at the level of mathematical reasoning as well as the level of visual aesthetics (Fukagawa & Rothman, 2008; Huvent, 2008). The key idea to solve it can be given through a small drawing, introducing the right-angled triangle CDE, which leads to use Pythagoras' theorem. This gives directly: $(R + r)^2 = (R - r)^2 + AB^2$ and then the result (Figure 3).

Here, we meet the spirit of the Japanese *Sangakus*, which are small masterpieces at both mathematical and aesthetic level. Jean Constant (2014) for example, made it an artistic speciality (Figure 4).

The next example illustrates a *sangaku* discovered by Hidetoshi Fukagawa, a contemporary Japanese mathematics teacher. The triangles being equilateral, what is the ratio between the two circles (Figure 5).

The inradius of a triangle is equal to the area of the triangle divided by its semiperimeter. This formula becomes easy to understand through a drawing: the triangle can be divided in three smaller triangles having the incircle centre as a vertex and two vertices of the large triangle as the others. Those small triangles have the same altitude, which is the inradius. Thus, the areas of those triangles are equal to the radius multiplied by half the length of the opposite side. Summing those areas introduces naturally the semiperimeter of the triangle (Figure 6).

Thus, to compute the two radiuses, we have to compute a certain number of lengths of Figure 5. The idea to do so comes easily when we forgot a part of the figure. Using the angles 60° and 45° in evidence, we find that the four triangles in red have angles 60°, 45° and 75°. Thus there are similar. For example, the ratio between the two triangles down is easy to compute; we find it equal 2. Thanks to the ratios of similarity and to the Pythagorean theorem, length measures are gradually computed, one of them (AC) having been chosen as the unit. The drawing is useful to follow the reasoning. The results fall like domino pieces. We've marked them on the drawing (Figure 7).

Figure 3. Hervé Lehning, the missing triangle
(© 2015, H. Lehning. Used with permission).

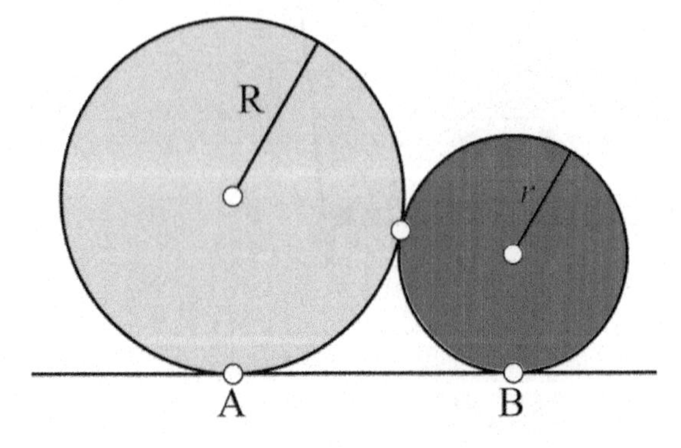

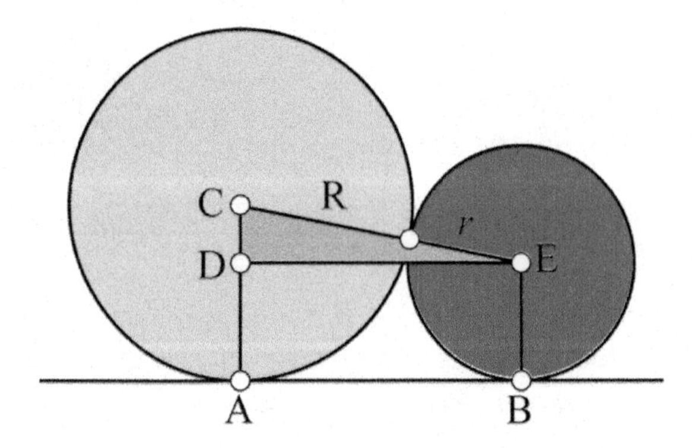

Figure 4. Jean Constant, Sangaku volume #9
(© 2009, J. Constant. Used with permission).

Figure 5. Hervé Lehning, the sangaku of Hidetoshi Fukagawa
(© 2015, H. Lehning. Used with permission).

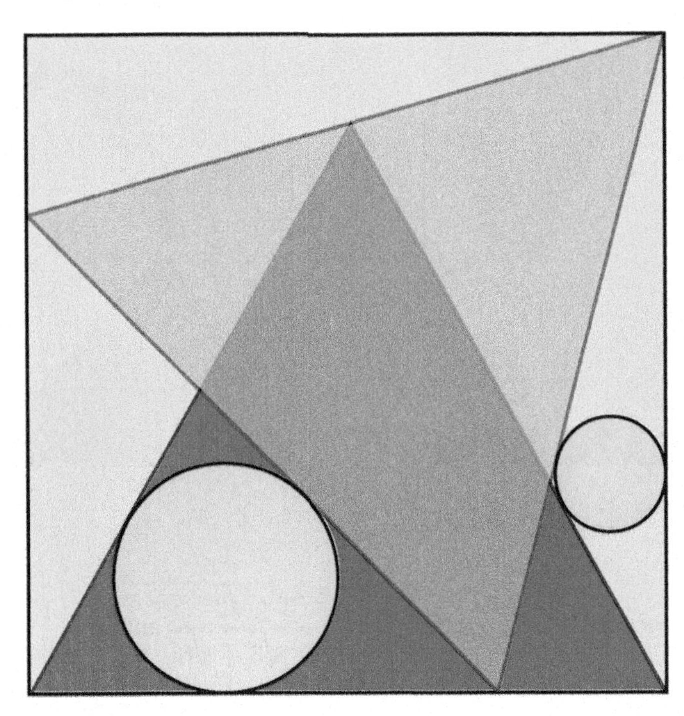

Figure 6. Hervé Lehning, the radius of the incircle
(© 2015, H. Lehning. Used with permission).

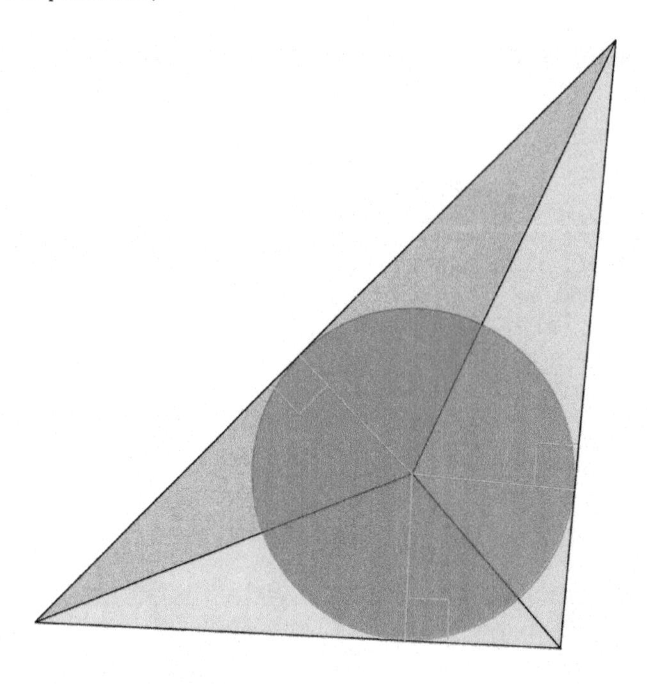

Figure 7. Hervé Lehning, diagram of a proof
(© 2015, H. Lehning. Used with permission).

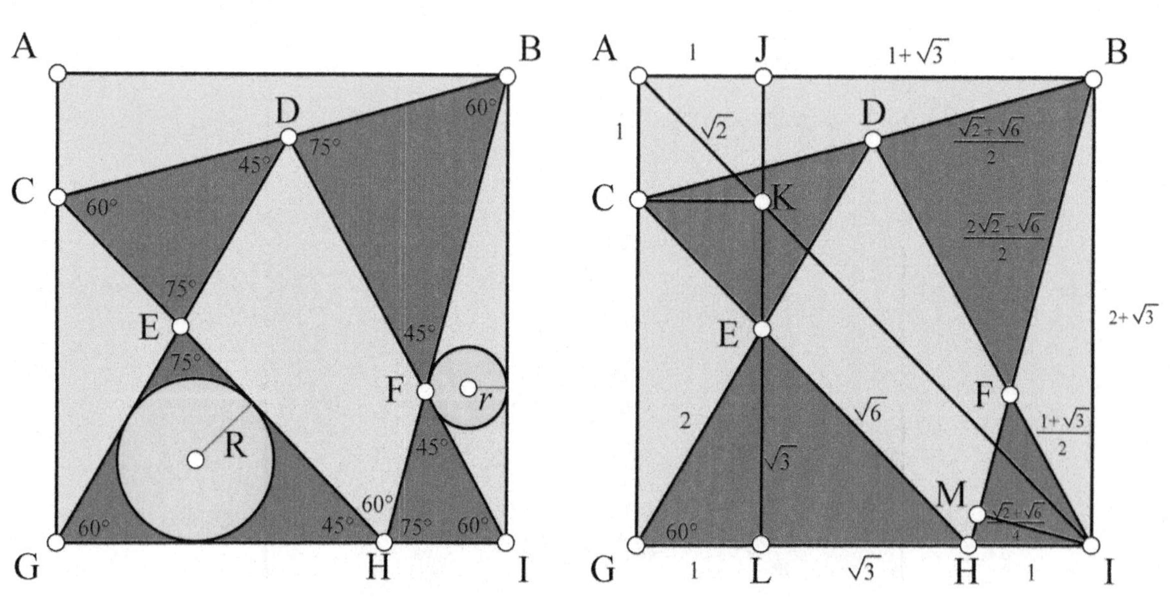

At the end, we find the values of the two radiuses: $R = \frac{3+\sqrt{3}}{3+\sqrt{3}+\sqrt{6}}$ and $r = \frac{1}{2}\frac{5+3\sqrt{3}}{5+2\sqrt{2}+3\sqrt{3}+\sqrt{6}}$.
Then, an algebraic computation allows us to show that: R = 2 *r*. For this last stage, no visualisation is necessary and we can do it with a Computer Algebra System (Lehning 1995).

We won't develop other examples of *sangaku* in such detail. The crucial thing is to realise that a good overview needs mathematical literacy, especially when a problem requires auxiliary constructions or even destructions. In fact, in the previous example, it is essential to "see" those four similar triangles, which allow us to compute progressively the lengths of the segments on figure 7, and thus the values of the two radiuses.

In some *sangaku*, the authors clearly focus on the aesthetics. For example, one of them represents a semi-circular fan spanning 120° and the question is: find the ratio between the radiuses of circles green and red. Here too, the crucial thing is to introduce the right points, which are not directly visible. Thanks to them, in a similar way as before, we find the number: $\dfrac{31-16\sqrt{3}}{2}$ (Figure 8).

ANAMORPHIC VIEWS

Beyond the simple visualisation met before, an anamorphic view, that is to say slightly distorted one, can reveal unexpected properties. That is the case of affine and projective transformations, which allow us to "see" ellipses as circles or even parabolas. In a simpler way, through an affine transformation, all triangles are identical. More precisely, a triangle may always be deduced from another one through rotations, translations, and dilatations in several directions, that is to say an affine transformation. Here again, the idea is visual. Let us start with two right-angled triangles. A translation followed by a rotation helps to match the right angles; a dilatation is enough to match them completely. To finish with the proof, we just have to show that any triangle is obtained of a right-angled triangle through an affine transformation. The proof needs to distinguish several cases, but the crucial point is given by a simple drawing (Figure 9).

From that, we deduce that all triangles share properties invariant in affine transformations that is to say concurrency, barycentre, ratio of lengths and areas. With this tool, we can derive from the concurrency of the medians of an equilateral triangle (which are also its perpendicular bisectors), the concurrency of the medians of every triangle! More generally, if an affine property concerns a triangle, it is

Figure 8. Hervé Lehning, the geisha fan
(© 2015, H. Lehning. Used with permission)

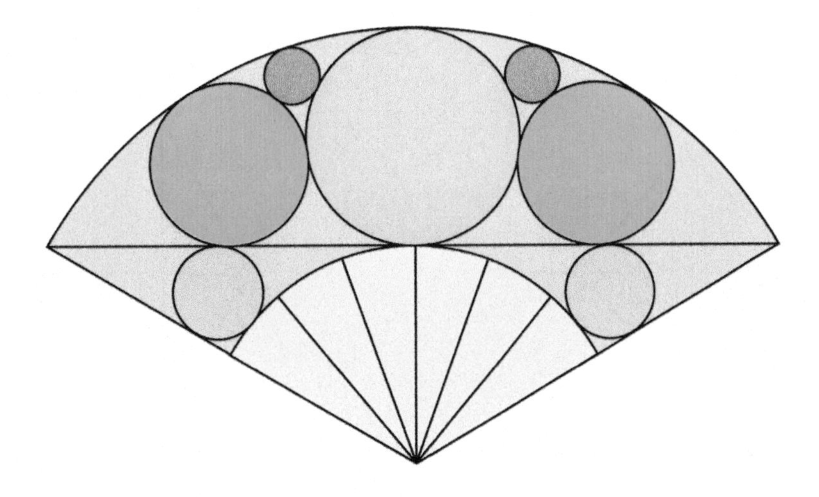

Figure 9. Hervé Lehning, an affine transformation allows to transform a triangle in another one (© 2015, H. Lehning. Used with permission).

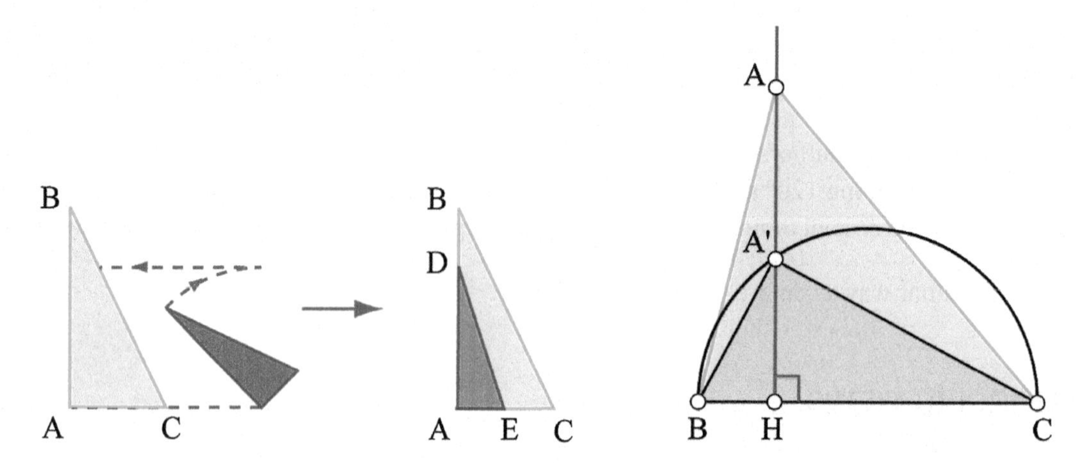

sufficient to prove it on a particular triangle to conclude that it is true for every triangle. It is the same for the affine properties of ellipses: it is enough to prove them for a circle (for example). To illustrate this point, let us consider the problem:

Given an ellipse E, find the locus of all points M for which the two tangent lines to E meet E in two points P and Q such that the centre of gravity of the triangle MPQ is on E.

The research starts with a simple observation: the notions used in the problem are all affine. Thus, it can be treated for a circle centred at O, using metric properties, a method that may seem almost absurd! Let us take a point M moving on a ray emanating from O, M starting on C and going to the infinite. During this process, the centre of gravity G of triangle MPQ remains on this same ray. It starts from the interior of C to go continuously to the infinite and its abscissa grows strictly during this process thus it exists one and only one point M such that G is on C. This point can be found intuitively: it's the point such that OM is the double of the radius. Indeed, in that case, OPH is equilateral, PHM isosceles, OPM right-angled. It follows that MPQ is equilateral. Thus H is its centre of gravity. So, the only point M such that G belongs to the circle C is the point M such that: OM = 2R (Figure 10).

Figure 10. Hervé Lehning, determination of a point of the locus on a ray in the case of a circle (© 2015, H. Lehning. Used with permission.)

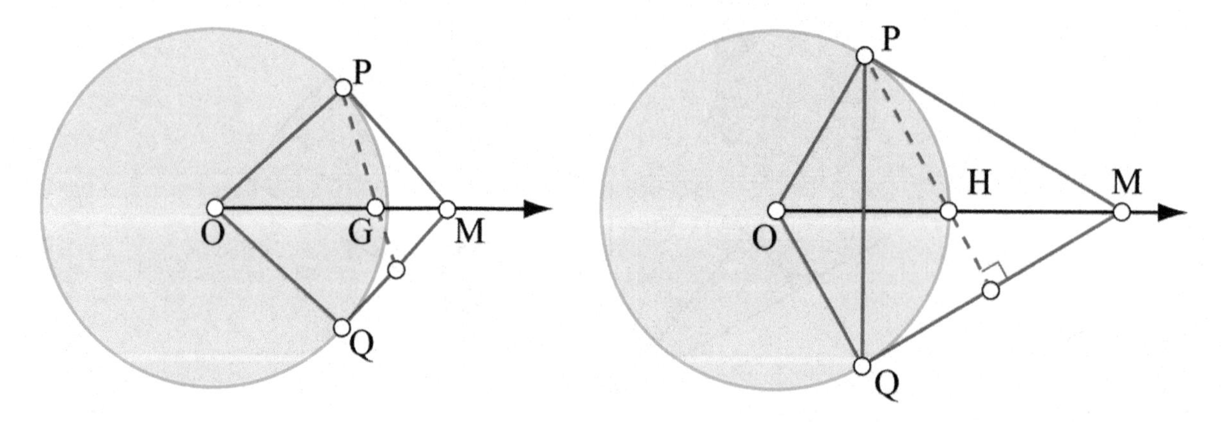

So, the locus C' we are looking for is the circle centred at O of radius 2R. It remains to transform this result in affine terms to transfer it to ellipses. For that, it is sufficient to remark that C' is the homothetic of C in the homogenous dilatation centred in O of ratio 2. Thus, the locus we are looking for is the homothetic of the given ellipse in the homogenous dilatation centred in O of ratio 2 (Figure 11).

A projective view of conics allows us to find other results. Indeed, the definition of conics as plane sections of a cone implies that there are all transforms of a circle in a central projection. As a triangle is the transform of an equilateral triangle through an affine transformation, a conic is the transform of a circle (or any other conic) through a central projection, that is to say an application of a plane P into another plane Q such that any point M of P associate the intersection M' of OM and Q where O is the centre of the projection (Figure 12). A central projection transforms points on a same straight line in points on a same straight line, and lines concurrent or parallel in lines concurrent or parallel.

Blaise Pascal used this result to prove that, if a hexagon ABCDEF is inscribed in a conic, it's opposite sides (AB and DE, BC and EF, CD and FA) intersect in three aligned points. To do so, he just showed the property for a circle and generalized the result via a central projection. He called this property the mystic hexagram because of the beauty of his proof (Figure 13).

Those two remarks on visualisation are at the centre of Erlangen program of Felix Klein (Hawkins, 1984), which classifies geometries according to the groups of transformations preserving properties: similarities for Euclidean geometry, affine transformations for affine geometry, and projective transformations for projective geometry. This idea of Felix Klein was at the centre of modern mathematics but, unfortunately, was little understood (Adler, 1972).

INFINITE DESCENT

Pierre de Fermat invented a method for proving in number theory based on a geometrical idea as its name evoke a never ending descent which is impossible in the set of natural numbers (Weil, 1984). This

Figure 11. Hervé Lehning, locus of all points M for which the two tangent lines to an ellipse meet it in two points P and Q such that the centre of gravity of the triangle MPQ is on E

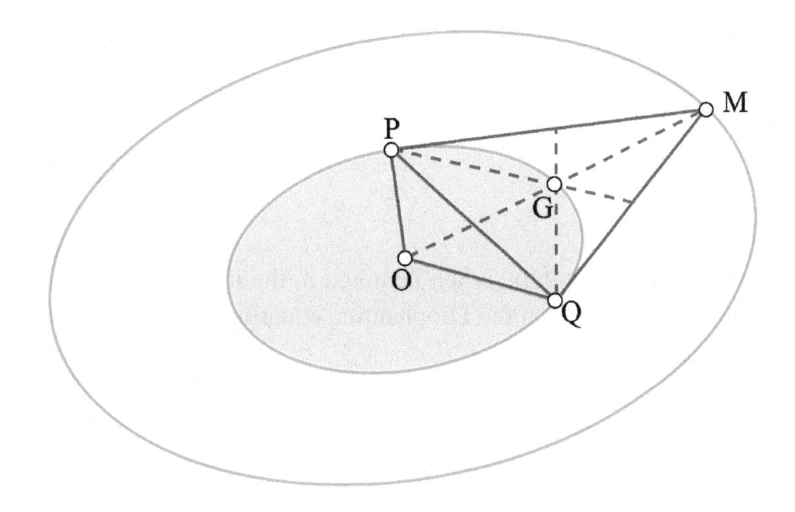

Figure 12. Hervé Lehning. The definition of conics shows that they all can be deduced form a circle through a central projection
(© 2015, H. Lehning. Used with permission).

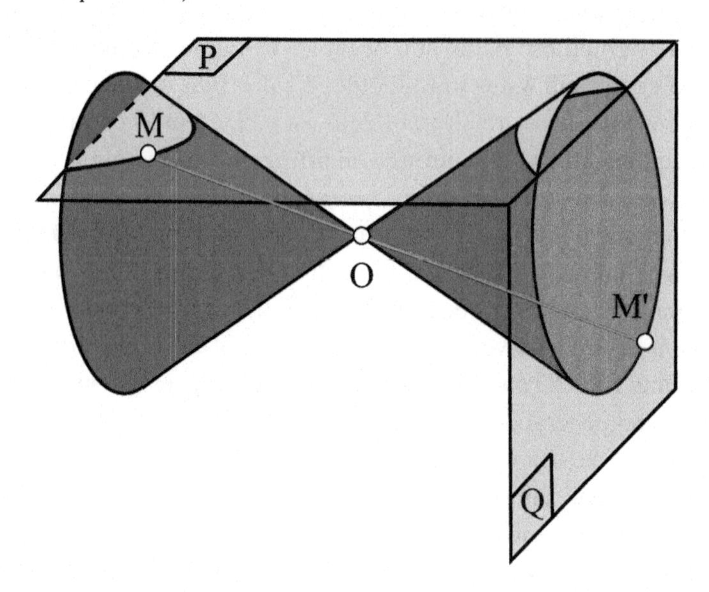

Figure 13. Hervé Lehning, Pascal's mystic hexagram in an ellipse and in a hyperbola (© 2015, H. Lehning. Used with permission)

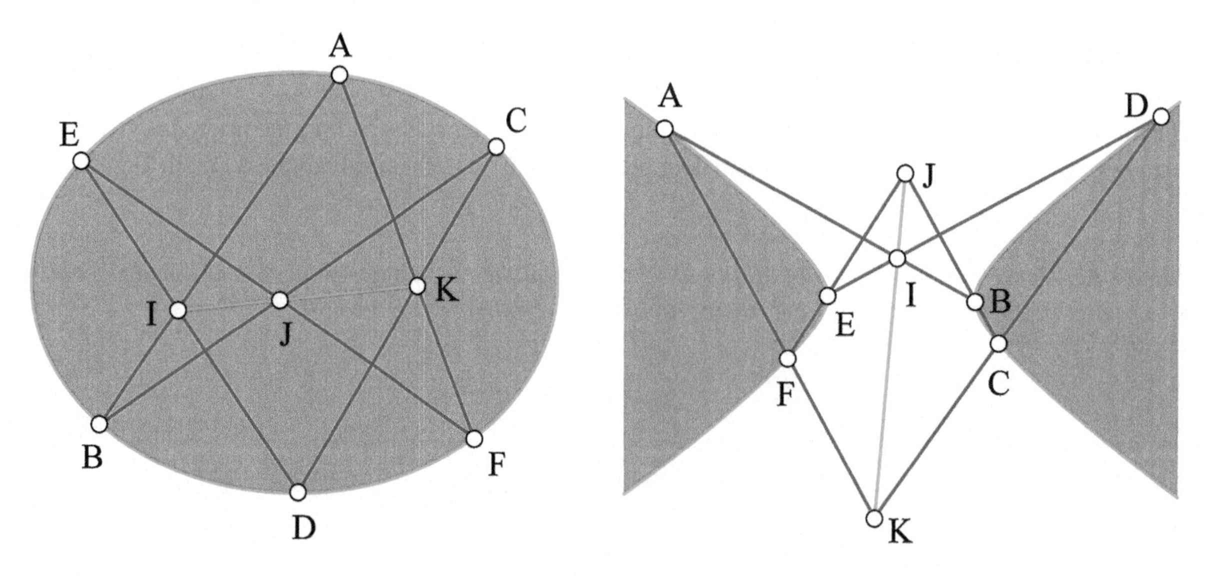

method is possibly the one he had in mind when he noted in the margin of its copy of *Arithmetica* by Diophantus, in front of the discussion on the Diophantine equation (that is to say in natural numbers): $x^2 + y^2 = z^2$ (3, 4, 5 is a solution):

"But it is impossible to divide any power beyond the square into like powers. I have discovered a truly marvellous proof of this, which this margin is too narrow to contain".

As its theorem was finally proved at the end of the XXth century by Andrew Wiles, it's improbable that Fermat's proof covered all cases. However, let us see how his method works to prove the irrationality of the square root of 2, which affected the world of Pythagoras. For that, we use a proof by contradiction and suppose that $\sqrt{2}$ can be written as the quotient of two natural numbers p and q, which leads to: $p^2 = 2\,q^2$ and to the existence of an isosceles right triangle with natural numbers as side lengths. In this case, we can find a smaller triangle with the same properties, and so on indefinitely (Figure 14).

Considering the equation: $x = \sqrt{2}\,y$ which represents a line D and the points with integer coordinates on it leads to another proof. From a point M on D, we construct his homothetic M' in the homogenous dilatation centred in O of ratio $k = 3 - 2\sqrt{2}$. As k belongs to (0, 1), M' is nearer O than M. Furthermore, if the coordinates of M are natural numbers, it is the same for the coordinates of M' ($p' = 3p - 4q$ and $q' = -2p + 3q$ because $p = \sqrt{2}q$), so the descent is infinite (Figure 15). This idea also works to find positive results, for example, to solve the Diophantine equation: $x^2 - 2\,y^2 = 1$. The previous computations lead us to think that if M is a solution of this equation, M' is another one. A simple calculation shows that is the case:

$$p'^2 - 2\,q'^2 = (3\,p - 4\,q)^2 - 2\,(-2\,p + 3\,q)^2 = p^2 - 2\,q^2.$$

From a solution M, we define another one (M') and then, iterating the process, a sequence. The descent is now along a hyperbola H (Figure 15).

The difference is that the descent stopped at the point A with coordinates (1, 0). Inversely, ascending back the descent from A, we obtain all the solutions of the equation. More precisely, we have to go

Figure 14. Hervé Lehning. Infinite descent of isosceles right triangles
(© 2015, H. Lehning. Used with permission).

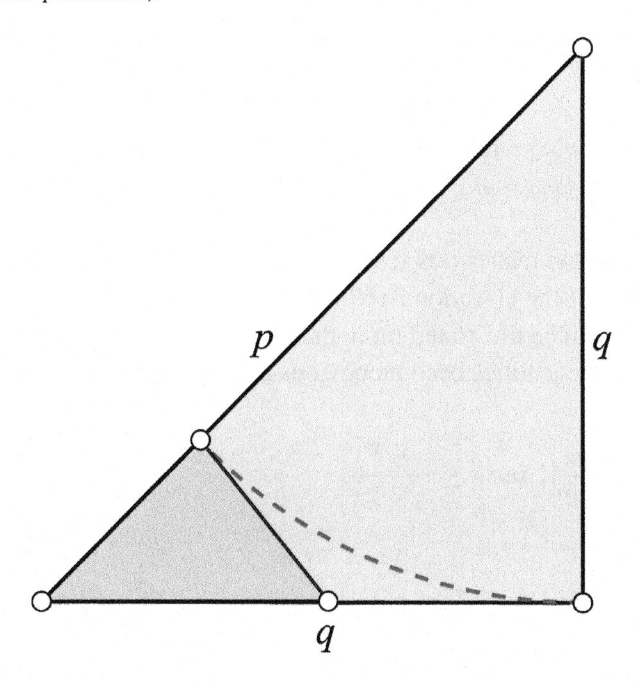

Figure 15. Hervé Lehning. Infinite descent along a straight line or a hyperbola to solve a Diophantine equation
(© 2015, H. Lehning. Used with permission).

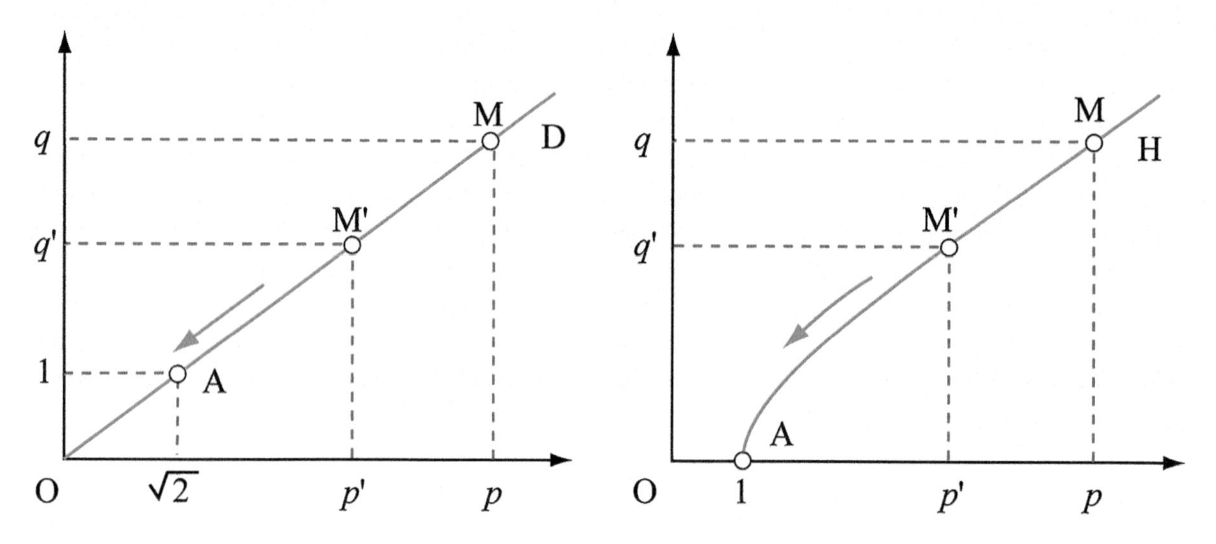

from M' to M ($p = 3p' + 4q'$, $q = 2p' + 3q'$) which allows to find the solutions. Starting from A and applying this transformation, we obtain: (1, 0), (3, 2), (17, 12), (99, 70), (577, 408), (3 363, 2 378), (19 601, 13 860), *etc.*

ANALYSIS

In the same way, in analysis, several ideas have visual origins. The most obvious case is probably the intermediate value theorem:

Given f a continuous function on an interval [a, b] and m a number between f (a) and f (b), there is a number $x \in (a, b)$ such that: $f(x) = m$.

A visualisation in mountain makes this result obvious: to climb a summit situated at the elevation 3172 metres from a refuge at the elevation 3169 metres, we have to pass through all the intermediate elevations. The same idea can be illustrated more mathematically and aseptically (Figure 16).

In the same way, some inequalities become obvious through a small drawing as, for example:

Show that, for every $x \in \left[0, \frac{\pi}{2}\right)$, $\tan x \le \frac{\pi x}{\pi - 2x}$.

The functions sine and cosine are concave on the interval $\left[0, \frac{\pi}{2}\right]$ thus the graph of sine is below its tangent at 0 and the graph of cosine above its chord between its extremities (Figure 17), that is to say:

Figure 16. Hervé Lehning. Grande Ruine, Roche Méane pic and graphical interpretation
(© 2015, H. Lehning. Used with permission).

$\sin x \leq x$ and: $\cos x \geq 1 - \frac{2x}{\pi}$ so: $\frac{1}{\cos x} \leq \frac{\pi}{\pi - 2x}$. We obtain the result by multiplying these inequalities.

Likewise, graphic visualisation of an integral allows us to find some results. Here is an example:

Find the value of the integral: $\int_0^1 \left(\sqrt[5]{1 - x^3} - \sqrt[3]{1 - x^5}\right) \mathrm{d}\, x$.

Figure 17. Hervé Lehning. Concavity and inequality
(© 2015, H. Lehning. Used with permission).

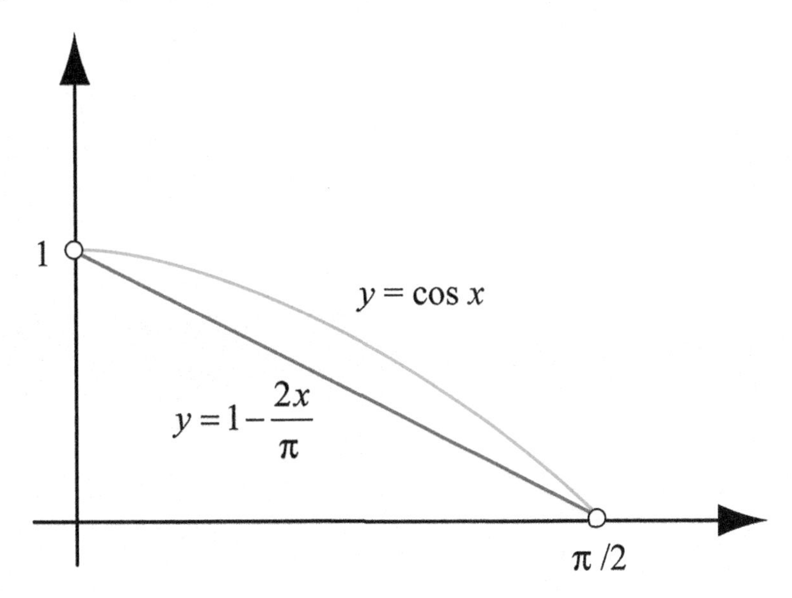

A numerical computation (made by a Computer Algebra System or otherwise) leads us to think that the value is 0, that is to say: $\int_0^1 \sqrt[5]{1-x^3}\,\mathrm{d}x = \int_0^1 \sqrt[3]{1-x^5}\,\mathrm{d}x$. As integrals mean areas, an idea is to try to show directly that it is the same area. Let us consider the set D of the (x, y), such that $0 \le x \le 1$ and $0 \le y \le \sqrt[5]{1-x^3}$. The domain D is limited by the axis and the curve of equation: $y^5 = 1 - x^3$ which can be written as well as: $x^3 = 1 - y^5$. In other words, $y = \sqrt[5]{1-x^3}$ is equivalent to: $x = \sqrt[3]{1-y^5}$.

The area of D equals the first integral: $\mathrm{A} = \int_0^1 \sqrt[5]{1-x^3}\,\mathrm{d}x$, which represents the sum (for x from 0 to 1) of the infinitesimal areas $\sqrt[5]{1-x^3}\,\mathrm{d}x$ of the vertical rectangles of $y = \sqrt[5]{1-x^3}$ altitudes and of infinitesimal width dx. By cutting the same domain in horizontal rectangles of $x = \sqrt[3]{1-y^5}$ width and of infinitesimal altitude dy, we show that the same area can be written: $\mathrm{A} = \int_0^1 \sqrt[3]{1-y^5}\,\mathrm{d}y$. In the integral, the name of the variable has no particular meaning: y can be changed in x. Thus: $\int_0^1 \sqrt[5]{1-x^3}\,\mathrm{d}x = \int_0^1 \sqrt[3]{1-x^5}\,\mathrm{d}x$ (Figure 18).

ALGORITHMIC

A relevant visualisation is equally useful in algorithmic. An interesting question on this point of view was conceived as a puzzle by its inventor, a French mathematician, Édouard Lucas (1842 – 1891), there is the tower of Hanoi (Hofstadter, 1985). This game consists of three rods, and a number of disks of different sizes, which can slide onto any rod. The puzzle starts with the disks on the left rod in ascending

Figure 18. Hervé Lehning. Calculations of an area
(© 2015, H. Lehning. Used with permission).

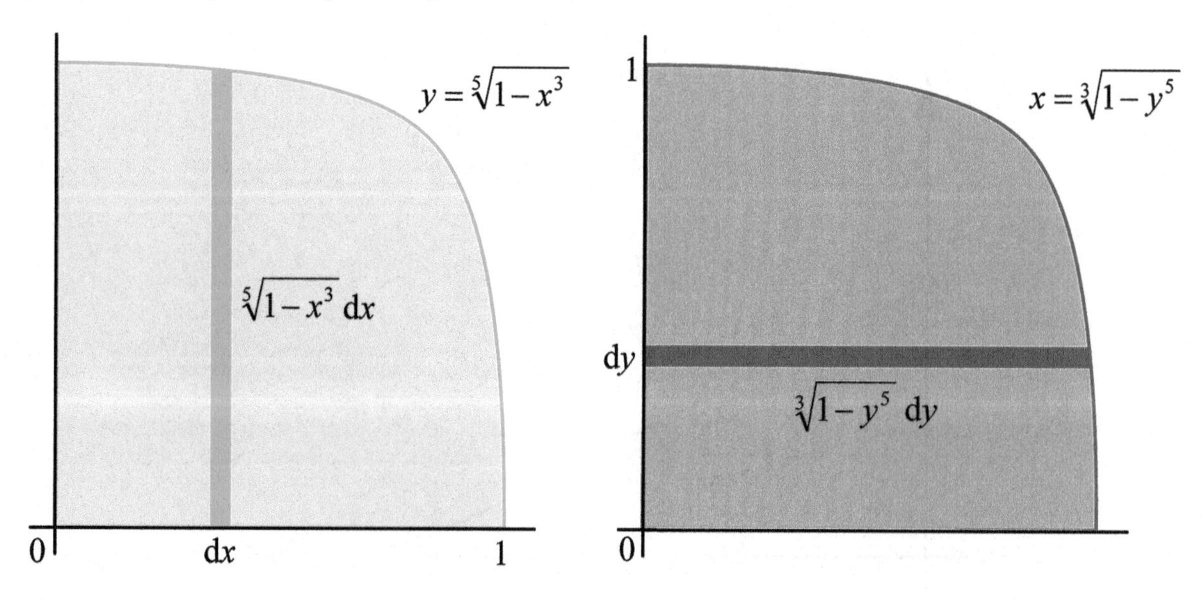

order of size. The objective of the game is to move the entire stack to the right rod, obeying two rules. First, only one disk can be moved at a time. Second, each move consists in taking the upper disk of a stack and placing it on another stack on a bigger disk. The goal is to reach the initial configuration but on the right rod.

As the question becomes more complex when the number of disks increases, we will start with two disks! A few trials lead us to a sequence of three moves. The case of three disks can be seemed as far more complex. The following idea can be seen as a magic trick but, of course, it isn't. We consider the two disks on the top as a single disk. Then we have just two disks and then we know how to solve the puzzle. *A priori*, this solution is illegal but it allows us to find a legal solution as the move of the two upper disks is possible according to the solution of the game with two disks. In fact, we have to add intermediates moves, for example, between the first and the second step of the magic trick. Thereby, the transformation I is realised at bottom-left and the transformation II at bottom-right. By linking all these moves together, we obtain a solution in seven steps for three disks (Figure 19). For n disks, we obtain a solution in $2^n - 1$ steps. Thus, the puzzle is unworkable for a human being if we exceed seven disks but it gives a good exercise in programing.

STUDIES OF DYNAMICAL SYSTEM

Several Fields medals, the most prestigious price in mathematics, including those of Maryam Mirzakhani and Artur Ávila in 2014, have been attributed for outstanding contributions to dynamical systems; that's indicative of the importance of the subject and of its difficulty too. But even here, drawings help a lot to understand the question. What is a dynamical system? *A priori*, it's about the evolution of a system over time according to a deterministic law. For example, the solar system is a dynamical system. Abraham

Figure 19. Hervé Lehning. The tower of Hanoi
(© 2015, H. Lehning. Used with permission).

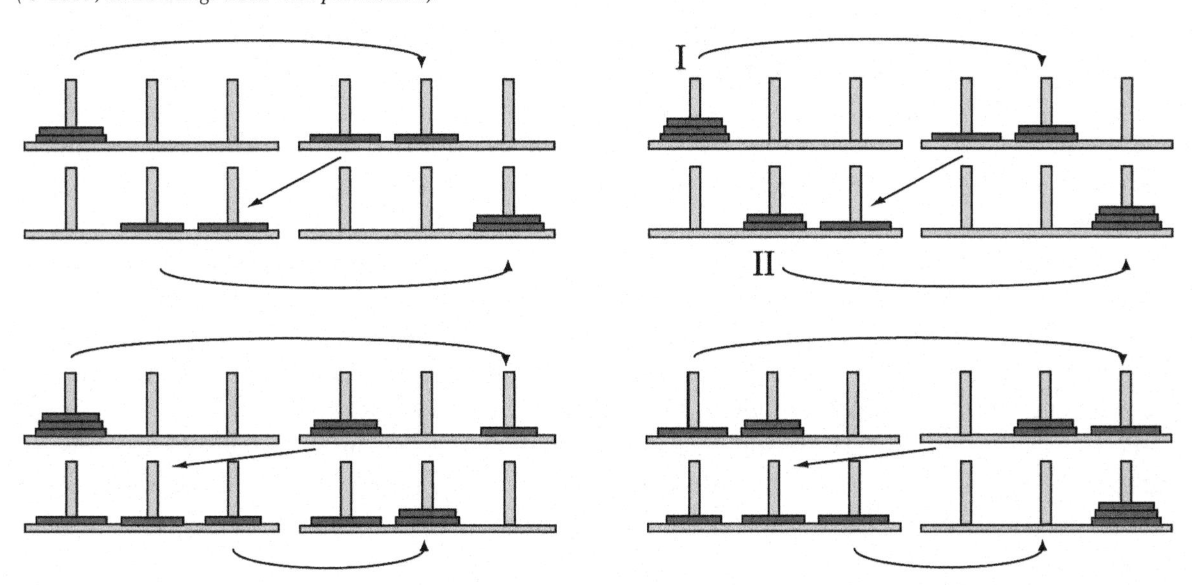

Newton showed that, if we know its state at a given instant, we can determine it at each instant later. The climate and the evolution of a population are also dynamical systems. As these questions are complex, generally we study elementary models of them.

The simplest law of evolution of populations is the exponential growth model where the population of a year is equal to the population of the previous year multiplied by a certain rate k. It corresponds to the function: $f(x) = kx$ and the sequence u_n defined by:

$$\begin{cases} u_0 & = a \\ u_{n+1} & = f\left(u_n\right) \end{cases}.$$

By induction, $u_n = a\,k^n$. According to the value of k, the sequence converges to 0 (if $k < 1$) or the infinite (if $k > 1$) with the exception of $k = 1$ (where u_n is constantly equal to a). This sequence was used by Thomas Robert Malthus (1766 – 1834) to argue for birth control. However, this function is not realistic and has been replaced by the logistic map introduced by Pierre François Verhulst (1804 – 1849). It consists of making the rate to vary with x; it leads to the function: $f(x) = kx(1 - x)$ where we have limited the variation of x to the interval [0, 1]. The parameter k varies between 0 and 4 so that f remains valued in [0, 1] (the maximum of f is obtained for $x = 1/2$ so it's equal to $k/4 \leq 1$), that allows the sequence to have a sense if the initial value a is taken in [0, 1]. The sequence is now more complex than before, but we obtain a closed formula if we use the composition of functions and define f^n in the following iterative way:

$$\begin{cases} f^1 = f \\ f^{n+1} = f \circ f^n \end{cases}.$$

With this definition, $u_n = f^n(a)$. Thus the graphs of the functions f^n allow us to visualise the behaviour of the sequence u_n. For small values of k ($0 \leq k < 1$), the population varies in a very regular way: it converges to 0 for every initial condition a. For $k = 1$, the evolution is slower. For $k > 1$, the behaviour of the sequence varies. For $k = 4$, the sequence has even a more complex behaviour. For $a = 0.5$, it seems very simple as we obtain: 0.5 1 0. For a slightly different initial condition ($a = 0.49$), we obtain a very different sequence: 0.49 0.9996 0.00159936 0.006387208192 0.02538564705 0.09896486392 0.3566832785 0.9178412692 0.3016346950 0.8426048232 0.5304877404 *etc*. The sequence seems to sweep the interval [0, 1]. A visualisation of the graphs of functions f^n on [0, 1] allows us to have the intuition of the behaviour of the sequence. For a value of a, we can see it on the vertical of equation: $x = a$. The convergence corresponds to a crushing of these values on a single one. An oscillation between two or several values corresponds to an oscillation between two or several graphs. More complex behaviours correspond to a smudging of the square $[0, 1]^2$ (Figure 20), where we draw the graphs of functions f^5, f^6 and f^7 in blue, ochre and red in this order for the values 1, 2, 3 and 4 of k).

For $k = 4$, this totally deterministic sequence seems to have a random behaviour. We call this sequence chaotic. If the behaviour of the sequence is regular (convergence or oscillation between several values), the corresponding parameter is said to be regular, elsewhere it is non-regular. Every parameter

Figure 20. Hervé Lehning. A visualisation of the logistic sequence.
(© 2015, H. Lehning. Used with permission).

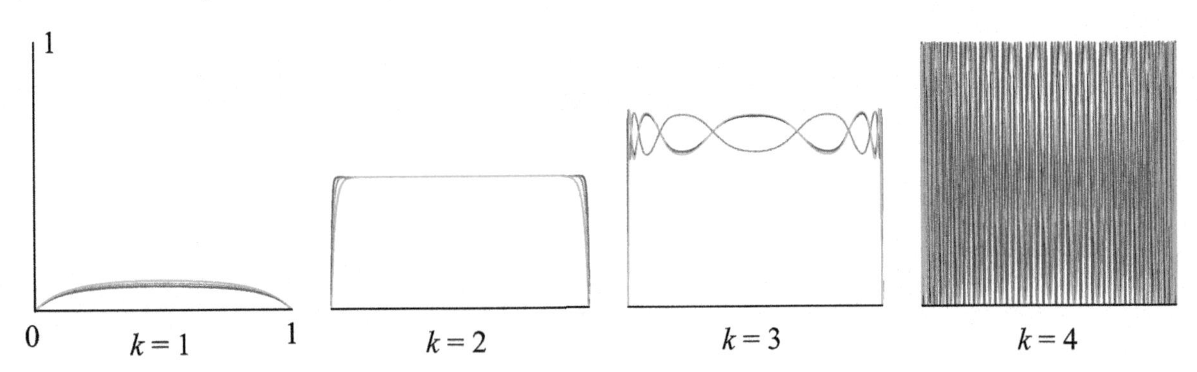

between 0 and 3 is regular and 4 is non-regular. Artur Ávila has shown that, for almost every non-regular parameter, the dynamic of the logistic function is chaotic, that is to say that if you take a non-regular parameter randomly, the corresponding sequence is chaotic.

The results of Artur Ávila, proved with Mikhail Lyubich, Welington de Melo, and Carlos Gustavo Moreira concern a larger class of functions, those of unimodular functions having a graph shape similar to the logistic function. Visualisation is essential to realise that.

MATHS LEADING TO ART

As we have seen it with the Japanese Sangaku, mathematical visualisations have an artistic side. A mathematical question can lead to a photograph, a painting or a sculpture. Let us give an example, which leads me to produce several paintings. In a square, let us imagine four parts limited by lines. They are neighbours if they have a frontier line in common. On this only criterion, on how many ways can they be assembled? To solve this topological question, let us introduce a graph linked to each solution. If a solution is given, let us place a point in each part and link this point to the neighbouring parts. We obtain a graph with four vertices. To know the number of possible parts, it is sufficient to count the number of graphs linking four points. For that, the best thing to do is to organize them in a square. The minimum number of vertices is three and the maximum, six (Figure 21).

The same graphs can be drawn in other form (Figure 22).

Thus, we have six possible arrangements, which lead to six paintings. When adding the condition: the sets have equal areas, we obtained several possible paintings (Figure 23).

Figure 21. Hervé Lehning. Graphs of four neighbouring sets, first version
(© H. Lehning. Used with permission)

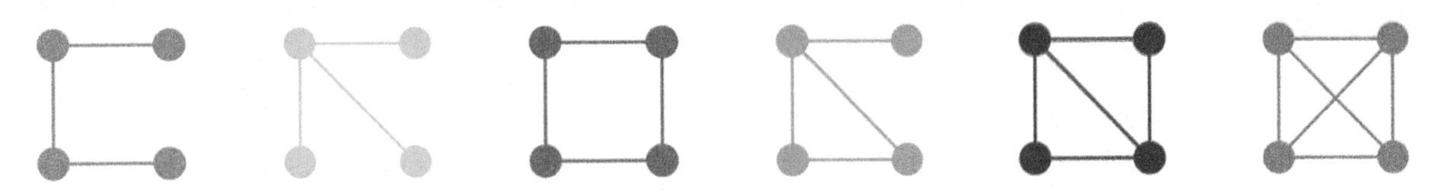

Figure 22. Hervé Lehning. Graphs of four neighbouring sets, second version
(© H. Lehning. Used with permission)

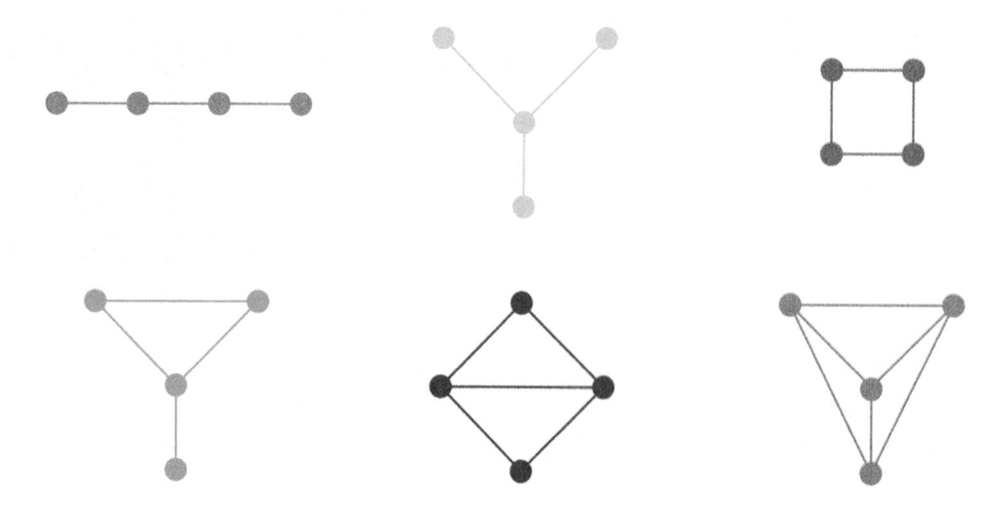

Figure 23. Hervé Lehning. Triptych of four neighbouring sets
(© H. Lehning. Used with permission)

In a way we would not discuss, strange forms evoking topology can be found in nature. In particular, we have found them in ice rocks melting on Nordic beaches (Figure 24).

These strange forms came from ice melting and from the violence of nature (Figure 25).

CONCLUSION

To conclude with, we see that at the levels of popularisation teaching and research, visualisation is not only a simple illustration. It can help the understanding of some concepts but also can bring new ideas.

Figure 24. Hervé Lehning. An ice rock melting on a beach of Greenland
(© H. Lehning. Used with permission)

Figure 25. Hervé Lehning. The violence of Nature at work on an ice rock on a beach of Iceland
(© H. Lehning. Used with permission)

Most of the figures are absent of research articles and even of teaching manuals but they appear on blackboards in a lot of lectures. We can find them on videos but rarely in papers (Klarreich, 2014). In addition, these ideas of visualisation give artistic ideas where *sangakus* are one of the most famous axes.

REFERENCES

Adler, I. (1972). *The new mathematics*. New York: John Day Company.

Constant, J. (2014). *Wasan geometry*. Retrieved from http://hermay.org/jconstant

Fukagawa, H., & Rothman, T. (2008). *Sacred Mathematics: Japanese Temple Geometry*. Princeton, NJ: Princeton University Press.

Hawkins, T. (1984). The Erlanger Programm of Felix Klein: Reflections on Its Place. In the History of Mathematics. *Historia Mathematica, 11*(4), 442–470. doi:10.1016/0315-0860(84)90028-4

Hofstadter, D. (1985). *Metamagical Themas*. New York: Basic Books.

Huvent, G. (2008). *Sangaku. Le mystère des énigmes géométriques japonaises*. Paris: Dunod.

Klarreich, E. (2014). A Tenacious Explorer of Abstract Surfaces. *Quanta Magazine*. Retrieved from https://www.quantamagazine.org/20140812-a-tenacious-explorer-of-abstract-surfaces/

Lehning, H. (1995). Learning mathematics with CAS.*Proceedings of the World Conference on Computers in Education VI*. London: Chapman & Hall. doi:10.1007/978-0-387-34844-5_78

Weil, A. (1984). *Number Theory: An approach through history from Hammurapi to Legendre*. Boston: Birkhäuser.

ADDITIONAL READING

Bishop, A. J. (1989). Review of research on visualization in mathematics education. *Focus on Learning Problems in Mathematics, 11*(1), 7–16.

Images des mathématiques, website of Centre National de la Recherche Scientifique (CNRS).

Lehning, H. (2011). *Questions de maths sympas pour M. et Mme Toutlemonde*. Bruxelles: Ixelles.

Lowrie, T., & Kay, R. (2001). Relationship between visual and nonvisual solution methods and difficulty in elementary mathematics. *The Journal of Educational Research, 94*(4), 248–255. doi:10.1080/00220670109598758

KEY TERMS AND DEFINITIONS

Algorithm: The mathematical and rigorous version of a recipe.

Anamorphosis: A distorted visualisation.

Computer Algebra System: A piece of software designed to do numerical and formal computations.

Convergence: The notion of convergence refers to the notion that some sequences or functions approach a certain limit. There are several notions of convergence: pointwise convergence, uniform convergence, *etc*.

Dynamical System: A dynamical system is a way of defining a sequence of numbers, points, functions, or curves through a fixed deterministic rule. It is widely used in mathematical modelling.

Chapter 7
Collage Strategy:
A Robust and Flexible Tool for Knowledge Visualization

Dennis Summers
Strategic Technologies for Art, Globe, and Environment, USA

ABSTRACT

In this chapter the author analyzes and defines collage in some of its many forms and media. He introduces three terms (the gap, the seam, and contested space) necessary to characterize the unique aesthetics of collage. Via a review of specific artists and art historical movements he creates taxonomy that typifies three distinctive collage strategies. He extends this review into other media including artists' books, cinematic film, and digital media. In the second part of the chapter he describes the work of three artists (including the author) and their relevance to this theory of collage and scientific visualization. Following that, he reviews the use of digital software and the pedagogical implications of collage.

INTRODUCTION

Collage enables us to experience everyday life in such a way that its disparate and idiosyncratic fragments resist coalescing into a unifying whole, which philosophers Gilles Deleuze and Félix Guattari (1983) refer to as 'disjunctive synthesis'. Instead of a totalizing body of knowledge, the composition of collage consists of a heterogeneous field of coexisting and contesting images and ideas. Its cognitive dissociation provides the perspectival multiplicity necessary for critical engagement. Dialectical tension occurs within the silent, in-between spaces of collage, as it's fragments, its signifying images and ideas interact and oppose one another. Such complexity and contradictions represent the substance of creative cognition and cultural transformation (Garoian & Gaudelius, 2008, p. 63).

As many have commented, collage is the art form of the 20th and 21st Century (Ulmer, 1983; Durant, 2002; Kohler, 2012). The word collage will be used broadly and inclusively within this chapter. Strictly speaking, collage refers to the gluing of elements historically considered to be outside of the realm of painting, onto paintings or simply onto a flat surface. Assemblage extends this into three dimensions;

DOI: 10.4018/978-1-5225-0480-1.ch007

photomontage consists of multiple images combined in a single photograph, and montage refers to the cuts between film clips in motion pictures. Collage will stand in comprehensively for these and the multitude of other "–ages" that exist, in addition to other words such as cut-up, mashup, etc. In many cases installation art and performance art can also be a collage. A more complete definition of collage will be found in the Background section of this chapter, and numerous examples will follow throughout.

Collage strategies are uniquely suited for visual presentations of information. However, there are few scholarly resources that really grapple with the theoretical implications of collage and the strategies used to create them. In 1975, art critic Harold Rosenberg (1989) lodged this complaint, and unfortunately little has changed in the 40 years since. In both educational settings and in too many fine art environments, collage is taken to be simply the combination of different elements on a page (Garoian & Gaudelius, 2008). A notable example of the debasement of this word includes a Groupon advertisement called "A Collage Of Our Best Deals: This Collage Is Crafted With Care," which was simply a web page grid of links to merchants (sent via email September 12, 2013). This confusion over collage is extensive and often begins in public schools. According to the authors Charles Garoian and Yvonne Gaudelius (2008) in their book *Spectacle Pedagogy, Art, Politics and Visual Culture*,

> *A visit to a public school would give us a learning environment replete with social studies collages, arithmetic collages, language collages, health collages, and even physical education collages, in addition to using this genre in art classes. Such ubiquity notwithstanding, we find little evidence that the aesthetic dimension and disjunctive narrative of collage is understood at any depth in schools (Garoian & Gaudelius, 2008, p. 4).*

This chapter is an effort toward resolving some of these shortcomings. Digital tools in particular have made the creation of collage easier and more affordable than ever before – well within the reach of most students. The downside of such ease of use is that students (and others) can easily fail to recognize the strategic and cognitive possibilities inherent in recombining various components. "Left to a 'cut-and-paste' mentality, the conceptual profundity of its [collage] narrative is easily misunderstood as a pastiche of essentialized images and ideas" (Garoian & Gaudelius, 2008, p. 68).

One of the challenges facing educators and students today is the gulf between the necessary expertise in visual communication and familiarity with technical content (Ursyn, 1997; Kosera, 2007). In spite of resources such as the series of books by Edward R. Tufte (1983, 1990, 1997) on design methods for knowledge visualization, it is rare to find people with skills in design, graphic software, and deep understanding of specific scientific concepts. At the commercial professional level it usually takes a team of at least two people to craft a visual experience, irrespective of media, that can accurately communicate the details of some aspect of scientific research. Furthermore, there is often a minimization of the skills or values of one group by the other. Of course, there are exceptions on both sides of the so-called two cultures divide as originally described by C. P. Snow (1993) in 1959. In spite of numerous reports as to how much artists and scientists have in common, it is critical to remember that there are fundamental differences between the goals and methodologies of scientists and those of artists. Science is about facts and falsifiability; art is about everything else.

Additionally, there are fundamental differences between the goals and methodologies of the quite distinct fields of "commercial arts" and "fine arts." Although there is obvious overlap between visual judgment skill sets, the pedagogy and intention of both is significantly different (Kosera, 2007). In a typical university setting shared courses between students in both disciplines might include color theory and basic drawing. After those freshmen level courses the two cohorts will split out into courses with names such as "Typography" or "Intro to Visual Communication" for graphic designers; and "Mark Making" or "Intro to Painting" for fine artists. Although it is likely that the global lessons of good composition or visual psychology will be taught to both cohorts, the context of delivery is certain to be entirely different. This includes the required history classes, which for the former would be "History of Modern Design" and for the latter "Contemporary Art History." The design student *may* be given a specific project to visualize a scientific concept. It is highly unlikely the same would be given to a fine arts student.

It will be posited in this chapter, however, that the fine arts *can* support knowledge visualization, assuming that one is willing to consider alternatives to the typical commercial demands of unambiguous intelligibility. Additionally, as acceptable and appropriate content in a fine arts project can be more inclusive, concepts from science can be connected to larger domains of knowledge, psychology, and emotion. Collage in particular has a long and interesting history within the arts. Collage strategies can be uniquely productive for knowledge visualization. Furthermore, this approach can be useful in a number of educational environments for students of any age. The prevalence of commonly used digital software creation tools has made this simpler than it might have been in the past. The necessary skills are: how to do research; how to capture salient aspects; how to translate them into compelling visual components; and how to combine them into a compelling visual experience. This is not to say that any of these skills are necessarily easily taught, learned, or assessed. However, for some percentage of people, learning from this process will be more enjoyable and retainable then other traditional methods (Ursyn, 1997, 2015; Ursyn & Sung, 2007).

This chapter will be divided in two sections. The first, *Theory* will review the history of collage strategies in a variety of visual media from painting to cinema to artists' books. These strategies will be defined taxonomically and their theoretical implications explored, in order to lay a foundation for the second section. That section, *Practice* will begin by describing and analyzing three contemporary works of art using the approaches described in the first section, and conclude with the pedagogical implications of collage and methods for manifesting different types of scientific knowledge. Throughout this section different aspects of using digital software will be addressed.

BACKGROUND

Some preliminary definitions must be introduced. This author has developed a *collage taxonomy* that systematically and consistently identifies categorical characteristics that can be determined within collage and mixed media artworks. This taxonomy is created by four different sets obtained by the combination of two terms.

There have been several authors that have suggested different defining characteristics of collage; they are overly simplistic, ambiguous, or incomplete (Janis & Blesh, 1967; Wescher, 1971; Wolfram, 1975; Waldman, 1992). For something that everyone can identify in a moment, why has it been so difficult to come up with a consistent descriptive theory? And for a creative strategy that extends through multiple media in addition to the visual arts, including but not limited to literature and music, there is relatively little scholarship (Garoian & Gaudelius, 2008). Surprisingly, for an art that hasn't always been taken very seriously, there is much more written about cinematic montage, which can be quite useful in this context, but nonetheless, is a subset of collage.

By far the most common term applied in discussions of collage is "juxtaposition" (by everyone, perhaps popularized originally in 1961 by Shattuck, 1992, p. 126). Although this term is necessary, it is by no means sufficient, and on further inspection it really says nothing at all about the complex interrelationships between elements within a collaged structure. More subtlety is required and can be found within these two taxonomical terms: *gap* and *seam*. Gap has been used by some writers on this topic (Shattuck, 1992 and many others). Seam has received little attention. Here are dictionary definitions (American Heritage Dictionary, 1969):

Gap: 3. A suspension of continuity; interval; hiatus. 4. A conspicuous difference; disparity. 5. Electricity. A space traversed by an electric spark: a spark gap (p. 542).

Seam: 1. a. A line of junction formed by sewing together two pieces of material along their margins. b. A similar line, ridge, or groove made by fitting, joining, or lapping together two sections along their edges. d. A scar (p. 1170).

The literary theorist Brian McHale (1991, p. 13) uses an expression "ontological discontinuity" in describing the distance between the real and the fictional within a narrative. The phrase can capture some of the sense of the word gap. In addition to the perception of discontinuity what is interesting is that this space is filled with one's imagination.

Bert M-P. Leefmans (1983) is a writer who describes the gap quite well: "Let us say that collage is the art of the *space between* or of *crossing boundaries* ... the *open* space left, upon which our imaginations may work" (p. 220). When comparing poetry to collage he wrote,

It is the leaps across the gaps that constitute the action, the dynamic, of the poem. These gaps, these bondings transform the elements they link but at the same time become autonomous and significant in themselves – and they are gratuitous in that they are neither caused nor limited by any laws but those of the imagination as it is affected by associations. It is thus the "content" of the gap that becomes the source and power of the new, and it is also upon it that selection is free to work as poetry itself evolves (Leefmans, 1983, p.193).

There are often literal seams in collages, but here it is also used metaphorically. Note too, the idea of seam as a scar. This is evidence of history and a kind of stitching together.

Another term needs to be introduced. It is not part of the taxonomy, but it is a critical concept. It is *contested space*. Building on the idea of the gap's content as described above, when two elements either in time or space are juxtaposed there is a conflict as the first one to be perceived influences one's psychological anticipation of meaning, which is then altered by the influence of subsequently perceived elements.

Although most histories of collage begin with Georges Braque and Pablo Picasso near the beginning of the 20th century, there are some sources that refer to collage-like structures beginning in the mid-19th century (Seitz, 1961) particularly with reference to photography (Fineman, 2012) and even earlier. The use of cutting and pasting different photographic elements and mixing them with paint was a common procedure in order to achieve aesthetic and interpretive goals. Because of the overlap between this and the creative and conceptual processes involved in crafting digital work, this topic will be postponed until the end of the *Theory* section. This section, by necessity, leaves out art movements, genres, and numerous practitioners who advanced collage in exciting ways. It focuses on styles that can be seen as taxonomic exemplars, along with some practices that are included for their relevance to the themes of this chapter.

1. THEORY

1.1. Cubism and Collage

There is some controversy over whether or not the first "fine art" collage was *Fruit Dish and Glass*, 1912 (Figure 1) by Georges Braque (likely to be the case according to Wolfram, 1975, p. 16 and others), or *Still Life with Chair Caning*, 1912 (Figure 2) by Pablo Picasso.

Both artists by this time had created numerous cubist works where they explored abstractions of space and time in decidedly non-illusionistic paintings and drawings. In the case of the Braque charcoal and gouache drawing, the artist has pasted pieces of wood patterned wallpaper; in the case of the Picasso oil painting the artist has added a piece of wallpaper with a chair caning pattern, along with a length of rope as a "frame." Below is a quote from Picasso (as cited in Gilot & Lake, 1964) that provides a good introduction to understanding their motivation:

The purpose of the papier collé was to give the idea that different textures can enter into a composition to become the reality in the painting that competes with the reality in nature. We tried to get rid of "trompe l'oeil" to find a "tromp l'esprit" ... If a piece of newspaper can become a bottle, that gives us something to think about in connections with both newspapers and bottles, too. This displaced object has entered a universe for which it was not made and where it retains, in a measure, its strangeness. And this strangeness was what we wanted to make people think about because we were quite aware that the world was becoming very strange and not exactly reassuring (Gilot & Lake, 1964, p. 70).

Picasso explains several things here. The first is that not only is there a conflict, as there had always been between the reality of a painting and the reality of the world outside of painting, but that it can be exhibited within the painting itself. That is, one object can now represent two realities. The second is that when two things are conflated, for example a bottle made of newspaper, a dialogue is created between those two entities as concepts. But additionally a dialogue is created between these two entities as formal visual and material elements – one illusory and one real: that is, the representation of a bottle and the reality of newspaper. Finally, that injecting a real element into an illusionistic world independently of any content is in itself an act that carries meaning. These dialogues are created within the gaps by the spectator. For example, within the contested space between the chair caning and the painted elements, one's thoughts go to the conflict between the two physical materials; the conflict between different modes of representation; and the conflict between different realities and histories.

To follow up on the formal aspects of collage, art historian Christine Poggi (1992) describes *Glass and Bottle of Bass*, 1914 this way:

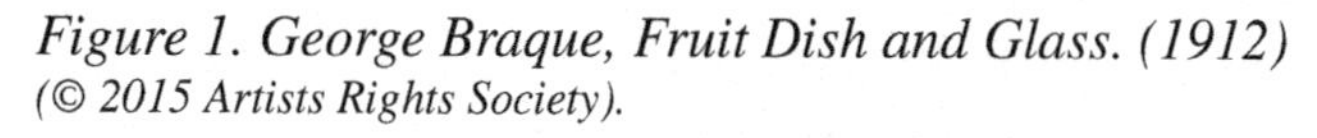

Figure 1. George Braque, Fruit Dish and Glass. (1912)
(© 2015 Artists Rights Society).

Figure 2. Pablo Picasso, Still Life with Chair Caning (1912)
(© 2015 Estate of Pablo Picasso/Artists Rights Society).

Picasso created a mock frame by pasting a wallpaper border to the four sides of his picture. But this frame fails to function convincingly because the paper has been crudely cut and glued (scissor marks and overlapping are visible), the orientation of the pattern alternates around the four edges, and most crucially, a section is missing from the upper right corner. Picasso filled this gap with a frame drawn in pencil directly onto the cardboard ground, without, however, making any attempt to imitate the wallpaper pattern. Further emphasizing the difference of this hand-drawn section from the wallpaper frame, Picasso made it cast an illusionistic shadow to the right, as if only the drawn frame had volumetric presence. Yet, because of its isolated and fragmentary character, the shadow cannot be confused with a real shadow and thus calls attention to itself as an illusion. The function of the frame as an enclosing border is also negated by the extension of the cardboard ground beyond the perimeter marked by the (inner) frame, causing the literal and framed edges of the collage to diverge. The wallpaper border thus appears as a (badly rendered) picture of a frame. The small bit of paper bearing Picasso's name is similarly paradoxical. It functions in relation to this picture both as a literal nameplate of the type (if not the material) frequently found in museums and as an ironic imitation of such identifying labels (Poggi, 1992, p. 82).

The conclusions that can be drawn from this are as follows: although the cubist painters, irrespective of collage, often played with space and perception of space, collage lends itself to an extra layer of complexity that can not be achieved in paint alone. Unfortunately, this quite interesting aspect of collage has rarely been examined with the same complexity and subtlety as had Picasso. Note too, that this picture is filled with seams both literal and metaphorical.

The final consideration is Picasso's use of newspaper text as content. Most of the critical writing on cubist collage considers the use of newsprint as simply a formal issue of pattern and reference. But at least for Picasso, from the beginning, collage was a method to bring text and its associated content into art, thus initiating a long tradition. Art historian Patricia Leighten (1985, p. 653) describes in convincing detail how the newspaper clippings of the collages from 1912-14 not only were included for specific political content, but that as they are reviewed chronologically a significant number of them document the evolution of the Balkan War prior to World War I.

It can be seen then, that both gaps and seams are important carriers of meaning within cubist collage.

1.2. Futurism and Collage

Collage created by futurists tends to get very little coverage in the scholarship. For example, Brandon Taylor's (2004) otherwise excellent book on the history of collage ignores them entirely. The futurists stand in sharp contrast to both the cubists and the surrealists from which the two major threads of collage through the 20th century are developed. Additionally, it will be noted that this underappreciated futurist collage thread is consistent with feminist approaches to constructing collage.

William Seitz (1961) in the catalog for the pivotal exhibition at the New York Museum of Modern Art, *The Art of Assemblage,* puts his finger on a key element of the problem, "futurism's key words are 'interpenetration' and 'synthesis' rather than [the cubist's] 'interval' and 'juxtaposition'" (p. 26). But, perhaps more importantly, futurists noticeably expanded the material possibilities of collage. Picked up by the dadas and becoming especially important after World War II, the futurists explored the use of *any* material to create art. In 1914 futurist Carlo Carrà (as cited in Poggi, 1992) wrote,

We Futurists also believe that 'painting does not lie in Lefranc tubes.' If an individual possesses a pictorial sense, whatever he creates guided by this sense will always lie with the domain of painting. Wood, paper, cloth, leather, glass, string, oil-cloth, majolica, tin and all metals, colors, glue, etc. etc., will enter as most legitimate materials in our present artistic constructions (Poggi, 1992, p. 165).

Poggi, one of the few writers to carefully assess futurist collage, discusses how their use of outside materials had an entirely different agenda to that of the cubists. Instead of accentuating contrasts between paint and non-paint realities, their goal was to fuse outside elements into paintings in order to achieve a new kind of dynamism. This minimizes the gaps between different realities, as within cubist painting, although the seam remains. But this seam is a subtler one; it is seam in the sense of scar. The outside elements have healed within the body of futurist painting. For example, Poggi (1992) describes the painting called *Still Life with Glass and Siphon* (Figure 3) by Umberto Boccioni probably created in 1914:

Figure 3. Umberto Boccioni, Still Life with Glass and Siphon (1912)
(Public Domain).

Unlike Picasso and Braque, Boccioni did not tend to assert the identity of each pasted element by allowing its boundaries to remain visible. In Still Life, for example, he covered almost the entire surface of his canvas with various kinds of paper and cardboard, so that the glass and siphon seem embedded within a textured ground. Because Boccioni painted over the layered materials glued to his canvases, Herta Wescher was led to claim that these materials served no particular expressive purpose. This is patently mistaken: they provided Boccioni with a plastic means of suggesting the 'molecular' interpretation of the depicted object and its environment. Just as Boccioni had once rejected a too rigorous application of separate strokes of pure color, he now rejected a collage technique based on the play of clearly defined edges and on the relations between distinct elements (Poggi, 1992, p. 181).

Thus Boccioni found in collage an ideal means of allowing the expressive properties of different materials to suggest the dynamism of all things, including inorganic substances (Poggi, p. 11).

Conclusions to be drawn from the futurists are that collaged elements may be continuous with the overall visual field. The distinctions between real elements and illusory painted ones create their own dynamics, but some juxtapositions are palimpsestic which will be discussed in greater detail further on. Their work was not about creating conceptual gaps between entities, and the seams present are seams that are knitted together, but conceptually potent.

1.3. Dada and Collage

Dada is included here, not because it leads to one of the taxonomic categories, but because of its subsequent influence across media and genre. Following somewhat on futurist strategies of incorporating a wide range of non-traditional media into painting, the dadaists wanted to integrate lived lives in all their turmoil and uncertainty. This led to new kinds of multi-media performance events. Dadaist Richard Huelsenbeck (1981) wrote in 1920,

The word Dada symbolizes the most primitive relation to the reality of the environment; with Dadaism a new reality comes into its own. Life appears as a simultaneous muddle of noises, colors and spiritual rhythms, which is taken unmodified into Dadaist art, with all the sensational screams and fevers of its reckless everyday psyche and its brutal reality (Huelsenbeck, 1981, p. 244).

Writer and founder of surrealism, André Breton adds:

[Dada] is the marvelous faculty of attaining two widely separate realities without departing from the realm of our experience, of bringing them together and drawing a spark from their contact; of gathering within reach of our senses abstract figures endowed with the same intensity, the same relief as other figures; and of disorienting us in our own memory by depriving us of a frame of reference reality (Huelsenbeck, 1981, p. 15).

Dada artist Kurt Schwitters is known for two separate types of work that he created throughout his career, and which influenced subsequent generations each in their own way (Figure 4). The first of these were his small collages made of paper and other items literally picked up off the streets, with paint added as necessary. The second were the mixed media sculptural structures that were built into the rooms of his houses.

Figure 4. Kurt Schwitters, doremifasolasido (1930)
(© 2015 Artists Rights Society).

So unlike the previously mentioned artists, Schwitters' starting point was with the outside elements – in itself a significant change of methodology and emphasis. His work illustrates another direction of abstract art that collage can take. However, Schwitters can be contradictory in his own statements. He claimed explicitly that items were not included for their semantic signification but for formal reasons only (Schulz, 2010). Nonetheless, the signification remains and is meaningful. There is a difference between a painted blue square and a blue bus ticket. As Isabel Schulz (2010) explains,

In the context of the work, the "chunks of everyday refuse" that Schwitters applied to his collages surrender their original function to be sure, but not all their semantic meaning. As used objects, discarded materials distinctly relate to the social reality of their time, and as vestiges of modern civilization, they become metaphors of a society increasingly shaped by industrial production and consumerism, advertising, and the media. The unresolved discrepancy that emerges between their "literary" content, rooted in time, and their abstract form places a limit on Schwitters's endeavor to exclude all manner of symbolism (often by way of overpainting) in favor of a work's absolute autonomy (Schulz, 2010, p. 52).

In contradiction to Schwitters' claims against semantic content, and that his art was meant to be self-enclosed, he also said that his motto was to "create connections if possible between everything in the world" (as cited in Schulz, 2010, p. 61).

Several dada artists were instrumental in popularizing photomontage – collages created by pasting clipped fragments of photographs into single, often quite complex compositions. In contrast to the almost entirely abstract collages of Schwitters, Hannah Höch (Figure 5) created collages filled with specific referential content. One example created in 1919-1920 is *Cut with the Kitchen Knife Dada through the last Weimar Beer-Belly Cultural Epoch of Germany.*

Visual studies professor, Maria Makela (1996) explains this work:

This large and complex photomontage unites representatives of the former Empire, the military, and the new moderate government of the Republic in the "anti-Dada" corner at the upper right, while grouping Communists and other radicals together with the Dadaists at the lower right. These mostly male figures are paired with photographic fragments of active, energetic women – dancers, athletes, actresses, and artists –who animate he work both formally and conceptually. The newspaper fragment at the lower right identifies the European countries in which women could or would soon be able to vote, including Germany, which granted women suffrage in its 1919 constitution. By placing the clipping in the corner she normally reserved for her signature and including a small self-portrait head at the upper-left edge of the map, Höch identified herself with the political empowerment of women, who, she envisioned, would soon "cut" through the male "beer-belly" culture of early Weimar Germany (Makela, 1996, p. 25).

Owing to the fact that dada was more of an attitude than a coherent art style, numerous dada artists created collages in widely differing methods and forms. Thus, dada as a movement does not fit into any of the taxonomic categories. However, as seen in the next section, Max Ernst, originally a dada artist, helps to define one of the categories.

Figure 5. Hannah Höch, Cut with the Kitchen Knife Dada through the last Weimar Beer-Belly Cultural Epoch of Germany (1919-1920)
(© 2015 Artists Rights Society).

1.4. Surrealism and Collage

Surrealism introduces the third thread of collage typology. Indeed, this is where juxtaposition is strongest. But surrealism also exposes a paradox that some have called the *collage aesthetic* in order to distinguish it from a collage of differing materials. As has been suggested earlier, and will become explicit subsequently, it is more appropriate to simply extend the meaning of the word collage. Max Ernst (Figure 6) began as a dadaist, and without really changing his style became a surrealist. His categorization had more to do with the art politics of the time than anything else. Consider the technical craft of Ernst's collages. As Waldman (1992) tells us,

Figure 6. Max Ernst, image from the book Une semaine de bonté ("A Week of Kindness"). (1934)
(© 2015 Artists Rights Society).

Ernst took the process of collage one step further than anyone had previously. Rather than disposing the paper fragments as separate elements upon the picture plane, as had the Cubists, or exploiting the shock value of their social and political content, as had the Futurists, the Russians, and other fellow Dadaists, Ernst created from them a seamless image and identified them as a single entity on the picture plane. … The psychological ramifications of this procedure were enhanced by Ernst's skillful technique of manipulating his material so subtly that it became difficult to discern whether the work was, in fact, a collage. In accordance with this fastidious method, he cut and pasted as little as possible. Eventually he improved on his technique by reproducing his images photographically, so that the cut edges were no longer visible. It was this invention that constituted his major contribution to Dada collage (Waldman, 1992, p. 124).

Critic Werner Spies (1982) supplies more insight into the images themselves,

The conflict between the overall image and the constantly interfering interpretation of its units, gives rise to the strange mood that pervades our encounter with Max Ernst's work –elements that are comprehensible in detail turn ambivalent on the level of composition.

This ambivalence is disturbing. In order to escape it, our mind attempts to analyze each detail and to fashion an image, as it were synthetically, from this knowledge. Yet the image is not simply the sum of its details – it is like a force field in constant flux, irreducible and disquieting. The image as a whole resists logical solution (Spies, 1982, p. 95).

This disturbing ambivalence has much to do with one's expectations when reviewing the "scene" of an artwork. One is surprised by juxtaposed elements when they don't correspond with either what is expected in artwork, or expected in the real world. This is fundamental, but not necessarily unique, to collage. It allows for the special frisson that occurs between the collage itself and the world to which it references.

When reading the two quotes through the gap and seam filter, it is clear that the work may be seamless, but the gap can be enormous. This applies to surrealism in general, and as shall be seen is especially relevant within digital environments where "cut and paste" are only metaphors, and all seams – visible or not – are illusory. In fact, looking next to the paintings of Salvador Dali (Figure 7) the defect in a too literal definition of collage can be driven home.

What's at issue here is the pure technical virtuosity of Dali. Since he could paint anything and make it illusionistically indistinguishable from a cutout image reproduction he can easily create a collage effect in paint alone. For example, compare two paintings. The first called *The Accommodations of Desire* done in 1929 is a collage in the strictest sense.

The lion head is taken from a book of advertising illustrations. The second called *Partial Hallucination. Six Apparitions of Lenin on a Grand Piano,* 1931 (Figure 8) could certainly be a collage, but according to the strict definition, is not.

Anyone who insists on seeing a difference between the two is simply being obstinate, contentious, or missing the point. As early as 1948 Margaret Miller (2007, as cited in Fergonzi) recognized this point, and in the press release for a MOMA collage exhibition wrote,

Figure 7. Salvador Dali, The Accommodations of Desire (1929).
(© 2015 Artists Rights Society).

Figure 8. Salvador Dali, Partial Hallucination. Six Apparitions of Lenin on a Grand Piano (1931)
(© 2015 Artists Rights Society).

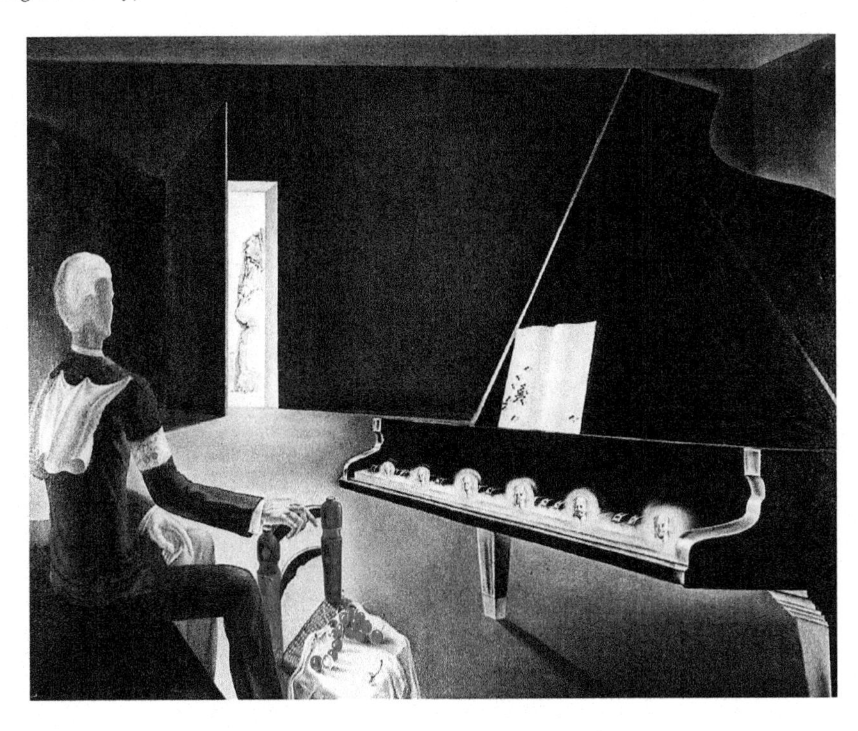

Collage cannot be defined adequately as merely a technique of cutting and pasting, for its significance lies not in its technical eccentricity but in its relevance to two basic questions which have been raised by 20th century art: the nature of reality and the nature of painting itself (Fergonzi, 2007, p. 332).

1.5. First Review

At this point three major threads of collage typologies can be reviewed (Figure 9). Cubism supports the first one, where both the gap and the seam are observed. Futurism, the second, where the seam is more salient than the gap. Surrealism, the third, where the gap is more salient than the seam. It is important to note that in spite of their historical origins these threads have nothing to do with content, they are solely a heuristic designed to categorize the structure of collages. It must be stressed, as the lines on this chart indicate, that each of these typologies exists on a continuum, one to the other. The previously referenced artists are exemplars of each typology – other artists can and have blurred these distinctions.

1.6. Collage since 1950

Everything changed after World War II. Collage began taking on entirely new physical forms, and within only 20 years became one of the major creative strategies for artists, and within 40 years became *the* major creative strategy for artists. Taylor (2004) writes that,

The story of collage tells us that the real tension within modern art of the post-war period was between the procedural licenses of Surrealism (chance, and the strangeness of the image) and the formal inheritance of Cubism (surface articulation, material construction, even abstraction). (Taylor, 2004, p. 133).

Figure 9. Dennis Summers, Collage Taxonomy 1
(© 2013, D. Summers. Used with permission).

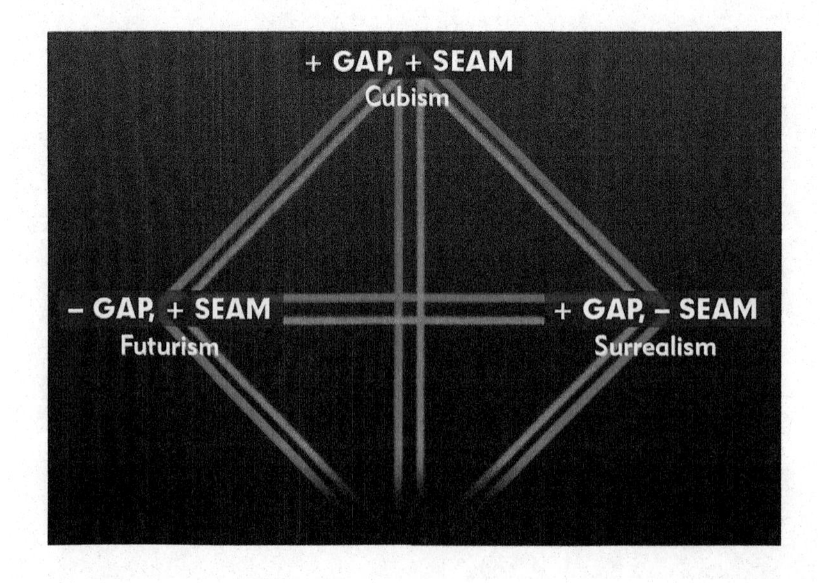

To which, as seen, the third source of tension is the less recognized legacy of futurism (healing, synthesis, interpenetration). At this point one of the best known collages in the history of art will be explored. Its troublesome nature has implications for digitally created collage.

1.7. Richard Hamilton

In 1956 the English artist Richard Hamilton created a small traditional collage made of images cut out of magazines. It was called *Just What Is It That Makes Today's Homes So Different, So Appealing* (Figure 10).

Many people point to it as a progenitor of pop art. But there is something else as interesting. Look closely at the image. It has almost nothing in common with preceding collages and much with succeeding ones. There is no confusion of the picture plane; little ambiguity of space and perspective; no abstract combination of elements picked up off the street; no combination of separate realities of media; no unusu-

Figure 10. Richard Hamilton, Just What Is It That Makes Today's Homes So Different, So Appealing (1956) (© 2015 Artists Rights Society).

ally surprising juxtapositions; no dreamlike imagery. It is nothing like collage up to that point aside from the fact that it is glued paper elements taken from different sources. It is mostly a photomontage, but one entirely different than the disturbing and visually complex photomontages of dada artist Hannah Höch from a generation earlier. With only a couple of exceptions, Hamilton's collage could have been set up as a photo-shoot and look quite similar. This artwork reveals that collage can on the one hand be a much simpler visual experience, and on the other hand can *mirror* – not just include elements of – the actual world. In some ways this turns collage on its head. It is a return to a more or less illusionistic image but not in paint. This also exposes the importance to collage of *expectations*. Yes, it may be unusual to see a muscle man holding an oversized tootsie-pop, but not as disturbing or mysterious as the juxtapositions seen in an Ernst collage. For the most part there is not much in the gaps between its constitutive elements. The seams are obviously there, but not really in a disjunctive way. This piece can be problematic to fit within the schematic, but would lie somewhere along the continuum between futurism and surrealism.

Its real importance lies in how easily this image could be created today in Photoshop. In fact, so much digitally collaged imagery resembles *Just What Is It ...* that it can be seen – in a way that cubism, futurism and surrealism cannot – as the prototype for digital collage.

1.8. Feminist Collage

Feminist collage was mentioned earlier in the discussion of futurism. The idea of fusing elements in order to create synthesis as opposed to juxtaposition is a shared ideology. This is not to suggest that more recent artists considered futurist collage as an influence on their own creative strategies (not that they would have been ignorant of futurism). But, their respective content is quite different. As early as 1987 the critic Lucy Lippard (1995) wrote that: "The feminist 'collage aesthetic' – putting things together without divesting them of their own identities – is a metaphor for cultural democracy" (p. 209). And Collage is born of interruption and the healing instinct to use political consciousness as a glue with which to get the pieces into some sort of new order" (p. 168). As has already been seen, the first quote isn't unique to a feminist collage aesthetic. And the second quote is not applicable to all collage. However, the shift of emphasis toward this particular metaphoric use is significant. Note the use of the word healing, and recall the scar definition of seam. In 1998 Gwen Raaberg (1998) developed this premise:

Works of feminist collage created during the 1990s continue to deconstruct cultural representations and discourses but also seek to expand their concerns to include, along with gender issues, a full range of social, political, and cultural concerns. In many works, we may find a reconstructive impulse based not on a totalizing perspective but on a collage strategy that utilizes fragmentation, discontinuity, and dialectical opposition to stage multiple, fluid relationships. Moving beyond a postmodern emphasis on fragmentation that results in a pastiche of 'distinct and unrelated signifiers' these feminist collages provide a mediating site that suggests new ways of connecting multiple open identities and perspective in a multitude of possible relationships. ... Although we cannot conclude that collage is particularly feminist, we may speak of 'feminist collage,' mindful that this body of work includes as number of theories and practices in various media arising from different historical and social contexts, which have been guided by and have furthered feminist goals (Raaberg, 1998, p. 169).

1.9. Mixed Media

The careful reader will note that the fourth category within the collage taxonomy has yet to be addressed (Figure 11). Briefly this refers to what is called mixed media. Although too casually and problematically included in surveys of collage (to take one example, Waldman, 1992), some mixed media artwork does not participate in the collage strategies that have here been delineated. As the 20th century progressed, artists no longer felt that they need be constrained by reductionist attitudes toward appropriate use of media. Alternatively, although collage historically was often mixed media, this is no longer the case – especially in the domain of digital collage. Nonetheless, thinking that all mixed media is collage is a serious error in analysis. In such work, there is no gap and no seam. For example, perceptual work, like the installations of Robert Irwin, although made of multiple materials, have nothing to do with collage. Additionally, – GAP, – SEAM would refer to any standard illusionistic representation in painting, photography, or film. That is, artwork that is not collage.

1.10. Second Review

Many artists prior to World War 2 were responsible for creating a number of different styles of collage ranging from the representational to the non-representational and the purposeful blurring between the two; from real world elements incorporated into paintings to paint incorporated into real world elements; from 2D to 3D to 4D. Only selected highlights have been discussed here, but the interested reader can find many more especially in *Collage: The Making of Modern Art* (Taylor, 2004). The first major exhibit of sculptural (i.e., 3D) collage, *The Art of Assemblage* (1961) at the Museum of Modern Art, New York, curated by William Seitz in 1961 was controversial, and the tone of the accompanying catalog is somewhat defensive. Shortly after this exhibition, and perhaps influenced by it, mixed media collage

Figure 11. Dennis Summers, Collage Taxonomy
(© 2013, D. Summers. Used with permission.)

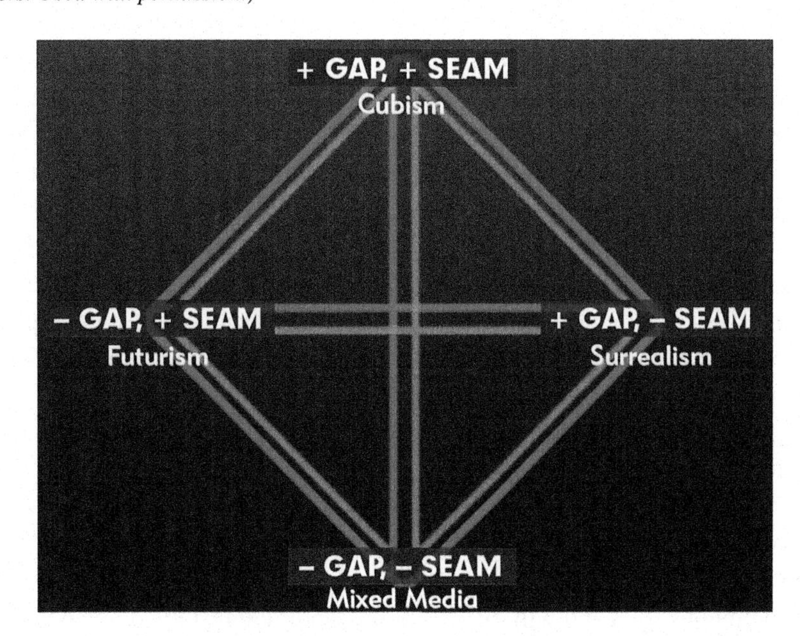

would become commonplace and uncontroversial. This chapter cannot cover the breadth of the many different forms that collage has taken since. However, as will be seen, the four typologies described here remain surprisingly reliable in classifying the full range of collage in many media and disciplines.

1.11. Some Other Forms of Collage: Introduction

This review of the history of collage has covered only a few emblematic examples of visual artwork or "static art." There are two more media, which must be discussed as they extend collage into time and are relevant to the potential pedagogical use of college for scientific visualization. They are film (or more appropriately to the purposes here – moving pictures – that can be created with video technology, manually, or entirely synthetically), and artist books. Both of these art forms can be especially useful in an educational setting.

The addition of time raises new issues that must be explored and developed. Static imagery is reviewed by its audience iteratively. One grasps the image as a whole quite quickly, and then focuses on details in order to add depth to experience, subsequently connecting those details to one another in order to recreate the complete image. Although it can be possible to do this with time-based media, such as film, typically understanding is gained sequentially and retrospectively over durational time. This allows collage to take on different forms of presentation with different implications. Generally, this is the same approach taken with literature, such as novels, although poetry might be more atomized. Like poetry, artists' books fall into a grey zone where they are likely to be experienced both iteratively *and* sequentially. Additionally, static imagery requires the viewer to move in space around the object of interest (which has prompted numerous investigations into the relationship between the body and art). Whereas accessing time-based media generally requires the viewer to be static. This distinction identifies a key difference between a collage created in Photoshop and one in After Effects.

Writing in the 1950s and 60s, early film theoretician Jean Mitry (1997) echoes this: In a painting "spatial relationships are immediately perceptible within the organic structure of the *whole*. Now, a film, like a symphony, is not an *immediate* entity, but a series of relationships, which gradually take shape. Moreover, relationships of time … are not perceptible to our eyes as they are to our ears" (p. 149). Film-maker Vsevolod Pudovkin adds, "When we linger over an image in life, we have to make an effort and spend some time, moving from the general to the particular, concentrating more and more until we begin to notice and appreciate the details. Film saves us this effort through editing" (as cited in Mitry, p. 132).

As mentioned at the beginning of this chapter the word collage is used to stand in for numerous other closely related terms including "montage." Montage is traditionally the term used in describing a film editing method. Loosely speaking it describes different film clips or scenes cut together. Thus, they are images that are combined sequentially over time. These cuts can be more or less apparent to the viewer. Practically speaking, film *is* montage. However, montage has a secondary meaning describing a less frequently used technique. This is the combination of different still or moving images – either juxtaposed or superimposed – within a single segment of film.

1.12. Cinematic Film and Montage

Every time there is a "cut" between two strips of film then a montage has occurred. The nature of this montage, and where it falls along a continuum from "seamless," "continuous," or "invisible," to "disjunctive," or "discontinuous" can and will differ amongst individual films or even within a single film.

The style of montage employed usually depends on that of the film's director. Different editing styles can be categorized and will loosely parallel that of the collage taxonomy classified here.

Theorizing on this topic may not have begun with filmmaker Sergei Eisenstein, but throughout his career he wrote extensively about montage. In 1929 he wrote,

in my view montage is not an idea composed of successive shots stuck together but an idea that DERIVES from the collision between two shots that are independent of one another (the "dramatic" principle). ("Epic" and "dramatic" in relation to the methodology of form and not content or plot!!) As in Japanese hieroglyphics in which two independent ideographic characters ("shots") are juxtaposed and explode into a concept (Eisenstein, 2004, p. 26).

The incongruity in contour between the first picture that has been imprinted on the mind and the subsequently perceived second picture – the conflict between the two – gives birth to the sensation of movement, the idea that movement has taken place.

The degree of incongruity determines the intensity of impression, determines the tension that in combination with what follows, will become the real element of authentic rhythm (Eisenstein, 2004, p. 27).

Eisenstein uses the term "a leap" to describe the "collision" or juxtaposition of clips in montage (as cited in Bogue, 2003, p. 51). Leaps, of course, imply gaps. Additionally, these "conflicts between the two" should echo what was written earlier regarding contested space. Eisenstein (1949) also created taxonomy of different types of montage, that is to say different psychological effects of cutting various film images together. He sets a pattern here that continues in subsequent critical writing on cinema.

One of the first writers to take cinema seriously as deserving philosophical attention was the critic Andre Bazin. Writing in the 1940s and 50s, Bazin (2004) was no fan of montage, but he does define its implications. He writes that montage creates

a sense of meaning not proper to the image themselves but derived exclusively from their juxtaposition. The well-known experiment of Kuleshov with the shot of Mozhukin in which a smile was seen to change its significance according to the image that preceded it, sums up perfectly the properties of montage (Bazin, 2004, p. 26).

Montage suggests "an idea by means of a metaphor or by an association of ideas" (p. 26). As can be seen, these comments are entirely consistent with those regarding static image collage.

It is Mitry, however, whose four types of montage most closely approximate the system described within this chapter. They are "narrative," "lyrical," "constructional," and "intellectual." Each of these has an analog in the taxonomy of collage.

According to Mitry (1997), the only purpose of narrative montage is to "ensure the continuity of action. ... It can be said to be invisible for the reason that it never violates the logic of the concrete" (p. 129, 130). This is montage that is subsumed to simply storytelling. This would be equivalent to the - GAP, - SEAM quadrant. An example would be the movies of Howard Hawks, described here by Sam Rohdie (2006): "The fragments, characters, looks, movements, expressions, voices, actions, all belong to different registers, but all so perfectly linked and harmonized, so naturally motivated, that the fragments are effaced for the unity that they constitute and to which they belong and which they return" (p. 87).

Of course, any montage could be said to have seams. Unless the film consists entirely of one camera shot for its entire duration, it will have cuts. But as these quotes point out, the intention is clearly that the viewer's impression of those cuts be so negligible as to be "invisible."

Lyrical montage, Mitry (1997) writes, ensures

narrative or descriptive continuity, exploits the continuity in order to express ideas or sentiments which transcend the drama. ... It's purpose is to inform but also and more especially to sublimate and magnify. The most significant scenes are broken down into a series of extreme closeups, in such a way that the editing can present the scene from every angle (Mitry, 1997, p. 130).

This is employed by directors such as Pudovkin in the movie *Mother*, and would be equivalent to – GAP, + SEAM.

Constructional montage is about editing clips together at the expense of traditional storytelling. The director Dziga Vertov, well known for the film *Man With A Movie Camera*, would be the exemplar of this type. Vertov

used his art above all to order selected documents and assemble them in such as way that a new idea should spring up from several objective and independently interrelated facts. ... The editor juxtaposes facts which are clearly authentic but have no signification other than that which they verify. An idea thus expressed –but one which obviously does not exist except though this relationship (Mitry, 1997, p. 133).

Rohdie (2006) actually uses the word gap in describing Vertov's film in this quote, "one shot does not 'answer' another ... but instead functions to highlight the gap and interval between them. ... This gap or interval is the center and source of movement and interest and energy in the film" (p. 81). Rohdie uses the word gap in just the same way as it is used by this author. This would be equivalent to + GAP, + SEAM.

Lastly, intellectual montage according to Mitry (1997) is "less concerned with ensuring the continuity of narrative than with constructing it and less with expressing ideas than with *determining* them dialectically" (p. 129). It is "concerned with expressing and signifying through image relationships rather than a purely cumulative continuity" (p. 135). A good example of this is the cut in the Eisenstein movie *Strike* from an ox being slaughtered to the workers being shot down by police. Rohdie tells us,

Nearly every image in Strike, certainly every sequence, brings together unharmonious, contradictory forms that correspond to no reality or at least no coherent continuous one. ... The central montage strategy of Eisenstein is a montage of correspondences whereby elements distant in time and space and from different realities are brought together... (Rohdie, 2006, p. 36).

Although entirely different in intent, one can see that in construction this is not dissimilar to Surrealism. This is equivalent to + GAP, – SEAM.

These parallels should not be overstated. The aims of Mitry are quite different than the ones here. However, they are relevant to any artist creating moving picture collage. These differing strategies, more or less, comprise the options available when crafting a movie. As with the visual arts described above, this system is a heuristic, and individual films and videos can fall along a continuum between these categories. Or different categories can be applied within the same movie for creative effect (see Figure 12).

Figure 12. Dennis Summers, Collage Taxonomy with Montage
(© 2013, D. Summers. Used with permission).

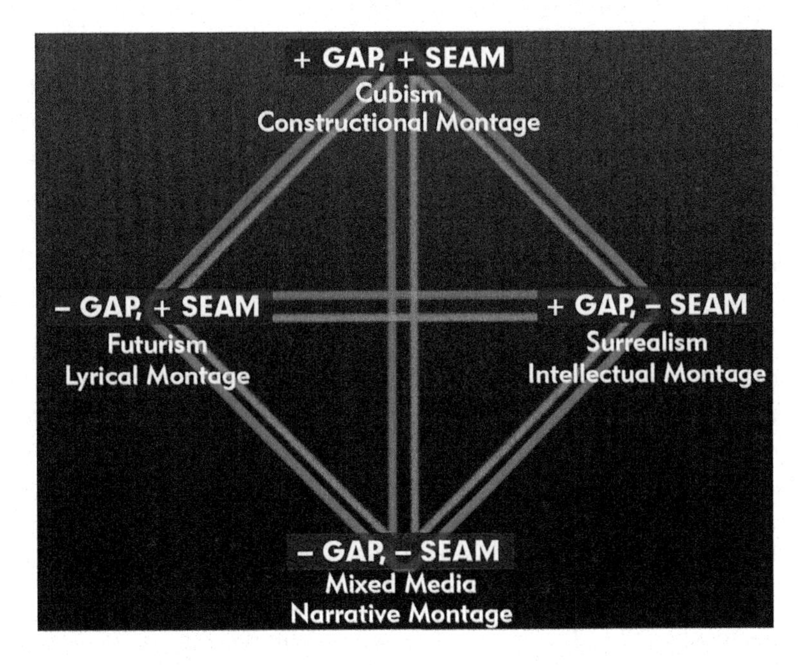

The key point is that narrative montage is similar to perceiving events as if they are continuous and traditionally novelistic. This is the construction for most film and television that comes out of Hollywood. In response to Mitry's categories, it is clear that there *are* alternatives – narrative is not the only option. For creative projects, even or especially those created for knowledge visualization, other choices are likely to be useful. As author William Burroughs (as cited in Miles, 2012) has said: "Life is a cut-up. As soon as you walk down the street your consciousness is being cut by random factors. The cut-up is closer to the facts of human perception than linear narrative" (p. 32). "Cut-up" is the word Burroughs used for collage. The bricolage of lived experience can also be constructed lyrically, constructionally, or intellectually. Recognizing this and acting upon it completely transforms perception, ontology, and understanding.

Montage in cinematic film by no means exhausts the possibilities of moving pictures. Video has a distinct aesthetic of its own, and many artists have created a wide range of collage work in this media. Perhaps the best known might be the couple Steina and Woody Vasulka who have been "abusing" video since the late 1960s. Interested readers can also check out the excellent book of essays edited by Doug Hall and Sally Jo Fifer (1990) called *Illuminating Video: An Essential Guide to Video Art.* Regardless of whether the capture format is video, digital video or film, in the past 20 years or so it has become commonplace to see "artists' cinema" in museums. These artists' work can range from that which is similar to traditional film to that which deconstructs traditional film. A good reference for this is *The Place of Artists' Cinema: Space, Site and Screen*, by Maeve Connolly (2009).

In order to draw out the implications of montage consider the film *Last Year at Marienbad* directed by Alain Resnais from a screenplay by Alain Robbe-Grillet. Note that this film would fall in the + GAP - SEAM quadrant, which when compared to Eisenstein gives an idea of just how robust these categories can be. From a visually constructed standpoint it appears to be like most other traditional movies. But creating a continuous and coherent narrative across scenes is not only challenging but actually *not* possible. *Last Year at Marienbad* makes the retrospective nature of cinema explicit. There are characters remembering and perhaps creating a past within the movie. Aside from beauty and mystery, what makes the movie valuable is that there are at least 2 and possibly 3 incommensurate realities presented within the movie – the behavior and memories of each of three characters. This is one of the things that make collage so philosophically useful. Remember from Picasso and Braque: collage is capable of representing multiple realities within the same place. This is in contrast to most styles of illusionistic painting, and Hollywood film making, which can be considered "Cartesian" in their rationalized structure. Robbe-Grillet (1962) addressed this distinction in reference to the movie:

Two attitudes are then possible: either the spectator will try to reconstitute some 'Cartesian' scheme – the most linear, the most rational he can devise – and this spectator will certainly find the film difficult if not incomprehensible; or else the spectator will let himself be carried along by the extraordinary images in front of him ... and to this spectator, the film will seem the easiest he has ever seen: a film addressed exclusively to his sensibility, to his faculties of sight, hearing, feeling (Robbe-Grillet, 1962, pp. 17–18).

If the word incommensurate reminds the reader of Thomas Kuhn (1970) and *The Structure of Scientific Revolutions*, this is intentional. One can see collage in the numerous culture clashes throughout the twentieth century. As mentioned earlier, incommensurate realities can be productively generative when collaged together. Such combinations of realities have creative and pedagogical implications.

1.12a. Montage of Superimposition/Palimpsest

Philosopher Gilles Deleuze (2013) wrote extensively on film. One topic is particularly relevant. He is one of the few people to address in any detail *montage of superimposition*. His thoughts are also consonant with static image collage and digital media.

In referring to the use of montage by director Abel Gance, in the film *Napoleon*, Deleuze wrote,

By superimposing a very large number of superimpositions (sixteen at times), ... the imagination is, as it were surpassed, saturated, quickly reaching its limit. But Gance relies on [this effect] which presents to the soul the idea of a whole as the feeling of measurelessness and immensity. ... In short, with Gance the French school invents a cinema of the sublime. The composition of movement-images always presents the image of time in its two aspects: time as interval and time as whole; time as variable present and time as immensity of past and future (Deleuze, p. 47, 48).

The take-away is this: there is something sublime in superimposition. Superimposition is a special type of collage and not just in cinema. For example, it can be seen in some of the art of Robert Rauschenberg, where paint is combined with silk-screened images. More specifically a superimposition is a palimpsest. A palimpsest, in turn, is "A written document, typically on vellum or parchment, that has been written upon several times, often with the remnants of earlier, imperfectly erased writing still visible, remnants of this kind being a major source for the recovery of lost literary works of classical antiquity" (American Heritage Dictionary, 1969, p. 944). Thus a palimpsest collapses history, that is time, within itself. Perhaps this ought to be the lens with which to understand Deleuze's comment on the image of time in its two aspects. Unlike other forms of collage, which *connect* concepts across spatial and/or temporal gaps, a palimpsest *aggregates* concepts across spatial and/or temporal gaps.

There is a history of painters creating palimpsests. For example, many of the painting by Cy Twombly demonstrate this attribute. However, constructing palimpsests digitally in either Photoshop or After Effects are so much easier to do than in other prior media. Via the (often overused) function of blending "layers," this technique has perhaps become the hallmark of digital collage. These blended elements occur via levels of transparency and specific algorithms for color mixing. These algorithms offer possibilities previously unavailable. A Google image search for "Photoshop collage art" will reveal that roughly one half of the image results display this feature.

1.13. Artists' Books

Artists' books, of course, by definition *are* books. They may expand and challenge what is thought of as a book along a continuum from something that is almost indistinguishable from a standard book to something that might casually be considered sculpture, or with advances in digital technology might even resemble a movie. This fluid character puts artists' books at the heart of this analysis. A good place to begin is with a quote from Fluxus artist Dick Higgins (1996),

A book in its purest form, is a phenomenon of space and time and dimensionality that is unique unto itself. Every time we turn the page, the previous page passes into our past and we are confronted by a new world. ... The only time a text exists in a solid block of time is when we are no longer reading it, unlike for example, a single painting which is all present before us when we consider its presence physically. In this way a book is like music, which is only experienced moment by moment until it, too, is past and remembered as a whole (Higgins, 1996, p. 103).

And, in this way, a book is like film.

Artist and writer Johanna Drucker (1995) confirms: "the codex form is described as a continuity with continual interruptions. This is quite similar to film, where the viewing experience is in relation to those breaks whether they are emphasized or repressed beneath the illusion of a continuous image" (p. 131). This critical point has already been noted in the theory of film. It should be clear that these "interruptions" are analogous to the term seam. Drucker, referring to books, discusses this in more detail, but it can be extended to collage in general,

This use of the edge of the page to manipulate the tension between continuity and discontinuity is so fundamental to the book form that it shows up in many works. In every book a decision has to be made about how to either emphasize, ignore, or overcome the fact that the openings are discreet units, separate spaces, each from the next and yet part of a continuous whole (Drucker, 1995, p. 175).

One artists' book that pushes the form was created by artist Isidore Isou in 1960 and titled *Le Grande Désordre*. As described by Drucker,

The envelope contains more than the cover for the book – it also contains its content. These are elements which spill forth as so much trash and debris: matches, cigarette butts, theater tickets, canceled stamps, [and much more]. ... The text of the "novel" is to be constructed from this mass of disorganized mate-

rial. There is no set sequence. There is no structure, order or framework. And yet the empty cover of the codex form serves as the major object according to which the rest of the elements gain their identity (Drucker, 1995, p. 126).

This artwork draws attention to the performative aspect of artists' books, which like static visual art, multimedia installation, and some new media; but unlike literature and traditional cinema, can be a corporeal experience actively created by the reader in collaboration with the artist. Engaging with Isou's book, by extension, makes us aware of our performance within – and engagement with – all of the elements of the world. Reading this one should think both of the dada artists, and of contested space.

A final quote by Clive Phillpot (1987) to confirm the concordance between artists' books and both static image and durational collage,

Reading sequences of pages in book works sometimes has an affinity with the way in which one reads a painting or a photograph, rather than a novel, in that it is a non-linear, quasi-random process. Reading page by page might be likened to traversing the surface of a collage or montage in which the eye experiences disjunctions between discrete sections of the work. It can also be likened to one's experience of a movie, in that the visual images are sometimes juxtaposed in time instead of space and cumulatively create an experience (Phillpot, 1987, p. 129).

Artists' books then, can be an ideal way for students to create projects of knowledge visualization. The convenient mix of text and imagery can help those who find it difficult to think solely in images. Additionally, text sometimes can capture a concept "just right." Finally, in spite of the virtual world experienced by today's students, the viable integration of digital and traditional media into a single tangible object can be quite satisfying.

1.15. Digital Collage or Compositing

In 2002 *The Language of New Media* was published. This highly influential book written by Lev Manovich examined montage in detail. Unfortunately, what he wrote on this topic is both internally inconsistent and unsupported by other scholarship including the information presented in this chapter. For this reason, and because of the importance of digital media to the practice of both collage and within educational settings, these misconceptions must be addressed. Manovich (2002) writes,

In computer culture, montage is no longer the dominant aesthetic, as it was throughout the twentieth century, from the avant-garde of the 1920s up until the postmodernism of the 1980s. Digital compositing, in which different spaces are combined into a single seamless virtual space, is a good example of the alternative aesthetics of continuity; moreover, compositing in general can be understood as a counterpart to montage aesthetics. Montage aims to create visual, stylistic, semantic, and emotional dissonance between different elements. In contrast, compositing aims to blend them into a seamless whole, a single gestalt (Manovich, 2002, p. 144).

This quote will be unpacked piece by piece. For the moment, accept *his* definition of montage. Manovich offers no support, here or elsewhere, for the claim that either montage was the dominant aesthetic "from the avant-garde of the 1920s," or that it ended with postmodernism in the 1980s. Or even that postmodernism has ended. By montage does he mean that which is specifically cinematic, or more broadly that which has been referred to as collage in multiple other media. This is unclear because throughout his book he sees all cultural artifacts through the lens of cinema. And yet, certainly here, the implication is that the word is being used in the broader sense, as the huge majority of films made (especially in Hollywood, the global engine of film) during this period would clearly fall into the category of narrative montage, which does *not* aim to create dissonance between elements. Assume then, that he is using the word in a broader sense to include for example the tradition of the visual arts as described earlier in this chapter. Montage (collage) was neither as monolithic a presence as he claimed during this period, nor did it end in the 1980s. A quick review through any art history book will expose numerous other art movements during this period, and show that montage (collage) remains a viable force today.

The word "compositing" is currently the word used by digital artists and editors to refer to combining different visual elements into a single image or film sequence. By "different spaces" I believe he means visual elements that may have originated from different capture media sources or may have been constructed synthetically, i.e. 3D animated models and scenes. Compositing, digital or otherwise, can be traced to the earliest decades of photography from the 1850s onward. Manovich discounts these photographs out of hand, but numerous others do not (Fineman, 2012). Additionally, compositing can be seamless, but is not technically obligated to be so. In most Hollywood movies it *is* seamless, but this is by no means determined by post-production software. It is determined by the aesthetics of continuity editing. Manovich asserts a distinction between physically manipulated photographs from the 19th century, and digitally manipulated imagery from today. Digital manipulation of photographs is fundamentally no different than physical manipulation. As Professor Tom Gunning (2004) wrote, "No question digital processes can perform these alterations more quickly and more seamlessly, but the difference between digital and film-based photography cannot be described as absolute" (p. 41). Furthermore, artist Mark Alice Durant (2002) makes clear that seamlessness is not a unique characteristic of digital media compositing: "[Using Photoshop] we can recreate our visual past without the violent telltale tears or cuts from an old pair of scissors. Our collage is seamless with edges softened, airbrushed, and blended-in with the new background" (Durant, 2002, p. 26).

Return now to Manovich's definition of montage, which cannot be supported by most scholarship. Certainly at one extreme there exists a montage of dissonance. But it should be clear that montage and collage include a wide range of approaches and results. A standard definition of montage might be this one by Rohdie (2006): "Montage simply is the joining together of different elements of film in a variety of ways, between shots, within them, between sequences, within these" (p. 1). As Rohdie and other writers on this topic explain, montage can emphasize or minimize discontinuities between film clips. It's a question of style and intent.

As mentioned, continuity editing refers to filmmakers who minimize psychological or perceptual disjunction by the viewer. The aesthetics of continuity can certainly be created via digital compositing, but this is by no means something new. It has been seen that the collages of Ernst were produced to just the same aesthetic of seamlessness. And should anyone argue that in the case of Ernst there is a psychological dissonance between elements, even if it's visually seamless; it is suggested that an image of dinosaurs and human beings in the same place is just as psychologically dissonant. And, in fact that certainly was the point of the composited scenes of *Jurassic Park*, which Manovich (2002, p. 142) refers to as an example of the perfect blending of digital compositing.

Manovich (2002) makes the startling claim that "Computer multimedia also does not use any montage" (p. 143). He explains that, *it follows the principle of simple addition. Elements in different media are placed next to each other without any attempt to establish contrast, complementarity, or dissonance between them. This is best illustrated by Web sites of the 1990s that typically contain JPEG images, QuickTime clips, audio files, and other media elements, side by side (Manovich, 2002, p. 143).*

As Manovich certainly knew the composition that he described here is a limitation of technology, not of aesthetics. Furthermore, even if such a web site construction is accepted, there's no reason at all to assume that images and clips side by side are incapable of establishing contrast or dissonance. In fact they are more likely to do so than not.

But even more surprising, later on in the book he completely contradicts this earlier statement. There he favorably defines his own new terms of "spatial montage" and "temporal montage" without the "postmodern" connotation. Contrast the previous quote to this:

In general, spatial montage could involve a number of images, potentially of different sizes and proportions, appearing on the screen at the same time. This juxtaposition by itself does not result in montage; it is up to the filmmaker to construct a logic that determines which images appear together, when they appear, and what kind of relationships they enter into with one other (Manovich, 2002, p. 322).

(His definition of spatial montage is simply the same one that has been used for collage throughout this chapter. His use of the term temporal montage describes the same structure for which most others simply use the word montage).

Finally, this last quote from Manovich,

Digital compositing exemplifies a more general operation of computer culture –assembling together a number of elements to create a single seamless object. ... As a general operation, compositing is a counterpart of selection. Since a typical new media object is put together from elements that come from different sources, these elements need to be coordinated and adjusted to fit together. Although the logic of these two operations –selection and compositing – may suggest that they always follow one another (first selection, then compositing) in practice their relationship is more interactive. Once an object is partially assembled, new elements may need to be added; existing elements may need to be reworked (Manovich, 2002, p. 139).

Can this be more stunningly obtuse? Every artist in every media works precisely the same way, especially collage artists.

1.16. Digital Collage or New Media

"New Media" can be notoriously difficult and controversial to define. As I use it, new media is likely to include at least two of the following: it is digital; it is interactive; it is virtual, or simulated; it is multi-, i.e. it incorporates more than one component of what would be considered old media. Additionally, it almost certainly takes advantage of some sort of programming code.

It is worth noting, that the word collage has been utterly debased via software apps found on the internet that allow users to combine their own photos and imagery according to different simplistic templates. I'm not exactly sure how or why this particular word appears to have lost its meaning, but the internet clearly has something to do with it. Collage that for the Dadaists was a pointed gun has become a tool for "Dummies." I think that one reason for the misuse of the word collage in these sorts of examples is that people have latched on to one rudimentary characteristic of collage, and that is "making connections." The fact that these connections are unsophisticated at every level is apparently beside the point. And making connections is something that digital applications do very well. At the heart of this are hyperlinks.

1.17. Hypertext and Hyperlinks

Writer and computer guru Ted Nelson coined the words hypertext and hyperlink in the mid-sixties. Although his full intentions for this technology were lost in the creation of the World Wide Web, for most of us the Web has defined these terms. Nelson himself envisioned a much more robust use of linking. Nonetheless, hypertext links, such as they are, have permeated our experience of virtual data for the past 25 years. George Landow (1999), a writer on digital culture, has made the case that hypertext is a kind of "digital collage-writing." He writes,

[I] am more interested in helping us understand this new kind of hypertext writing as a mode that both emphasizes and bridges gaps, and that thereby inevitably becomes an art of assemblage in which appropriation and catachresis rule. This is a new writing that brings with it implications for our conceptions of texts as well as of reader and author. It is a text in which new kinds of connections have become possible" (Landow, 1999, p. 170).

Two conclusions can be drawn here. The first is that web and standalone hyperlinked projects are definitely collage and their interactive nature offers possibilities that are difficult if not impossible in other media. The second is that the distinction between art as created by the author and art as created by the participant is blurred. This resonates with some of my previous comments regarding the way that collage has a way of expanding outward from itself to then incorporate other foreign elements. It should be pointed out that not all hypertext projects are collagic. The act of simply linking texts is not enough, there needs to be some sort of gap and/or seam, and contested space, otherwise the reading experience is simply unusually robust but relatively linear.

Of course hyperlink constructions are not limited to hypertext novels. These can include websites such as cameronsworld.net, which includes a myriad of words, images and gif animations taken from archived webpages on Geocities, a web hosting service that closed in 2009. These elements in turn are links that open new windows. Scrolling down through the main page is an experience like the rings of Dante's Inferno. The top "level" consists of outer space and lower levels include a "fantasy realm," an ocean filled with fish, a hell level and several more. The site reveals a wide range of amateur design and personal content created between 1994 and 2009. It also exposes the often ephemeral nature of such art. Searching for net-art on the internet can be an exercise in frustration, as many sites depend on out of date plugins and lost links.

Net-art can include static-image collages that change over time, or links where the participant creates a durational montage effect similar to film and video. This is one example by Jim Gilmore called *Black Cloud* {Gilmore, uncertainmachine.com} where the participant clicks on a button to expose short texts that tell a story as strange creatures float across an image of typical suburbia. An extraordinary interactive net-art piece that participates in both types of collage, and one that explores possibilities unimagined by Lev Manovich, was completed by David Clark in 2008 and is titled *88 Constellations for*

Wittgenstein (to be played with the Left Hand), (http://88constellations.net/). This project was created using Flash which offers possibilities unavailable in standard HTML. The following quote is taken from the Electronic Literature Collection website with no attribution.

Exquisitely designed with a confident, understated visual vocabulary relying on icons and degraded images, 88 Constellations at times can feel like a really well-made independent documentary, but one which swerves from seemingly normative biographical reportage into visual puns, fantastic associations, and quirky digressions. Infused with the paradox, playfulness and occasional paranoia of the philosopher's life, this is a massive work with circles-within-circles logic that would take several hours to exhaust. ... 88 Constellations is a tour-de-force interactive, multimedia essay ... (http://collection.eliterature.org/2/works/clark_wittgenstein.html)

This piece is carefully designed to make connections between the content of each constellation. However, the order of selection can subtly determine one's interpretation of that content. Although this exists to a slight extent in static-image collage, it does not exist at all in cinema where our viewing order is predetermined. Interactive collage is durational, yet not pre-determined. This gives such new media collages –when done well which is often not the case– *variable* gaps, seams and contested spaces which can be quite powerful.

1.16. Conclusion to Section 1

Taxonomy of collage can be created and summarized here, using the concepts of the gap and the seam. To reiterate: within these categories the gap is entirely metaphoric, but the seam can either be literal or metaphoric.

1. Cubist type collages and their descendants. Imagery that concentrates on formal visual complexity, spatial configuration, and ambiguity; on illusion versus reality; and on the inclusion of outside elements for purposes of visual and conceptual juxtaposition. Art that draws attention to the gaps between materials, images and concepts; and art that makes both the metaphoric and the literal seams of the collage explicit as part of its visual strategy. This would also include constructional montage as described by Mitry. These would be + GAP and + SEAM artworks.
2. Futurist type collages and other entirely different, sometimes feminist collages, where outside elements are used in such a way that the contrast between them and the rest of the structure is minimized, synthesized or unified, both visual and conceptually. That is the gap between elements is minimized in order to communicate transcendent or socio-political harmony. However, each element retains its identity, and thus the seams – either metaphoric or literal – can be seen. This would also include lyrical montage in film as described by Mitry. These would be – GAP and + SEAM artworks.
3. Surrealist type collages and their descendants. This imagery may or may not consist of actual outside elements, as the focus is on the conceptual disjunction of components, and surprising and unpredictable combinations. The gaps between these components may or may not reflect material differences, but always reflect conceptual ones. However, in many cases the seams between ele-

ments are minimized or completely absent in order to create an overall formal visual unity. This would also include intellectual montage in film as described by Mitry. These would be + GAP and – SEAM artworks.

4. Mixed media structures where the goal is formal and conceptual unity. Attention is not drawn to the variety of sources, elements, materials or their relationships to the outside world, or their juxtapositions with one another. Even if the elements are, strictly speaking, from different cognitive realms they are put together in such a way that the context reflects external and/or internal coherence, and juxtapositions yield no surprises. There's no discontinuity, ontologically or formally. This would also include narrative montage in film as described by Mitry. These would be – GAP and – SEAM artworks.
5. Collage is transferred into the realm of time with artists' books and cinema. It becomes durational instead of static. Regardless of media, the collage taxonomy remains applicable. Superimposition or the palimpsest is a formal strategy in both moving pictures and static ones. It becomes perhaps the major strategy in digital media.
6. Static image collage always *implies* movement across time and space. This is evident in cubist collage with its attention to different perspective. Static image collage requires the perceiver to move within time and space in order to fully apprehend it. Thus, we perform the art. In contrast, in regards to moving pictures, the perceiver is generally static; the movement occurs within the media itself, and thus the art is performed for us. In addition to the iterative/sequential binary there is a corporeal/ephemeral one.

2. PRACTICE

2.1. Introduction

Most of my own artwork has been inspired by scientific concepts; about linking these ideas to others from distant intellectual domains; and crafting a visual environment where this may be understood more or less intuitively depending on the background of the participant. The first digital artwork I produced was called *Crosslinked Genome: Or Data Fugue* (1996) an interactive CD created in Oracle Media Objects. This was a collage somewhat similar to the artists' book *Le Grande Désordre*. Selecting one of three interfaces, the user picked texts and images via titles that then popped into different windows on the screen. Extremely primitive by today's standards, it is mentioned here because in some way every digital collage I've created since, in any presentation format has been ever more sophisticated approaches to the same essential aesthetic challenge: How to create a rich visual environment that includes images and texts, and addresses a wide range of content allowing the participant to make their own connections between data elements. The project that will be outlined here is not interactive – it is a series of digitally created videos – yet is impelled by the same motivation.

The project is called *Slow Light Shadow Matter* (*SLSM*). Following a project description, the theory portrayed in the first section of this chapter will be used to analyze the work. Following this, the work of two other artists will be described and analyzed. The section will conclude with implications and suggestions for adapting these approaches to an educational setting.

2.2. Project Description

Slow Light Shadow Matter is a long-term project comprising 13 short digitally created animated videos: twelve chapters and a prologue. The title itself refers to two concepts from physics. In 1999 scientists "used a new state of matter called a Bose-Einstein condensate ... to bring light down to the speed of a bicycle and then later to a dead halt" (Perkowitz, 2011, p. 8). This is called slow light. Shadow matter refers to a hypothetical material "that interacts with ordinary matter, only by gravity, which means hardly at all" (Thomsen, 1985, p. 296). The imagery in *SLSM* consists of complex motion collages, combining modified representational and non-representational elements, text, music and voice. Most visual elements are synthetic, that is created entirely via software. Each chapter is inspired by an artist-scientist dyad; each of these pairs includes an artist and a scientist who were approximate contemporaries. The chapters are united by multiple themes. For example, one narrative thread is based on the Greek god Hermes. The physical nature and history of light (or more generally electromagnetism), that of force and fields, and by extension technologies of communication or information transfer are investigated. I am weaving into the chapters a cross-referential structure of systems that include alchemy and biology in order to draw out related conceptual connections. In addition to a free-improv soundtrack, there is a voice-over recounting the biography of a musician based on jazz saxophonist Ornette Coleman.

The scientist/artist pairs are listed here:

- Prologue (El Greco)
- Isaac Newton/JanVermeer
- Michael Faraday/William Turner
- James Clerk Maxwell/Claude Monet
- Rosalind Franklin/Ana Mendieta
- Ernest Rutherford/Paul Cezanne
- Niels Bohr/Wassily Kandinsky
- David Bohm/Sol Lewitt
- Albert Hofmann/Pierre Soulages
- John S Bell/Jasper Johns
- Murray Gell-Mann/not yet determined
- John Archibald Wheeler/Agnes Martin
- James Lovelock/Richard Long

The scientists are almost all physicists who have worked with some aspect of electromagnetic or quantum theory. There are three dissimilar scientists who represent a sub-theme of biology. The work of the artists usually relate to light and/or color. Here too, divergent members are included in order to develop diverse ideas. The connections drawn between the artists and scientists are of course entirely aesthetic and based on my own intuition. The resulting video is exceedingly dense with superimposed imagery and texts. It may take viewers some time to parse out all the components and their relationships.

Each chapter is a roughly 3-minute video with components that move, scale, rotate, and fade in and out. Thus the aspect of motion will be compromised by the inclusion here of only still images. But enough can be represented to give the reader some sense of how information is visualized and aesthetically combined.

SLSM: Chapter 2, the one inspired by Turner and Faraday, will be examined. This high definition video is projected on 3 separate screens, two narrower ones to the right and left, and a square one in the center. There is a small gap between each screen. This is represented here by two vertical black lines, as seen in Figure 13.

This video was composited in Adobe After Effects (AE), and included elements were either created in AE, Autodesk 3ds Max, Adobe Photoshop or E-on Vue. The video opens with digitally created ocean waves covered by an orange fog of clouds. Superimposed on the top of all components throughout the video are vertical colored bars, which the viewer might recognize as the emission spectrum of Helium. There is a pattern of small light blue dots, which represent the constellation Taurus. This slowly rotates 360 degrees on the z axis throughout the duration of the video. Also seen is a pattern of embossed "spots". Each of these is actually a 2D representation of a 3D model of a radiolarian called *Heliosphaera actinota* based on an illustration by 19th century scientist Ernst Haeckel. These spots randomly pop on and off throughout the video. Soon several elements begin to appear, as seen in Figure 14. These include two bodies of texts or text fragments, and a third reveals at about 1.5 minutes. Each of these is taken from the writings of Michael Faraday, and they slide across the screen in different directions.

A "blobby" pattern representing that seen in electrical filings around a magnet appears and is seen throughout the video. At about 15 seconds, emerging from the center and slowly moving toward the viewer, is a model of the Egyptian sun-god Ra as represented by the sun atop a solar barge of a serpent with reflective wings, as seen in Figure 15.

As this leaves the screen a close-up view of a neutrino detector appears (Figure 16), fading off about 20 seconds later.

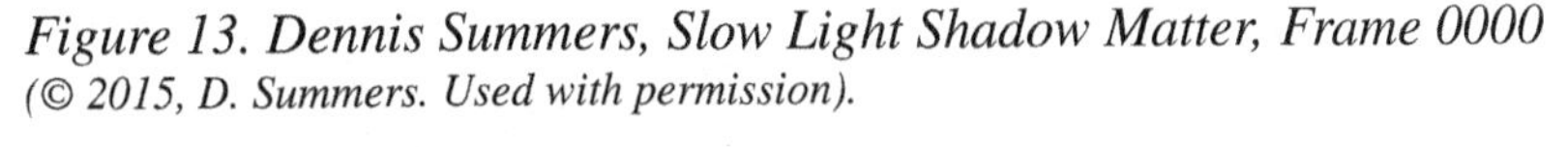

Figure 13. Dennis Summers, Slow Light Shadow Matter, Frame 0000
(© 2015, D. Summers. Used with permission).

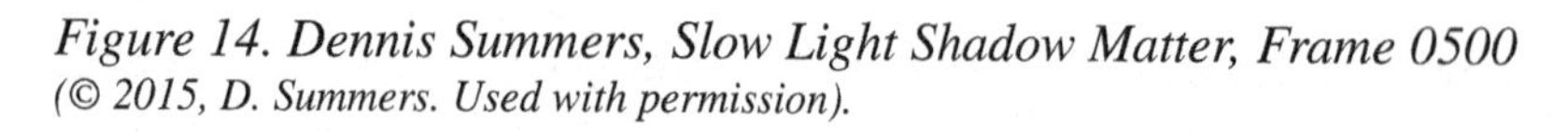

Figure 14. Dennis Summers, Slow Light Shadow Matter, Frame 0500
(© 2015, D. Summers. Used with permission).

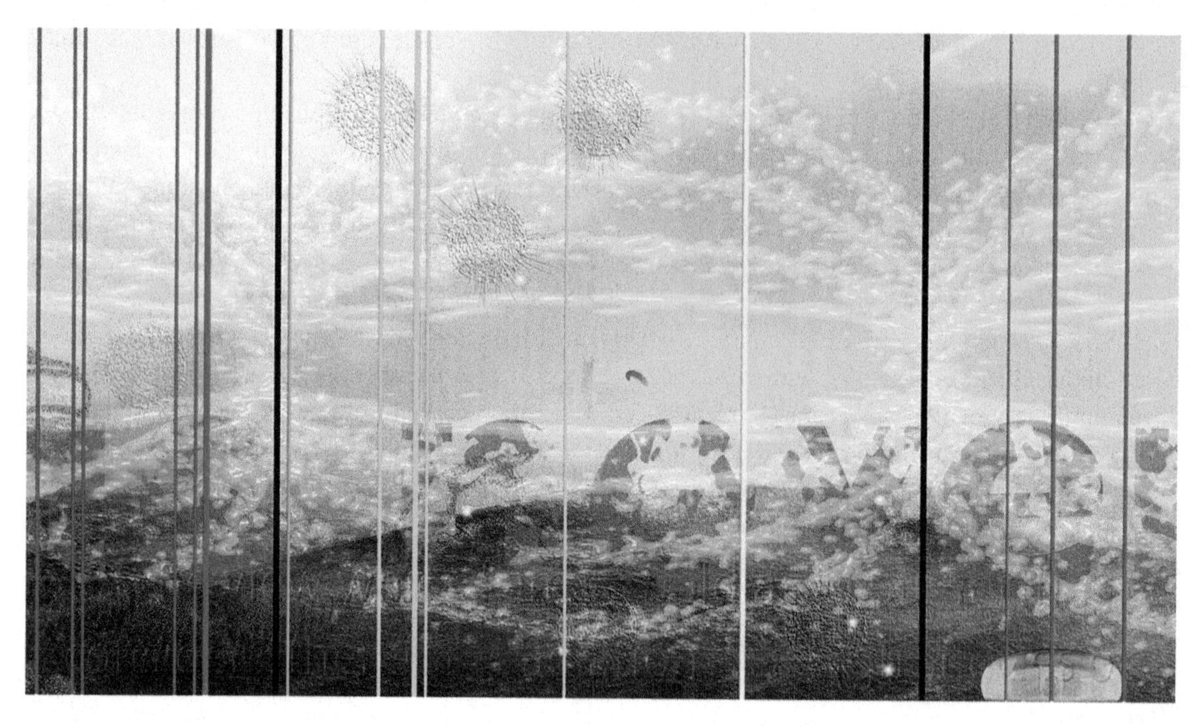

Figure 15. Dennis Summers, Slow Light Shadow Matter, Frame 1000
(© 2015, D. Summers. Used with permission).

Figure 16. Dennis Summers, Slow Light Shadow Matter, Frame 2000
(© 2015, D. Summers. Used with permission).

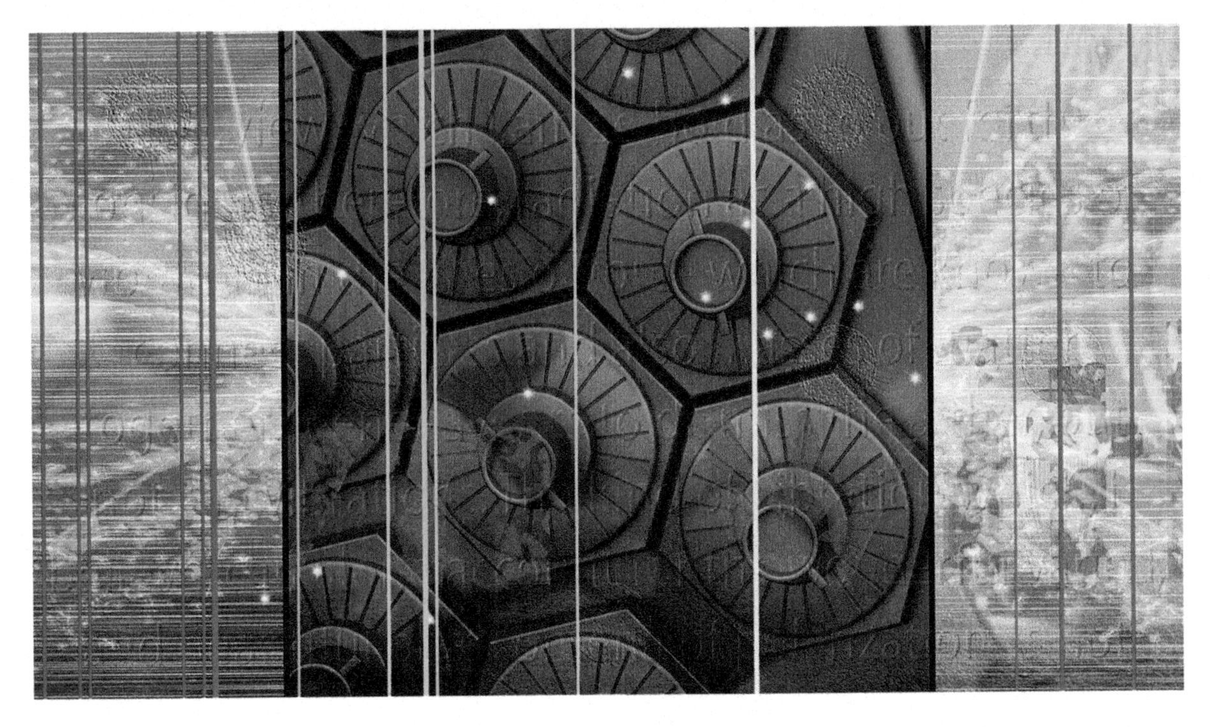

While all this is occurring, on the right panel a self-illuminating jellyfish is slowly moving vertically upward. At about 1.5 minutes a figure can be seen climbing a stylized representation of the DNA double helix (Figure 17). On closer examination it can be seen that the figure is a half-man/half-woman hermaphrodite wearing a crown like the similar symbol within the visual history of alchemy.

During this period dragonflies can be seen flying across the top of all 3 screens. As this ends a spinning earth with glowing green loops in the pattern of the earth's magnetic field fades on, as birds fly across all screens from bottom right to top left (Figure 18).

Subsequently moving vertically downward on each side panel is a close-up shot of a rotating jade figure of the Minoan Snake goddess (Figure 19).

The video ends with a fade up of a stylized representation of the Greek god Hermes (his hat, staff, and boots) along with the tortoise shell lyre that he made and subsequently gave to Apollo (Figure 20).

There are several more visual components present within this video that will not be described in order to keep this relatively manageable. From the still images and this description one should be able to get a sense of the visual density along with the depth of layering of superimposed images.

2.2a Exposition

One of the aesthetic goals is to create a manifold visual space similar (in form, not content) to paintings like those by Robert Rauschenberg, Anselm Kiefer, and many others. Yet adding motion into the experience further complicates the effort to keep the total imagery coherent for both the artist and the

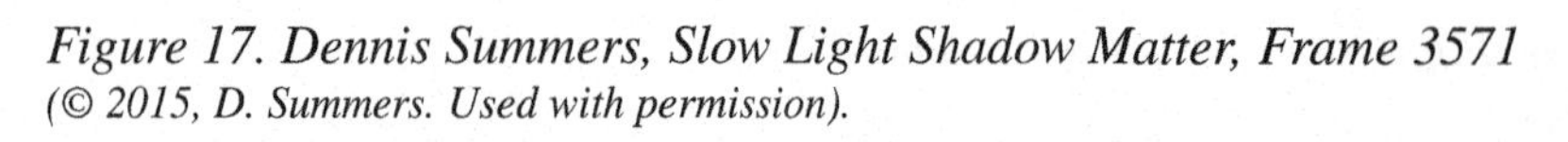

Figure 17. Dennis Summers, Slow Light Shadow Matter, Frame 3571
(© 2015, D. Summers. Used with permission).

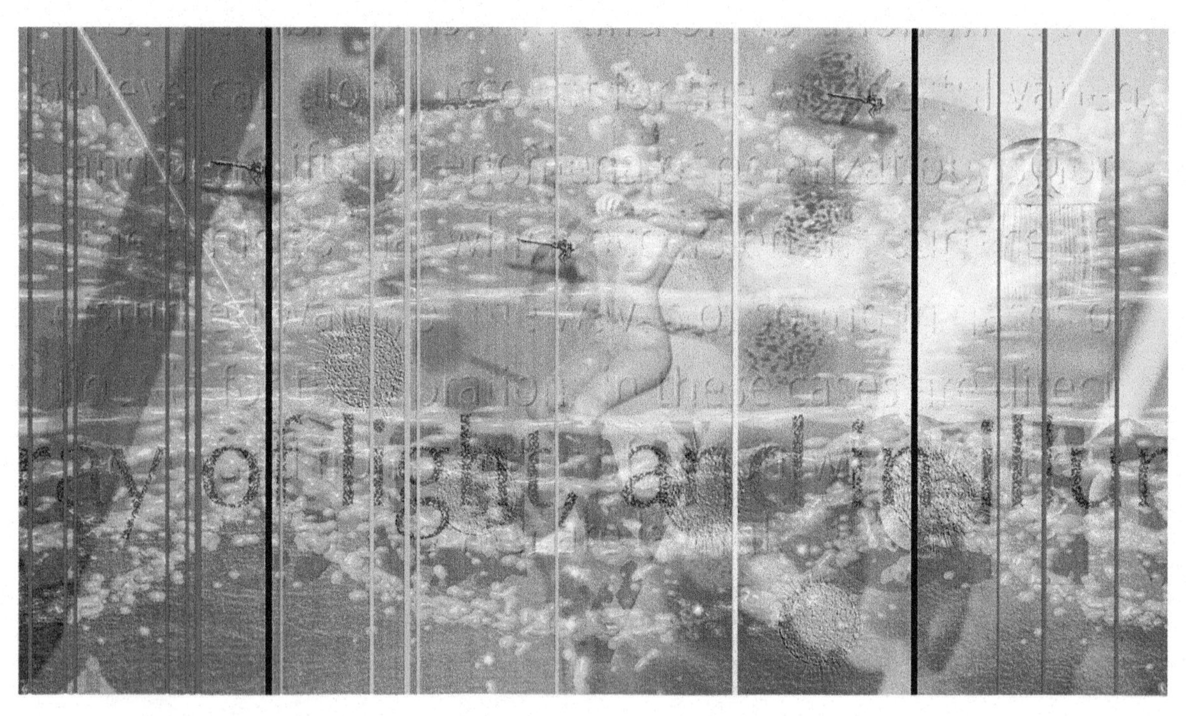

Figure 18. Dennis Summers, Slow Light Shadow Matter, Frame 3750
(© 2015, D. Summers. Used with permission).

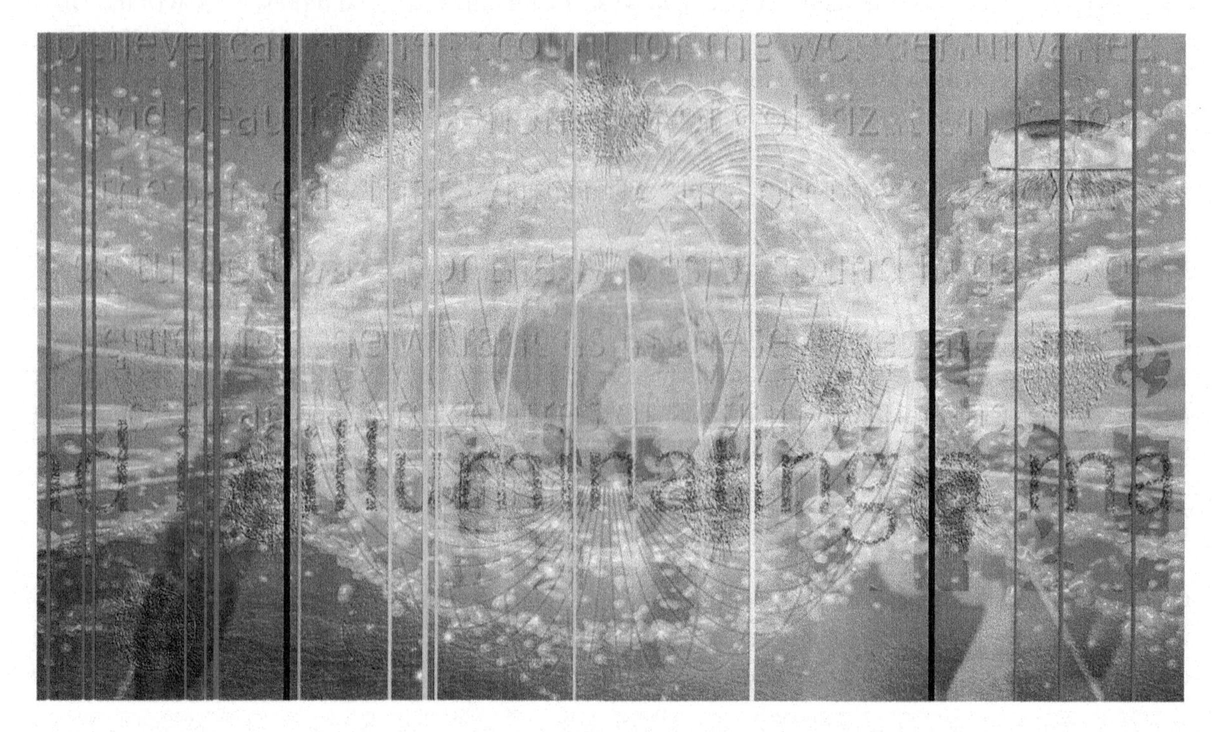

Figure 19. Dennis Summers, Slow Light Shadow Matter, Frame 4600
(© 2015, D. Summers. Used with permission).

Figure 20. Dennis Summers, Slow Light Shadow Matter, Frame 4922
(© 2015, D. Summers. Used with permission).

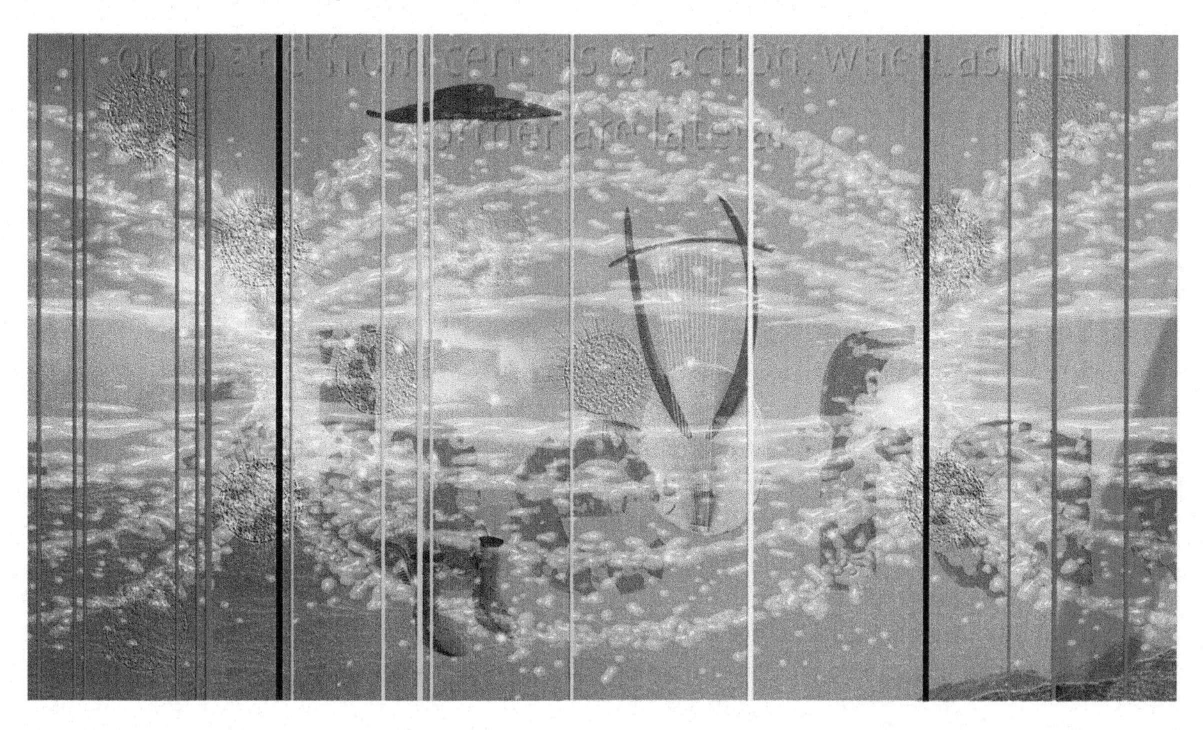

viewer. This is meant to require the viewer's iterative and sequential attention as referred to in Section 1. Each chapter is a variation on a theme, much like jazz. Many of the same components are used in each chapter, but in different styles and configurations guided by the respective pairs of individuals. It ought to be clear that this project is all about knowledge visualization but it differs from the objective style as described by Tufte. It is more poetic, evocative. The difference between this sort of knowledge visualization and the more conventionally explicit will be addressed further on.

As an artist I work intuitively, guided by what "feels" right both formally and conceptually. I don't normally have to articulate or explain any of the details, either to myself or others. In some cases I've forgotten information about specific components; sometimes I rediscover connections between them. For the sake of elucidating relationships between this video and collage strategies I will suggest possible signification.

Each chapter includes the emission spectra lines, and the element that they represent matches the number of the chapter. So for example, in this case *SLSM: Chapter 2* shows Helium; *Chapter 3* would show Lithium. Taurus is the second constellation of the zodiac and thus present in the second chapter. These are alternative numbering systems that have arbitrary correspondence. The relevance of the magnetic components to Faraday should be clear, and the ocean and fog speak to the paintings of Turner. A perhaps subtler point is the reference to the birds and the earth's magnetic field. Recent research has suggested that migrating birds orient themselves via magnetically sensitive particles within their heads.

There are numerous connections of commonality, particularly surrounding the concept of light. Turner was famous for painting light and the effects of light – sunsets, skies, reflections on water. Faraday discovered the connection between electricity and magnetism, fusing them into one force: electromagnetism, and made many discoveries about the characteristics of polarized light. Light, of course, is one form of electromagnetism. One of the quotes from Faraday in the video is this fragment: "magnetizing and electrifying a ray of light." Certain species of jellyfish are self-illuminating which in turn connects to the water, the sun, and the stars. However, these are not the interesting parts of this collage. As mentioned earlier, it is the conflicts and contested spaces between juxtaposed elements that make collages interesting.

2.2b The Gap

Although there is some overlap between Turner and Faraday there are also differences. Both men were analytical and visual thinkers. Turner was an almost unintelligible public speaker; he communicated with pictures. Faraday was known to be a spellbinding public speaker and cogent writer. Faraday was humble to a fault, not so Turner. But more than simple personality differences, the contrasting goals of an artist compared to a scientist cannot be underestimated. Turner may have wanted to represent the many possible variations of light; Faraday wanted to explain the underlying physical structure of light.

The hermaphrodite in alchemic tradition represents the union of opposites, spirit and matter, and opposing forces, which can be taken as a symbol here for collage itself. The gap between that figure and the double helix on which it climbs could speak to issues such as gender differentiation and identification. There are some species of jellyfish that have both male and female reproductive organs.

The characteristics of Hermes are many, and many are relevant to this project. This allows for a certain flexibility regarding associations both close and distant that can be made between Hermes and other elements within the collage. One attribute is his aspect as the messenger god, swiftly traveling between humans on the earth and the gods who live in the sky. People that believe in astrology believe that forces related to the stars control their lives. Birds occupy a literal space between the earth and the sky, and in many cultural traditions occupy a similar mythic space. Hermes is the patron of travelers, and this relates to the migration of birds. What characteristics distinguish humans, gods, and animals? What of the messages carried within DNA? How do these messages compare to literal ones? Neutrinos connect humans to every other point within the whole of the universe. How is this both a metaphor for the video itself, and the relationships of components within it?

These are just a small sampling of potential juxtapositions. Viewers are likely to raise more questions. The richness of their experience depends on their questions and their answers.

2.2c The Seam

Digital media has made compositing disparate elements in entirely fictional spaces commonplace. This is one such fictional space, in spite of the fairly realistic ocean environment that underlays the video. In fact much like futurist collage these components are thoroughly blended in order to purposely obscure which layers are on top of which. "Blended" here of course has a specific technical meaning within AE. In contrast to typical additive paint mixing, AE offers many different ways of combining color values of pixels from one image with those of another. This range of options allow for robust and complex combinations, many of which are used within this project. Additionally, this blending – as in the painting referred to earlier by Boccioni – is about creating texture and "suggesting the 'molecular' interpretation of the depicted object and its environment" (Poggi, 1992, p. 181). Real, that is tangible, texture is of course quite difficult to achieve in digital media, and is a personal frustration in creating these videos.

Ambiguous space is no longer as surprising in contemporary art as it was in Picasso's time. However, looking closely at Figure 17. The reader will note that the dragonflies not only cast a shadow but they are atop the emission spectrum lines which throughout the video act as a "cage" containing all of the other components. Thus, this shadow "cannot be confused with a real shadow and thus calls attention to itself as an illusion" (Poggi, 1992, p. 82). It paradoxically creates a kind of illusory realism not present in the other elements. The seams between the 3 screens are real, and most components respect the extent of the implied frames, and yet some components, like the birds, stitch the screens together.

Although commonplace now, the use of text as content in artwork, as mentioned earlier, can be traced to the newsprint used in Picasso's collages. Much like dadaist art, *SLSM: Chapter 2* purposely shows life "as a simultaneous muddle of noises, colors and spiritual rhythms" (Huelsenbeck, 1981, p. 244). As mentioned earlier there is a free-improvisational musical soundtrack, which mimics the density and complexity of the visuals, along with the voice-over, which further tangles the audio experience.

Although not informed by the same psychological concerns as surrealism, the combination of components within *SLSM: Chapter 2* does not conform to any normal illusionistic space. Nor do they conform to any consistent conceptual space. The visual environment includes "elements that are comprehensible in detail [that] turn ambivalent on the level of composition" (Spies, 1982, p. 95).

However, much like feminist collage *SLSM: Chapter 2* moves "beyond a postmodern emphasis on fragmentation … [and] suggests new ways of connecting multiple open identities and perspective in a multitude of possible relationships" (Raaberg, 1998, p. 169).

It is important to note that in this chapter there are no cinematic cuts, yet there is some overlap with constructional montage. Components are assembled "in such a way that a new idea should spring up from several *objective* and *independently* interrelated facts. … An idea thus expressed – but one which obviously does not exist except though this relationship" (Mitry, 1997, p. 133). Of course what is most relevant to this chapter and permeates my thinking about this project is that of the superimposition or palimpsest. Although perhaps difficult to tell from a series of still images, the intention of this video is to present "time as interval and time as whole; time as variable present and time as immensity of past and future" (Deleuze, p. 47, 48).

2.3 Nelson Smith and Interactive Collage

Nelson Smith has created art in a variety of media over the past several decades. These include paintings, drawings, installation art, and performance art. Most of this work follows collage strategies. His paintings are a perfect example of a collage with no outside material, just paint. And yet the juxtaposition of constituent items would be recognized as collage. He reorients and reconsiders painting in the multi-media installations will be reviewed here instead.

This project called *Round Tower Cosmic Teacup Network* was created in 2014 and first exhibited in 2015. Smith was participating, as an artist-in-residence producing paintings at the Ballinglen Art Foundation in Ballycastle North Mayo, Ireland. While eating lunch at a local cafe a memorable thing happened. A small group of people sat down at a table and ordered tea. When the cups arrived they each began stirring the liquid. This created a chorus of different sonic patterns, reminding Smith of the music of composer Steve Reich. This striking event led him to record the teacup-stirring patterns for possible inclusion in some as yet undefined artwork.

During this residency he discovered something else. This was a "round tower" site in nearby Killala. He learned that towers like this one have been built by monks throughout Ireland in the sixth and seventh centuries for unknown purposes. He found a book in the Ballinglen library titled *Nature's Silent Music: A Rucksack Naturalist' s Ireland* written by Dr. Philip Callahan (1992). Callahan's book explores his experiences in Ireland during World War II and applies his expertise in infrared radiation to the mysteries surrounding the round towers. According to Callahan, the round towers share the same proportions of insect antennae, designed to communicate in the infrared spectrum. In addition, Callahan (1997) aligned the positions of the known round towers in Ireland to the constellations in the night sky above Ireland.

One of the most challenging and provocative books ever written is Finnegans Wake by James Joyce. This book is practically a collage of words, sentence fragments and languages, that combines Irish mythology, science, communications, popular art, sociology, and much more into a complex and seemingly endless stream of internal and external references. Smith's interest in Ireland included pursuing his understanding of Finnegans Wake and incorporating that into his own work. He was particularly interested in the process of embedding layers of history and mythology into a cyclic narrative.

2.3a Exposition

As these apparently unrelated preoccupations fermented in Smith's brain, they began to entangle and he started to see the juxtapositions that would become *Round Tower Cosmic Teacup Network.*

As seen in Figure 21, the interactive sculpture consists of a wooden tabletop painted with a layering of diagrams, five teacups each with an embedded speaker, sound module, and motion sensor. The tabletop rests on saw-horse-styled legs fabricated to the same general shapes of the round towers, or insect antennae. Each teacup has a distinct recording of a stirring pattern, which is triggered by someone moving a spoon. Each spoon covers a motion sensor that is hidden inside a teabag on the saucer (Figure 22).

The tabletop painting consists of the superimposition of a diagram indicating the locations of the round towers, and one of matched constellations found in the night sky over Ireland. Also superimposed

Figure 21. Nelson Smith, Round Tower Cosmic Teacup Network, installation view
(© 2015, N. Smith. Used with permission).

Figure 22. Nelson Smith, Round Tower Cosmic Teacup Network, detail
(© 2015, N. Smith. Used with permission).

is a reformatted version of a diagram of Finnegans Wake created by the artist Laszlo Moholy-Nagy. The final layer of information is an interpretive schematic created by Smith that seeks out connectivity between the layers below and the sounds coming from the teacups. The drawings used for the tabletop painting can be seen in Figure 23.

Smith (N. Smith, personal communication, September 28, 2015) writes that:

Circles and oval forms have three connotations for me – 1) as portals or entry points of some kind; 2) as implied motion like a cyclic or rotating motion; and 3) as container, embracing multiple things. This has been the case for a long time so Joyce's use of cycles in Finnegans Wake fed into something I have been interested in for a long time. As a result, I converted Moholy-Nagy's Wake chart into a kind of bulls-eye format of extending circles. I had no real expectation for how the interaction of the Moholy-Nagy diagram would intersect, but like most of my chance superimpositions I was amazed at how they created unlikely relationships that I would not have thought of otherwise.

Figure 23. Nelson Smith, Round Tower Cosmic Teacup Network, drawings
(© 2015, N. Smith. Used with permission).

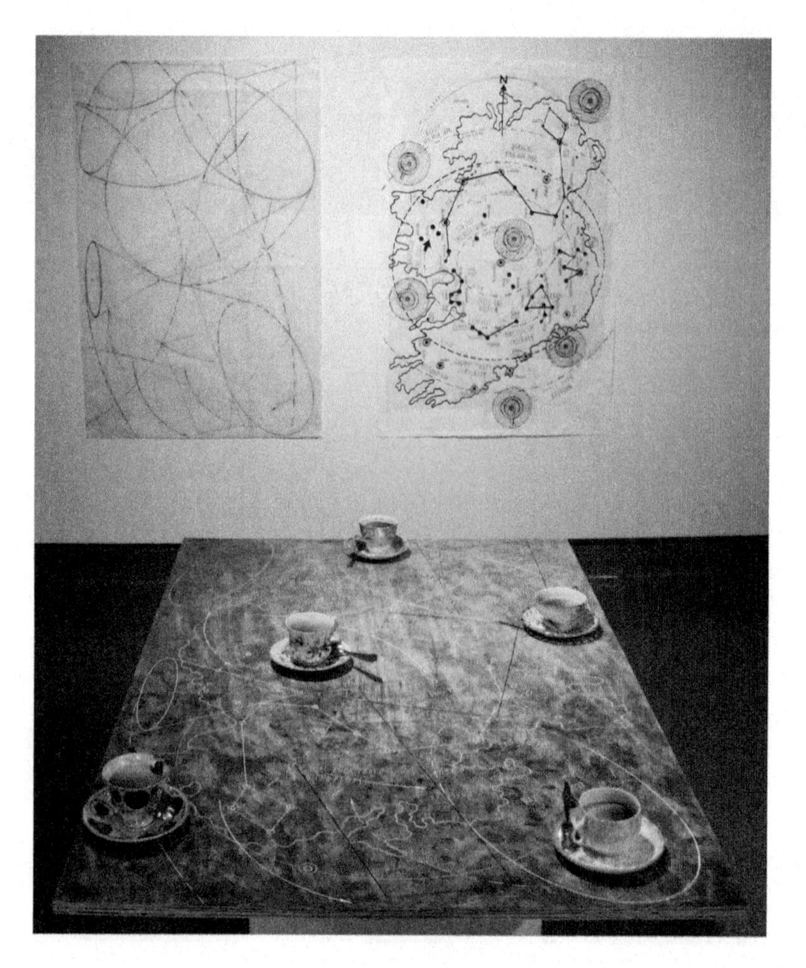

Artists in all media and formal structures frequently refer to the role of chance and coincidence in their work. Whether purposely pursued or accidental, such correspondences in collage seem to have special meaning, in part because collage so often causes such unexpected juxtapositions.

Networks, like collages, are about linking components. Smith sees the round towers as a network, not unlike that of television antennae. The interlocking table legs are another kind of network, and a kind of upside-down representation of the towers. The global system of radio telescopes designed to capture information from the stars can be extrapolated to the round towers on the map of Ireland and their correlated constellations. Referring to the teacups as "cosmic" reinforces the idea of a teacup network participating somehow with these other ones. Like bells, all five together sound as if they are signaling. This can be interpreted as a nearby event, like school bells, or farther like African talking drums, or distant – to the outer reaches of space.

All of this in turn resonates with the infrared communication networks of insects. Some insect communication is of course audible, which has additional parallels to the teacup chorus. As Smith writes,

Since the insects use this communication largely for finding food and forming their communities, it seems a better fit than I might have imagined at first. The use of only five teacups and stirring patterns on a table that echoes the scale of dining, allows for people to make this kind of connection. While my positioning of the teacups is based mostly on aesthetics, they are positioned to be "played." I have imagined using the piece as an instrument in a performable sound score. The final layer of schematic that I painted onto the table top attempts to speak to the "logic" of the teacups sound design. Some of the configurations appear to be wormhole-like or audio speaker-like; they make a leap from antennae to the audio speakers embedded in the teacups. Much of this configuration is also formal, using my design fundamentals to make a cohesive visual work and unify the underlayers of information as a composition while creating a relationship with the physical teacups sitting on the surface of the painting. (N. Smith, personal communication, September 29, 2015).

Like all good collage the connections and contrasts between entities radiate outward from the artwork itself into the larger world, and in this case into the rest of the universe. This work can be seen as identifying and representing specific concepts from science. But it also can be seen as creating new scientific structures.

2.4 Susan Gold

Susan Gold has produced artwork in a wide range of media over the past several decades. She's created numerous site specific installations inspired by biology and ecology, one of which is *OBSERVATIONS IN,* created in 2013. This project was located in the Weldon Library of the University of Western Ontario, and created as an extension of her nearby gallery exhibition. One of her objectives in choosing this site was to comment on the relationship between the artwork and texts, animal objects, and systematic classification methods. More specifically, at its heart was a text titled *Systema Naturae*, written in 1758 by Carl Linnaeus, Swedish botanist, physician, zoologist, and father of modern animal taxonomy.

The artwork consisted of three display cases containing a variety of objects paced atop distorted photocopies of the Linnean texts, as seen in Figure 24.

Items in addition to that text included: a variety of preserved horned larks (a species of bird); a manifest of the bird collection; assorted preserved bugs; assorted eggs; some small stones; raku fired

Figure 24. Susan Gold, OBSERVATIONS IN, installation view
(© 2013, B. Lambert. Used with permission).

ceramic objects; and plastic Petri dishes containing transparent photocopies of enlarged portions of the text. One case contained the birds and the Petri dishes; one contained boxes of collected eggs; and the third contained insects, stones, and Petri dishes. Within all of the cases were the sheets of manipulated photocopied text.

These transformed text images were created by manually pulling the originals across the copier plate during its reproductive action. The horned larks, eggs, and insects came from the collection housed in the University's Biology Department. Each bird was tagged with place, date, and a signature of the original collector. Stones collected by the artist came from an area around the municipality Sanikiluaq located on the Belcher Islands in Hudson Bay. Engraved on them are circles that appear to be manually instead of naturally produced. The ceramic pieces were made by the artist in the 1980s, as basic primal shapes designed to represent "almost nothing." Each Petri dish was filled with a text fragment copied on transparent acetate. This combination gave the illusion of a magnifying glass when placed over the larger sheets of paper.

2.4a Exposition

This project allowed the artist to place materials affiliated with ideas of classification and knowledge storage into the appropriate context of a library. The library, a place of classification and stored information can also be likened to a book. The installation can be seen as a book presenting its contents physically. Such a book becomes interactive with its space and the viewer. This entanglement of the artwork with the context and content of its location, – always inherent in collage – creates a poetic space where questions, reflections, inspirations, humor, magic, memory, and loss can come into play (Figure 25).

Figure 25. Susan Gold, OBSERVATIONS IN, detail 2
(© 2013, S. Gold. Used with permission).

The work of Linneaus as both content and methodology are the substructure of this project. The text included a section outlining and explaining his binomial system of classification. Another section described the necessity of including poetics, histories, traditions, and stories that relate to the natural organisms under consideration. Modern systematic scientific practice has eliminated this approach thereby separating arts understanding from scientific understanding.

OBSERVATIONS IN can be seen as an effort to redress this partition. Linneaus traveled widely for botanical specimens, and used his original taxonomies for organizing and displaying them in cabinets. At that time, his collection was considered one of the finest in the world. His goal was to create consistent and predictable systems of classification for natural organisms based on specific visual relationships. These relationships of resemblance were aesthetic as much as scientific in a period of history when distinctions between the two were not as explicit as they were to become. In 1754, he said (as cited in Koerner, 1996) that "the earth is then nothing else but a museum of the all-wise creator's masterpieces, divided into three chambers." He created a display in his own home that "was a microcosm of that 'world museum'" (p. 153). The parallels between this, collage in general, and the installation described here should be apparent. Analogously to Linneaus, Gold describes her creative process with the words "Collect. Sort. Align." (S. Gold, personal communication, September 27, 2015). This, of course, can also be used to describe the process of creating a collage.

The pages of warped text imagery act as a substrate in supporting, reflecting, and refracting the other elements within the display cases. Much as in the collages of Picasso, the text has a dual presence as both content and as image. Gold describes her production method this way:

Pulling material across the copier plate is a practice of mine to jar loose the content of the material, to see it in a different way. It is a performative act and it harnesses the operations of chance. Chance and error are productive for me in my creative work. Words are smeared to illegibility, but the process exposes embedded alternative meanings. Certain words come forward; the composition of the text is exposed. When pulling back and forth across a digital copier plate very strange mirrored images occur. These can perhaps be deciphered but more likely they are felt experiences. I am startled into an alternative view of the quotidian. Startled into a poetic space. (S. Gold, personal communication, September 27, 2015).

The insects were chosen for their plainness in order to resemble a mark on the page. An additional possible association here can be to software "bug." Such programming flaws have a parallel in the text image distortion. Birds eat insects, insects eat paper, and this appears to overlap with the phrase that one "consumes books." The stones appear to be inscribed with messages from non-human nature, this is juxtaposed with the inscriptions on the various components within the cases, and the books within the library (see Figure 26).

What is it talking to whom? The insects and birds once had mobility: did they carry messages? What does it mean to classify? Classification locks in the range of possible interpretation. Linnean classification was originally based on visual resemblance. Occasionally, DNA sequencing will clarify that an animal

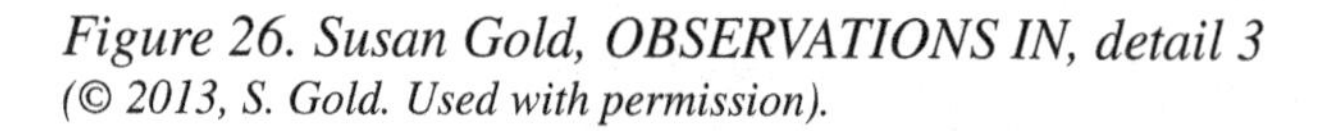

Figure 26. Susan Gold, OBSERVATIONS IN, detail 3
(© 2013, S. Gold. Used with permission).

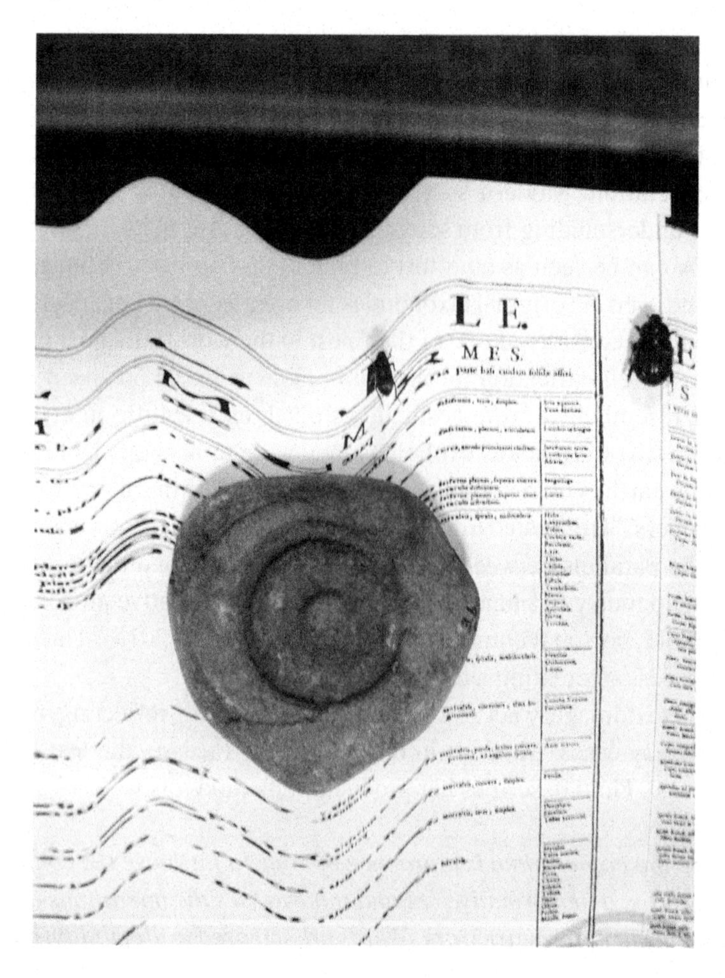

was mistakenly included in the wrong group. How does this relate to organizing visual information? Isn't collage about organizing visual (and other sensory) information? What are the distinctions between non-human nature and human culture? What of the contested space between the dead birds and the cases within which they are trapped? (Figure 27).

Gold adds these comments,

Bird bodies can be anthropomorphized into individualized readings of death or pain or they can appear as a scientific collection revealing their similarities and their objectness, named and labeled; plastic petri dishes contain transformed text on floating transparent acetate, reminding the viewer of the scientific eye and laboratory process. Is text an object for scientific observation? Natural stones or manmade objects confound the viewer as does the morphed text. I brought these objects to the space hoping not so much to create questions with my assembly but to create a state of confoundedness. The objects are meant to speak to each other and confound each other as well as the viewer in their juxtaposition or in the gaze. They are of course mute, as is the relationship of the viewer to object, but their illogic or strangeness breaks down boundaries of linear thought. (S. Gold, personal communication, September 27, 2015).

Figure 27. Susan Gold, OBSERVATIONS IN, detail 1
(© 2013, S. Gold. Used with permission).

As described earlier in this chapter, collage is inherently irrational. The conflict between the artwork itself and its presentation as a pseudo-scientific classification method creates a contested space that brings into question the differences and commonalities between contemporary and historic art and science.

2.5 Solutions and Recommendations for Knowledge Visualization

In this section several aspects of the use of collage in pedagogical settings will be addressed. These include the use of software; the importance and value of using collage strategies; assessment of –and suggestions for– student projects. The reader should note that the purpose of this section, and indeed the complete chapter is to create a preliminary framework for a deeper understanding of collage than has been previously available, and for the employment of collage as a creation strategy within and without the classroom. Although some of the following may strike the reader as somewhat under developed, this author hopes that enough information has been presented to allow readers to open it up in directions appropriate to their own individual skill-sets and interests. Collage has always been uniquely expansive, inclusive, flexible, and deeply personal. Take the following as an invitation and a challenge.

2.5a Collage and Software

Unlike sophisticated complex scientific visualization requiring advanced software tools, popular and omnipresent software is available to most students. In practical settings one can expect some familiarity with Photoshop, perhaps less with compositing and effects applications, and even less with 3D applications. Engaging collages can be created with fairly minimal software experience. The tools necessary in Photoshop can be as limited as selection, image adjustment options, and the use of filters. Components created in Photoshop can relatively easily be incorporated into an After Effects (AE) comp. Although AE can be a little more challenging, simply learning how create *keyframes*, and using the Transform options: Position, Scale, Rotation and Opacity, can be enough to create moving collages of surprising sophistication. 3D components are by no means necessary, however, the robust and free 3D application Blender is available to anyone. Although not necessarily appropriate for all projects, a wide range of free 3D Blender models can be found on the Internet. In addition to using software to create components, almost everyone today has a phone that can take photographs, and of course the Internet is a vast resource of texts and imagery, which can be incorporated into projects under "fair use" laws. And as noted in the work of Susan Gold, even a photocopier can be expedient.

2.5b Pedagogical Implications of Collage

As mentioned at the beginning of this chapter, scholarship on the philosophy of collage is sorely lacking. The one stellar book, *Spectacle Pedagogy, Art, Politics and Visual Culture* (Garoian & Gaudelius, 2008), reviews the principles and implications of collage pedagogical and otherwise. It is limited however, by a narrow definition of collage. Notwithstanding that constraint, it is especially applicable in this section.

Considering its importance in twentieth century art, why has collage not received the kind of critical pedagogical attention that it deserves in the field of art education? What is the principle of collage? How is it constructed? What are its cognitive operations? What is the significance of its disjunctive form? What is the epistemology of collage and how does it function pedagogically? (Garoian & Gaudelius, 2008, p. 91).

But more importantly this is what they have to say about the value of teaching collage:

Because the postmodern condition is pervasively mediated by visual culture, our awareness of its dominating assumptions and our ability to expose, examine, and critique the spectacle of visual culture make the critical pedagogy of collage, montage, assemblage, installation and performance art all the more imperative. When students understand the critical and paradoxical relationships between their art-making activities and the habitus of institutionalized schooling, between the images and ideas that they create through art and the spectacle pedagogy of visual culture, then a liminal in-between space opens that enables the potential of art-making for transgressive and transformative experiences (Garoian & Gaudelius, 2008, p. 39).

In order to do this, students need to be within an educational environment that allows for "paradoxical space[s] where an oppositional tension exists" (p. 71).

For that paradox to exist, the discipline specific content of art should be juxtaposed with the other academic disciplines in schools and the images and ideas of visual culture so that the dialectical tension between them yields multiple critiques, interpretations, understandings, and applications (Garoian & Gaudelius, 2008, p. 71).

This point goes to the heart of *Knowledge Visualization and Visual Literacy in Science Education.* Despite the very real differences between the arts and the sciences, it is the arts, specifically collage that can capture and explore the complexity and contradictions of what science means for human beings.

Such boundary-breaking education is transdisciplinary because it enables unlikely creative and intellectual associations. It is transpersonal because it recognizes a diversity of learning abilities and allows for students' expressions and performances of subjectivity. It is transcultural because students' memories and cultural histories are recognized as significant content in the classroom and allowed to interplay and intersect with one another as they expose, examine, and critique the commodity fetishism of visual culture and the academic knowledge taught in schools. Hence the critical pedagogy of collage enables students to transgress and transform academic and institutionalized assumptions into new cultural understandings and representations (Garoian & Gaudelius, 2008, p. 72).

2.5c The Difference Between STEM and STEAM

There are consequential differences between the goals of STEM (science, technology, engineering, and mathematics) and those of STEAM (science, technology, engineering, arts, and mathematics). If the math is wrong then the bridge will fall down. If the art is "wrong" then the bridge will be ugly. These are not trivial distinctions. "Correct" art can transform a bridge from being utilitarian and disregarded into one that becomes a tourist attraction. These differences are relevant to the use of collage in knowledge visualization insofar as they are recognized as requiring disparate assessment criteria. Aesthetic beauty may be one criterion, but a special kind of "verity" can be another. There is not and should not be an "answer," much less a correct one. The complexity and stimulating uncertainty of response is the necessary outcome.

The images and ideas that are radically juxtaposed in these visual art genres constitute a disjunctive collage narrative that is 'apprehended' rather than 'comprehended' through a fugitive epistemological process in which the interconnectivity of its disparate understandings is indeterminate and resistant to synergy ... In the in-between spaces of the fragments of collage, where knowledge is mutable and undecidable, opportunities exist for creative and political intervention and production –a kind of educational research that exposes, examines, and critiques the academic knowledge of institutionalized schooling (Garoian & Gaudelius, 2008, pp. 91-92).

For those in disciplines with different expectations this will make assessment problematic. There may not and should not be one rationalized interpretation of students' work. The assessment should reflect the depth and range of the student's understanding, and of the questions raised within their work, not the falsifiability of "conclusions" drawn from that work.

2.5d Assessment

The method commonly used in the visual arts is the public critique. This can be problematic for other disciplines for at least three reasons. The first is that they are time consuming; the second is that if done well they are difficult; and the third is that there may not be institutional support for potentially atypical assessment methodologies; Ursyn (2015) addresses some of this. Multiple-choice exams are significantly faster, easier, and cheaper to score.

Assuming that the collage project will be critiqued, there are techniques that can encourage success. Too often a visual arts critique devolves into the student supplying a linear narrative: "this means this, which in turn is succeeded by that meaning that." The professor then responds either with unrelated pontification, or disagrees "that does not mean that," therefore the student must be mistaken or confused. Aside from the general ineffectiveness of these exercises in learning acquisition, such critiques would be particularly inappropriate for the types of collage projects described here. The objective ought to be the following: can the student articulate the specific meaning of components used, i.e. "this is an example of an electron "s" orbital," and can the student express the questions arising between juxtapositions found within the work. Looking for answers, which is commonplace in most disciplines would be useless here.

It is important for readers to recognize that incorporating any of the ideas or suggestions presented here in an educational setting require recognition of how easily it can all go wrong. Garoian and Gaudelius list several examples of bad student collage concluding with this:

Students' science projects often consist of didactic panels whose compositions are loaded with texts, images, and found materials to illustrate the elemental processes found in nature. ... Mathematics students' use images and texts from magazines and newspapers to visualize quantification and the logic of abstract equations. In each of these pedagogical instances, the understanding sought among the fragmented, disparate remnants of collage is a teleological one that conforms and tames its radical aesthetic to naturalized, academic, and logical outcomes. (Garoian & Gaudelius, 2008, p. 90).

Without an active critical approach on the part of students to the strategies employed in combining components, the full potential of collage will be lost to them. These possibilities include but are not limited to the real world implications of science and technology. For example, a collage on the biology of food production that leaves out genetically modified organisms would be incomplete. It is important to examine closely the complexities and sophisticated possibilities of collage. Otherwise, projects can too easily become haphazard agglomerations of stuff.

2.5e Possible Student Projects

As this chapter was being written, there was an article in the New York Times about the assemblage artist Sarah Sze, that captures perfectly the idea of visualizing science in a fine arts context. Sze: "also makes references to science – "You want to have all of the information, then you want to titrate it." The computer on the desk is connected to the NASA site that measures in real time the distance between Earth and the Voyager 1 spacecraft" (Pogrebin, 2015, p. C1). This is apt as the possibilities for student projects are explored. Keeping in mind the caveats just mentioned, the suggestions listed here must be approached with careful attention to the capacity of collage.

- **Chemistry:** An element from the periodic table could include many visual components: electron orbitals, molecules built from that element, isotopes, chemical reactions, photographs of the natural physical form of the element, different examples of how the element is used in science and technology, historic, artistic and folkloric representations. Additionally, the sociological impacts of the element should be considered: lithium and psychiatric applications; uranium and plutonium as nuclear waste; mercury poisoning in fish. Economic aspects should be recognized, pollution and social disruption from mining; the use of rare earth metals in computers, post-consumer waste. There are political implications to the geographic location of certain elements.
- **Biology of Food:** This can include the biological transition from wild grasses to cultivated grains. Additionally, fungal diseases such as wheat rust can be explored, along with the biology of antibiotic use in farm animals. This topic can range from the origins of domesticating livestock and cultivation of grains to the vast industrial farms of today. The former led to population centers, governments and written languages; the latter to the associated problems of monocultures, poisoned environments, loss of jobs, and patented corporately owned seed. The genetics of breeding plants and animals and genetically modified organisms can be explored. The "green revolution" in farming, which has extended arable land and led to decreases in hunger can be contrasted to improved global economies that have led to an increase in the consumption of meat, which is one cause of climate change.

The methods for visualizing some of these abstract concepts can be quite challenging. This is where the "art" comes in. For example, one could show a tomato with a fish inside of it to communicate the idea that GMO tomatoes have been created that include fish genes within them. The contested space between these two organisms can be provocative. Of course there are alternative methods of visualizing the same idea. One could include an image of a genome and symbolize the difference between organisms with graphic methods such as color or design styles. I'm sure the reader can imagine even more approaches.

These two examples from quite different realms of science should be enough to suggest the extent of possible uses for collage in science visualization. There are advantages to creating science informed collages by students in the arts *and* by students in technical fields. Those in the arts may discover inspiration that may have gone undeveloped otherwise. Too often young art students really have nothing to say. Their lives aren't all that differentiated or interesting and their artwork can be indistinguishable. However, having to research topics outside of their realm of comfortable expertise is both challenging and engaging. For technical students (and their instructors) the challenge is how to conceptualize hard data that they are comfortable with in totally different forms – some of which they may not even consider to be legitimate. Taking advantage of the inherent nature of collage, students will link concepts from distant epistemological worlds or even opposing tenets. This can teach a flexibility of thought that might be missing in more explicit disciplines. They will learn to critically assess those relationships, which can, and should extend beyond the specific scientific topic into its sociological implications.

As remarked earlier, it should be clear that digital software is an ideal tool nowadays for students to exploit the many visual forms within which such knowledge can be found. They can access still and moving images and texts from the internet, scan images from print resources, draw and paint their own images on a computer, and easily make their own videos of real elements for inclusion in their end product. This end product can take the form of a digital collaged image or video. However, it is worth mentioning that one valuable form their work can take is that of the artists' book. As observed, artists' books are uniquely flexible in that they retain aspects of both static and durational presentation. Additionally, although perhaps seeming somewhat outdated in the virtual world of many students, this author believes that there is something to be gained from crafting a tangible object. Students will experience visual space differently when manipulating objects in physical space. It is easy to print out imagery, text and compositions that have been created digitally, however, artists' books can also include authentic elements taken from the real world. Much of the Theory section of this chapter has explained that such elements retain a quality that refers to reality in a way that is missing from virtual objects.

CONCLUSION

Successful collages must operate at both the visual and the conceptual level. For a student to create a collage based on their studies in, for example, physics or biology, they must not only understand how to visualize a concept, they must understand what connects, *or separates*, two or more realities. This link is not necessarily analogous –collage allows for the juxtaposition of antagonistic concepts. It is not necessarily logical – collage allows for metaphoric poetic relationships. And it is not necessarily linear – collage allows for networks that span time and space. But these associations are no less valid, no less scientific – the history of science is filled with metaphoric leaps that led to significant discoveries and insights.

The implications of understanding relationships in this way are many and can be applied to seemingly distant phenomena. I would suggest that extractive cultures such as industrialized western societies view the world as isolated unrelated components that can be harmlessly removed, taken out of context, and recombined as temporary consumer goods destined for the trash. This neglected understanding of the basic truths of ecology obviously results from the adverse view that the lives and materials of the world are isolated –not a coherent collage of elements but an agglomeration of stuff. According to the ecopsychologist Laura Sewall,

In The Voice of the Earth, Theodore Roszak presents a provocative theory that the roots of our collective [ecological] misbehavior can be found in the historic and conceptual split between 'in-here' and 'out-there.' This dichotomy manifests as the large and despairing gap we feel between ourselves and nonhuman nature. In response, Deep Ecology and progressive psychology have begun to flesh out a conception of an ecological self, in which the division between inner and outer worlds becomes an arbitrary and historical distinction. In contradiction to an identity in which the mature self is culturally defined as fully individuated and possessing intact, absolute, decisive, and divisive boundaries, the ecological self experiences a permeability and fluidity of boundaries (Sewall, 1995, p. 202).

One should easily see the parallels between this quote and the description of collage strategies presented here. This fluidity of the border between what is collage and what is not collage, and our perception of it within both art and reality poses interesting questions. When we experience the world with our senses, we are aware of only small bits of the data that is present, and we focus on even less. As we go through our day we may choose to examine something closely, paying attention to ever more levels of detail. Additionally, as we experience the world sequentially our attention may jump from one thing to another; from one sense to another; and our mind retrospectively ties these events together into a seamless experience. We don't generally remember our day as numerous discontinuous jumps of focus. Cinematic montage, which every student is at least subliminally aware of, organizes these jumps of focus. All of these strategies of attention are collage strategies. Becoming aware of this transforms not only the way we understand the world but the way we understand our role within the world.

If the twentieth century has taught us anything, it is that fragmentation, reassembly, and hybridity are the life forces of a vibrant culture; calls for purity and a return to wholeness are naïve and dangerous. Collage is our natural environment. Once it is understood that we are hybrid culturally, psychologically, esthetically, in our foods, traditions, beliefs, and art, it becomes possible to understand relations as reciprocal. Meaning and identity are constructed vis à vis, in the relationship between things. The impulse toward collage is no longer one of fracture and violence but one of vitality, recognition, and temporary resolution (Durant, 2002, p. 28).

Collage artists in all media teach us to create realities that wouldn't otherwise exist.

REFERENCES

American Heritage Dictionary. (1969). Boston, PA: Houghton Mifflin.

Bazin, A. (2004). *What is cinema?* (2nd ed.; Vol. 1). Oakland, CA: University of California Press.

Bogue, R. (2003). *Deleuze on cinema*. New York, NY: Routledge.

Callahan, P. (1992). *Nature's silent music: A rucksack naturalist's Ireland*. Kansas City, MO: Acres U.S.A.

Callahan, P. (1997). *The Enigma of the Towers*. Retrieved September 28, 2015, from http://whale.to/b/callahan.html

Connolly, M. (2009). *The place of artists' cinema: Space, site and screen*. Bristol, UK: Intellect.

88 . constellations for Wittgenstein (to be played with the left hand). (n.d.). In *Electronic Literature Collection Volume Two online*. Retrieved from http://collection.eliterature.org/2/works/clark_wittgenstein.html

Deleuze, G. (2013). *Cinema 1 the movement-image*. New York, NY: Bloomsbury Academic.

Drucker, J. (1995). *The century of artists' books*. New York, NY: Granary Books.

Durant, M. A. (2002). Some assembly required: Ten fragments toward a picture of collage. In T. Piché (Ed.), *Some assembly required: Collage culture in post-war America* (pp. 19–28). Syracuse, NY: Everson Museum of Art.

Eisenstein, S. (1949). *Film form; essays in film theory* (J. Leyda, Trans.). New York, NY: Harcourt, Brace.

Eisenstein, S. (2004). The dramaturgy of film form. In L. Braudy & M. Cohen (Eds.), *Film theory and criticism: Introductory readings* (6th ed.; pp. 23–40). New York, NY: Oxford University Press.

Fergonzi, F. (2007). Episodes in the critical debate on collage, from Apollinaire to Greenberg. In M. M. Laberti & M. G. Messina (Eds.), *Collage/Collages from cubism to new dada* (pp. 322–335). Milan, Italy: Mondadori Electa.

Fineman, M. (2012). *Faking it: Manipulated photography before Photoshop*. New York, NY: Metropolitan Museum of Art.

Gap. In (1969). *The American Heritage Dictionary of the English Language*. New York, NY: American Heritage Publishing.

Garoian, C., & Gaudelius, Y. (2008). *Spectacle pedagogy, art, politics and visual culture*. Albany, NY: State University of New York Press.

Gilot, F., & Lake, C. (1964). *Life with Picasso*. New York, NY: McGraw-Hill.

Gunning, T. (2004). What's the point of an index? or, faking photographs. *Nordicom Review, 1-2,* 39-49.

Hall, D., & Fifer, S. J. (1990). Illuminating video: An essential guide to video art. New York, NY: Aperture in association with the Bay Area Video Coalition.

Higgins, D. (1996). A book. In J. Rothenberg & D. Guss (Eds.), *the book, spiritual instrument* (pp. 102–104). New York, NY: Granary Books.

Huelsenbeck, R. (1981). Collective dada manifesto. In R. Motherwell (Ed.), The dada painters and poets: An anthology (2nd ed.; pp. 242-245). Cambridge, MA: The Belknap Press of Harvard University Press. (Reprinted and translated from Dada almanach, 1920)

Janis, H., & Blesh, R. (1967). *Collage: Personalities, concepts, techniques* (Rev. ed.). Philadelphia, PA: Chilton Book.

Koerner, L. (1996). Carl Linnaeus in his time and place. In N. Jardine, J. A. Secord, & E. C. Spary (Eds.), *Cultures of natural history* (pp. 145–162). Cambridge, UK: Cambridge University Press.

Kohler, T. (2012). Manifesto collage: Collage in arts and sciences. In C. Salm (Ed.), *Manifesto collage: Defining collage in the twenty-first century* (pp. 9–10). Nürnberg, Germany: Verlag für Moderne Kunst.

Kosara, R. (2007). Visualization criticism – The missing link between information visualization and art, *Proceedings of the 11th International Conference on Information Visualisation (IV)*, (pp. 631–636). doi:10.1109/IV.2007.130

Kuhn, T. (1970). *The structure of scientific revolutions* (2nd ed.). Chicago, IL: University of Chicago Press.

Landow, G. (1999). Hypertext as collage-writing. In P. Lunenfeld (Ed.), *The digital dialectic: New essays on new media* (pp. 150–171). Cambridge, MA: MIT Press.

Leefmans, B. M.-P. (1983). Das undbild: A metaphysics of collage. In J. P. Plottel (Ed.), *Collage* (pp. 189–228). New York, NY: New York Literary Forum.

Leighten, P. (1985). Picasso's collages and the threat of war, 1912-13. *The Art Bulletin*, *67*(4), 653–672. doi:10.1080/00043079.1985.10788297

Lippard, L. (1995). *The pink glass swan: Selected essays on feminist art*. New York: New Press.

Makela, M. (1996). *The Photomontages of Hannah Höch*. Minneapolis, MN: Walker Art Center.

Manovich, L. (2002). *The language of new media*. Cambridge, MA: MIT Press.

McHale, B. (1991). *Postmodernist fiction*. London, UK: Routledge.

Miles, B. (2012). The future leaks out: A very magical and highly charged interlude. In C. Fallows & S. Genzmer (Eds.), *Cut-ups, cut-ins, cut-outs: The art of William S. Burroughs* (pp. 22–31). Nürnberg, Germany: Verlag für Moderne Kunst.

Mitry, J. (1997). *The aesthetics and psychology of the cinema*. Bloomington, IN: Indiana University Press.

Palimpsest. In (1969). *The American Heritage Dictionary of the English Language*. New York, NY: American Heritage Publishing.

Perkowitz, S. (2011). *Slow light invisibility, teleportation and other mysteries of light*. London, UK: Imperial College Press.

Phillpot, C. (1987). Some contemporary artists and their books. In J. Lyons (Ed.), *Artists' books: A critical anthology and sourcebook* (pp. 97–132). Rochester, NY: Visual Studies Workshop Press.

Poggi, C. (1992). *In defiance of painting: Cubism, futurism, and the invention of collage*. New Haven, CT: Yale University Press.

Pogrebin, R. (2015, August 23). Sarah Sze aims for precise randomness in installing her gallery show. *The New York Times*, p. C1.

Raaberg, G. (1998). Beyond fragmentation: Collage as feminist strategy in the arts. *Mosaic: A Journal for the Interdisciplinary Study of Literature, 31*(3), 153-171.

Robbe-Grillet, A. (1962). *Alain Robbe-Grillet's last year at Marienbad: a ciné-novel* (R. Howard, Trans.). London, UK: John Calder.

Rohdie, S. (2006). *Montage*. Manchester, UK: Manchester University Press.

Rosenberg, H. (1989). Collage: Philosophy of put-togethers. In K. Hoffman (Ed.), *Collage: Critical views* (pp. 59–66). Ann Arbor, MI: UMI Research Press.

Schulz, I. (2010). Kurt Schwitters: Color and collage. In I. Schulz (Ed.), *Kurt Schwitters: Color and collage* (pp. 51–63). Houston, TX: Menil Collection.

Seam. In (1969). *The American Heritage Dictionary of the English Language*. New York, NY: American Heritage Publishing.

Seitz, W. (1961). *The art of assemblage: Catalogue of an exhibition held at the Museum of Modern Art, New York.* New York, NY: Museum of Modern Art.

Sewall, L. (1995). The skill of ecological perception. In T. Roszak, M. E. Gomes, & A. D. Kanner (Eds.), *Ecopsychology: Restoring the earth, healing the mind* (pp. 201–215). Berkeley, CA: Counterpoint Press.

Shattuck. (1992). The art of assemblage: A symposium (1961). In J. Elderfield (Ed.), *Studies in modern art, no. 2: Essays on assemblage* (pp. 118-159). New York, NY: Museum of Modern Art.

Snow, C. (1993). The two cultures (Canto ed.). London, UK: Cambridge University Press.

Spies, W. (1982). *Focus on art*. New York, NY: Rizzoli.

Taylor, B. (2004). *Collage: The making of modern art*. London, UK: Thames & Hudson.

Thomsen, D. (1985). Shadow matter. *Science News, 127*(19), 296. doi:10.2307/3969495

Tufte, E. (1983). *The visual display of quantitative information*. Cheshire, CT: Graphics Press.

Tufte, E. (1990). *Envisioning Information*. Cheshire, CT: Graphics Press.

Tufte, E. (1997). *Visual explanations: Images and quantities, evidence and narrative*. Cheshire, CT: Graphics Press.

Ulmer, G. L. (1983). The object of post-criticism. In H. Foster (Ed.), *The Anti-aesthetic: Essays on postmodern culture* (pp. 83–110). Port Townsend, WA: Bay Press.

Ursyn, A. (1997). Computer art graphics integration of art and science. *Learning and Instruction, the Journal of the European Association for Research on Learning and Instruction, 7*(1), 65-87.

Ursyn, A. (2015). Cognitive Learning with Electronic Media and Social Networking. In A. Ursyn (Ed.), *Handbook of Research on Maximizing Cognitive Learning through Knowledge Visualization* (pp. 1–71). Hershey, PA: IGI Global. doi:10.4018/978-1-4666-8142-2.ch001

Ursyn, A., & Sung, R. (2007). Learning science with art.*Proceeding, SIGGRAPH '07 ACM/SIGGRAPH 2007 educators program.*

Waldman, D. (1992). *Collage, assemblage, and the found object*. New York, NY: Abrams.

Wescher, H. (1971). *Collage*. New York, NY: Abrams.

Wolfram, E. (1975). *History of collage: An anthology of collage, assemblage and event structures*. New York, NY: Macmillan.

ADDITIONAL READING

Garbagna, C. (2007). *Collage/collages: From cubism to new dada*. Milan, Italy: Electa.

Teitelbaum, M., & Lavin, M. (1992). *Montage and modern life: 1919 - 1942*. Cambridge, MA: MIT Press.

KEY TERMS AND DEFINITIONS

Collage: The gluing of elements historically considered to be outside of the realm of painting, onto paintings or simply onto a flat surface. Assemblage extends this into three dimensions; photomontage consists of multiple images combined in a single photograph, and montage refers to the cuts between film clips in motion pictures. Collage will stand in comprehensively for these and the multitude of other "–ages" that exist, in addition to other words such as cut-up, mashup, etc. In many cases installation art and performance art can also be a collage.

Collage Taxonomy: A heuristic method for analyzing artwork via the combinations of two descriptive aspects of collage, the gap and the seam. 1) + GAP and + SEAM, Cubist type collages and their descendants. Imagery that concentrates on formal visual complexity, spatial configuration, and ambiguity; on illusion versus reality; and on the inclusion of outside elements for purposes of visual and conceptual juxtaposition. Art that draws attention to the gaps between materials, images and concepts; and art that makes both the metaphoric and the literal seams of the collage explicit as part of its visual strategy. This would also include constructional montage in film; 2) – GAP and + SEAM, Futurist type collages and other entirely different, sometimes feminist collages, where outside elements are used in such a way that the contrast between them and the rest of the structure is minimized, synthesized or unified, both visual and conceptually. That is the gap between elements is minimized in order to communicate transcendent or socio-political harmony. However, each element retains its identity, and thus the seams – either metaphoric or literal – can be seen. This would also include lyrical montage in film; 3) + GAP and – SEAM, Surrealist type collages and their descendants. This imagery may or may not consist of actual outside elements, as the focus is on the conceptual disjunction of components, and surprising and unpredictable combinations. The gaps between these components may or may not reflect material differences, but always reflect conceptual ones. However, in many cases the seams between elements are minimized or completely absent in order to create an overall formal visual unity. This would also include intellectual montage in film; and 4) – GAP and – SEAM: Mixed media structures where the goal is formal and conceptual unity. Attention is not drawn to the variety of sources, elements, materials or their relationships to the outside world, or their juxtapositions with one another. Even if the elements are, strictly speaking, from different cognitive realms they are put together in such a way that the context reflects external and/or internal coherence, and juxtapositions yield no surprises. There's no discontinuity, ontologically or formally. This would also include narrative montage in film.

Compositing: Combining visual elements from different sources of origin into a single image or film sequence.

Contested Space: When two elements either in time or space are juxtaposed there is a conflict as the first one to be perceived influences one's psychological anticipation of meaning, which is then altered by the influence of subsequently perceived elements.

Chapter 8
How We Hear and Experience Music:
A Bootstrap Theory of Sensory Perception

Robert C. Ehle
University of Northern Colorado, USA

ABSTRACT

This chapter examines occurrences and events associated with the experience of composing, playing, or listening to music. First it examines virtual music, and then recounts an experiment on the nature of pitch and psychoacoustics of resultant tones. The final part discusses the prenatal origins of musical emotion as the case for fetal imprinting.

INTRODUCTION

It is typical of us to assume that when we perceive things with our senses, we just take in things as they are, and we understand them using their nature as a basis. We just perceive things as they are and then we work with them, and lead our lives with them. The problem with this concept is that the nature of the things we perceive never reaches our brains. Our senses convert incoming sensations into neural impulses and then the neural impulses carry information about the world to our brains. There are various codes that stand for the characteristics of things, and the neurons are set up to detect things and then send information to our brains by means of these various codes. Neural impulses that travel up to our brains resemble the pulses that travel around in computers. They are not the same, however, because they are not digital in the sense that they do not code for numbers. They code for various things: edges, shapes, colors, pitch, loudness, saltiness, etc.

Learning before birth and also immediately after birth is traditionally called Imprinting, to distinguish it from the intellectual type of learning that will take place years later. Imprinting has been extensively studied in animals and birds and has been extensively documented. Konrad Lorenz (1937), probably the best-known researcher on imprinting, defined imprinting in his classic studies on graylag geese and other animals as the rapid learning occurring in early stales of life. Obviously, animals, and humans too,

DOI: 10.4018/978-1-5225-0480-1.ch008

are capable of learning some things around the time of birth and before. This type of learning is usually said to be subcortical because it takes place in lower parts of the brain than the cerebral cortex, which is rather undeveloped at this stage of life.

This chapter tells about the ways we experience music, how our brain perceives pitch, and discusses the role of its early development in the perinatal period.

BACKGROUND

At the end of the 19th century, Ernst Haeckel summarized the biogenetic law in a phrase, 'ontology recapitulates phylogeny' and posed that the fetal growth and development goes through stages resembling the stages of animals in evolutionary history. In his 1871 book, *The Descent of Man* Charles Darwin (2004) proposed a view, confirmed later by the evolutionary developmental biology, that early embryonic stages resemble embryonic stages of previous species but not the adult stages of these species. Charles Darwin's theory has been extensively discussed and attacked.

In 1960s Paul McLean (1990) formulated the triune brain model of the vertebrate forebrain evolution and behavior. According to then acclaimed model, the triune brain comprises three sequentially evolved structures. First was the primitive reptilian complex, and then followed the paleomammalian complex including the limbic system consisting of a number of separate components: the thalamus, the hypothalamus, the hippocampus, and the amygdala, among them. Further on, the neomammalian complex (neocortex) was added to the forebrain. The parts of Maclean's triune brain develop sequentially in the human fetus or child; specifically before birth, a human being does not yet have a developed cerebral cortex but is able to learn general sensory and emotional things in their limbic system. The limbic system along with the sense organs is said to have begun to function in the third trimester of gestation and so is available to do this task. Jaak Panksepp and Lucy Biven (2012) presented in *The Archaeology of Mind* the neural mechanisms of affective expression in mental processes, brain functions, and emotional behaviors characteristic of all mammals to locate.

There were controversies related to the theories about the prenatal imprinting, with a great number of research conducted, mostly at the beginning of this century. Imprinting is primarily determined by basal forebrain structures (including the cerebral hemispheres, the thalamus, and the hypothalamus), to which the hypothalamus is integral (Keverne, 2015). Developmental changes occur in neocortical (concerned with sight and hearing) forebrain. After birth, development of the neocortex lets free the child's behavior from hormonal mechanisms and the dependence on pheromonal cues. Nicolaïdis (2008) examined the role of under- or over-nutrition in the pregnant mother in postnatal regulation of feeding preferences and fat reserves in offspring, predisposing the offspring, in case of over-nutrition, to later development of obesity and the type 2 diabetes mellitus. Merlot, Couret, & Otten (2007) suggested that prenatal and early life events, such as stress experienced by the pregnant mother may cause future disorders of the child's immune system. Richard Parncutt (2011/1989) asserted in his *Music Theory, A Psychoacoustics Approach,* that a newborn child recognizes its mother's voice. Having extensive knowledge of sheep, Parncutt described how a newborn lamb could identify its mother's bleat, which seems to be a trait common in mammals.

At the TED Global 2011 Conference (TED – Technology, Entertainment, Design: Ideas worth spreading is a global set of conferences), Annie Murphy Paul (2011) delivered a talk, *What babies learn before they are born*, where she provided many cases of prenatal imprinting including the imprinting of tones in the mother's voice associated with her native language. She gave documented examples where a newborn child shown a preference for tones used by French speakers over German speakers, even though the child did not known French or German. A newborn child has yet no intellectual knowledge, simply because the cerebral cortex is undeveloped at that age. The implication was made that a lower portion of the brain has done the learning.

In his book *Why You Hear What You Hear: An Experiential Approach to Sound, Music, and Psycho-acoustics* Eric Heller (2012) refers to the sounds we hear as "executive summaries." I take this to mean that they are illusions created by the brain but they are related to experience in a convoluted way. Other research has also suggested that we hear what we expect to hear, or that the brain provides something for us to hear that conveys a maximum amount of useful information (in a survival context) for us to make successful choices in dangerous situations. As such, it would be of maximum usefulness if the brain were to provide well-known experiences to consciousness.

VIRTUAL MUSIC

Perhaps not everybody knows that in addition to the usual world of music that we study every day and teach in music theory classes, there is a second world of music, a virtual world. Like a reflection of a mountain range in a lake, the virtual world is generated from and depends upon the real world of music, yet it is separate, different, and affects us differently. It can be independently controlled and employed, too.

The world of virtual music is somewhat evident to us in acoustic phenomena. A composer and music theorist Georg Andreas Sorge, who was a fiend of Johann Sebastian Bach described first the psychoacoustic phenomenon of combination tones, which are perceived when two real tones are played at the same time. However, combination tones are usually ascribed to the famous 17th century violinist Giuseppe Tartini, and a manifestation of this phenomenon in difference tones have come to be called "Tartini Tones." The combined tones are usually called difference tones because their frequencies are the difference between two sounding notes.

As described by the Encyclopedia Britannica (2015), "Combination tones are heard when two pure tones (i.e., tones produced by simple harmonic sound waves having no overtones), differing in frequency by about 50 cycles per second or more, sound together at sufficient intensity." Two varieties of the combination tones are the difference tones whose frequencies result from the difference between the frequencies of real tones, and the sum tones whose frequencies can be found by adding the frequencies of two real tones. There has been some speculation about summation tones too but little confirming evidence. However, there is little confirming evidence about sum tones.

Periodicity pitches are the resultant tones that arise from the linear mixing of acoustic waves. They are phantom pitches in that they have no energy at their frequencies, but many writers claim that we can hear them, anyway. They are the result of additive and subtractive interference between pairs of sound waves. The equivalent in water waves may be seen when the wakes of two boats interact on the surface of a river. Where the wakes cross, new wave patterns will arise, relatively permanent and static.

Brass players and also didgeridoo players use difference tones to generate complex chords. They can play triads, even whole chorale chord progressions, or, in the case of the didgeridoo, the sounds of wild animals. The French 18th century composer Jean-Philippe Rameau used them to show the true roots of chords in inversion. Guitarists and other musicians use similar difference tones when they overdrive an amplifier producing distortion or when they leave out the root of a chord. Many players use beats, a difference tone phenomenon to play in tune.

Acoustical phenomena as listed above are some of the well-known parts of virtual music. The less known parts of this phenomenon relate to psychoacoustics, as they occurr in the inner ear and the brain. These events include virtual pitches (Terhardt, 1974), subjective tones (Lewis, & Larsen, 1927), missing fundamentals (Schnupp, Nelken, & King, 2011), Schouten's residue pitches (Schouten 1940; Schouten, Ritsma, & Cardozo, 1962), and others.

In the realm of the psychoacoustics, we have rootless voicings of jazz musicians. We also have quartal chords such as Aleksandr Nikolaevich Scriabin's *Mystic* chord, and Ferde Grofe's Painted Dessert music in the *Grand Canyon Suite* that seem to have virtual chord functions.

This entire arena has been somewhat known for centuries, but has been dismissed as the junk of music for most of that time. Recently though, experiments with the new medical technology machines (MRI, fMRI, PET, MGG, MEG, CAT, etc.) have implicated the presence of these phenomena in musical emotion. Showing that the amygdala, the brainstem, the cerebellum, and other lower parts of the brain light up in scans is an evidence of emotional response in the centers of emotion in the brain. The implication is that these parts of the brain are emitting neurotransmitters into the nervous system and thus changing states of feeling. Most of this takes place subliminally, subconsciously, and unknowingly to musicians.

Today, several European music theorists use the term Virtual Pitch for such phenomena, for example Ernst Terhard (1974) and Richard Parncutt (2011). Americans (John R. Pierce (1990), Max Mathews (Mathews, & Moore, 1970), John Backus (1969) seem to prefer the term Periodicity Pitch. I would suggest that there is a wide range of phenomena and experiences, some of which are acoustical and some psychoacoustical, yet the key component is the ability of these phenomena to modulate emotion. There is a deep emotional component in the music of some composers that results from these unwritten and unaccounted-for pitch phenomena, both acoustical and psychoacoustical.

The subjective emotional charges that some pitch collections and voicings are able to carry comes from primitive voice recognition functions in the brain. The brain has mechanisms for monitoring the moods of other people: major means calm, minor means sad or distraught, and dissonant (containing tritones and seconds) means angry. There is also a mechanism for memorizing people's voiceprints that we use every day on the telephone. We recognize people from their voices. This is similar to the mechanism that allows us to recognize musical instruments from their tone colors. The Pavlovian conditioning psychology can be referring to these identifications, so we experience a learned emotional response every time we hear certain voices or sounds.

Human beings or quasi-human beings have existed on this planet for hundreds of thousands or even millions of years without metal tools, living in similar conditions as the ancient Native Americans, the Yannomamo of Brazil, or the tribes of central Borneo. They could not modify their environment very much but had to learn to live within it. The sense of vision might be dominant, so they would deal with the things they saw, but the auditory sense was supportive, doing what is called auditory scene analysis Bregman, 1990). These people would be constantly analyzing their environment with the auditory sense looking for threats or opportunities. Thus, the auditory sense would become finely honed. Such people would be able to recognize friend or foe, family member, voracious animal, or other important things a mile away, from a wisp of sound carried on the wind. These things carried strong emotional meanings, because they indicated life or death matters. Such is the legacy we own and deal with. It is mostly out of place in our modern world where we have modified our environment to make it more comfortable and safer, and where danger can come so fast that you never see or hear it coming, but we were not designed for the modern world, we are products of the primitive world.

I have always thought and felt that concerts, recitals, performances, etc. are fine, but that the meaning of music is not contained in them. The meaning of music comes from primitive responses to primitive sounds that can be embodied in various musical compositions. Such music is powerfully meaningful, no matter how you hear it. By contrast, other music may lack meaning, no matter how it is performed or presented.

How to Play or Write Virtual Music

To be sure, this is mysterious stuff. This is a fertile territory for composers, arrangers, and improvisers, and it works spectacularly well on the piano. Here are some rules or at least suggestions:

1. Leave out the roots and most of the bass notes.
2. Use one tritone in some sonorities. A tritone implies 3 up to b7, or 8 up to #11, and can generate a subjective root in your inner ear.
3. Use one or two minor seconds in a chord. These imply #11 up to 12, or 6 up to b7 in a chord, and can generate subjective roots, too.
4. Use several whole steps in a row. These imply b7, 8, 9, 10, and #11, and will generate subjective roots.
5. Play or write for piano with the middle pedal depressed. On most pianos the lower strings will pick up resonances from the notes you play. (On a few concert grand pianos, capture a thirteenth chord in the key in which you are playing.) These resonances can provide bass notes and roots.
6. Practice imitating various natural sounds with your voice, on the piano, or in compositions.
7. Most of all, be aware that like the reflection of the mountain in the lake, there are notes being produced by these processes. If you write or play the root and/or bass, your played notes will dominate. It is only by leaving these notes out that you can hear the virtual ones provided by nature.

AN EXPERIMENT ON THE NATURE OF PITCH: PSYCHOACOUSTICS OF RESULTANT TONES

When two musicians perform together on stage (playing a flute and a clarinet, for instance) their notes may sometimes interact but usually they do not audibly interact. Brass instrumentalists frequently play one note, sing another, and complete the chord by the resultant tone. There are three places where tones might interact: in the instruments themselves, in the air or other transmission medium such as an audio system, and in the human auditory system.

Musical acoustician John Backus described three types of interactions: difference tones, subjective tones, and beats. For example, resultant tones (difference tones, subjective tones, and beats) are all a part of the sound of the Australian didgeridoo. Musical instruments produce an audible difference tone, which might be incorporated into the harmony. Jazz musicians often play rootless voicings in which the root or bass note is a subjective resultant tone. Sometimes two players will use beats to help them tune.

Physics and acoustics books frequently discuss interference. The two categories of interference are the additive and subtractive interference, and the resultant wave patterns are called interference patterns or interference waves. The wave interference patterns are frequently displayed as diagrams and photographs presenting water waves, light waves, as well as sound waves.

However, many aspects of the complex subject of interference remain controversial, and discussions that give frequencies and calculate difference or sum frequencies are rare. Studies on musical acoustics usually include frequencies because we hear these frequencies as musical tones and under some circumstances we can hear difference tones. This leads us to pay attention to the frequencies. There are two or more types of difference tones, one being f1-f2, and the second one being 2f1-f2, in both cases f1>f2. These are sometimes called quadratic and cubic difference tones. Sum tones (f1+f2) might be also included, although their existence has been debated in musical acoustics. Some musicians claim to be able to hear sum tones as the resultant tones.

How We Hear Pitch

Over the years, many people have speculated about how human beings and other animals hear pitch. Suggested answers have included genetic factors, harmonic series factors, learning theories, environmental concepts, and no doubt other approaches (Stevens, & Warshofsky, 1981). Speculations have included Fourier analysis by the Place process, autocorrelation analysis in the auditory system, time-based analysis, harmonic templates, volley processes, and many others. The 1961 Nobel Prize laureate Georg von Bekesy explored the Place process. He detected under the microscope an undulation sweep over the basilar membrane when a sound was introduced into the cochlea. Von Bekesy found that the high-frequency tones were perceived near the base of the cochlea and the lower frequencies toward the apex.

One may ask, what if the brain does not really hear pitch at all. The inner ear detects pitches, but immediately converts them into neural impulses that travel up the auditory nerve system to the primary auditory cortex (PAC) in the middle level of the brain. What if the PAC records not the pitches but the nerve activity on which the impulses arrive? The auditory nerve system is tonotopically organized, meaning that the basilar membrane in the cochlea in the inner ear always sends an impulse to the PAC on the same nerve when it receives a specific frequency. Thus, when the PAC receives an impulse on a particular nerve, that nerve represents a particular pitch even though no remnant of the pitch itself might ever be received in the brain.

It is helpful to think of the auditory system as a survival mechanism that evolved according to the Darwinian survival strategies. Its task is to recognize danger and safe havens. This is done by creating templates in the brain that correspond to the voice elements of the mother's voice, other family members voices, and familiar domestic sounds that represent safety for the young child. Also, it is necessary when encountering danger to create templates in the brain representing that danger so it can be avoided in the future. However, these templates might represent only the nerve on which a neural impulse was received. They are not necessarily harmonic templates, and the brain knows nothing of harmonics as such. If the young child is bombarded with harmonic sounds, the child will appear to have imprinted the harmonic series, but actually it only imprinted the neurons that were stimulated by sounds that had a harmonic series aspect to them.

Because, according to Ernst Haeckel, Charles Darwin, and others ontology recapitulates phylogeny, the development of the brain takes a course that parallels the evolution of previously existing brains: upward from the brainstem to the mid brain and finally to the upper or cerebral part of the brain. Thus, in a very young child the mid brain may be mature enough to function, while the upper or cerebral part of the brain is too immature to do so. Thus, the mid brain can imprint sounds heard before birth and immediately after birth by creating templates in the PAC. However, it cannot analyze the structures of sound patterns because it does not have that capacity. The analysis of sound patterns must await the cerebral development that makes it possible, and that initiates the natural language learning phase of human development. This takes place after a year or more and phases out the sound-imprinting phase that took place in the perinatal period.

The templates that are created represent the nerve patterns that are stimulated by each of these sounds, perhaps along with visual and other (tactile, odor, taste, etc.) sensations. It is important to note that the pitch need never be recorded or experienced. There are 30 to 50 thousand nerves in the auditory nerve system and so the frequency scale is divided into 30 to 50 thousand small steps. These represent the frequencies detected by the Place process in the inner ear but do not carry any vestige of pitch representation with them. In fact, by the time the nerve impulses reach the PAC, all vestige of pitch can be discarded and forgotten. The PAC looks at these nerve impulses solely as a pattern that represents danger or safety. If danger is encoded then the PAC sends messages of fight or flight to the appropriate brain centers. If safety is detected, the PAC sends messages of pleasure and contentment to other brain centers.

The consequences to this imprinting process are that we will always like and prefer sounds like the ones we heard in our mother's voice and our prenatal and perinatal environment and that these preferences will stay with us for life. Also, people living in a close-knit group will share certain vocal sounds that everyone will know in that area (accents, dialects, pidgins, Gullah language, etc.). Thus, there will be auditory communities that share certain vocal sounds. Everyone from another community will automatically be identified as a stranger and a potential source of danger. Note that the voices themselves are not recorded in the PAC, only the pattern of nerve endings sending impulses when a certain sound is heard is recorded. Because the PAC's operations are largely subconscious, we have little knowledge of this taking place. Note that this sound imprinting takes place in the middle brain and that the cerebral cortex is so undeveloped at this period as to be incapable of involvement. Also note that only isolated sounds are imprinted. The learning of sound structures will come later in the natural language learning phase, which this period precedes. For this reason, I like to refer to this phase as the phonological learning phase.

It should be apparent now that all sound is relative. Only the nerve patterns are recorded. Nature knows no number system, so the nerves are not numbered. The nerves are known only by their order from low to high. Nature has no need to record the sounds themselves, nor any need to know the specific pitches (or frequencies) involved. Nature's sole need is to establish a basis for safe conduct for the yet-to-be-born child. To do this, it needs a sound detection and analysis system, an emotional and behavioral control system, and a simple system of values (or weightings) to connect the detection system to the control systems.

The most mysterious part of the system is the illusion that we are actually hearing sounds. This illusion is generated in the PAC and is sent to consciousness, that means to the cerebral cortex. According to some recent research, we hear what we expect to hear! Thus, sounds that are outside of our life experience are not heard. This seems absolutely bizarre and impossible. It is most likely that colors, scents, tastes, and other world qualities are similarly synthesized for us in our brains!

This is the mysterious part of the process and no one seems to have much knowledge of how it works. We do know that our experiences of sense data do not correspond with the realities of the sensory data received. For example, the sounds of musical instruments are single gestalts that have very little to do with the harmonic construction or non-harmonic construction received by the ear. A violin is "warm," an oboe is "spicy." A clarinet is "hollow." Chili is "hot." To some extent, there is logic and a general agreement to this, but it also seems to reinforce the idea that we hear or experience what we expect to hear or experience. This suggests that we are unlikely to experience sensory patterns that are completely new. It is as if an agency in the brain looks at what we know and what we expect, and then gives us that thing as a sensation. The experiences we had in the perinatal period from several months before birth up to several years after being born are all likely contributors to this experience.

During the period when the nervous system is growing it is capable of configuring itself in conformity with incoming nervous system patterns. Once it is completed there is no possibility to do this anymore, and learning takes place through the acquisition of weightings in neural networks. Thus we reach the stage of learning that we all know and use. It is fundamentally different from the system of learning (imprinting) that takes place in the earliest months or years of life.

Otoacoustic emission (OAE) is a sound produced by the outer hair cells in the cochlea. Otoacoustic occur spontaneously or can be evoked by acoustic stimuli (Farlex, 2012) indicating that this capacity exists in the cochlea. Low-intensity sounds are transmitted through the middle ear apparatus to the ear canal (Glattke, & Kujawa, 1991). Research into otoacoustic emissions has identified a category called distortion product otoacoustic emission (DPOAE) that do this process of creating difference tones. Research results suggested that DPOAEs might serve as an objective indicator of frequency discrimination. The difference tones are generated, it is said, due to non-linearities in the alignment of frequency across the basilar membrane. Once they are generated they travel to their proper place on the basilar membrane where they might be heard, and then feedback through a process called the cochlear amplifier causes them to be regenerated by the outer hair cells and then they can be picked up by ultra-sensitive microphones placed in the auditory canal. Two major categories are quadratic difference tones: f2-f1, and cubic difference tones: 2f2-f1).

What all of this suggests is that if you were not exposed to difference tone effects as periodicity pitch waves in the perinatal years, you will be unable to hear them as an adult because consciousness has no experience on which to base its synthetic sensory experience. As a result, a percentage of the population will hear them and another percentage will not. If no templates have been created in the PAC for neural patterns representing a certain experience, then that experience does not exist for that person. Later in life, the synthesis process depends on pattern templates to create its version of reality. A good example is that of absolute pitch (or perfect pitch) – the rare ability to name or produce a note of given pitch in the absence of a reference note (Deutsch, 2013). This ability may be related to the critical periods in perceptual and cognitive development, the brain substrates of specialized abilities, and the role of genetic factors in perception and cognition. Later in life the conscious brain can create specific pitch experiences with names only if an earlier process imprinted neural patterns representing pitch effects with an unambiguous identity. That imprint does not have to be a musical one, but it must be strong and unambiguous as a result of having been reinforced a great many times in early life. Once that has been done, the brain will have acquired a neural pattern representing a specific pitch and then later, consciousness can use that template to create additional specific pitch experiences.

The current thinking is that you learn pitches from aural experience before or immediately after birth probably identified by the emotions accompanying them, and then these pitches acquire musical or other meanings later in life, mostly by association. In other words, there is nothing intellectual about acquiring perfect pitch because it happens before the intellectual portions of your brain develop. In later years the assigning of names to pitches can be intellectual and can be carried out in a complex way that introduces other elements: transposing instruments, movable clef signs, Baroque pitch and so on.

Sometimes this can become so complex that the perfect pitch breaks down. You may hear people say that playing Baroque violin killed their perfect pitch. This is because Baroque pitch is usually about a half step lower than modern pitch but it can also have different pitches altogether, and some Baroque players played higher than modern pitch. The violinist Franz Biber apparently did so, and he must have used a particularly short violin to get these higher pitches tuned on the open strings.

The Nature of Pitch: An Experiment

This experiment is derived from the musical experience but is not typical of physics. We are specifically interested in the musical results that we can hear. The experiment that I run in my laboratory at the University of Northern Colorado had to do with the nature of periodicity pitch and by extension, pitch in general. I connected three sin wave generators to a mixer, an oscilloscope, an amplifier, and a loudspeaker. First, I generated two frequencies and then three frequencies using the sin wave generators.

In the three-pitch experiment, I used the previous two frequencies and added the frequency E3 at 165 Hz. The objective was to see if a stable display on the oscilloscope could be achieved with each combination. I consider a stable display to be indicative of the existence of a common denominator to the frequencies. The common denominator is the sweep frequency to which the oscilloscope must synchronize, if a stable single-trace display is to be achieved.

First, in the two-frequency experiment I generated frequencies A3 at 220 Hz and D4 at 293.3333 Hz. The oscilloscope was adjusted to show a single trace synced to sub-multiple of the frequency 220 Hz. On the oscilloscope screen a wave was shown, which was the composite of the two waves: a 220 Hz wave with an amplitude change at the frequency of the largest common factor: 73.3333 Hz (D2). This is a periodicity pitch; it can be shown on the face of an oscilloscope, thereby proving that periodicity pitch has a physical basis and is not just an artifact of the hearing process.

I then played the tone pair into a spectrum analyzer (iSpectrum running on an Intel Macintosh computer). The original pair of tones shows up very nicely on the screen, but the periodicity pitch at the frequency of the largest common factor is not seen on the screen of the spectrum analyzer. I take this result to indicate the following: a periodicity pitch, as it is commonly called, is an acoustical phenomenon but it differs from the waves produced by oscillators (and musical instruments) in that there is no power at that frequency. The periodicity pitch is a wave phenomenon produced by the additive and the subtractive interference. It appears that any time two waves are linearly mixed in a common medium (like the air, electrical waves in a wire or air in the middle ear), the periodicity pitch will appear, and it can be displayed on an oscilloscope screen. An implied result is that when the two test frequencies are mixed in the perilymph of the inner ear, the periodicity pitch interference frequency will also appear, and that it might be audible, at least to a listener who has been trained to hear it.

For the three-pitch experiment, I added a third oscillator running at the frequency of E3. This three-oscillator group produces what is commonly called a quartal chord of E3, A3, and D4. The smallest common factor is C minus 1 at about 8 Hz. A slightly higher common factor is D minus 1 at about 9 Hz. This is a common chord existing in the 20th century classical music, and it is a widely used left hand jazz piano chord in modern jazz. I wanted to hear if this chord would be heard as a D chord or a C chord. What I perceived was that I could hear it either way depending on context. I think that this is a common experience in hearing periodicity pitches. They can be heard, but they can be heard in different ways depending on context.

If C is the root, D is the ninth of the chord, A is the thirteenth, and E is the third. If D is the root, E is the ninth, A is the fifth, and D is an octave; of course, all at multiple octaves up. But I really wanted to see if I could get a stable single-trace oscilloscope display for this 3-frequency-quartal chord. I found that I could do so with a sweep frequency of around 18 Hz. Because my oscilloscope has a lowest sweep frequency of 14 Hz, I must conclude that I was sweeping the wave at twice the fundamental frequency. The display did stabilize but did not show a simple waveform. By adjusting each of the frequencies very carefully I could get an unchanging, stable single-trace display indication of a common factor on which the oscilloscope could synchronize.

My point in running these experiments has been twofold: first, I wished to show that periodicity pitches are a natural acoustical phenomenon, not just a psychological one; and second, I wished to show that there was no power present in the periodicity pitches, but that they were produced by a natural additive and subtractive wave interference process. I am speculating that if such process takes place in the air or electrically in a wire, then it will occur naturally in the perilymph of the cochlea as well, and that it might be perceived, at least with the proper training or experience. The research published in the Heidelberg University article suggests that some people do hear such periodicities and others do not, hearing instead, the present frequencies separately.

Conclusion and Recommendation

If my hypothesis that the only pitch receiving process is the Place process occurring in the cochlea, then it is necessary to consider the various pitch phenomena that have been reported, to see if they can be performed by the Place process in the cochlea. Such phenomena include periodicity pitch, phantom tones, residue pitch, some sort of autocorrelation-like sorting of pitches into more and less commonly recurring categories, missing fundamentals, rippled noise and pitch sensations from noises such as staircase pitch, and fast echo effect pitches. The cochlea is a transducer like no other and its full capabilities are not known. I think it is fair to say that there is a good chance that the cochlea can produce all of these effects without invoking any additional nervous system or brain system of processing. A point to be made is that while all sounds differ from each other in small or large ways, all sin waves are, by definition, the same.

Again, all sin waves having the same frequency and amplitude are exactly the same. Thus a neural impulse representing that frequency and amplitude carries all the information that is necessary for a complete description of that sin wave. If all sounds can be represented by sin waves, the neural patterns are capable of carrying all the required information for all sounds. The auditory neurons are said to have a refractory period (recovery time after firing) of around 3 milliseconds before they can fire again. Thus they cannot fire on every period of common pitches for sin wave frequencies above about 350 Hz. An advantage to vibrato, tremolo, reverberation, early reflections in the reverberant field, echo, pitch spreading (as in violin jitter), chorus effect, and other effects is that they can recruit more neurons and thus produce a stronger sensation because more neurons will be coaxed to fire.

A critical stage in the processing is the early childhood (the perinatal period) stage of imprinting neural impulse patterns in the PAC. If the child should be extensively exposed to harmonic sounds at this time, the impulse patterns that get reinforced will be harmonic and it will appear that the person has harmonic templates. If the reinforced experiences are specific pitches (as with native Mandarin Chinese speakers) then the templates will appear to be of specific pitches. If the reinforced experiences are a specific type of intonation, specific noises, specific linguistic intonations, or any other sonic phenomena, then it will appear that the person has acquired templates for that phenomenon. In actuality, according to my theory, all that has been recorded by the memory section of the PAC is that the nerve-ending pattern of the auditory nerve from the cochlea has been recorded and reinforced. Thus it is totally non-sound specific and may be redirected to other sonic activities at any time.

It is widely agreed that different people hear sound and particularly music in different ways and that they have different preferences. This research suggests that the executive summaries of sounds that we hear are based on early experiences that are replicated in current experience. In other words, if one was not exposed to a certain sound or sound producing process in early life (the perinatal period) the executive summary process will have no way of synthesizing it later in life. There will be no basis for it. Specifically in the context of this chapter, if one was not exposed repeatedly to periodicity-pitch producing sonorities early in life, one will have no mechanism for the executive summary process to create them for you to hear later in life. The same applies to absolute pitch, complex harmony (like quartal chords) and even the experience of pitch, itself. One has to have created neural templates in the PAC that can represent an auditory phenomenon or one will be unable to perceive it later in life. The executive summary process will have nothing to work with. Most likely, this applies to all percepts.

Since there are two classes of acoustic waves: powered waves and resultant tones that have no power at their characteristic frequency, then perhaps other types of wave energy would exhibit this duality as well: water wave, light waves, and esoteric forms of waves such as quantum wave or gravity wave. Researchers should look for non-powered resultant waves of these types.

PRENATAL ORIGINS OF MUSICAL EMOTION: THE CASE FOR FETAL IMPRINTING

We typically associate emotion with things or events; we say, "the boy was sad because his toy was lost" or "the girl was sad because her mother left." It is almost as if we cannot imagine disembodied emotion that lacks some object in things or experiences. We do recognize disembodied emotion occasionally as melancholy or joy, but when pressed we will ascribe it to some object such as a rainy day or a sunny sky.

It can be assumed that the centers of emotion in the brain are prepared long before birth, and we do experience disembodied emotions that we get from our mothers. Before birth, we are able to pick up mother's emotions from her voice, her movements, perhaps through hormones and neurotransmitters. In this way we experience the emotions of life vicariously, before we have had a chance to live it. This is a way to get us prepared for the experiences we will eventually have.

These emotions may eventually be employed by artists to portray all the feelings of life, even those the artist has not experienced. This can apply to all aspects of aesthetic expression including such things as the taste of various foods, the sounds of musical passages, the aesthetic attributes of various colors, and even feelings such as faith and hope. This is why it is sometimes said that an artist can create a work that allows us to experience all the emotions of life, even those we have never experienced in life. We can experience emotions vicariously, which is seemingly a very mysterious thing.

Strange though it may seem, we have these things imprinted from our mother, who had them imprinted from her mother. Thus they are passed on for generations to make up the fabric of culture and life as it is experienced. In this way there is continuity in experiences and feelings passed on from generation to generation.

However, this mechanism can break down in a culture that is a fusion of many cultural traditions. For example, a child can be imprinted with a traditional culture, yet after birth can be faced with one or more other cultures that present new emotional experiences. The child can reach out to these cultures and old cultural models can be lost, at least temporarily. I would say that this particularly describes the American experience with its cultural fusion.

The place where I live, Greeley, Colorado is one of the new places on the planet. Two hundred years ago a thinly distributed Native American nomadic population occupied this region. Since then it has seen an influx of people from all over the globe. The primary language is English, but the population contains many Hispanics, Native Americans, Europeans, Asians, South Americans, and other people from everywhere. This set of circumstances represents a cultural fusion that characterizes our time and particularly newly settled areas like Greeley, Colorado.

The fascinating question is why there are so many different musical styles in the world, while every person shows a general fondness for the music of their homeland that they heard in their youth. One may assume that the prenatal fetal imprinting takes place before birth, so the unborn child imprints the sounds of the mother's voice along with her emotions, and this becomes the bedrock of one's musical taste later in life.

A BOOTSTRAP THEORY OF SENSORY PERCEPTION

When the brain receives these impulses, the meanings of various codes are only understood with respect to the built-in interpretation systems. Many of these systems are genetically programmed. My theory, which I call a bootstrap system of sensory perception, is that in the perinatal period, which stretches from several months before birth until several months after birth, the subcortical brain records sensory perceptions along with the mother's emotions that accompany them. The emotions are recorded directly from the mother's body (through the action of neurotransmitters, hormones, and physical behavior plus the mother's emotions) before birth and by a close physical connection after birth. These sensations are stored in a subcortical part of the brain and are available throughout the entire lifespan of the individual. It is believed that the stored emotional responses to sensory inputs become a bootstrap, which the individual will use to understand his or her perceptions of the world throughout their entire life. They become a bootstrap by which the world is understood.

We start recording safety and danger messages at least three months before we are born and an undetermined period after birth, but probably no more than a year, and use this mechanism to encode our mother's voice. We also encode our mother's emotions in conjunction with her voice so that we will be able to detect our mother's responses to our actions when we enter the world. The pleasure we receive from music is simply the pleasure we receive from safe sounds like the ones we heard in our mother's voice. The emotions are built in to our brain and nervous system and have nothing to do with specific pitches or sounds, but rather are simply the patterns (templates) of nerve impulses received. It is important that there be pre-natal imprinting because at that stage our mother's voice and emotions are separate from every one else's and we take advantage of that fact.

Many popular concepts may be explained by the bootstrap theory, such as the concept that the color red is a warm color and the color blue is a cool color. It is believed it may come about because before birth the color red is the most common color that would be perceived. It filters through the mother's body and the unborn child would receive the mother's emotions connected with light and heat, hence they would be perceived as being warm. The color blue, on the other hand would for the most part not be perceived until after child is born; it would carry little mother's emotion, hence it would be perceived as cool. However, brain neurons that process vision develop in mammals after birth, when they receive signals from the eyes, which was confirmed in classic research conducted by the Nobel Prize laureates Torsten Wiesel and David Hubel (1963).

Another example may be the perception of the major keys and chords as happy and the minor keys and chords as sad. The lowest harmonics in the harmonic series are major and higher ones are minor (5, 6, 7, 8, and 9). When the mother's voice is at rest and happy, it stresses the lowest harmonics of the vocal tract; hence happy gets imprinted with major and perfect intervals. But when the mother is sad, she raises her voice stressing the higher harmonics in her vocal tract. Hence sad gets imprinted with minor intervals. People intuitively know all of this and, although it is mostly unconscious, they practice using it during all their lives. Nursery rhymes and baby talk, for example, are designed to exercise this pitch identification system.

Food preferences and food (tastes) to be avoided are conveyed to the young child by this means (see Annie Murphy Paul's TED video, *What We Learn Before We Are Born*). Faith, FEAR, hope, and numerous other feelings are conveyed in the same way, being recorded before birth in the subcortical parts of the brain (before the neocortex is mature enough to do anything) and then they are available throughout life as a bootstrap system by which we understand the incoming sensations. We learn to understand the world with reference to our mother's feelings.

In this way we learn our mother's perceptions directly, but we also learn the standards of our culture that are transmitted from generation to generation by this means. Later in life we may maintain our culture's transmitted ideals or we may rebel against them, but we always know what they are because we got them down and documented in our subconscious mind before birth. The importance of the prenatal imprinting phase is that we are still in an intimate direct contact with our mothers, so that we can get her emotions imprinted. After being born, this connection is less direct.

My assertion is that once the cerebral cortex is sufficiently developed (beginning about the second year of life) the human is capable of learning many things including language. Before that, the learning takes place in other structures such as the limbic system, and it is not intellectual, having no logical structure. Items acquired in the prenatal and the early post-natal periods are not characterized by names, pictures, or other intellectual constructs such as physical descriptions. They are however characterized by emotions they evoke, as the limbic system has been called by Ramon y Cajal (1988) and other neurohistologists the center of emotion in the brain.

Methods

The goal of the described project concerning music theory is to identify emotional responses. I want to identify evidence of emotions induced by music. As music theorist, most of the time I have been using my own emotional responses as points of departure. There is no guarantee that others will have the same emotional responses as I do. In fact, a major part of my hypothesis is that they will not. My hypothesis is that the emotional basis for each person's musical taste lies in the sound imprinting that took place in their limbic system before they were born, and that this imprinting was created by the mother's voice.

Methodology adopted in this project involved collecting the accounts of musical emotion experienced in specific places from specific pieces of music. Physical evidence of such emotions might result from oral reports or from accounts of chills, goose bumps, tears, or other physical responses. In cases where music is written and scores can be obtained, the methodology is to circle specific notes, chords, and passages in the music that produce these emotions, and then to perform a thorough analysis of these sounds based on music theory and acoustics with a goal of categorizing them in specific ways.

My work on this process over the years has suggested that harmonic series voiced sonorities are particularly powerful. I call these effects the Harmonic Series Voicings. I have written extensively about harmonic series voicings present in the Stravinsky's *The Rite of Spring* and in modern jazz. In cases where no scores are available, the analysis will be performed aurally from the available recordings and descriptions of the general or specific places where these effects have been felt.

Results

Example 1

The first example is taken from Nikolay Andreyevich Rimsky-Korsakov's composition *Scheherazade Op. 35.* A passage that I have identified as having a particular emotional impact is found near the beginning of the first movement after the violinist plays the introduction to the movement. This passage, intended to describe the ocean waves, consists of the famous motto theme played by the trombones accompanied by four chords. These four chords may be analyzed as follows: I, French sixth on the tonic, ii half-diminished seventh on the second scale degree and a V chord. The passage is heard as modulating, not just progressing to the dominant chord, and thus the passage is repeated, this time in the dominant key and so on, around the circle of fifths.

Rimsky-Korsakov could have chosen a simpler and more conventional chord progression, such as I, V7/IV. V4/2/V, V. It is clear that he wanted to achieve the tonicization of the dominant chord, in order to create a modulation, and that these more conventional chords would have accomplished this. The question is why he chose using a French sixth chord in place of the V7/IV, and a half-diminished sonority in place of a secondary dominant. The answer to this question goes to the heart of my research.

I would claim that the French sixth and the half-diminished chord are harmonic series voicings. The French sixth chord is found in the harmonic series at harmonics 7, 8, 9 and 11. The half-diminished seventh chord is found in the harmonic series at harmonics 5, 6, 7, and 9. These are higher in the harmonic series than the usually employed sonorities, particularly the major triad (harmonics 4, 5, and 6) and the dominant seventh chord (harmonics 4, 6, 7, and 9). The point is that a person elevates their voice when they are emotionally driven. The elevated voice falls in a higher harmonic series range of the harmonics of the larynx, and thus represents higher emotion. The formula is that higher partials represent elevated emotion.

Example 2

The second example consists of the first five chords at the beginning of the last movement of Pyotr Ilyich Tchaikovsky's *Symphony No. 6, Pathétique, Op. 74.* The first two of these chords are half-diminished seventh chords. The third chord is a minor seventh chord, the fourth is a half-diminished 11th chord, the fifth chord is a 13th chord, and the last is an augmented dominant chord that becomes a dominant triad. All these chords can be found in the fourth and fifth octaves of the harmonic series.

Tchaikovsky is often accused of wearing his heart on his sleeve, displaying his feelings openly and habitually. He actually does so in fact, but what people should ask is, how is it that you can wear your heart on your sleeve in abstract notes. What is it about notes that allow a composer to express his deeply felt emotion? Tension and release; dissonance and consonance? Of course, but that's not enough for Tchaikovsky's brand of emotionalism. There has to be something more. Prenatal imprinting of mother's emotions is the answer.

Example 3

The third example is *Prelude No. 1* from the book of 24 Preludes by Frédéric Chopin. Each of the measures of this work consists of an arpeggiated triad plus one or more non-harmonic tones. The first measure for example, contains an arpeggiated C-major triad plus a neighboring tone. The second measure contains an arpeggiated G major triad in first inversion plus a neighboring tone. The third measure is like the first, and the fourth measure contains an arpeggiated C-major triad in first inversion plus an accented passing tone, etc. Some of the no-harmonic tones are appoggiatura, and some are escape tones, but throughout this small piece all of the measures may be analyzed employing traditional 18th -century harmonic analysis techniques. In each case, though, each measure may be alternatively heard implying an extended tertian chord that may be mapped onto the third and fourth octaves of the harmonic series, which may be heard as triggering subconscious emotion. For example, the first measure may be verticalized into a I+6 chord, the second measure into a V6+9 chord and the fourth chord into a I major-seven chord in first inversion and so forth. Every measure may be analyzed in these two ways, but each measure may also be heard in these two ways depending on your aesthetic preference. Triad plus non-harmonic tone or a tall tertial chord, the difference depends on your preference, but your preference depends upon your sonic prenatal imprinting. If you imprinted tall chords then you will hear them, but if you just imprinted triads, then Chopin gives you an alternative way to his music. Gustav Mahler does the same thing.

Example 4

The fourth example is the second chord of Franz Liszt's *Hungarian Rhapsody No. 2*. This famous piece is well known, partly because it has been often turned into commercial music in radio, television, and the movies. The second chord, occurring immediately in the first moments of the piece is a half-diminished seventh chord in a place where a simple dominant seventh chord would work well.

This example further highlights my basic methodology. I ask why it is that many famous composers in their most famous compositions substitute some other sonority for I, IV, or V chords or the dominant seventh chord. The answer to this question could simply be a desire for greater variety, and I am sure there is some of this. Still, I like to point out that the substitute sonorities can generally be mapped on to the harmonic series in the upper fourth and fifth octaves, and I think that this is significant. My claim is that we get to know these higher harmonics in the prenatal period of our lives, and that they carry emotional meaning for us, acquired by imprinting of our mother's emotions.

Sonorities frequently encountered as substitute sonorities include the half-diminished seventh chord (5, 6, 7, 9), the minor-major seventh chord (6, 7, 9, 11), the augmented seventh chord (7. 9. 11, 13), the French sixth chord (1, 5, 11, 7), the minor seventh chord (5, 6, 15, 9), the so-called dominant 9th chord (1, 5, 3, 7, 9), and so on. The various types of ninth, eleventh, and thirteenth chords most frequently encountered may be so mapped as well.

The so-called ninth chord with sharp 9, flat 5 may be found in the harmonic series at 5, 6, 7, 9, 11 and the thirteenth chord is 1, 3, 5, 7, 9, 11, 13. Beethoven used this chord occasionally, as did other older composers, probably because it is so similar to a first -inversion minor chord (1, 3, 13), when some of the pitches are omitted and the 13th chord is rarely presented with all seven of its pitches sounded (there is a famous example of all 7 pitches sounded in Mahler's Tenth Symphony).

The movie composer John Barry commented about the precious half-diminished seventh chord. I think that the half-diminished chord is the poster child for emotionally heightened chords because it is low enough in the harmonic series that many people would have had a chance to imprint on it in their mother's voice, yet high enough that it can represent heightened emotion. The major triad (1, 5, 3) represents no emotion other than rest or contentment. The dominant seventh chord (1, 5, 3, 7) represents a need for action and the minor chord (6, 7, 9 or 10, 12, 15) represents sadness or melancholy, as is well known. The reason this is so is that you would have heard those harmonics in your mother's voice, when it was in an elevated pitch state while she was expressing these types of feelings. The association of the major triad with happy feelings and the minor triad with sad feelings goes back to prenatal imprinting from the mother's voice. These sonorities are imprinted in subcortical regions of the brain, and they remain available for life. They are not normally noticed however, and might be thought of as subconscious, although they may be noticed with practice.

In my own case, a thirteenth chord produces a strong physical jolt that I observe and then after the fact can comment on. I can say that I just heard a 13th chord because I felt a jolt. On going back and examining the music I just heard, the 13th chord will be found. This will normally be an incompletely voiced thirteenth chord (1, 7, 5, 13 or 1, 3, 7, 5, 13, or occasionally 1, 3, 7, 9, 13). I might encounter 1, 3, 11, 7, 9, 13, as well. Actually all combinations are possible, but most composers delete many of the members of a thirteenth chord simply because of the amount of dissonance that is built up. The closer to the modern era (20th century) the more chord members might be included, until Mahler used all 7 of the chord members in his modernistic 10th Symphony. Note that the 3rd harmonic is the fifth of the chord, and the fifth harmonic is the 3rd of the chord.

The general rule is that the lower in the harmonic series the pitches are found, the greater is the number of listeners who will have imprinted an emotional response to the sonority. Thus higher groupings of more pitches are just dissonances, and no imprinted emotion exists other than horror. Still, a composer needs to express horror occasionally, and a minor second cluster 13, 14, 15, 16, 17, 18, etc. will do it.

CONCLUSION AND RECOMMENDATION

It seems to be pretty clear that some people have imprinted on the higher partials of the harmonic series in conjunction with specific emotions, while other people have not. I would guess that nearly everyone imprints on something though, just depending upon their mother's specific vocalizations. Perhaps it might be non-pitched vocalizations, rhythmic patterns, or melodic configurations. While this chapter focuses on classical music, there might be certain patterns found in popular music of various types that could be imprinted. The evidence is to be found in some people liking and deriving specific feelings from the sounds of different types of sounds, as found in a variety of music.

It seems to be true that everybody likes some sorts of music or sounds that bring back good feelings imprinted from their mother at the time these sounds were being heard. So, can you learn before you are born? Well, we hear of the Mozart Effect, so obviously people think some things are possible. However, there is a big error in the so-called Mozart effect; it is assumption that the music will produce the learning effect alone. This is definitely not the case. It is not an intellectual thing. It is an emotional thing and the key to prenatal learning involves the mother's emotions. The subcortical place in the brain where prenatal learning occurs is the center of emotions in the brain.

There's plenty of general evidence that one can learn before one is born. de Villers-Sidani, Chang, Bao, & Merzenich, (2007) summarized the findings as follows: "Like many areas in the neocortex, the functional properties of the adult primary auditory cortex (A1) are highly dependent on the sounds encountered early in life." And, "The change is persistent in that it lasts throughout life." If this is correct, then there is a vast amount of music and musical sounds in the world to be explored and documented. There is considerable evidence that this is possible, a lot of it coming from native Mandarin Chinese language speakers. It turns out that Chinese people are much more likely than westerners to have perfect pitch because their language is tone based. The speakers of Chinese use the same pitches over and over again and their infants hear these same pitches repeated constantly. They will learn typical phonemes from their native language as spoken primarily by their mother's voice and they will use that learning to improve their language performance later on. The assignments of the tones to music will come later, too.

The conclusion is that emotion is the nature's way of getting us to do something. The emotions we feel when we listen to the music or sounds imprinted in our brains during our perinatal period come not really from that music or those sounds. The emotion we feel seems to be an evolutionary trick that evolved for the purpose of promoting social cohesion over multiple generations. In other words, we should not be fooled. In my case, it is not the music of Tchaikovsky or Mahler that produces my emotions per se. It is caused by the fact that my mother (and my grandmother before her) liked the late Romantic music, listened to it, played it, and sang it in full voice with full professional piano accompaniment during the critical periods in my earliest development. I have imprinted this music and I now receive emotions from it. For other people who grew up in different times and in different places other types of music and sound patterns will produce these emotions. I have learned that I should not allow myself to be fooled.

REFERENCES

Backus, J. G. (1969). The Acoustical Foundations of Music – Musical Sound: a Lucid Account of its Properties, Production, Behavior, and Reproduction. W. W. Norton.

Bregman, A. S. (1990). *Auditory scene analysis*. Cambridge, MA: MIT Press.

Darwin, C. (2004). The Descent of Man. Penguin Classics.

de Villers-Sidani, E., Chang, E. F., Bao, S., & Merzenich, M. M. (2007). Critical period window for spectral tuning defined in the primary auditory cortex (A1) in the rat. *The Journal of Neuroscience*, *27*(1), 1809. doi:10.1523/JNEUROSCI.3227-06.2007 PMID:17202485

Deutsch, D. (2013). *The Psychology of Music* (3rd ed.). Elsevier Inc.

Encyclopedia Britannica. (2015). *Combination tone*. Retrieved October 27, 2015, from http://www.britannica.com/science/combination-tone

Farlex. (2012). *Otoacoustic emission*. Farlex Partner Medical Dictionary.

Glattke, T. J., & Kujawa, S. G. (1991). Otoacoustic Emissions. *American Journal of Audiology*, *1*(1), 29–40. doi:10.1044/1059-0889.0101.29 PMID:26659426

Heller, E. J. (2012). *Why You Hear What You Hear: An Experiential Approach to Sound, Music, and Psychoacoustics*. Princeton University Press.

Keverne, E. B. (2015). Genomic imprinting, action, and interaction of maternal and fetal genomes. *PNAS, 112*(22), 6834–6840. Retrieved October 25, 2015, from www.pnas.org/cgi/doi/10.1073/pnas.1411253111

Lewis, D., & Larsen, M. J. (1927). The Cancellation, Reinforcement, and Measurement of Subjective Tones. *Proceedings of N.A.S., 23*(7), 415–421. doi:10.1073/pnas.23.7.415 PMID:16588176

Lorenz, K. (1937). Imprinting. *The Auk, 54*(3), 245–273. doi:10.2307/4078077

MacLean, P. D. (1990). The Triune Brain in Evolution: Role in Paleocerebral Functions. Springer.

Mathews, M. M., & Moore, F. R. (1970). GROOVE – a program to compose, store, and edit functions of time. *Communications of the ACM, 13*(12), 715–721. doi:10.1145/362814.362817

Merlot, E., Couret, D., & Otten, W. (2008). Prenatal stress, fetal imprinting, and immunity. *Brain, Behavior, and Immunity, 22*(1), 42–51. doi:10.1016/j.bbi.2007.05.007 PMID:17716859

Nicolaïdis, S. (2008). Prenatal imprinting of postnatal specific appetites and feeding behavior. *Metabolism: Clinical and Experimental, 57*(Suppl 2), S22–S26. doi:10.1016/j.metabol.2008.07.004 PMID:18803961

Panksepp, J., & Biven, L. (2012). The Archaeology of Mind: Neuroevolutionary Origins of Human Emotions (Norton Series on Interpersonal Neurobiology). W. W. Norton & Company.

Parncutt, R. (2011). *Harmony: A Psychoacoustical Approach*. Springer.

Paul, A. M. (2011). What babies learn before they are born. *TED Global 2011*. Retrieved October 25, 2015, from https://www.ted.com/talks/annie_murphy_paul_what_we_learn_before_we_re_born/transcript?language=en

Pierce, J. R. (1970). *Telstar, A History*. SMEC Vintage Electrics.

Ramon y Cajal, S., & DeFelipe, J. (1988). Cajal on the Cerebral Cortex: An Annotated Translation of the Complete Writings. Oxford University Press.

Schnupp, J., Nelken, I., & King, A. (2011). *Auditory Neuroscience*. MIT Press.

Schouten, J. F. (1940). The residue and the mechanism of hearing. *Proceedings of the Koninklijke Akademie van Wetenschap, 43*, 991–999.

Schouten, J. F., Ritsma, R. J., & Cardozo, B. L. (1962). Pitch of the residue. *The Journal of the Acoustical Society of America, 34*(9B), 1418–1424. doi:10.1121/1.1918360

Stevens, S. S., & Warshofsky, F. (1981). *Sound and Hearing* (Revised Edition). Time Life Education.

Terhardt, E. (1974). Pitch, consonance, and harmony. *The Journal of the Acoustical Society of America, 55*(5), 1061–1069. doi:10.1121/1.1914648 PMID:4833699

Wiesel, T. N., & Hubel, D. H. (1963). Effects of visual deprivation on morphology and physiology of cell in the cat's lateral geniculate body. *Journal of Neurophysiology, 26*(6), 978–993. PMID:14084170

ADDITIONAL READING

Cope, D. (2004). *Virtual Music: Computer Synthesis of Musical Style*. The MIT Press.

Heller, E. J. (2012). *Why You Hear What You Hear: An Experiential Approach to Sound, Music, and Psychoacoustics*. Princeton University Press.

Novak, D., & Sakakeeny, M. (Eds.). (2016). *Keywords in Sound*. Duke University Press.

Sun Eidsheim, N. (2015). Sensing Sound: Singing and Listening as Vibrational Practice. Duke University Press Books. doi:10.1215/9780822374695

Tomlison, G. (2015). *A Million Years of Music: The Emergence of Human Modernity*. Zone Books.

KEY TERMS AND DEFINITIONS

Acoustic Waves: Longitudinal waves with the same direction of vibration as the direction of their traveling in a medium such as air or water. Linear mixing of acoustic waves results in forming periodicity pitches of the resultant tones.

Pitch: A sound quality describing the highness or lowness of a tone, defined by the rate of vibration that produces it.

Sound Waves: Sinusoidal waves characterized by their frequency, amplitude, intensity (sound pressure), speed, and direction. Pairs of sound waves may reveal additive and subtractive interference.

Tonotopic: Organization means the arrangement of spaces in auditory cortex where sounds of different frequency are processed in the brain. Tones close to each other frequency are represented in topologically near regions in the brain.

Tritone: An interval of three whole tones with an augmented fourth. For example, between C and F sharp.

Section 3
Computing and Programming

This section delves into selected methods aimed at assisting the learners in acquiring computing and programming skills with the use of video tutorials and metaphorical visualization.

Chapter 9
Using Video Tutorials to Learn Maya 3D for Creative Outcomes:
A Case Study in Increasing Student Satisfaction by Reducing Cognitive Load

Theodor Wyeld
Flinders University, Australia

ABSTRACT

This study tracked the transition from traditional front-of-class software demonstration of Autodesk's Maya 3D to the introduction of video tutorials over a five-year period. It uses Mayer and Moreno's (2003) theory of multimedia learning to frame the analysis of results. It found that students' preference for the video tutorial increased over the course of the study. Students' preference for video tutorials was correlated with a reduction in cognitive load, increase in satisfaction with the learning experience and subsequent reduction in frustration with the software. While there was no apparent change in measurable outcomes, students' satisfaction rating with the video tutorial in preference to other learning media suggests more efficient learning was achieved. As a consequence of the findings, the traditional demonstration was discontinued. Overall, the introduction of video tutorials for learning Maya 3D reduced frustration and freed up time for more creative pursuits – the primary purpose for learning the software.

INTRODUCTION

Traditional forms of software tutorial instruction include front-of-class demonstration (see Figure 1), books, online documents and digital resources such as files that can be opened in software, edited and reconfigured in different ways to achieve various creative outcomes (McNeil & Nelson, 1991). Increasingly, online video tutorials are providing an alternative learning medium. They can be sourced from online repositories such as: YouTube, *Digital Tutors*, and software online help websites such as Autodesk, among many other online sources. Video tutorials step through sequences of actions towards a prescribed goal – usually to demonstrate a particular function of a group of tools within the software. Anecdotally,

DOI: 10.4018/978-1-5225-0480-1.ch009

learners are increasingly accessing online video tutorials as they prefer the apparent easier step-by-step procedural learning they provide. However, little research has been conducted into why learners may show a preference for online video tutorials over their traditional text-based alternatives.

This study attempts to address the need to provide improved learning experiences for early learners of Autodesk's Maya 3D software package. In so doing, it provides a better understanding of the preference by early learners of video tutorials over text-based equivalents. It was found that more satisfying learning experiences could be achieved using video tutorials than for text-based tutorials and that this was largely due to the reduced cognitive load video tutorials provide. It is worth noting, however, that not all learners chose to use video in preference to text-based tutorials and others continued to use both in conjunction. The remainder of this report discusses the shift from text-based tutorials to video in the classroom and its impact, as well as provide a theoretical explanation for the overall more satisfying learning experiences reported by participants.

Autodesk's Maya 3D

Autodesk's Maya 3D is a modelling, animation and visual effects software package. It is widely used in the animation, games and visual effects industry. It is used by animation companies such as Pixar,

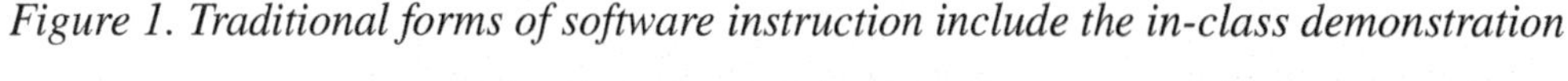

Figure 1. Traditional forms of software instruction include the in-class demonstration

Dreamworks and Blue Sky Studios; games companies such as Blizzard, EA, Polyphony; and, visual effects companies such as Pixomondo, Double Negative, Industrial Light and Magic. It is not an easy product for early learners to master (Park, 2004).

Traditional Maya 3D Instruction

Traditional software instruction, such as Maya 3D, follows the front-of-class demonstration process – the teacher performs operations using the software whilst its interface is projected onto an adjacent wall (see Figure 1). Students attempt to follow the teacher's instructions while the teacher is demonstrating. Although the teacher may moderate their demonstration pace to allow time for the students to catch up, some students are unable to match the pace of the teacher and may stop to take notes instead. For other students, the teacher often needs to interrupt the demonstration to assist them when problems are encountered that are not addressed in the demonstration. As such, a 20-minute tutorial exercise can take more than an hour to complete as a demonstration (McNeil & Nelson, 1991). An online PDF guideline may supplement the in-class demonstration. With the software package open, learners can open examples and manipulate the various options.

Autodesk's Maya 3D Software

Autodesk's Maya 3D software is not designed as a simple tool for early learners. It has been designed by engineers as an industry tool with few compromises for ease of use. It is a powerful 3D modelling and animation software package. It has a complex and intimidating interface (Park, 2004) (see Figure 2). It is particularly intimidating for those who simply want to use Maya for creative pursuits. The complexities of the software make it very difficult to begin to be productive without first learning how to use the tool in detail. It is necessary to amass a considerable legacy of knowledge about its functions and procedures for achieving specific results before any real production can proceed. This is typical of many software packages in the screen graphics industry. Hence, a method for making it easier for early learners to achieve productivity more quickly is needed. Students expressing short-term frustration with the software are unlikely to be focused on long-term outcomes. More generally, overwhelming frustration with learning how to use the tool can lead to class attrition.

Early Learners of Maya 3D

Early learners of Maya 3D, in the context of these studies, can be defined as those with little or no prior experience with the software but do have a general aptitude for computer programs such as the Microsoft suite of Office tools, Adobe graphics tools, and video games. Observations from this study and online forums dedicated to early learners of Maya 3D (see forums on digitaltutors.com; cgsociety.org; simplymaya.com) suggest they struggle to comprehend the complexity of the Maya 3D software. They struggle not only with the introduction to modelling in three-dimensions but also with the complexity of the interface. The software interface is organised in a non-intuitive way. Some core tools may be alternately represented by textual descriptions, icons, symbols or text input. The same tool may be present in multiple forms. Tools can be embedded within other tools, and multiple versions of the same tool may be found in different parts of the interface, including slightly different forms of the same tool. This can lead to confusion due to the overwhelming complexity of the software, compounded by the

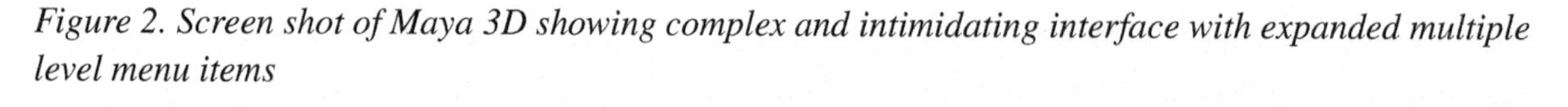

Figure 2. Screen shot of Maya 3D showing complex and intimidating interface with expanded multiple level menu items

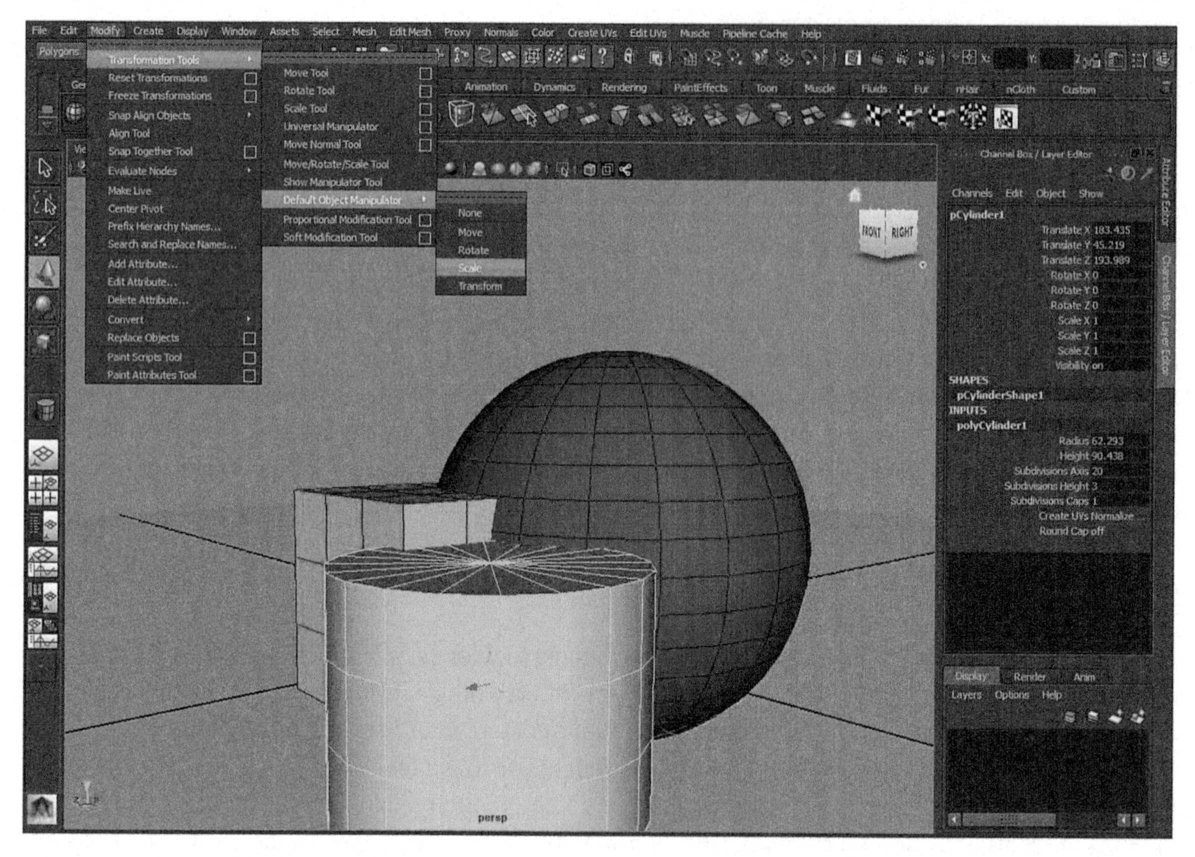

need to understand a new way of manipulating three-dimensional objects on a screen via mouse and keyboard. Collectively, these confounds often lead students to despair whether they will ever be able to master the software. Their primary goal is to be productive and creative with the tool. However, the difficulty in learning how to use the tool and understand the 3D visual medium itself means, for many early learners, they struggle to comprehend even the basic potential of the software and achieve only limited creative outcomes, such as the rudimentary construction of three dimensional objects used in a keyframed animation.

For example, creating a simple sphere in Maya 3D may require the user to alternatively select the 'Polygon' menu from the six top-most dropdown menu options, then, under 'Create' select 'Polygon Primitives', 'Sphere', ▣ (which opens the 'toolbox' with multiple parameters to choose from), or type 'polySphere;' in the MEL (Maya Embedded Language) textfield, or select the sphere icon from the 'shelf'. Hence, student frustration can be increased in an environment that includes multiple methods for achieving the same end result (see Figure 2). Alleviating student frustration and improving the quality of the learning experience was the goal of the current study.

INTRODUCTION OF VIDEO TUTORIALS: CASE STUDIES

Overview of Case Studies

The series of case studies reported here focused on the transition from traditional front-of-class software demonstration with supplemental online PDF tutorials to narrated screen-capture video tutorials. The PDF tutorials continued to be provided in a format that could be displayed on screen or printed. The video tutorials were streamed. The same material was covered in both the PDF and video tutorials. The topic of the tutorials was procedures in Autodesk's Maya 3D modelling and animation software. The study was conducted over five years from 2011 to 2015 as part of the annual introductory 3D Animation and advanced 3D Effects class offerings at Flinders University in semesters 1 and 2 respectively, and as part of an offshore intensive teaching program. This research reports on five case studies over five years. The timeline for development of use of video tutorials included the following case studies:

1. 2011, introductory video tutorial of 'basic walkcycle' in the 3D animation class – then available in 2012 and 2013 classes;
2. 2011, introductory video tutorial for 'paints' in the 3D Effects class – then available in 2012 and 2013 classes;
3. 2011, test performed on 3D Effects students using a 'smoke' tutorial to compare the efficacy of the PDF and the video versions. The class was divided into two equal groups. The video tutorial was then available in 2012 and 2013 classes;
4. From 2014, all tutorials in 3D Animation and 3D Effects were available in video and PDF format – in-class surveys were conducted; and
5. From 2015, all tutorials in 3D Animation were available in video and PDF format for the offshore teaching program – in-class surveys were conducted.

In case study 1, a single video tutorial was produced to assist students' better understanding of a difficult procedure in the 3D Animation class – the basic walkcycle. The basic walkcycle involves modelling legs and animating them in a 12 second walkcycle. As an introductory tutorial covering many core competencies in 3D Animation and Maya 3D more generally, students often struggled with the strict linearly sequenced procedures necessary to achieve the desired outcome. Although the demonstration and PDF tutorials followed a linear format, the video tutorial was found to be less ambiguous due to its overtly linear viewing nature.

In case study 2, as in case study 1, a single video tutorial for a difficult procedure in the 3D Effects class was produced – 'paints'. The paints tool in Maya relies on difficult procedures, depending on what type of object is being painted: 2D, 3D, polygonal or NURBS (Non-Uniform Rational B-Spline) objects. Using the 'paint' tool involves 'painting' colours or textures onto objects in a 2D or 3D scene. The tutorial was used until 2014 after which it was replaced with an updated version. Using the paint tool in Maya requires strict adherence to certain environment settings and modelling typologies. Whilst this is explained in the traditional demonstration and PDF tutorials, the sequential instruction was clearer in the linear format of the video tutorial.

Case study 3 is a follow-up to case study 2. It outlines an in-class study of the use of a video and PDF tutorial for creating 'smoke' in Maya 3D. Creating smoke in Maya employs the physics simulation engine and keyframing. Numerous steps are required before a renderable result can be achieved for simulated 'smoke'. 'Smoke' can be generated in the form of a collection of spheres or other types of objects (which grow in size) 'emitted' from a point or plane in a 3D scene over time. This tutorial was used until 2014 after which it was replaced with an updated version. Unless certain environment settings are established at the beginning of the physics simulation process errors will occur or the end product may not be usable. Whilst this was explained in the traditional demonstration and PDF tutorials it was hoped the sequential instruction would be clearer in the linear format of the video tutorial. In this study, the class was divided into two groups of equal numbers of students. One group of students only had access to the video tutorial while the other group only had access to the online PDF version. There was no in-class demonstration.

Case study 4 outlines the process and end results of providing all tutorials in 3D Animation and 3D Effects in both online PDF and video versions from 2014. Front-of-class software demonstrations were no longer provided for either class from 2014. Case study 4 focuses on the students' choice of the online PDF or the video tutorial or their combination. Which type of tutorial or their combination was chosen, and user satisfaction were recorded.

Case study 5 is a follow-up study to case study 4. It outlines a study focussing on the cross-cultural implications for the use of video tutorials in the 3D Animation course alone. It helps clarify learners' reported experiences across all groups. The same 3D Animation and 3D Effects courses are offered as apart of a top-up degree in collaboration with Finders University and the Chinese University Hong Kong (CUHK). Hence, case study 5 also discusses the process and results of providing the 3D Animation course tutorials as videos and PDFs to EAL (English as an Alternative Language) students (the 3D Effects course used different material to that investigated in case study 4 and was not included in this study). Case study 5 provided some interesting insights into perceptions of how information was understood and interpreted by EAL students.

In summary, the 5 case studies reported here set out to track and compare the relative efficacies of front-of-class demonstration, text-based tutorial guides and their video corollary. The subject of the tutorials was about learning particular procedures in Autodesk's Maya 3D. Most of the students had little or no prior experience with the software. However, as all students were either enrolled in an IT or Digital Media Bachelor Degree it can be assumed they had an aptitude for working in a computer environment. The tutorials provided instructions on how to perform predetermined functions in the software. Creative production was the goal of each tutorial. The text-based PDF tutorials were contained in a single, continuous, sequential document segmented by sections and subsections. Each video tutorial included from three to seventeen discrete segments derived of the sections and subsections in the PDF version. The video segments were designed to be viewed in sequence. In-class demonstrations delivered the same material and in the same sequential order as the PDF document and video tutorial.

Video Tutorial Production

In the 3D Animation course there are nine tutorials sectioned or segmented into 62 2-7 minute videos. In the 3D Effects course there are seven tutorials sectioned or segmented into 25 2-9 minute videos. The average tutorial video length was 5 minutes. Each tutorial was recorded using Freez Screen Video Capture (www.smallvideosoft.com) – a freeware video screen capture software tool. This was used to capture the mouse movements on screen while the software was in use – including drop-down and pop-up menus, draft renderings and so on. While screen images were captured, it simultaneously recorded the narrator's voice describing the actions performed on screen. Each video was produced in an uncompressed AVI format in the first instance. This was converted to the mpg4 format in Quicktime Pro for editing and subsequent streaming.

Procedure for Video Production

There are 9 steps in the production of a video tutorial:

1. Each video tutorial is storyboard scripted according to its PDF version.
2. The narrator then completes the tutorial based on the PDF version storyboard to ensure there are no anomalies.
3. This is repeated and video captured with the narrator explaining what they are doing whilst completing the tutorial.
4. This initial capture is scrutinized for any errors, which may require a complete re-shoot, or segments re-shot. This process is repeated several times until all anomalies are removed or details added.
5. Any redundant parts from a final shoot are edited out (such as waiting for a render to complete or tool to open).
6. The final continuous shoot is broken down into logical segments of 2-7 minutes in length based on the logic of the PDF sections and subsections.
7. The segments may be exported at a lower resolution than the original to provide for faster streaming.
8. Segments are uploaded to the university streaming server running Apache.
9. RSIDs are provided by the streaming service which are used to link to a webpage that students can access the videos from within the university's content delivery MOODLE platform.

Following this process, a 30 minute video tutorial can take 8-12 hours to produce (see also Bork & Gunnarsdottir, 2012; Wang, 2009; Martinovic etal, 2012).

Process for Case Studies

Each case study investigated a different aspect of the transition from front-of-class demonstration with supplemental PDF tutorials to video tutorials. For case study 1 and 2, what media type was chosen or combination of media types, how long it took to complete the tutorial, problems encountered, user satisfaction and comments were recorded. For case study 3, how long it took to complete the tutorial, problems encountered and user satisfaction were recorded. For case study 4 and 5, only media format choice, user satisfaction data and comments was collected. See Table 1 for participant demographics by year and case study.

For case study 1 and 2, 2011, all students had access to both the PDF and video tutorial. An in-class demonstration was conducted at the same time. The study was conducted in a single two-hour session (the normal class time allocated).

For case study 3, 2011, students were organised into two groups of equal numbers. They were physically separated by a continuous bench in the centre of a long room with computer stations either side of it. On one side, students only had access to the video tutorial at the same time as students on the other

Table 1. Demographics for each case study

		Study 1	Study 2	Study 3	Study 4		Study 5
		Semester 1 2011	Semester 2 2011	Semester 2 2011	Semester 1 2014	Semester 2 2014	Offshore 2015
	Course	3D Anim	3D Effects	3D Effects	3D Anim	3D Effects	3D Anim
	Topic	Walkcycle	Paints	Smoke	all	all	all
	Male	24	17	17	38	25	18
	Female	3	3	3	8	5	16
	Totals	27	20	20	46	30	34
Age	Min	19	20	20	18	20	21
	Max	38	30	30	23	30	26
	Average	20.8	21.6	21.6	20.6	21.1	22.5
	Degrees	BCA(DM)	BCA(DM)	BCA(DM)	BCA(DM)	BCA(DM)	BCA(DM)
		BSC(DM)	BSC(DM)	BSC(DM)	BIT(DM)	BIT(DM)	
		BA	BA	BA	B.Media	B.Media	

side only had access to the online PDF version. There was no in-class demonstration. This study was conducted in a single two-hour session (the normal class time allocated).

For case study 4, in 2014, students were free to choose a format of tutorial from the online PDF or streamed video or both. All tutorials for both the 3D Animation and 3D effects courses were available in either format. No in-class demonstrations were performed. The study was conducted across two semesters. More than half of the students from the 2014 3D Animation class were also enrolled in the 2014 3D Effects class.

Case study 5, in 2015, was conducted with a group of students who were part of an off-shore intensive teaching program. Students were free to choose a format of tutorial from the online PDF or streamed video or both. All tutorials were available in either format. No in-class demonstrations were performed. The study was conducted across a one week intensive teaching program. The offshore students report English as an Alternative Language (EAL).

For case study 4, in each of the three settings, three surveys were conducted to identify tutorial format choice: at week 4; week 8 and week 12. How many chose which format and why was recorded. Student comments were solicited and recorded anonymously. The survey was conducted while the whole class was present and able to respond collectively and individually. This was repeated for case study 5 at the end of day 1, 3 and 5.

RESULTS FROM CASE STUDIES

Data collected for each case study was based on the specific investigative purpose of that case study.

Case study 1 and 2 investigated the introduction of a single video to assist with a difficult tutorial. Data collection included: student numbers, media format choice, time taken to complete tutorial, number of problems encountered, satisfaction rating and student comments. Case study 1 results indicate a preference for the video tutorial over other formats available. Case study 2 results indicate an increased preference for the video tutorial since case study 1.

Case study 3 included a controlled test comparing the PDF with the video tutorial in isolation. Case study 3 data collection included the same elements as for case study 1 & 2. Case study 3 results indicate the video tutorial could be completed more quickly than the PDF with higher satisfaction but with the same number of problems for either format.

Case study 4 and 5 tracked the choice of media type after all tutorials in the course were made available as videos. Data collection included: student numbers, media format choice, satisfaction rating and student comments. Case study 4 results indicate a steady increase across the semester in the number of students choosing to use the video tutorial or its combination with the PDF. Case study 5 results indicate students' initial choices were maintained across the study period.

Case Study 1 and 2 Results

In case study 1 and 2, semester 1 and 2, 2011, a satisfaction rating was generated using a Likert scale of 1 to 7, strongly agree to strongly disagree to the question "I was able to complete the tutorial easily based on the information available to me". The data is collated in Tables 2 and 3.

From Table 2 we see that the video tutorial alone achieved a higher satisfaction rating than other forms. This was equaled when it was used in conjunction with the PDF. The video-only cohort encountered the least problems and completed the tutorial more quickly than the other cohorts. More students chose to follow the demonstration than other formats. The next most popular was the video format, PDF and then the combined video-PDF format. Only a few used the PDF in conjunction with the demonstration and only one student used the video format in conjunction with the demonstration, or all three.

Table 2. Data collected from 2011 case study 1

	Demo	PDF	VT	PDF + VT + Demo	PDF + VT	PDF + Demo	VT + Demo
Students	7	5	6	1	4	3	1
Time ave (hrs:mins)	2:21	1:33	1:08	2:27	1:22	2:11	2:01
Problems ave	6	4	1	5	2	7	8
Satisfaction ave	5	4	2	3	2	5	7

Table 3. Data collected from 2011 case study 2

	Demo	PDF	VT	PDF + VT + Demo	PDF + VT	PDF + Demo	VT + Demo
Students	2	3	8	0	7	0	0
Time ave (hrs:mins)	2:18	1:41	1:17	-	1:57	-	-
Problems ave	4	4	2	-	0	-	-
Satisfaction ave	4	4	2	-	2	-	-

It was observed that the students who used the video tutorial in conjunction with the demonstration, or all three formats available, struggled with the fundamental concepts covered. Hence, these two students are not representative of the overall cohort.

Typical students comments included:

- "I like to follow… [the demonstration] because I can [keep up and] finish in class."
- "The PDF has everything I need and I can do it in my own time [or at my own pace]."
- "The video tute was great! I could see what I needed [to do straightaway] and do it [at the same time]."
- "I tried to follow… [the demonstration] but gave up. I used the PDF [instead] and looked at the video, but I just got [more] confused."
- "I could find the missing stuff in the PDF [after watching the video]."
- "I had the PDF open when you were... [demonstrating]."
- The demonstration "…was good, but its different to the video?"

From Table 3 we see that student attitudes to the video tutorial had shifted since its introduction in the previous case study. There were more students who used the video tutorial alone or in conjunction with the PDF than other options. The video-only cohort completed the tutorial more quickly, encountered fewer problems and indicated a higher satisfaction than those who chose the other formats.

Typical student comments included:

- The demonstration "…would be better if you could go slower."
- "The PDF should have more screenshots."
- "Its hard to see the numbers [as you type them in the video tutorial]."
- "I had the PDF open but mostly used the mpg [video tutorial]."
- "I can't see the point of… [the demonstration – because I can watch the video]."

Case Study 3 Results

As a follow up to case study 2, case study 3 outlines an in-class study of the 'smoke' tutorial. Less data was collected for this case study as it was a controlled test (see Table 4).

In case study 2, most students chose to focus on either the video tutorial or PDF or their combination rather than the in-class demonstration and combinations including the demonstration. As a control, case study 3 investigated the relative efficacy of either the video tutorial or the PDF in isolation. There was

Table 4. Data collected from 2011 case study 3

	PDF	VT
Students	10	10
Time Ave (hrs:mins)	1:26	0:44
Problems Ave	4	4
Satisfaction Ave	4	1

no in-class demonstration. From Table 4 we see that those students who only had access to the video tutorial completed the tutorial in significantly less time with a higher satisfaction but encountered the same number of problems as those students who only had access to the PDF tutorial.

Typical student comments included:

- "Its hard to follow the PDF when I am in the program [because it is on my iPad and I can't look at both screens at the same time]."
- "I just listened [to the video tutorial] and did what it said."
- "The videos are cool!"

Case Study 4 Results

With the success of the 2011 controlled studies, from 2014 all tutorials were made available in both PDF and video format. Students could choose which format or combination of formats they preferred to use. Case study 4 has two parts: results from the semester 1 3D Animation course survey and results from the semester 2 3D Effects course survey. While comprehensive data was collected for the semester 1 course, it was deemed much of this would be duplicated in the semester 2 course. Hence, the survey conducted at the end of the semester 2 course included only the gross numbers of students, media format choice and satisfaction data. The case study 4 data collected for semester 1 has been graphed to show trends (see Figure 3).

From the semester 1, 2014, 3D Animation course survey data graph we see a steady shift from PDF-only choice to video-only or combined video-PDF. The week-by-week spikes can be attributed to the intrinsic difficulty of individual tutorials. The rapid shift from PDF to video tutorials after the first week was due to the video tutorials not being available during the scheduled tutorial time for the first week (due to technical difficulties). There is a slow decline in the numbers of students who chose to use the video tutorial in conjunction with the PDF until mid-semester when it begins to incline. There is a spike in preference for video-only in the final tutorial. Overall, there is a steady decline in the PDF-only cohort numbers, with a corresponding steady incline in the video-only numbers.

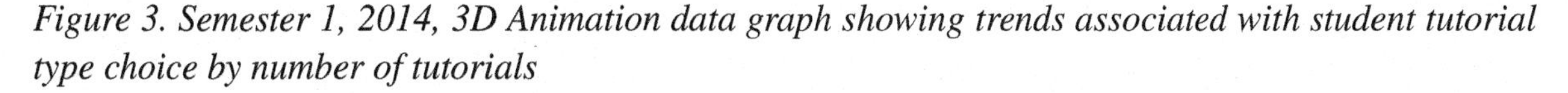

Figure 3. Semester 1, 2014, 3D Animation data graph showing trends associated with student tutorial type choice by number of tutorials

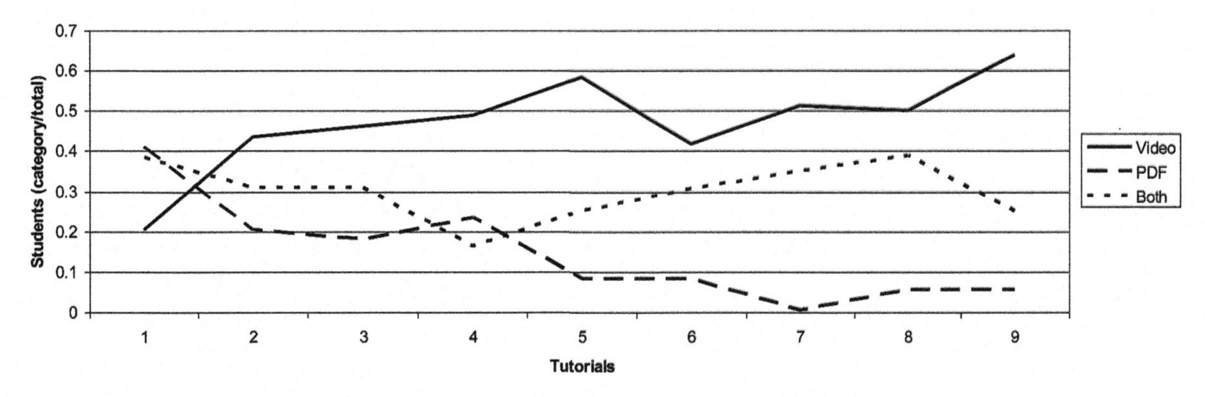

Typical student comments included:

- "Seeing you do it [in the video] helped."
- "I don't miss anything – I can rewind [the video]."
- With the PDF I can "...go at my own pace – text form [is] easier to understand."
- "I could follow both [the PDF and video] and get detailed info from the PDF."
- "I like to read [the PDF] and then watch [the video]."
- Initially "I used the video but later [in semester found I] needed the PDF" also.

Less data was collected from the semester 2 3D Effects course survey. From semester 2, 2014, all tutorials were available in video or PDF format. A short survey was conducted at the end of semester 2. Of the nineteen students in the class:

- Six chose to use the video tutorials only;
- One student chose to use the PDF tutorials only; and
- Twelve students chose to use both in conjunction.

This follows the trend established in semester 1. Although, there has been a greater shift towards using the video in conjunction with the PDF tutorials than in semester 1. More than half of the students enrolled in the semester 2 3D Effects course were also enrolled in the semester 1 3D Animation course.

Typical student comments included:

- "I changed from video to the PDF because the video lags at home."
- "If I needed to know something specific [that I couldn't find in the video] I could look it up in the PDF."
- "After doing... [the tutorial from the PDF] I could check if it was right in the video."

Case Study 5 Results

Case study 5 is a follow up and direct cross-cultural comparison to the semester 1, 2014, 3D Animation course survey in case study 4. Students from an offshore teaching program were invited to comment on the efficacy of the same video and PDF tutorials. Similar to the conditions for case study 4, all tutorials were made available in both formats. Students could choose a format or combination of formats. The data has been graphed to show trends (see Figure 4).

The graphed results for case study 5 are distinctly different to those for the Australian cohort for the same course and conditions in case study 4. The initial low count for the first video and subsequent high count for the PDF tutorial formats was due to the video tutorials not being available during the scheduled tutorial time for the first week (due to technical difficulties). After this initial anomaly, it appears students made a choice of what format they would use and largely did not deviate from this choice for the remainder of the course.

Typical student comments included:

- "I look at PDF before video and it seemed easier [that way]."
- "I just look to the PDF for details [I am not sure of in the video]."

Figure 4. Offshore program, 2015, 3D Animation data graph showing trends associated with student tutorial type choice (divided by total number of students) by number of tutorials

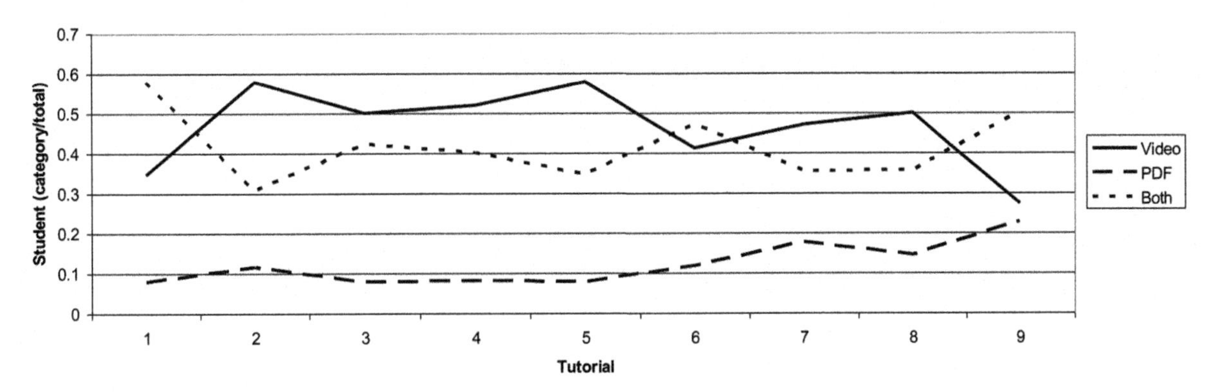

- "I look at PDF first then go to video to see dropdown menu and actions [which are only described in text in the PDF]."
- "I may use PDF as well for tutorial 6 because it looks difficult."
- "Its easier to set my own pace with PDF only."
- "I am good at Maya [I have used this program before] so I can guess what the text [you are typing in the video] is."

Overall Satisfaction Ratings for Case Study 4 and 5

As case studies 4 and 5 were controlled studies with many variables it was not possible to record time taken and numbers of problems encountered for individual tutorials. At the end of the course, a satisfaction rating was generated from the same Likert scale used in case study 1, 2 and 3 (see Table 5).

From Table 5 it is clear that those who chose to use the video format only expressed a higher satisfaction than other formats. However, the other 2 options also achieved a positive satisfaction rating.

Summative Assessment Results

For the courses under investigation, summative assessment included marks for skills acquisition and application of those skills in a major self-directed project. Marks for skills acquisition was based on competent completion of a tutorial. Components assessed included: accuracy, mastery and aesthetic quality. Each were assigned explicit requirements in the assessment guide and marked accordingly. Students either

Table 5. Satisfaction rating data collected from case study 4 and 5

	2014 Australia		2015 Offshore
Format	**3D Animation**	**3D Effects**	**3D Animation**
Video	1	1	1
PDF	3	2	2
Both	3	2	2

demonstrated that their work was: accurate, inaccurate or incomplete; mastered, competent or unskilled; realistic, basic or absent. Following this schema, each component of an individual tutorial was allocated a raw score of 2, 1 or 0 respectively. The marking schema was aggregated to generate an overall mark for individual tutorials and the final project, resulting in a final summative mark and subsequent grade for the student. Given this scenario, any improvement in comprehension, mastery and aesthetic application of the material covered in a course should be reflected in a trend towards higher overall marks for successive cohorts across the five years this study was conducted. The summative assessment results for the five years included in this study have been graphed to show any trends (see Figure 5).

From Figure 5 we see that there appears to be no significant change in summative assessment results for the period the courses surveyed in this study were taught (2009-2015). This indicates that the experimental introduction of the video tutorial to these courses has not resulted in any *measurable* improvement in overall comprehension, mastery or aesthetic application of the material covered. This is consistent with reporting by others (see Veronokis & Maushak, 2005; Hoffler & Leutner, 2007; Keefe, 2003; and, Boster etal, 2006, 2007). However, from the individual case studies, it is clear that students prefer to use the video tutorials and their combination with PDFs to the traditional front-of-class demonstration. Indeed, the removal of the front-of-class demonstration from 2014 did not result in any change in measurable outcomes. Nonetheless, anecdotally, the author of this chapter would suggest the overall quality of the productions submitted for assessment in these courses improved after the video tutorials were introduced. The difficulty is in how to measure this apparent improved quality? This is discussed later in this chapter.

DISCUSSION OF RESULTS

From this study, the traditional front-of-class demonstration appears redundant when compared to the combined PDF and video tutorials. Taking much longer to deliver, the primary difficulty with the demonstration appears to be that students struggled to keep pace – delivery was either too slow or too quick. This reduced the benefits of its synchronous delivery mode. By comparison, the video tutorial provided the same or more material in a form that students had control over its speed of delivery – they

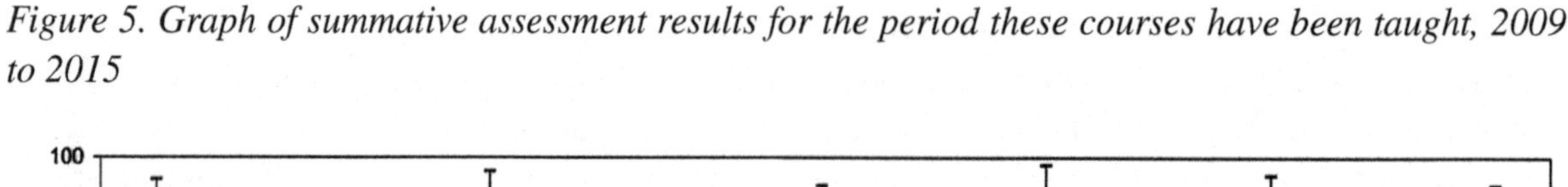

Figure 5. Graph of summative assessment results for the period these courses have been taught, 2009 to 2015

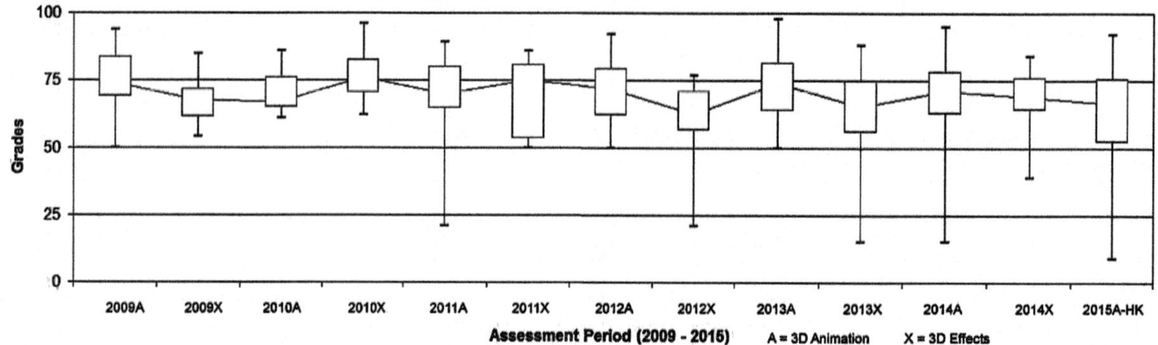

could stop, rewind, or revise as needed – with the PDF tutorial providing back-up detail. From 2014, the traditional demonstration was discontinued. No negative impact from this change was evidenced from student comments, ability to complete tasks, or the summative assessment results. From the results, while similar problems and number of problems were encountered by either group, the video tutorial was completed more accurately, quickly, and with greater satisfaction than those who chose to use the online PDF version only.

In case study 1, the video tutorial achieved a higher satisfaction rating than other forms. Nonetheless, it is interesting to note that more students still chose to follow the demonstration than other formats. This may be indicative of a reluctance to change from a prior learning mode. From the student comments, it appears students became confused when they tried to mix media types. Those who used a single media type, or at most two, tended to show greater understanding, satisfaction and encountered fewer problems. This was particularly evident for the video-only and video-and-PDF cohorts.

In case study 2, while the same procedures and concepts were explained in the traditional demonstration and PDF tutorials, the sequential instructions appeared to be clearer because of the linear format of the video. More students chose to use the video tutorial in case study 2 than in case study 1. This may be attributed to the fact that more than half of the students enrolled in this course had access to the walkcycle video tutorial from the previous semester. The increased shift to the video tutorial appears also to have divided the class between those who used it exclusively or in conjunction with the PDF and those who relied on the demonstration or PDF alone. This second group were mostly comprised of those who had not previously accessed the walkcycle video tutorial. Thus, it appears they may have preferred a more traditional method, as highlighted in case study 1. Despite its greater popularity, from the student comments, it is clear that the video tutorials could be further refined (updated versions of the video tutorials and PDFs were released in 2014). More importantly, it is clear that students still required access to more than one type of media: demonstration, video and/or PDF tutorial.

In the controlled test conducted in case study 3, students were divided into 2 equal groups having access to only the video or PDF tutorial. Again, the results indicated that the video tutorial was the preferred format. This may be because, although the selection of students for each group was random, the video-tutorial-only group was simply more capable than the PDF-only group. However, this is not borne out by the final assessment results, suggesting the quicker completion time was indeed due to the apparently clearer video instructions. That the satisfaction levels for the video tutorial cohort are higher than that of the PDF group may also be because of peer-group pressure and the perception that the video-tutorial group were somehow advantaged – confirmed by the quicker completion times. The PDF group may have artificially adjusted their satisfaction rating to compensate. However, that both groups encountered a similar number of problems suggests the intrinsic difficulty of completing the tutorial by either method was similar. From student comments and general observation of class behaviour, some in-class rivalry was evident. This may have skewed the results. Dividing the class into two equal groups and assigning them specific learning methods is an artificial imposition. To remove this apparent bias, in the follow-on case studies, students were able to choose which media format or combination they preferred.

In case study 4, all tutorials were available in either video or PDF format. Front-of-class demonstrations had been discontinued. Across the semester, there was a general increase in video-only use and a correlating decrease in PDF-only use. That there were spikes in use of the different media across the semester may be due to the more complex and/or detailed nature of individual tutorial material – requiring students to clarify values from the PDF document – as demonstrated by the resumption in preference for video-only in the final tutorial (a more straightforward tutorial). This suggests students explored the different media formats available before committing which to use, based on the intrinsic difficulty of each tutorial. This appears to be confirmed by spikes in the graph and from student comments such as: initially "...I used the video but later [in semester found I] needed the PDF" also.

In the second part to case study 4, it appears there was an even greater shift to using the video tutorials or its combination with the PDF in preference to PDF-only than in semester 1. This may be because of the more detailed information needed to complete the more advanced 3D Effects course (larger concepts to comprehend, more input values and optional extensions) than in the introductory 3D Animation course. From the student comments, it appears students found a new sophistication with the video and PDF tutorials since their introduction in semester 1. They were able to navigate both with equal facility – using the benefits of one to clarify the other. In this way, they appear to be better able to understand what they were doing and overcome frustration with the software.

Case study 5 attempted to investigate if there were any cross-cultural implications for the overall study. The case study 5 cohort comprised 34 Hong Kong (HK) students as part of an offshore intensive teaching program hosted by the Chinese University Hong Kong (CUHK). From the graphed data, it appears that the English as an Alternative Language (EAL) students tended to persevere with their initial choice of media format rather than explore other options. This was surprising as the series of tutorials vary in their degree of intrinsic difficulty and detailed instruction. That the offshore cohort did not show a similar trend to the Australian cohort, in terms of an overall shift towards use of the video and combined video and PDF tutorials across the study period, suggests there may be other mitigating factors involved. One of the comments suggests a possible reason: "...it is easier for me to read English than listen to you [speak it in the video]." From their comments, the offshore cohort appears to have naturally divided into 3 groups, those who: were confident with spoken English (video-only); were confident with written English (PDF-only); and, paid close attention to what they saw on the video but relied on the PDF for clarity (PDF and Video). Nonetheless, from their comments we see also that many of the same concerns were being expressed by the offshore and Australian cohort, such as: "I just look to the PDF for details [I am not sure of in the video]"(HK); "If I needed to know something specific [that I couldn't find in the video] I could look it up in the PDF"(Aust). But, they approached the issues raised in different ways. For example, instead of exploring the different media to find the information they needed or ask the teacher, they tended to instead persevere with what they could glean from their initial media format choice. However, it was observed that there was more in-class problem solving between students than that for the Australian cohort. Hence, their overall reliance on the video, PDF or combined formats for problem solving was less than that for the Australian cohort. This is quite different to the Australian students who tended to work in isolation.

Satisfaction ratings were collected for all case studies. In all cases, the video format or its combination with the PDF achieved the highest satisfaction rating. Of particular interest is the comprehensive data collection in case studies 4 & 5. In these case studies, we see that all formats achieve a high satisfaction rating. This may be due to the following reasons:

- A media format was chosen based on prior experience hence a positive outcome was anticipated;
- Students were reluctant to report a negative experience due to their choice of media format as they may have self-reflected on this as poor judgment (the inverse occurred when choice was artificially imposed, as in case study 3);

- The quality of the learning material was the same and of a high standard regardless of the media format and so success was achieved despite potentially varying difficulties in synthesising the information provided between formats, leading to positive satisfaction reporting.

Despite the positive comments and satisfaction rating, the summative assessment results collated for the period the courses were taught – from 2009 to 2015 – appear to show no correlation between the introduction of the video tutorials in 2011 and overall assessment results. This suggests that a preference for a particular media format does not necessarily result in improved outcomes or correlate with satisfaction ratings. However, this may also be a function of the marking method – as outlined in the summative assessment results section.

Two key questions arise from this. While preference for, and satisfaction with, the video tutorials is demonstrated in the results, *why* this is so remains unanswered. The section, Developing a Theory for Findings, attempts to address this question. What knowledge is available to the students and how it is visualised by the various media formats is addressed in the next section.

Knowledge Visualisation Modes

The three different media formats (demonstration, PDF, video) visualise the knowledge contained in the tutorials in different ways. The visualisation mode tends to dictate how much and what type of knowledge is available according to which media format is used. For example:

- While the flexibility of the front-of-class demonstration allows for detailed explanation, this often didn't occur because of overriding time constraints;
- The PDF also allows for detailed explanations but it tended to be more open to misinterpretation and misunderstanding;
- The video provides the most comprehensive coverage of procedures but the narrator had to be very thorough with his explanations as, unlike the demonstration, the user could not ask questions.

Table 6 outlines what knowledge was available to the user and how it was visualised, accompanied by the teacher's reflections on the mode of delivery.

While the knowledge available to the students was consistent across all three modes, its practical access was limited by its visualisation mode. Listening to a demonstration, watching a large screen and then turning to a local monitor to repeat the steps is not the same as working from the same monitor alone, be it reading text or listening and watching a video.

Comparing Modes of Knowledge Visualisation

To understand the students' practical access to the knowledge in the tutorials we need to explore each visualisation method. The following is a narrativised account of part of one of the tutorials which describes how to use the Maya physics engine to simulate smoke. It demonstrates the different knowledge visualisation modes, content and effectiveness.

Table 6. Outline of knowledge visualisation and reflections on delivery mode

Mode	Knowledge Visualisation	Teacher Notes
Demo	Knowledge: techniques, procedures, locations (where to find information – different types and formats – on the user interface), outcomes, problems and solutions, options, explanations, settings. Visualisation: linear but disjunct – can stop and start (but can't rewind), projector, mouse cursor, walk up to screen and point at, gesture, repeat, stop and move about class to help students catch up (point on their screen, encourage, advise).	Rushed (always conscious of time), need to accommodate slow and fast learners, need to stop to assist students to catch up, too much tutorial time dedicated to the demo, not enough time for one-on-one assistance, students need to look at the screen, remember procedure and then repeat steps from memory whilst looking at their own monitor.
PDF	Knowledge: procedures, notes, extras, history, purpose, overview, step-by-step instructions. Visualisation: narrative, tables, lists, screen grabs, diagrams, examples, links to external resources, references, self-directed assessment criteria.	Open to misinterpretation, misunderstandings, students tempted to jump forward, miss steps as eyes move from PDF to screen and back again (in particular, small steps, embedded in larger steps), limited screen grabs, teacher makes decisions about what to include and what to leave out – whereas, with the video or demo everything is included – thus, absolute guarantee of success – can see the whole process unfold, students have the PDF open next to Maya.
Video	Knowledge: techniques, procedures, locations (where to find information – different types and formats – on the user interface), outcomes, problems and solutions, options, explanations, settings. Visualisation: screen, mp4, streamed, linear format but can stop, start, rewind, mouse cursor, voice over.	Opens up possibilities to be more creative, not rushed, more content (than demo or PDF), reusable, can provide examples of extended tutorials, students have video open next to Maya.

Front of Class Demonstration Method

The front-of-class demonstration is conducted in a room which is longer than it is wide with rows of computers opposite each other on a long bench in the centre of the room with more benches and computers lining the side walls. At one end of the room is the screen for a projector and the teacher is stationed at the other end of the room. The Maya software is manipulated by the teacher at a console. The same view the teacher sees on his monitor is projected onto the screen at the far end of the room. There is insufficient time for a historical introduction on the use of physics in 3D animation and effects (included in the PDF version). This particular demonstration begins by launching the program and opening the settings and preferences. The teacher verbalises his actions as they are performed. The unit measure needs to be set to millimetres as the physics engine is sensitive to scale. On the screen, drop-down menus can be seen to be opened and navigated. A particle emitter is created with its toolbox open to adjust the parameters. The teacher waits for students to catch up. The parameters are adjusted to generate a colour ramp. This involves a series of complicated left and right mouse button clicks. This launches a series of interconnected dialog boxes. It is easy for the students to miss a step. Despite the teacher's slow pace, by this stage, at least a third of the class has stopped trying to keep up and are simply taking notes. Small points can be seen to emanate from the centre of the screen (particles from the emitter). Parameters are changed again to change the particles from points to spheres. But before these can be rendered the renderer's settings need to be adjusted. This involves many nested commands, some 4 or 5 mouse clicks deep. Students cannot see the teacher's hand when he operates the mouse – only the result. The rendered result is impressive. More parameters are changed to make the effect more realistic – colours, duration and random seeding of sphere sizes. Only half of the class has been able to complete this part of the task. The teacher moves around the class assisting students to catch up. This, normally five-minute tutorial has taken over 45 minutes. Hence, although the front of class demonstration provides flexibility, contiguity, completeness and time for the teacher to stop and start as needed to assist the students to keep up, the overall time taken to complete the demonstration prevents more knowledge being communicated. The demonstration does not cater well for different learning paces.

Using the PDF Tutorial

The PDF tutorial is accessed online. Students tend to have the PDF open on one side of their monitor and Maya on the other. When moving from one to the other they make comparisons between the values and screengrabs they can see in the PDF with those in Maya. The PDF includes a short introductory history of how and why the Maya physics engine is used in 3D animation and effects. It describes the different types of physics effects – particles, forces, soft and rigid bodies; dust, smoke, sparks, lightening, bullets and so on; how the particles can interact with each other – bounce, collide, fall, fade; and, types of fields: gravity, wind and turbulence. The procedural format for students to follow uses symbols, acronyms and icons. For example, to change the particles from points to spheres the greater-than sign (>) is used to indicate the steps: 'In the Attributes Editor: particleShape1 > Render Attributes > Particle Render Type > Spheres'. This expression indicates that, 'under the particleShape1 tab in the Attributes Editor there will be an expandable subtab called Render Attributes within which a function called Particle Render Type has a drop-down menu which includes Spheres as an option – select this'. The second step is to left-mouse-button click (LMB) on Current Render Type and change the Sphere Radius to 0.5. However, from the above, it can be seen that the greater than sign infers much more than the direction steps must be taken. Hence, the second step is often missed. This is because, after navigating to the Spheres option the drop-down menu closes giving the impression that the procedure is now complete. Without this secondary step the particles on the screen do not update to spheres (they remain as points) and do not render. The various terms, acronyms and icons (stamps derived of cropped screen grabs) used in the PDF to describe steps graphically include: LMB (left mouse button), RMB (right mouse button), MMB (middle mouse button), < (go back), > (go forward), (open the toolbox), (material channel), (keyframe), (render), (snaps), and so on. Hints are also included throughout the PDF tutorial to provide extra information to assist understanding of the underlying principles and how to resolve common issues that may arise. For example: '…if you RMB click on the RGB PP text field again you should see the tag popup: < - arrayMapper1.outColorPP > Holding down the RMB, move the mouse cursor over < - arrayMapper1.outColorPP > and select Edit Ramp. This lets you change the colours for your ramp'. Here, the greater-than and less-than signs are used as parentheses by Maya, leading to possible further confusion. Despite the need to communicate all the information graphically (as text and pictures) in an overtly linear manner commensurate with the linear nature of Maya more generally, many steps are still missed or misunderstood leading to errors. And, although students can progress at their own pace, the information contained in the PDF tutorials remains less than other formats, as it is not always possible to communicate all the procedures visually or by symbolic narrative – students still need to interpret the coded nature of the symbolic narrative. For example, it is surprising how many students do not understand what LMB stands for.

Using the Video Tutorial

The narrativised video is most closely related to the demonstration. However, it can be longer, more thorough and explore more alternatives than the demonstration or the PDF. Nonetheless, the limitations of the streaming server which hosts the videos means some information is not included – such as the introductory history. As for the PDF tutorials, students tended to divide their screen between the video and Maya. This particular video tute was divided into 5 clips or segments: settings and preferences and creating the emitter; changing emitter type from points to spheres; adding a colour ramp to the 'particles' object; setting the hardware renderer up to render the effect; and, making adjustments such as timespan, colours and random seeded sphere sizes. These 5 clips make it easier to remember the order of each critical step. It also reinforces the overall linear nature of Maya. In the first clip the mouse cursor slowly moves over the dropdown menu items while the narrator explains what is going to happen next and why this is necessary. For example, adjusting the default settings from metres to millimetres. Then the array of menu types is changed to dynamics (where the physics functions are accessed). The different types of particle emitter (omni, directional, surface and so on) are explained and where to access them are shown. A directional emitter is created with direction and spread. Total frames are set to 200 and the initial animation is run showing the particles emitted from a point at the origin. What will be done in the next video clip is briefly described at the end of the previous clip. In the second clip, the animation is run about halfway, stopped, and the resultant particle cloud is selected. The particle type is changed to spheres. In the third clip, colour is added to the particle cloud with careful attention paid to the use of the RMB and all of the nested dialog boxes that pop up and the values needed to ensure a useful result. In the fourth clip, how to set up the renderer to render the result and what happens if the settings are incorrect is explained. With a renderable result, the final clip shows how to make adjustments to the particle cloud to refine the result and allow for some personal creativity. It includes a typical result and an example of a more realistic, extended, result and briefly how to create it. While the video clip opens up opportunities for the teacher to be more creative with the content and examples, its production is considerably more time consuming than other modes. Added to this is the need to update the video tutorials at least every 3 years due to software upgrades. Unlike the demonstration and PDF which can be edited and revised, the video clip needs to be completely re-shot. This makes the video clip the least efficient to produce and maintain.

Summary of Knowledge Visualisation Modes

From these accounts, three core operational themes emerge: the role of the screen; how much control students had over the delivery of the content; and, the level of detail in the content, dictated by the media format. It was observed that the screen size and location affected how much and how accessible the content was; the mode of delivery (by teacher, PDf or video) affected how much control students had over delivery pace; and, duration (demonstration), need to mentally verbalise text (PDF), or restrictions of the streaming server (video) affected how much and how detailed the content was. Table 7 shows in brief the comparative effectiveness of each mode.

Overwhelmingly, students found having the Maya software open and either the video or PDF or both open alongside the most effective method of accessing the knowledge contained in the tutorials. So much

Table 7. Comparative effectiveness of visualisation mode

	Screen	Control	Content
Demo	Large but far away from and separate to the student's monitor	Teacher controls the pace, students can interrupt but no direct control otherwise	Detail limited by the time needed to complete the whole tutorial
PDF	Restricted to local monitor resolution, but can be open in tandem with Maya	Students have direct control over pace	Detail is limited by the reasonable number of screen grabs, diagrams and text included
Video	Restricted to local monitor resolution, but can be open in tandem with Maya – requires headphones	Students have direct control over pace but limited to start, stop, pause and rewind – all linear functions	Detail is limited by capacity of streaming server to support large files – but detail is generally greater than demonstration or PDF

so that the demonstration was made redundant. But, while the visualisation modes can be seen to affect how much and what type of knowledge was practically available to the students, a more general theory is needed to make sense of the students' choices.

Developing a Theory for Findings

In order to address the question, "why did most of the participants indicate their preference for, and satisfaction with, the video tutorials over the alternatives?", we need to understand the cognitive processes involved in this type of learning environment. The three types of teaching methods investigated – front-of-class demonstration, self-directed text-based PDF tutorials and pre-recorded video tutorials – all invoked different cognitive processes. While the use of audio-visual stimuli (video tutorials) in learning contexts has a long history (see Verhagen, 1994; McNeil & Nelson, 1991;), front-of-class software demonstration does not. However, the purpose of this study was not to identify the most effective teaching method, rather to address why the video tutorial generated such a high satisfaction rating among students. And, the current study focused on comparing the text-based PDF tutorial with its video equivalent. Hence, the efficacy of the front-of-class demonstration is not addressed in any depth in this study. Indeed, for the courses investigated, the clearly greater efficacy of the PDF and video tutorials deemed the demonstration redundant from 2014. For a review of research on the front-of-class demonstration see Majerich and Schmucklerv (2008). Therefore, the remainder of this section focuses on cognitive processes related to the PDF and video tutorials only.

Learning involves all of the senses. Typically however, learning in the classroom focuses on auditory and visual stimulation. Each type is subject to cognitive load – how much and how difficult it is to process information. We can use Paivio's (1986) dual coding theory (Clark and Paivio, 1991; Reed, 2006) and Baddeley's (1998) theory of working memory (Baddeley, 1992, 2003; Reed, 2006) to make sense of the cognitive loads for the auditory and visual information presented in a video tutorial. According to Paivio and Baddeley, these types of information are processed by separate channels. While Paivio's theory focuses on verbal (words) and non-verbal (pictures), Baddeley's theory focuses on visual (text and pictures) and audio (sounds and narration) – both agree on the processing senses: eyes and ears. Baddeley's theory extends Paivio's by suggesting the limited capacity of working memory can inhibit learning if there is a cognitive overload in either channel. This supports Sweller's (1988) cognitive load theory more generally (Reed, 2006; Sweller and Chandler, 1994). Sweller identified intrinsic, extraneous and germane sources of cognitive load. The limited capacity of cognitive load of Paivio and Baddeley's channels is subject to Sweller's (1988) cognitive load theory and Baddeley's (1998) working memory theory. Accordingly, reducing extraneous cognitive load increases germane or long term learning retention, moderated by the intrinsic difficulty of the task. Thus, meaningful learning only occurs when cognitive processing takes place simultaneously in both the verbal and visual channels (Wittrock, 1989; Mayer, 1999, 2002).

Applying Paivio, Baddeley and Sweller's theories to multimedia – such as video tutorials – the coordinated, simultaneous, stimulation of vision and sound appears to generate a lower cognitive load than the text and static images of the PDF tutorial used in the current study (Mayer & Moreno, 2003; Mayer, Moreno, Boire & Vagge, 1999). Moreno & Mayer (2000) also applied Paivio, Baddeley and Sweller's theories in proposing a new theory related to cognitive load when using multimedia. In Moreno & Mayer's theory, auditory instructions require a lower cognitive load than text, leading to better learning outcomes. In their studies, they compare visual stimulus with alternately synchronous and asynchronous text or auditory instructions. They found that synchronous auditory instructions with visual stimulus achieved the deepest learning outcomes.

Multimedia Learning Theory

In Moreno & Mayer's (2000) study they isolated animation, text and narration using synchronous and asynchronous delivery. The current study further isolates images and text from animation and narration by comparing static text and images with dynamic animated and narrated video tutorials. This completely separates the aural and visual channels: where the text-based tutorial relies entirely on visual stimulus (with the conversion of text to word sounds), the video tutorial separates aural from visual with narration and animation. This is illustrated in an adaptation of Mayer and Moreno's (2003) theory of multimedia learning flowchart (see Figure 6).

In adapting Mayer and Moreno's (2003) theory of multimedia learning flowchart to the current study, we see it also illustrates the comparative paths of the text and image based PDF tutorial with the narrative and video tutorial. Both forms communicate via the two channels: auditory/verbal and visual/pictorial, which are processed in working memory and integrated with and guided by prior knowledge from long-term memory. The ability of the learner to cognitively process this information is limited by

Figure 6. Adaptation of Mayer and Moreno's (2003) theory of multimedia learning flowchart to the current study

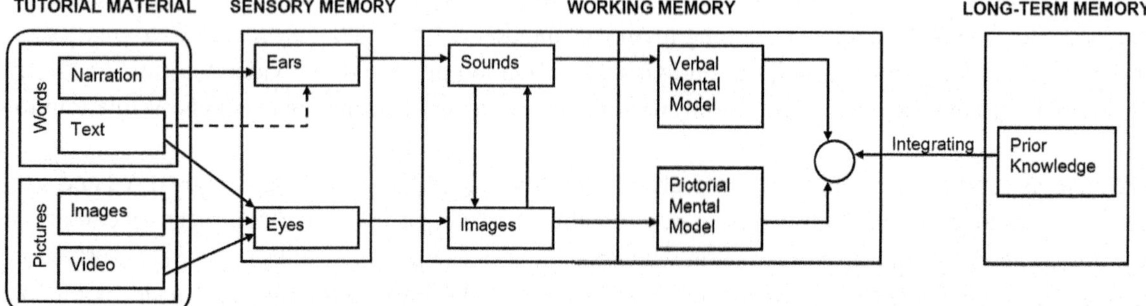

their short-term cognitive load capacity. However, the text and image based PDF tutorial relies almost entirely on the visual channel – textual information is mentally verbalised in working memory thus increasing the cognitive load when compared to its narrated corollary in the video tutorial. The video tutorial separates the auditory/verbal and visual/pictorial channels such that they are cognised synchronously, independently. In turn, this reduces the cognitive load, as the working memory processing of this information occurs in different regions of the brain which can act simultaneously (Baddeley, 1998). Hence, the text and image PDF path is a high cognitive load learning method compared to the narrative and video, low cognitive load path.

In adapting Mayer (1997) and Mayer and Moreno's (2003) theory of multimedia learning to the current study we see that a text and image based PDF tutorial is almost entirely mentally represented in visual working memory. As Moreno & Mayer (2000, pp13-14) point out, "although some of the visually-represented text eventually may be translated into an acoustic modality for auditory working memory, visual working memory is likely to become overloaded." Students are not able to pay attention to both text and images at the same time. They are "...not able to hold corresponding pictorial and verbal representations in working memory at the same time.... [Hence, students are] less able to build connections between these representations." Thus, their performance on transfer and retention tasks may suffer. On the other hand, for a video tutorial, containing narration and moving images, the moving images are mentally represented in visual working memory and the corresponding narration is mentally represented in auditory working memory. As a result the student "...can hold corresponding pictorial and verbal representations in working memory at the same time, [and so are]... better able to build referential connections between them" (Moreno & Mayer, 2000, p13). This process is illustrated as a flow diagram in Figure 7.

In short, while the video tutorial information is processed concurrently in visual and auditory working memory, the visual-only text and image tutorial is disjunct – as only one form of information can be processed at the same time. This leads to a higher cognitive load and greater chance for connections between text and image to be misplaced, misunderstood or missed all together.

In Mayer and Moreno's (1998) study they showed that students who learn with concurrent verbal auditory narration and moving images outperform those who learn with text and images alone. Critical to this process is the synchronicity of the delivery. Narration linked to moving images needs to be closely coordinated. Any spatial or temporal incongruence will overload one or the other working memory modality – auditory or visual. In other words, it may not be possible for the student to hold all of the narration related to a moving image sequence in working memory before that sequence is displayed. Hence, it may be necessary to chunk the auditory information with the moving image display so as not

Figure 7. Flow diagram showing the comparative cognitive loads and processes for a PDF and video tutorial respectively

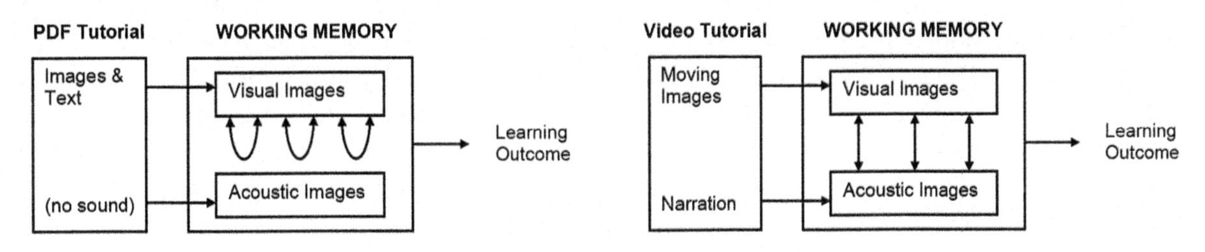

to overload their working memory. For example, narrating a procedure whilst performing that procedure but including segments which may not include any unnecessary narration and vice-versa: halting a moving image sequence to narrate a fuller explanation of what has just been seen, and so on (Mayer, Moreno, Boire & Vagge, 1999).

Applying Theory to Findings

Mayer and Moreno's (2003) theory of multimedia learning, supported by Paivio, Baddeley and Sweller's theories of cognitive load, provides a framework within which to analyse the current study's results. In the previous section we see how the cognitive loads can be distinguished for the text-based PDF and video tutorials respectively. However, before applying Mayer and Moreno's theory, beyond identifying cognitive load differences, it needs refining for the specific type of learning described in this study. Unlike Mayer and Moreno's (1998) studies, which focused on generalist declarative knowledge in the classroom, learning a software package such as Maya 3D is essentially a procedural-motor knowledge acquisition task (where to place the mouse cursor to find menu items); that is, it requires learning specific procedures for predetermined outcomes which can be later applied in different contexts. Hence, to better understand and measure the efficacy of the current study, we need to expand its theory base.

Hoffler and Leutner (2007) expand on Mayer (2001) and Mayer and Moreno's (2003) theory of multimedia concluding that animations or moving images are superior to static images when they focus on procedural-motor knowledge rather than declarative knowledge. Veronokis and Maushak (2005) went further by applying Mayer (2001) and Mayer and Moreno's (2003) theory to the learning of software. They used three different formats: text, static images and audio and animated screen capture with audio. Like Hoffler and Leutner, Veronokis and Maushak found that, while there was no significant difference in the outcomes determined by the format used, video presentations with a focus on procedural knowledge were preferred by the participants.

A problem with Hoffler and Leutner, Mayer and Moreno, and Veronokis and Maushak's studies is that they do not provide any evaluative metric for quantifying the outcomes. They provide heuristic analysis only. Moreover, Shepard's (2003) survey on research in the field of evaluating the use of streaming video to support student learning suggests there is little research in the field and what there is tends to focus on attitudes to the technology rather than quantifiable outcomes from its use (see also Keefe, 2003; Boster etal, 2006, 2007). Most report that video streaming frees up time for the teacher to focus on in-class problem solving rather than content delivery. This presents the current study with a dilemma: how to quantify the efficacy of using video tutorials when compared to their PDF corollary? The following section attempts to address this question.

How to Measure Learning Success?

To address the need to quantify the efficacy of video tutorials for learning outcomes we can look to innovation in education more generally. There is a long history of technological advances and innovation in education (Dunn, 2011; Hsu, 2007). The most obvious is simply how information is recorded by the student. For example, from chalk and slate (1800-1900) to pencil and paper (1900-1940 when the pencil was replaced by the ballpoint pen) to keyboard and mouse (since the late 1980s). The impact these technological innovations have had on the quality of the learning experience is clear. However, *measuring* the quality of the learning experience is less straightforward.

The increased capacity for recording information from chalk and slate to pencil and paper allowed students to contemplate larger chunks of information, to store these, reflect and correct them, but whether there was a measurable increase in knowledge and understanding as a result is not clear. The progression to computers and their exponentially increased capacity to store, share and facilitate collaboration has also not necessarily resulted in quantifiably greater understanding or knowledgeable scholarship.

The limited capacity to record words and illustrations on a slate with chalk would have been frustrating for those who wished to do more and grapple with greater chunks of information. Just as the fragility of paper and pencil – susceptible to fading in sunlight, disintegrating in water, destruction by fire – would have been frustrating to those who mastered the new medium. In turn, although the computer promised to overcome the fragility of pencil and paper and unchain the creative mind – with its seemingly infinite capacity to store and distribute recorded words and thoughts – what has been produced is not measurably of any greater scholarship than what came before. Across these innovations, the common liberating thread appears to be a simple reduction in frustration with the limitations of the prior technology; a frustration with the limited capacity for the technology to facilitate the recording of free thought at a pace commensurate with thought itself. In this vein, the video tutorial presents only an incremental technological improvement over illustrated text. Its main contribution is reduction in cognitive load and better coordination of information, and connections between information elements.

While some researchers suggest video tutorials are superior, or at least preferred, to their static text-based equivalents, few demonstrate actual improved outcomes (see Hoffler & Leutner, 2007; Veronokis & Maushak, 2005; Sheppard, 2003). And, while students are clearly more engaged, this does not necessarily translate to improved outcomes. Vogt etal (2013) are one of the few who claim students using video tutorials do better in an exam situation than those using text-based material. But, the procedural-motor learning of a new software package may be fundamentally different to the more generalist declarative knowledge learning that Vogt etal (2013) report on. In the current study there is no exam equivalent to quantify any learning efficacies as a result of the use of the video tutorials thus Vogt etal's findings cannot be verified here. Because the motivation for learning the software package was for creative production this also makes it doubly difficult to quantify any improvement in outcomes, as its quality tends to be subjectively measured. Similar to what Keefe (2003) and Boster etal (2006, 2007) found, the single most significant reportable change that students experience when using video tutorials in place of, or in combination with, the PDF documents was a reduction in frustration with the software.

Henceforth, students' attitude towards and preference for a particular or combined media format will be used as a measure of learning success. Although a qualitative measure, its reliability will be the distributive proportion of students reporting positive or negative experiences. In other words, as the increased choice of the video media format resulted in less frustration and hence more time on task this can also be used to measure the level of satisfaction. The source of the reduced frustration was a reduction in cognitive load provided by the video tutorials. This is explained in more detail in the next section.

Frustration with Computers

We often encounter frustration when using computers. The frequency, causes and level of severity of frustrating experiences differs depending on type of user (novice or advanced), their situation (work or pleasure) and the tools used (graphics or text intensive). The net result of user frustration is time off task. According to Ceaparu et al (2004, p334), frustration can arise from "…users' lack of knowledge, poor training or unwillingness to read instructions or take tutorials." Problems encountered can result from flaws in the computer software, network or interactions with the interface. These can be measured as user satisfaction or frustration. Three main sources of frustration were observed in this study: errors, confusion and unexpected events.

Errors can be broadly defined as those events out of the control of the user, such as hardware or software malfunction, and those that are due to user misunderstanding or incorrect input (Norman, 1983). Confusion can be broadly defined as the inability of the user to decipher the erroneous result of their apparently well-reasoned actions. Unexpected events can be defined as the appearance or disappearance of an element, tool, function of the software for no apparent reason. All of these events, regardless of their causes, are frustrating for users (Lazar & Norcio, 2000). Examples of the three main sources of frustration observed in this study include:

- **Errors:** User and hardware/software errors such as, accidentally tapping the capitals lock button on the keyboard which disables a number of commands – for which there is no error message (although later versions of Maya do now include a message that the caps lock is on), or not generating any render output because there is a conflict in the machine's configuration settings despite the option to render still being available from within the Maya program
- **Confusion:** Such as selecting a primitive object (e.g. NURBS sphere) to model with and expecting it to behave and respond to certain commands as it has in the past only to discover that, despite having a very similar tool icon, it is a different type of object (e.g. polygon sphere);
- **Unexpected Events:** Such as the disappearance of viewing aids (e.g. view cube) for no apparent reason. This is followed by the difficulty in finding the correct procedure for reinstating the missing aid or tool which may only be found under nested menu options.

Although frustration was not measured directly in the current study, measures of satisfaction were recorded. Satisfaction can be defined as the completion of a goal or task-directed behaviour aimed at the attainment of a need, desire, or want. Frustration occurs when there is an interruption, barrier to, conflict with, or inhibition of, the goal-attainment process (Dollard et al., 1939; Berkowitz, 1989; Shorkey & Crocker, 1981; Dennerlein et al, 2003). Satisfaction can be achieved despite the presence of frustration. This was witnessed when students persevered because their goals were important to them. Frustration, on the other hand, was observed in the way students demonstrated instances of help-seeking behaviours and sought clarification. Indicators of frustration include: yawning, aggressive behaviour, raised voice, tapping of pen/pencil on desk, strumming of fingers, heavy tapping on keyboard. Although related, frustration and anger should not be confused. Level of persistence can be operationalised as time on task and used as an indicator of frustration level.

To understand how frustration and satisfaction are related we can use Deci & Ryan (1985, 2000) and Ryan & Deci's (2002)*Basic Needs Theory* (BNT) of self-determination. According to Deci & Ryan, need satisfaction and need frustration are essentially equivalent measures. They claim the three innate psychological needs to be satisfied for well-being include: autonomy (self-determination), competence (ability) and relatedness (social connectedness). These are affective measures. They claim their BNT is universal across cultures and individual differences (Chen et al, 2015). In the context of the current study, the corresponding affective measures for a sense of well-being associated with working with computers and learning new software are:

- **Autonomy:** Self-sufficiency (sense of one's ability to overcome problems unassisted);
- **Competence:** Adequacy (sense that one's skills are improving over time); and
- **Relatedness:** Congruousness (sense that the computer and its software is a tool used to extend one's own capabilities).

From this we see that, although need frustration is a distinct construct, not just the absence of need satisfaction, it can be conceptualized as the deprivation of need. Sheldon and Gunz (2009) found need frustration was a driver of individual motivation to satisfy need. Hence, frustration and satisfaction appear to be at least affectively correlated.

With this in mind, in an educational context, the teacher should be attuned to a learner's sense of frustration. They should be able to determine if a learner is:

- Content to continue solving problems in a self-sufficient manner or they are becoming increasingly frustrated and likely to quit;
- Improving their skills over time leading to feelings of adequacy;
- Working in concert with the computer and its software or becoming increasingly overwhelmed by the apparent incongruities in the tool.

Indeed, addressing anxiety and frustration is an important part of successful learning experiences (Schank & Neaman, 2001; Kapoora et al, 2007). In the context of the current study, students responded to need frustration in a number of ways:

- Asking for help;
- Seeking solutions from online resources; and
- Investigating the software's included help tool.

The surveys conducted in the current study used a Likert scale to indicate levels of satisfaction. This can be used to operationalise frustration as indicated by increased time on task. In other words, frustration with the software, as observed in the current study, was defined as 'an inability to recognise problems as and before they arise and subsequently solve those problems using logical means'. For example, most procedures in Maya are linear in nature. This means, while there may be multiple steps in a procedure aimed at a predetermined outcome, each step must be performed in the correct sequence otherwise an error may occur. However, it is possible to skip some steps and get a result, albeit not what was the goal of the initial exercise. This makes it difficult to diagnose where the problem is in the sequence: it may be that a critical step was missed but its solution is not obvious from the results alone; or, it may not be immediately obvious to the student what step is being missed, as the order of the steps is not always obvious. Hence, the student may repeat the steps again with exactly the same erroneous result. Other procedures may not require the same strict sequential operations, or specific environment settings may need to be established beforehand. Again, this may not be obvious until many steps have been completed, only to find the result does not match what is called for by the tutorial. For example, the physics simulation engine is sensitive to environment scale. Hence, it is critical to set the unit measure for the environment in the project preferences before commencing a physics simulation. Without this critical step at the beginning it may not be possible to see any results when the physics simulation engine is executed, or the default scale may cause the physics engine to run slow or even crash. Trying to adjust the environment scale after setting up a physics simulation does not scale all of the elements evenly, compounding the opportunity for errors in the scene. Often, the only option in these situations is to start the tutorial again from the beginning. The linear nature of the video tutorial and its reduction of cognitive load in the learning process was found to facilitate a better understanding of the linear nature of the program thus reducing overall frustration.

Frustration as a Measure of Cognitive Load

As Hoffler & Leutner (2007), Veronokis & Maushak (2005) and Vogt etal (2013) report, where the video is superior to the static text-based tutorial is in the way it facilitates an explicit demonstration of the strict linear sequences needed to achieve the desired results for procedural-motor tasks, such as those for mastering Maya 3D. Students can be confident that if they follow exactly what the tutor has done in the video tutorial they should get the same result. The outcome is twofold. On the one hand, if the student follows the tutorial as it is presented they can expect to succeed. And, on the other hand, there is little ambiguity in the instructions that they must follow – there is little opportunity for the student to misinterpret what they must do to succeed. In turn, this boosts the student's confidence with a commensurate reduction in frustration with the software. According to Mayer and Moreno, as the video tutorial requires a lower cognitive load, success in completing the tutorial is more likely than if guided by other media formats. By contrast, the PDF requires a higher cognitive load leading to greater opportunities for misunderstandings and/or missed steps – increasing frustration with the software.

In short, the student who uses the video tutorial experiences a lower cognitive load in their learning and is more confident that they will be able to complete the task correctly with less frustration. This leads to more rapid mastery of the tool, such that their ultimate goal of a creative outcome is achieved earlier. Whether their creative outcome is of a higher quality as a result is then only contingent on the student's ability and diligence in applying their newly gained skill and knowledge to their creative pursuits. Nonetheless, because of the reduced cognitive load and expectations of success they suffer less frustration and report higher satisfaction with the video media than other media.

Evidence of Improved Learning Experiences

Students who used the video tutorial, or its combination with the PDF, were more likely to achieve their goals in less time, reporting higher satisfaction overall. This suggests there was an overall reduction in frustration. This is evidenced by them having more time and being more motivated to spend this time on applying their new knowledge in different contexts. They utilised the increased time to focus on their final creative productions. They did so in different ways. This included:

- Polishing
- Comprehension
- Problem solving
- Planning, and
- Refining

of their final productions – a short animation based on either a specific tutorial or as a final, self-directed project derived of the skills gained from all the tutorials in the course.

Prior to the introduction of the video tutorials, students often spent time cycling through problems until they were resolved or asked the demonstrating tutor for help. This took time away from, and compromised the potential quality of, their final productions. They now had more time to polish and refine their final productions.

They tended to better comprehend the procedures due to the success they found when rigidly following the video tutorial guides rather than their prior trial and error methods – as was often the case with the PDF-only and demonstration cohort. This is counter intuitive to the more typical learning method – whereby one learns by analysing and correcting their misunderstandings. When using Maya, misunderstandings resulting in errors tend to prevent learning. Instead, understanding how the system works and building a legacy of procedural knowledge establishes a more useful foundation for later exploration of potentially creative outcomes within the bounds of those procedures.

Many students who repeated a PDF tutorial did not always result in correctly identifying the source of a problem or error from their initial attempt. Sometimes they were not able to identify the problem in their second or subsequent attempts. Rigidly following the video tutorial guides, on the other hand, may have resulted in the same number of problems but students gained a deeper understanding of how important the linear procedures, inherent in the software, are to success. In turn, this also facilitated a different approach to problem solving. Simply revisiting the video tutorial and following it more closely often resulted in the original error being detected and strategies developed for avoiding it recurring in subsequent procedures – unlike the case with the PDF-only cohort, and not possible with the demonstration.

Planning before acting is essential to quality production outcomes. Prior to the introduction of the video tutorials, students did not have as much time to plan their productions before committing to them. This often resulted in overly ambitious goals and/or disappointing outcomes. With the extra time facilitated by the reduction in frustration with the software due to the use of video tutorials, students more thoroughly planned their productions before committing, resulting in more organised and legible productions.

The PDF-only and demonstration cohort often skipped planning and storyboarding, citing a lack of time. With the extra time available to the video tutorial cohort, these students were better able to refine all facets of the production – from planning, storyboarding, to rendering and editing. Anecdotally, this also manifest as an overall improvement in the quality of their productions. This quality is difficult to quantify however, as the assessment method measures only skills acquisition and the final productions are rated against their peers rather than against a pre-defined target. This means that, while the same level of skills acquisition was generally obtained by all cohorts of students, the numerical value assigned to their final productions tended to remain unchanged over time *across* cohorts. As the quality of the cohort's production work improved over time this was normalised to a common benchmark *within* cohorts. Thus, it is not reflected in a steady increase in overall marks across cohorts over time (as demonstrated in graph 3).

Following application of Mayer and Moreno's (2003) theory of multimedia learning, that the video tutorials were preferred suggests Hoffler and Leutner's (2007) conclusions regarding procedural-motor knowledge hold. More specifically, Veronokis and Maushak's (2005) findings regarding learning software also holds; that, while there was no significant difference in the outcomes determined by the format used, video presentations with a focus on procedural-motor knowledge are preferred by early learners.

And finally, Shepard's (2003) study on evaluating the use of streaming video to support student learning, which found that researchers tended to focus on attitudes, is consistent also with the findings in the current study (Keefe, 2003; Boster etal, 2006, 2007). Video streaming does indeed free up time for the teacher to focus on in-class problem solving rather than content delivery, and increases an overall sense of achievement resulting in higher reported satisfaction with learning experiences (although this remains difficult to quantify).

CONCLUSION

This study set out to track the transition from traditional front-of-class Maya 3D software demonstration with supplemental PDF tutorials to the introduction of video tutorials. Learning Maya 3D is not easy. The software has a complex and intimidating interface for the early learner. In-class software demonstrations are different to traditional scientific phenomena demonstrations. Software demonstrations are disjunct, time consuming and generate lots of problems for students due to misunderstandings. Similarly, the text-based PDF tutorial is open to misinterpretation – students can skip over steps causing errors leading to frustration. It is not always clear from the demonstration or PDF that linear procedures are crucial to successfully completing a Maya 3D tutorial. By contrast and nature, the video tutorial is explicit in its linearity. Step-by-step procedures contained in a video tutorial are unambiguous. It is clear to the student that if they follow the video tutorial they should expect to succeed. However, the PDF is still necessary for some detail and to accommodate different learning styles. Video tutorials are very time consuming to produce. But once produced, they are useful for up to 3 years (before annual updates to the Maya 3D software package dictate the tutorials need to be updated also). In the meantime, segments can be edited or replaced when needed. Without the need to demonstrate, the video tutorial frees up time for the teacher to assist in class with problem solving and students' creative pursuits.

The results from this study show that there was a progressive increase in the use of the video tutorial and its combination with the PDF across semesters and the 5 year study period for the Australian student cohort. The HK cohort showed a different result. This was due to EAL issues and more in-class student-student problem solving compared to the Australian cohort. However, in neither case did the uptake of the video tutorial translate to quantifiably improved outcomes. That the summative assessment over the 5 year period remained largely unchanged suggests either it is not possible to detect changes to outcomes due to the way marks are collated or there is no improvement to outcomes contingent on media format. Anecdotally however, observations from this study suggest otherwise. The general quality of participants' self-directed projects and tutorial exercises did seem to improve over the course of the study. This is possibly due more to the increased time students had to plan and execute them thus presenting more polished productions than any paradigmatic shift in understanding or application of new knowledge *per se*. But, as these projects were circumstantially rated against a benchmark set by the peers within a cohort, the numerical value placed against a production was normalised within the standard marking range – typically 45-85% (fail to high distinction) – obscuring any longitudinally measurable change across cohorts over the period of the study.

While there was no measurable change in outcomes, the traditional front-of-class demonstration fell victim to the success of the videos. It had been uncritically used for many years prior to this study. It became increasingly clear that it was not as effective as a learning tool when compared to the video tutorial. The introduction of the video tutorial brought into focus many of the traditional demonstration's failings – disjunction, uneven pace, and often asynchronous imagery and narration. The video tutorial's greater efficiency and ability to be played back at a pace of the students' choosing overcame many of the failings of the demonstration. The removal of the demonstration from 2014 went largely unnoticed by the students.

After analysing how the knowledge contained in the tutorials was practically visualised it became clearer which modes were effective. While the demonstration provided flexibility, it was constrained by the overall time available. The PDF allowed students to proceed at their own pace but steps could be missed. The video tutorial, on the other hand, leveraged the comprehensiveness of the PDF with the overt linearity of the demonstration and allowed the student to control the pace with less incidence of missed steps. This led to less frustration with the software and greater satisfaction with the learning experience. As a result, the demonstration was discontinued.

The remainder of the study focused on use of the video and PDF tutorials and their combination. A theory was needed to make sense of the results and observations. Clearly, students struggled less and reported higher satisfaction with the video tutorials. But, what was it about the video tutorials that evoked this response? There is little direct research on the use of the video tutorials in teaching. Mayer and Moreno's (2003) theory is perhaps the only comprehensive attempt to explain learning in the context of multimedia technology in the classroom. However, it was necessary to go back to first principles to understated how Mayer and Moreno developed and applied their theory to multimedia. This revealed the primary field applied related to the varying cognitive load experienced by learners using different media formats. From this, the synchronous audio-visual stimulus of a video tutorial presented the lowest cognitive load of the media formats investigated.

Using video tutorials, it was observed that students experienced reduced frustration with the software they were learning when compared to demonstration or PDF only methods. This suggests they were better comprehending the idiosyncrasies of the software, leading to more effective learning. The synchronous delivery of visual and auditory stimulus from the video tutorial seemed to remove the ambiguity of its text-based or teacher-demonstrated corollary. The strict linear nature of the procedures needed to learn how to use the software were not well supported by the demonstration or PDF tutorial alone. This increased understanding and reduction of frustration with the software appeared to be due to the reduced cognitive load afforded by the audio-visual synchronicity of the video. This is consistent with cognitive load and its relationship to audio-visual stimulus as researched by Paivio (1986), Baddeley (1998) and Sweller (1988) and applied in Mayer and Moreno's (2003) general theory of multimedia learning.

Changes in preferences observed in this study were also consistent with Mayer and Moreno's findings and others'. However, where the research reported here differed from prior research was in the type of learning involved. Learning software procedures is distinctly different to the more generalist declarative knowledge acquisition reported by others. Learning software procedures is a procedure-motor knowledge acquisition task. Hoffler & Leutner (2007) and Veronokis & Maushak (2005) addressed this mode of learning. However, their research, and an earlier survey by Shepard (2003), failed to provide a method for measuring the effectiveness of multimedia in learning software procedures. Instead, only attitudes and preferences were recorded. It was also beyond the scope of this report to formulate an efficiency metric directly. Hence, student attitudes and preferences for media format were used also as a measure of learning success. What evidence there was of improved learning outcomes was mostly anecdotal (as discussed earlier). These related to the increased time students had to focus on their final productions following the reduced cognitive load and frustration with learning the software procedures afforded by the video tutorials.

Evidenced by the results collated in this report, the introduction of the video tutorial was a success, in that it reduced student frustration and led to greater overall feelings of achievement and satisfaction with the course material due to reduced cognitive load. This showed the traditional demonstration to be redundant. However, the need for more than one type of learning material was still evident. The text-based PDF tutorial still plays an important role in providing detailed back-up information for the video tutorial and support for those students who prefer to read rather than listen to instructions – especially EAL students.

Going forward, the introduction of the video tutorials also opened up new opportunities for different methods of teaching altogether. They provided:

- More time for the teacher to move around the class and assist in problem solving;
- Time for being more creative in the production of the tutorials – whereas previously the time-consuming demonstration restricted how much could be covered and in what detail;
- The opportunity to introduce new material and advanced extensions to the standard tutorial; and,
- More generally, multiple learning paces and styles – introductory to advanced – better suited to the diversity of learning styles and ability within a cohort.

The reduction in cognitive load and subsequent reduction in frustration with the software freed students to explore creative outcomes rather than focus on learning the tool only. The primary motivation for learning the tool, creative outcomes, was achieved. More generally, understanding how Maya can be used to create 3D artefacts is critical in industries such as architecture, engineering, film production, games, animation and many others. The students' experience with Maya in these courses is transferable. Even if they choose not to pursue a technical role directly related to the use of 3D software their new-found skills will enable them to practically and conceptually direct others to create the 3D artefacts needed.

FUTURE DIRECTIONS

From the study of EAL students in HK this informed how the same material can be better produced for students of EFL (English as a First Language). Material was adjusted to benefit the EAL students but this tended also to make the same material even clearer for EFL students. For example, the removal of colloquialisms that EAL students might not be familiar with, correct enunciation when narrating, and the need for more detailed explanations and descriptions of what is expected when performing the tutorial assisted both cohorts. This highlights the impact and importance of cross-cultural considerations and understandings when preparing video tutorials more generally. From this, it will be interesting to see if over the ensuing years the HK cohort follow the Australian students' experience with more uptake of the video tutorial as a learning medium choice (see also Gunn & Kraemer, 2011). This would involve understanding better the HK students' pre-varsity learning modes, such as the extent to which rote learning might influence media format choices, and do the HK students access online video tutorials of their own volition as the Australian students do? It would also be interesting to explore the potential for establishing a metric to directly measure the efficacy of the different media formats discussed in this study. Typically, a series of before and after cognitive tests would be needed to measure the effectiveness of each type of media on skills acquisition and knowledge retention.

REFERENCES

Baddeley, A. (1992). Working Memory. *Science. New Series, 255*(5044), 556–559.

Baddeley, A. (1998). *Human memory*. Boston: Allyn & Bacon.

Baddeley, A. (2003). Working memory: Looking back and looking forward. *Nature Reviews. Neuroscience, 4*(10), 829–839. doi:10.1038/nrn1201 PMID:14523382

Berkowitz, L. (1989). Frustration-aggression hypothesis: Examination and reformulation. *Psychological Bulletin, 106*(1), 59–73. doi:10.1037/0033-2909.106.1.59 PMID:2667009

Bork, A., & Gunnarsdottir, S. (Eds.). (2012). *Tutorial Distance Learning: Rebuilding Our Educational System*. Springer Science + Business Media, LLC.

Boster, F. J., Meyer, G. S., Roberto, A. J., Inge, C., & Strom, R. (2006). Some effects of video streaming on educational achievement. *Communication Education, 55*(1), 46–62. doi:10.1080/03634520500343392

Boster, F. J., Meyer, G. S., Roberto, A. J., Lindsey, L., Smith, R., Inge, C., & Strom, R. (2007). The impact of video streaming on mathematics performance. *Communication Information, 56*, 134–144.

Ceaparu, I., Lazar, J., Bessiere, K., Robinson, J., & Shneiderman, B. (2004). Determining Causes and Severity of End-User Frustration. *International Journal of Human-Computer Interaction, 17*(3), 333–356. doi:10.1207/s15327590ijhc1703_3

Chen, B., Vansteenkiste, M., Beyers, W., Boone, L., Deci, E. L., Duriez, B., & Verstuyf, J. et al. (2015). Basic psychological need satisfaction, need frustration, and need strength across four cultures. *Motivation and Emotion, 39*(2), 216–236. doi:10.1007/s11031-014-9450-1

Clark, J. M., & Paivio, A. (1991). Dual coding theory and education. *Educational Psychology Review, 3*(3), 149–210. doi:10.1007/BF01320076

Deci, E. L., & Ryan, R. M. (1985). *Intrinsic motivation and self-determination in human behaviour*. New York: Plenum Press. doi:10.1007/978-1-4899-2271-7

Deci, E. L., & Ryan, R. M. (2000). The "what" and "why" of goal pursuits: Human needs and the self-determination of behaviour. *Psychological Inquiry, 11*(4), 227–268. doi:10.1207/S15327965PLI1104_01

Dennerlein, J. T., Becker, T., Johnson, P., Reynolds, C., & Picard, R. (2003). Frustrating computers users increases exposure to physical factors.*Proceedings of the 15th Congress of theInternational Ergonomics Association*.

Dollard, J., Doob, L., Miller, N., Mowrer, O., & Sears, R. (1939). *Frustration and Aggression*. New Haven, CT: Yale University Press. doi:10.1037/10022-000

Dunn, J. (2011). The Evolution of Classroom Technology. *Edudemic: Connecting Education and Technology*. Retrieved from http://www.edudemic.com/classroom-technology/

Gunn, M., & Kraemer, E. W. (2011). The agile teaching library: Models for integrating information literacy in online learning experiences. In S. Kelsey & K. St. Amant (Eds.), *Computer-mediated communication: Issues and approaches in education*, (pp. 191-206). Hershey, PA: Information Science Reference.

Höffler, T. N., & Leutner, D. (2007). Instructional animation versus static pictures: A meta-analysis. *Learning and Instruction*, *17*(6), 722–738. doi:10.1016/j.learninstruc.2007.09.013

Hsu, J. (2007). Innovative Technologies for Education and Learning: Education and Knowledge-oriented applications of blogs, Wikis, podcasts, and more. *International Journal of Information and Communication Technology Education*, *3*(3), 70–89. doi:10.4018/jicte.2007070107

Kapoora, A., Burlesonc, W., & Picard, R. W. (2007). Automatic prediction of frustration. *International Journal of Human-Computer Studies*, *65*(8), 724–736. doi:10.1016/j.ijhcs.2007.02.003

Keefe, T. (2003). Enhancing a face-to-face course with online lectures: Instructional and pedagogical issues. *Proceedings of the eighth annualMid-South Instructional Technology Conference*.

Lazar, J., & Norcio, A. (2000). System and Training Design for End-User Error. In S. Clarke & B. Lehaney (Eds.), Human-Centered Methods in Information Systems: Current Research and Practice, (pp. 76-90). Hershey, PA: Idea Group Publishing. doi:10.4018/978-1-878289-64-3.ch005

Majerich, D. M., & Schmuckler, J. S. (2008). *Compendium of science demonstration-related research from 1918 to 2008*. Xlibris.

Martinovic, D., McDougall, D., & Karadag, Z. (2012). *Technology in Mathematics Education: Contemporary Issues*. Informing Science Press.

Mayer, R. E. (1997). Multimedia learning: Are we asking the right questions? *Educational Psychologist*, *32*(1), 1–19. doi:10.1207/s15326985ep3201_1

Mayer, R. E. (1999). The promise of educational psychology: Vol. 1. *Learning in the content areas*. Upper Saddle River, NJ: Prentice Hall.

Mayer, R. E. (2001). *Multimedia learning*. New York: Cambridge University Press. doi:10.1017/CBO9781139164603

Mayer, R. E. (2002). The promise of educational psychology: Vol. 2. *Teaching for meaningful learning*. Upper Saddle River, NJ: Prentice Hall.

Mayer, R. E., & Moreno, R. (1998). A split-attention effect in multimedia learning: Evidence for dual processing systems in working memory. *Journal of Educational Psychology*, *90*(2), 312–320. doi:10.1037/0022-0663.90.2.312

Mayer, R. E., & Moreno, R. (2003). Nine ways to reduce cognitive load in multimedia learning. *Educational Psychologist*, *38*(1), 43–52. doi:10.1207/S15326985EP3801_6

Mayer, R. E., Moreno, R., Boire, M., & Vagge, S. (1999). Maximizing constructivist learning from multimedia communications by minimizing cognitive load. *Journal of Educational Psychology*, *91*(4), 638–643. doi:10.1037/0022-0663.91.4.638

McNeil. B.1., & Nelson, K.R. (1991). Meta-analysis of interactive video instruction: A 10 year review of achievement effects. *Journal of Computer-Based Instruction*, *18*(1), 1–6.

Moreno, R., & Mayer, R. E. (2000). A Learner Centred Approach to Multimedia Explanations: Deriving instructional design principles from cognitive theory. *Interactive Multimedia Electronic Journal of Computer-Enhanced Learning*, *2*(2), 12–20.

Norman, D. (1983). Design rules based on analyses of human error. *Communications of the ACM*, *26*(4), 254–258. doi:10.1145/2163.358092

Paivio, A. (1986). *Mental representation: A dual coding approach*. Oxford, UK: Oxford University Press.

Park, J. E. (2004). *Understanding 3D Animation Using Maya*. Springer.

Reed, S. K. (2006). Cognitive architectures for multimedia learning. *Educational Psychologist*, *41*(2), 87–98. doi:10.1207/s15326985ep4102_2

Ryan, R. M., & Deci, E. L. (2002). An overview of self-determination theory: An organismic-dialectical perspective. In E. L. Deci & R. M. Ryan (Eds.), Handbook of self-determination research, (pp. 3-33). Rochester, NY: The University of Rochester Press.

Schank, R., & Neaman, A. (2001). Motivation and failure in educational systems design. In K. Forbus & P. Feltovich (Eds.), *Smart Machines in Education*. Cambridge, MA: AAAI Press and MIT Press.

Sheldon, K. M., & Gunz, A. (2009). Psychological needs as basic motives, not just experiential requirements. *Journal of Personality*, *77*(5), 1467–1492. doi:10.1111/j.1467-6494.2009.00589.x PMID:19678877

Shephard, K. (2003). Questioning, promoting and evaluating the use of streaming video to support student learning. *British Journal of Educational Technology*, *34*(3), 295–308. doi:10.1111/1467-8535.00328

Shorkey, C. T., & Crocker, S. B. (1981). Frustration theory: A source of unifying concepts for generalist practice. *Social Work*, *26*(5), 374–379.

Sweller, J. (1988). Cognitive Load during problem solving: Effects on learning. *Cognitive Science*, *12*(2), 257–285. doi:10.1207/s15516709cog1202_4

Sweller, J., & Chandler, P. (1994). Why some material is difficult to learn. *Cognition and Instruction*, *12*(3), 185–233. doi:10.1207/s1532690xci1203_1

Verhagen, P. W. (1994). Functions and Design of Video components in Multi-Media applications: a Review. In J. Schoenmaker & I. Stanchev (Eds.), *Principles and tools for instructional visualisation*. Enschede: Faculty of Educational Science and Technology, Anderson Consulting - ECC.

Veronikas, S., & Maushak, N. (2005). Effectiveness of Audio on Screen Captures in Software Application Instruction. *Journal of Educational Multimedia and Hypermedia*, *14*(2), 199–205.

Vogt, N. P., Cook, S. P., & Smith Muise, A. (2013). A New Resource for College Distance Education Astronomy Laboratory Exercises. *American Journal of Distance Education*, *27*(3), 189–200. doi:10.1080/08923647.2013.795365

Wang, V. C. X. (Ed.). (2009). *Handbook of Research on E-Learning Applications for Career and Technical Education: Technologies for Vocational Training*. IGI Global. doi:10.4018/978-1-60566-739-3

Wittrock, M. C. (1989). Generative processes of comprehension. *Educational Psychologist*, *24*(4), 345–376. doi:10.1207/s15326985ep2404_2

KEY TERMS AND DEFINITIONS

3D Scene: A virtual world containing three-dimensional objects which can be rendered in perspective.

Autodesk's Maya 3D: Software package used in the industry for creating cartoon style animations, visual effects for movies, characters and environments for games, among other uses, such as in architecture and civil engineering.

Cognitive Load: The relative difficulty in processing different types of information, such as auditory, visual, tactile and so on.

Declarative Knowledge: Factual knowledge or information that is knowable or can be learned from books, teachers, and other forms.

Front-of-Class Demonstration: Teacher demonstrates software procedures via a projector and screen while students follow or take notes.

Frustration: Inability to achieve a designated goal.

Keyframe Animation: Similar to a classic Disney cartoon animation whereby individual images are organised in a sequence such that when 12 or more images are displayed in quick succession (minimum 12 images per second) objects appear to move across the screen. Each 'frame' or image is 'keyed' to a specific position in the sequence.

Long-Term Memory: Where information from prior knowledge is stored. For example, items from the short-term memory that require recognition are compared with similar objects in the long-term memory store.

NURBS: Non-uniform, rational b-spline, a mathematical description of a curve, grid or mesh in three-dimensional space which can be coloured and rendered to the screen.

PDF: A text-based document that can include images. Its small file size is suitable for accessing via the internet.

Polygon: Minimum definition has three points in 3-dimensional space forming a 'face' which can be filled with colour and rendered to the screen. Collections of polygons can be used to construct three-dimensional objects.

Procedural-Motor Knowledge: Knowledge that can only be learned by doing. For example, learning procedures in the Maya 3D software program requires motor actions such as moving the mouse around, using combination of keyboard controls and so on.

Satisfaction: The realisation that a goal has been achieved.

Video Tutorial: Simultaneous narrated demonstration of software procedures showing computer screen, mouse cursor and any actions performed with the software, such as drop-down menus, typing in textfields and moving objects across the screen.

Working Memory: Or short-term memory, where most operations are performed that require minimal detailed analysis, such as identifying how many items on one's desktop and so on.

Chapter 10
Metaphors for Dance and Programming:
Rules, Restrictions, and Conditions for Learning and Visual Outcomes

Anna Ursyn
University of Northern Colorado, USA

Mohammad Majid al-Rifaie
Goldsmiths, University of London, UK

Md Fahimul Islam
Queens College CUNY, USA

ABSTRACT

This chapter offers visual explanation on how to code using dance as a metaphor. This approach provides an overview of programming with ready to follow codes. It explores the implementation of restrictions and conditions in programming as compared to those ruling various dances. For those willing to learn or grasp the idea of coding for learning or acquiring better communication with co-workers, several programming languages are used to solve a similar task. Thus, similar codes are written in various languages while being related to the same topic. They delineate various dances and their rules, so the reader can compare and contrast the underlying principles for various environments. Then, exploration of invisible patterns created by movement of feet and aesthetics behind resulting patterns are presented, to highlight the dynamics behind the images generated by music, and subsequently the resulting movements of dancers according to various rules behind choreographies. The idea of randomness in coding, as compared to improvisation in dance is also investigated, when the dancers feel the music to create their own solutions to shape, space, and time, rather then following and obeying already designed rules.

DOI: 10.4018/978-1-5225-0480-1.ch010

INTRODUCTION

Many believe that particular skills or abilities are necessary to even begin thinking about coding and appreciating the outcome (Boden, 2016), as one needs a different mindset and training for coding and for creating art. At the same time, programmers and artists often cooperate having visual strategies at hand and developing on the go the new strategies, rules, patterns, and products. For this and many other reasons, there is a constant need for coders. Before software became easily available, artists eager to compute had to learn programming to be able to use dumb terminals. In the days of minicomputers and mainframes a dumb terminal consisted of a keyboard as an input and a display screen as an output, but it lacked the capability to process or format data. It was connected via phone line to the computer services, and was usually used by artists at night, when employees who agreed to share their machines were not using them for their daily tasks. Now, with electronic devices reaching new levels of usefulness and efficiency, new options open up and invite an interdisciplinary work involving collaboration and integration.

It may be not easy for an artist to enroll into a programming class, often without prior knowledge or prerequisites. At the same time, it is hard for a person to match the precision of a machine, and develop projects without expanding the capacities of mighty software packages. Also, many of us are curious and playful. We like to explore and welcome happy accidents, which when noticed in time can become a great source of inspiration. Curiosity and inspiration motivate our efforts; they often lead to collaboration, which in turn generates new ideas and solutions, and thus may advance progress.

Many visual artists need to acquire coding skills, while many coders play with visual outcomes. For example, Mohammad Majid al-Rifaie, one of the co-authors of this chapter investigated drawings created by two nature inspired, swarm intelligence algorithms (al-Rifaie, Bishop & Blackwell, 2011; al-Rifaie, Aber, & Bishop, 2012), and also served as a curator of an exhibition "A-EYE" for code based art, often using artificial Intelligence (AI) techniques. Another co-author, Fahimul Islam develops programs for music, while Anna Ursyn creates her art works based on programming.

The visual aspect behind constantly growing technology stimulates its users to become more knowledgeable, creative, explorative, open to ideas, and adventurous enough to meet the challenge set up by the presence of the computer. We can do it by creating an image in order to transform it into other dimensions: from two dimensional to three-dimensional rotational object, a time based image, or a virtual reality based scenario. The idea of robotics, computer generated poems, music, art, or literary works has been known for a long time. There are many images, games, apps, and other opportunities available on the web, and one might feel compelled to try one's own hand at them.

A metaphor indicates one thing as a representation of another, difficult one. Thus metaphors enable us to make mental models and comparisons. We usually choose for the metaphors concepts and objects that we hold to be easily understandable and familiar to our audience. Abstract images that resemble something through the metaphors may lower the learning curve. To some programmers or statisticians the coding or a statistical analysis may work like a cookbook: a chef follows some rules and uses ingredients presented in the recipe; if instructions are followed well, food resembles one that was done according to the grandmother's, grandfather's, or some other chef's recipe recorded in books, online, or by words. This may be also true in a case of coping traditional dishes. However, some ambitious restaurants will search for chefs who can do the research followed by their individual interpretations, which would not only be healthful for their audience but would thrill their palates. One can ponder in a similar way about performers of musical composition. Great performers mastered their own instruments so their music can sound according to the original score. However, the awards in competitions go often to those able to apply their own interpretation.

Dances, which are based on the rules set for each partner, also leave space for individual interpretations. A dance metaphor used for explaining programming aims to address the particular rules behind writing programs and guide the programming job through dances. If the steps, restrictions, and conditions are seen as the common basis that forms the rules, we might see some similarities between dancing and programming. It is a metaphor used for obtaining a bigger picture behind a subject, but also a way to examine the potential power behind graphics generated by the invisible lines created by the feet of dancers.

Auditory metaphors seem essential for music theory and appreciation. Metaphors are often applied to process complex information or non-physical, numerical data (such as network system data or stock market values). With the use of metaphors, such data often refers to naturally occurring objects and related events.

The aim of this chapter is to show how the restrictions used in dance might be relevant to those in coding. This means one could Google a description of a dance of ones own choice, and using a line and a shape attached to it, one would write a program that follows the rules for that particular dance. Then follow the explanations of the program-related texts and comments for each language/dance. This provides a quite simple idea allowing the reader to comprehend conditions and statements, and grasp the power of the 'if,' 'while,' 'else,' or 'transform and repeat' conditions.

The following text presents first a general discussion on music and dance, especially the ways music can be combined with words, visuals, and instructions coded by writing a program. The dance programming part is aimed at describing the rules for selected dances in different programming languages in order to apply the resulting dance related programs as metaphors for learning coding.

BACKGROUND INFORMATION: MUSIC, WORDS, AND PROGRAMS

Music may address human senses and thus convey a message without the use of words. Sound signals can be listened and understood by different groups of listeners, and the received messages may have a different meaning in particular tribes or social groups. Music that conveys fear or joy may be received and understood without the verbal means or previous education in music; animals often react to these messages. Musical messages may be previously encoded and then learned, thus providing communication between the composer and the audience.

Pitch

A musical scale, as a set of musical notes ordered according to basic frequency or pitch, brings about common understanding. Auditory messages may translate signals into ones enhancing information, perfection aimed exercises, and induce action readiness, preparedness, and even bravery. While there are courses that support building one's perfect pitch, no one can learn how to produce a new melody. Similar thoughts can be found about color in visual arts: one either possesses the absolute color memory or not. Figure 1 presents "A Flutist" – a sculpture by Anna Ursyn.

Figure 1. Anna Ursyn, A Flutist, a sculpture in bronze
(© 2009, A. Ursyn. Used with permission).

Rhythm

The first music making was done by hands and legs of a human, followed by mouth, and then by the human-created instruments. Some believe, mouth was the first instrument for music making. Rhythm created with a drum conveyed in all historical times messages on several levels and in different ways. Signals created by drummers were able to:

- Encourage various actions: body movement, festive dancing.
- Mobilize by inducing pressure and readiness to act.
- Communicate and inform.
- Warn or alert.
- Energize and stimulate enjoyment and fun.
- Elevate or decrease the aggression level.
- Elevate adrenaline level conducive to fighting an enemy.
- Cheer.
- Define the pace and rhythm geared toward supporting prayers, meditation, relaxation, exercises aimed at calming down and attenuating tension.

- Activate the release of regulatory substances such as hormones and mediators activating and signaling functions, thereby motivating the listeners or enhancing a feeling of belonging.

Music and Words

There may be many ways to combine music and words. They may include:

- Choral music, including a choir in Greek tragedies, liturgical church worship, chorale music, and musical ensemble of singers.
- Stage performances such as opera, operetta, and musical.
- Cantata, with a choir, vocal solos, and orchestra.
- Hymn.
- Recitative.
- Romantic songs based on ballads or poetry.
- Carol music.
- Lullaby.
- A score written and composed to accompany a film.
- Vocalization of musical themes consisting of melody without words; it makes the singer's voice an instrument, while such musical forms as an operatic aria or choral oratorio add a storytelling component to the musical one.
- Jazz music; one can associate it with the abstract forms of art.
- Rock music may often contain lyrics and thus the storytelling elements.
- Musicals.
- Special effects link voices with noise and sounds.
- Vocal music, with lyrics or singing without words where a voice is considered instrumental music.
- Songs with words set to music.
- Storytelling, which sometimes takes form of sound effects explaining and emphasizing an action.
- Music videos and animations combine many times verbal and visual messages.
- Songs and music videos that are often produced as part of commercials and advertisements.
- Elevator music and background music in social spaces often interlaces melodies with words.
- Phone ringtones.
- A Greek choir.
- Japanese theater NO and its contents explanation.
- Kabuki theater.
- Communication with the use of drums.
- Ritual music for life events (initiations, weddings, funerals).
- Music produced on a special occasion.
- Musical comment of social or political events, revolutions, for army, political cabaret (such as Guignol, Bread and Puppet, Shadow puppet theater) and many other forms.
- Silent movies with music supporting action understanding, acting as a non-verbal comment.
- Documentary films with music as a part of the data but also illustration.

Numerous forms of social entertainment involving music and words combined may include:

- Karaoke, where amateur singers perform along with recorded music and the displayed text.
- Couplets – a singing form from long ago comprising two rhythmic, rhymed lines.
- Sailor songs.
- "Happy birthday" songs conveying good wishes for a person.
- Sang recitation or declamation of poetry.
- Rap music with the rapidly, rhythmically recited texts often improvised without preparation.
- Singing telegrams invented in 1933 by George P. Oslin (Western Union, 1972); an artist delivers a message as a musical form, thus combining literary and musical presentation.
- Music played in silent films having no synchronized recorded speech may somehow substitute verbal storytelling.
- Ballads, as romantic songs, often based on narrative poems telling a story or expressing feelings. Also medieval romances telling about heroes and their chivalry combined storytelling with music. Medieval bards recited poems and epics supported with music. During High Middle Ages troubadours composed both music and lyric poetry to perform, in words of Dante Alighieri (2005), a rhetorical, musical, and poetical fiction.
- Rhythm in poetry becomes almost a musical form, and thus inspires composers to create music, songs, or music for films. Onomatopoeic qualities of a verse that refer to the phonetic properties of its words may inspire musicians to compose music. One may ponder whether onomatopoeia in poetry can be compared with the cases when music themes imitate sounds of nature.

Text in Music

Music may convey important matters shared by social groups. Words and musical tones may be seen as two constructs that may involve different cognitive capacities and mental capabilities. Words may serve to enhance a musical composition; on the other hand, music may support works such as a theatrical play, a film, an opera, or a television spectacle. Words in musical compositions may take form of songs, arias, or other vocal forms; they have to obey rhythm and agree with overall features of a composition. Text can be recorded for a musical work as written words that convey meaning with the use of a selected learned alphabet. Music in the visual and/or the text-based works may help control the viewers' understanding and feelings (for example, when the music becomes ominous or joyful). For example, working on drums may support conveying verbal messages. Techno poetry deals with notions and rhythms behind technological terms.

Literary elements of a work may influence perception of music by transposing psychological aspects of music into the attempted states of the listeners' minds. Performing rap music having a strong rhythm of the recited words may cause emotional reaction to texts, e.g., about the social pressure of privileged groups. Aural signals may indirectly evoke verbal connotations, such as by conveying a message with a Morse code. Rituals may comprise a message from the elders that is conveyed according to their knowledge to a whole group of inferior members of a tribe or a clan; such message may make them agree to sacrifice their children for a higher goal such as a better survival tactic. Words might be not necessary for a ritual; in specific cases music invoked a sacrifice of one's own child.

Education in Music

Many times the listeners do not have enough knowledge about the sound spectrum, pitch, value, and volume to perceive intentions of a composer/creator of music. However, music exists independent of the way it is created or recorded. Even making a rhythm by pounding a table with scissors may create music. Music education includes learning the music notation systems. They transform sounds and single tones of specific pitch made by human voice or musical instruments into written notes or symbols for software, which allow everybody to convey an intended message. Human creativity allows composing music without making a proper formal record. This makes appreciation of native or folklore music possible, but often causes a disappearance of valuable compositions. Many times compositions made without any record have strong spontaneity; this doesn't have to be correlated with education in music, connoisseurship, or the talent power. Many engineers feel a need to consider themselves artists, while many artists engage in music related media; all this is happening because the media became multisensory and related to more than one communication medium in the arts.

In semiotics, when someone sends a message, someone else in the audience (an addressee) decodes the message according to one's own cognitive framework and the semantic and syntactic rules related to particular social and cultural codes. A semiotician and writer Umberto Eco (1989, 1990) drew a distinction between the closed texts with one interpretation and the open texts that might have possibilities of multiple interpretations made by the readers. The notion of an open and closed message may somehow refer to the narrative and non-narrative music compositions (e.g., some contemporary symphonies and instrumental compositions). Also, an artist statement about a painting may somehow refer to verbal comments about a musical composition (if only by assigning a title). A music critic may, for example suggest events occurring in the Napoleon's campaign for the 3rd Symphony "Eroica" by Ludwig van Beethoven (1770-1827). Remembrance of the fields, sheep, and warm seasons may influence perception of the 6th "Pastoral" Symphony by this composer. Figure 2 presents "A Concert" by Anna Ursyn.

Elements and Principles of Design in Art and Music

One may ponder how the elements of a musical composition such as its structural parts (e.g., a four-movement structure of a symphony, or a cadence at the end of a piece of music), variation techniques, or virtuoso's interpretations may have formal counterparts in visual arts. Artists' inspiration for creating music or a literary work may have common sources driven from nature (such as clouds, weather), environment, or technology. It may come as thoughts (verbalized or not), spoken or written words (as in music with voice, including songs), or musical notes, but also as meditative forms, dreams with related feelings and emotions (such as fear or love). Concepts like talent, creativity, formulas, abstract symbols, and design are common to all artistic areas. They all take a creative mind to be good at it, see the spaces in between, and things that come out of them. Creativity is often considered a currency of the 21st century. For example, a brochure published for prospective students by the Ringling College of Art and Design (Make your Mark, 2015) gives strong emphasis to this direction, while big companies such as DreamWorks or Sony instruct educators to draw attention to the quality portfolio, because it is easier to train their newly hired employees in software after they were already taught how to develop creativity and support talent.

Elements of design in visual arts: color and value, shape and form, space, line, and texture refer to what is available for the artist/designer or anyone willing to communicate visually, while principles of

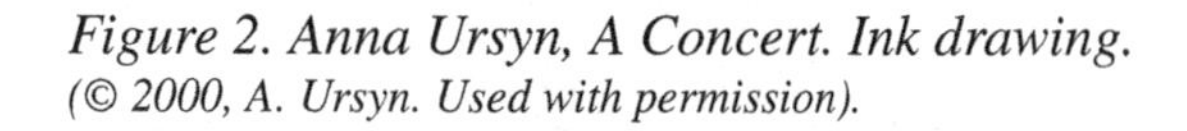

Figure 2. Anna Ursyn, A Concert. Ink drawing.
(© 2000, A. Ursyn. Used with permission).

design describe how the elements could be used and provide the rules by which an artist uses the elements of design. Principles of design that are most often used in visual arts are balance, emphasis, movement, variety, proportion, and unity. One can ask how a composer may affect perception and evoke emotion in viewers by applying the elements and principles of music making. Art and music, as well as science and mathematics may use the same principles of design.

Order can be seen in most of disciplines of music, art, and science. Both in music and in visual arts we use metaphoric representation to tell a story or create emotional response. Both music and art follow a preset line, but it can be distorted and abstracted earlier. Different terms have similar outcomes. Music has form and composition that can be symmetrical or asymmetrical (same as in visual arts), and they usually have balance. There are common elements in music, visual arts, and architecture such as form, shape, color, repetition, composition, and pattern. Most of them hold a rhythm. We can find symbolizing numbers in math or science and symbolizing notes in music. Visual arts and mathematics, especially geometry, are mostly spatial, while music is time based. Music involves science (physics of sound waves), and math (metered beats and measures, theories of music). One may see a difference in the use of unity and variety: in music, variety is offered by the use of altered rhythms, melody, and color. In visual arts, variety is offered by placement, theme, and color. Visual representations of concepts in science and mathematics are better understood with contrast. This is common in music and other arts – it evokes excitement and emotion. In music and other arts visual, societal, and life-oriented messages are communicated with beauty. There is a common need of simplifying their meaning. There are also aesthetic dimensions of science; elements and principles of design in the arts can be used to teach visually math and science. Music compositions and art works created in different, sometimes conflicting styles bear visible design characteristics according to the messages they carry. We can analyze how the design-related issues were solved in previous times and in contemporary arts, and how the style of an art work relates to the actions or behaviors of people, influences their thinking, keeps under control their emotions, and affects their reactions to the content of the image.

Musical composition has its own rules. According to a composer Aaron Copland, music has four elements or ingredients: rhythm, melody, harmony, and tone color. Music can be seen as a grammar, just like icons can serve as visual record notes. Understanding computer language in 3D software packages, such as Blender, Maya, and Python may make people who can do both art and coding ready to be hired in the creative industries (e.g. media, advertisement, gaming, etc.).

RULES AND RESTRICTIONS IN PROGRAMMING THROUGH A METAPHOR OF DANCE STEPS

In order to provide visual explanation on how to code we'll use dance as a metaphor, and show how applying restrictions in various dances might be relevant to those in coding. The images created by the "invisible" lines generated by the dancers' feet may form the visual representation of sound-based movement. Each dance is guided by some rules and adjusted to the type of music played, so it has a different tone, tempo, interval, pitch, volume, etc. Dancers must become familiar with the rules, absorb the music, and coordinate their movement with partners. When a pair is dancing, both partners are following the rules, yet each one would have somehow different interpretation, approach, or degree of freedom in the same dance.

In this chapter we offer codes written and explained in different languages, which delineate various dances and their rules. This way the reader might see codes related to the same topic. This approach makes possible to have an overview of programming available and ready to follow.

The reader willing to learn or grasp ideas of coding is introduced to coding in several languages used to solve a similar task. This way the reader gets a "bird's eye view" in coding, and can compare and contrast similar underlying principles set to work for various environments. The exploration of patterns created by movement of feet is set to explore the dynamics behind the image generated by music and resulting movements of the dancers.

In order to test this concept, we started with using software. Joshua Solomon, a student from the University of Northern Colorado was asked to work using software Adobe Illustrator to move a line of his choice attached to any shape so that it would follow the rules of a particular dance. He was asked to explore how a movement guided by music (with its rhythm, melody, intervals, pitch, or volume) could gain a visual representation, and perhaps its own line of aesthetics. Josh was asked to explore the transform and repeat function, where the lines and objects would change an angle, move, and then repeat the set, still honoring changes of an angle. It is an approach to a metaphor as a learning aid, though.

When we use a ready-made product, such as software, we rely on somebody else's work, often resulting from writing a program. Communication with software is usually visual, possible through the use of buttons having the iconic representations on them. Software is a result of coding with tasks assigned to visual icons representing particular actions, such as brush and pencil for painting or drawing, a note and staff for music composition, or trash for discarding unsatisfying files. Those are metaphors that are often referring to objects or actions from the physical world and later on transferred into the digital or analog type of representation.

Jacob Reuter, another student from the same university was asked to develop a similar pattern by coding in Processing.org. He was working on this project focusing on the transform and repeat function. He was learning the program while experimenting with still images and animations. Kasey Rees and Ben Fry, former students of John Maeda from MIT developed Processing.org. It possesses special learning shortcuts, allowing the user to load a drawing, and then alter it by tweaking a code, and/or do the same with animation, and even interactivity. The power behind this method lies in the visual aspect of learning by trial and error. The set of dances were developed as a visual approach to programming. Further recommendation for the reader could involve the Arduino board collecting sounds from the dancing partners, and then transferring them into an actual drawing with processing.org.

One of the co-authors, Mohammad Majid al-Rifaie wrote, as a professional programmer and lecturer, several programs with comments explaining each step, move, or action for the sets of movements recorded for various dances including the Tango, Foxtrot, and Rumba. The color turtles attached to lines follow the rhythms of the movement of the dancers.

The text and comments follow for each language/dance. Quite a simple idea allows the reader to grasp conditions and statements, and the power of the 'if,' 'while,' 'else,' or 'transform and repeat.'

At the moment, the colors are generated randomly and thus each time different set of colors is selected for the male and female dancers. The relevant code is copied and pasted below:

```
# a random colour is generated for the man
r = rand(0,255); b = rand(0,255); g = rand(0,255)
lM.pencolor(r,g,b)
rM.pencolor(r,g,b)
# a random colour is generated for the the woman
r = rand(0,255); b = rand(0,255); g = rand(0,255)
lW.pencolor(r,g,b)
rW.pencolor(r,g,b)
```

By replacing r (for red), g (for green) and b (for blue) in 'pencolor', a static color can be picked for the traces related to the foot of each dancer. In order to alter the colors, various combinations of values for R, G, and B could be applied. This means the visual representation of the image could be altered in order to satisfy personal taste of the image creator, or a particular viewer.

Al-Rifaie has also developed an algorithm to represent improvisation in dance. There is a notion of improvisation in music and the same applies to dance. When the explorer and mountaineer Sir Edmund Hillary (who with Tenzing Norgay first reached the peak of Mt. Everest) was asked why he'd climb Mount Everest, he answered simply, "Because it's there" (Bio, 2015). A dancer improvising her responses for various types of music, when asked how she knew when her performance looked good, replied that if it feels good, it looks good. When people improvise when dancing, this situation might bring the process of randomization in computing. The computer picks up numbers randomly and distributes them along an axis or a variable.

Randomization and improvisation absolutely go hand in hand. It's the restriction (or constraints) in stochasticity that allows *the freedom to be regulated* and as such a pattern would emerge, which while being loyal to the main principle of the dance, is not chaotic.

Another co-author, Fahim Islam created the Tango steps in C#. He explains how to secure a workspace, and how to approach this task step by step. He also explores the idea of randomness behind improvisation in dance, when the dancers prefer to feel the music and create their own steps and movements based on particular musical composition, rather then to follow and obey the pre-existing rules.

The concept of dance and its visualization provide us with a useful tool to expand on the idea of using simple rules, which are repeated over and over again. This loosely follows dance as a metaphor, where once the rules (composed of restrictions on a set of free moves, i.e. moving forward, turning, etc.) are learnt, all that is needed is the repetition of the rules while keeping in mind the restrictions.

With coding, a similar process is implemented. Take as an example a dance where for instance, initially the right foot 'moves forward' one meter and then the left foot 'moves forward' half a meter; afterwards, another set of rules are applied for each foot and so on and so forth.

Once all the rules are provided for a given dance, the process can be repeated as many times as needed, from various locations in the dance floor, and at varying tempo. Therefore, there is a great deal of freedom offered to the coder as they can inject various elements of flexibility in the dance by changing the step size, the tempo of the dance, etc.

An example of restriction relevant to this metaphor is to monitor when a dancer leaves the dance floor, at which stage they should stop and other dancers should be brought in and perform.

In order to make a stronger connection, the reader is invited to investigate the essence of programming. A programming language, among other elements, consists of three main building blocks:

- **Sequences:** Where commands are executed one after another (e.g. first moving the right foot, then moving the left foot, and so on.)
- **Conditions (IF Statements):** Where for example if a dancer is outside the dance floor, another dancer should be summoned to perform
- **Repetition (or Loops, FOR or WHILE Statements):** where, once a cycle of dance is completed, the same cycle could be repeated over and over again, following the same set of rules (consisting of sequences, conditions, and perhaps inner repetitions called nested loops)

These steps are present in the dance metaphor used in this work. While programming might be seen (by a non-programmer) as a tool to produce some rigid and inflexible outcome, it is possible to introduce a certain element of freedom or stochasticity, and thus make the outcome more unexpected. This is usually catered for by injecting some dynamically changing elements, for example, as mentioned before:

- Varying step size.
- An offset to keep the feet sliding away from where they should be exactly, making them more realistic.
- Introducing a different tempo randomly generated by the computer in each cycle, etc.

In order to examine how aural stimuli might become a visual outcome, we applied various strategies for a shape attached to a line and following the rules of various dances.

It's an integrative look at knowledge behind image generation, which is coming from sets of rules derived from another discipline. It might be a good place to talk about several aspects:

- To introduce the rules and conditions: in each dance and in each language as a translation into another language with the visual results in mind.
- To see if the invisible lines created by the footsteps that could create an interesting pattern and perhaps composition, when made visible. The color of the lines was coded for each dancer, and then some experiments with different color sets were applied for the sake of image quality.
- To have sets of programs for the user written and explained, so the new to programming reader can grasp the idea of coding by comparing and contrasting. It's like looking at the topic from the bird's eye view/perspective. Some information how to get the compiler and how to start the whole process is provided, so the user would be able to reconstruct, or modify each setup, analyze it, then understand it.
- A different degree of freedom/interpretation/personality of a particular dancer was introduced. This means the dancers have to follow the rules, but their actions can vary: the lines generated can create other interesting patterns and compositions. For example, one dancer might strictly obey the rules of a particular dance, while the other might apply their own interpretation, thus varying the lines of their foot.
- Dance improvisation: this way the coding conditions, (while, if/else, transform, and repeat), are explained as the essence of image generation. Dance improvisation obviously depends on the tempo, speed melody, rhythm, to name just a few, which makes the process even more interdisciplinary.

It is hard to negate that knowledge of coding is important. Although provocative philosophically, it helps to develop cognitive and abstract thinking, creativity, curiosity, and flexibility, to name just a few. It also helps the user in expanding many ready to use software packages. Knowing the coding creates opportunities for unexpected solutions going beyond what's known and expected. The audience of the HBO seasons of the "Silicon Valley" may share the same reaction, "I wish I could code like that." The programmer can change the state of usability behind the apps, services, security, and set new standards for whatever creative thoughts or experience might bring to one's mind. Looking at the potential of the code creates a new level of understanding what a "machine" could do, and how it could affect our lives on many levels and in many fields. This creates another layer of economical disputes.

PROGRAMS, EXAMPLES, AND VISUAL SOLUTIONS

This part presents solutions offered by students Josh Solomon and Jacob Reuter, and then by the co-authors of this chapter Mohammad Majid al-Rifaie and Md Fahimul Islam (see Figures 3-17).

Josh Solomon (Adobe Illustrator)

Tango

This design was made by taking a symbol (A): a triangular shape made out of two lines, and then repeating it in a pattern. The pattern was taken from the Tango steps seen below (B). The shapes (A) were flipped horizontally to differentiate between left and right footsteps in the dance. After creating the initial pattern (C), it was transformed and repeated three times to create a final, rotationally symmetrical pattern (D) (Figure 3).

Figure 3. Josh Solomon, Tango
(© 2015, J. Solomon. Used with permission).

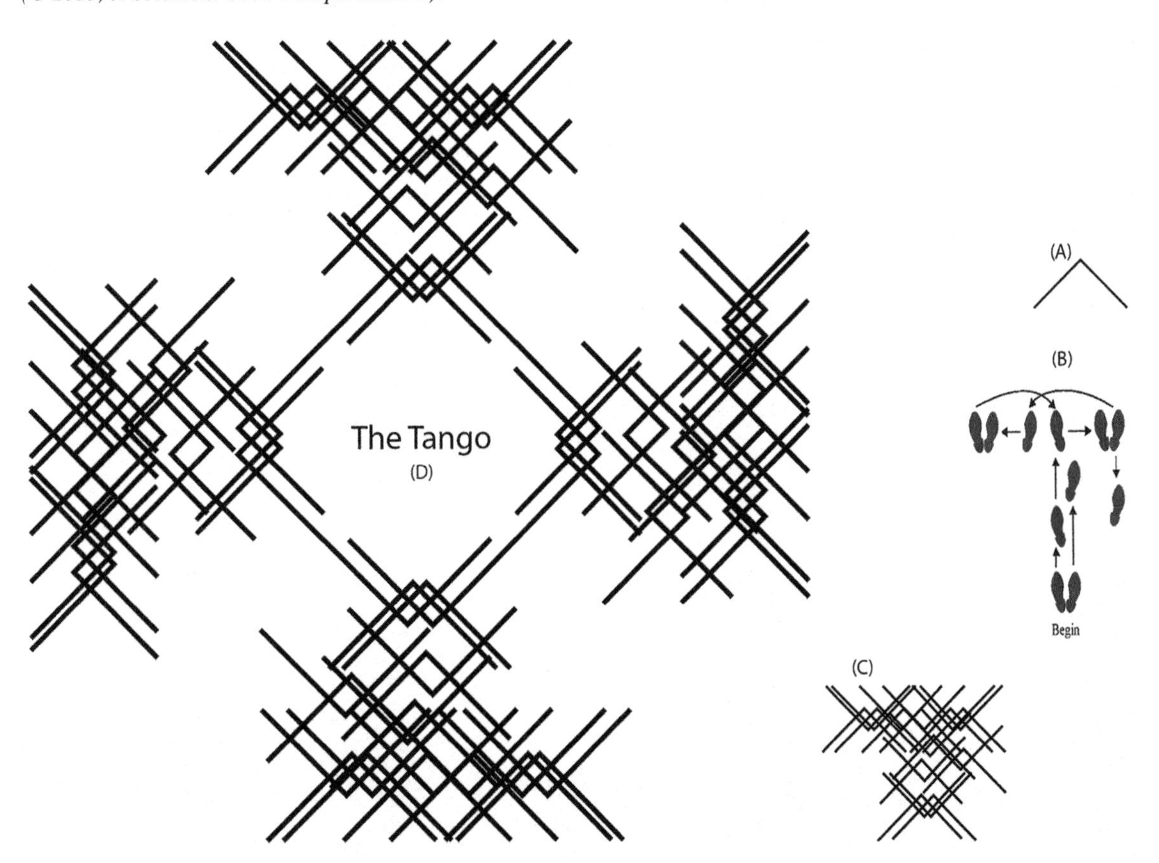

The icon made for this pattern involves several circles and is inspired with a water ripple effect. This icon was repeated and then rotated to create a pattern resulting from constraints of the dance diagram. To show the growth and motion through the composition of this figure, I transformed and repeated the pattern several times, so that it created a zig-zag shape (Figure 4).

Pastoral

This pattern was created using information and constraints from the dance steps sheet "The Pastorall" by M. Isaac, shown below. Previously, icons replaced footprints, but this pattern follows the path made by the movement of dancers. The icons resulted from combining shapes and symbols found on the M. Isaac's dance steps sheet. They were put in a shape of the dance, and then repeated and scaled to add interest and depth (Figure 5).

Swing

In this project the dance steps (a) created the path of motion for where the feet would travel (b). Then I used a section (d) of a picture showing a splash of water (c) to trace it over the path in spots where their line direction matched up. The resulting composition seems nonrepresentational and lacks closure (Figure 6).

The graphic (b) was made using the icon/symbol (a) – a graphic representation of a foot with nonrepresentational qualities. The icon was placed in a circular pattern according to the dance rules. I recycled the four shapes shown in the middle of the graphic (b) to recycle them into another pattern. This pattern was repeated and placed over the footprints of a circular dance (c) to make a new pattern derived from this dance (d) (Figure 7).

Figure 4. Josh Solomon, Tango
(© 2015, J. Solomon. Used with permission).

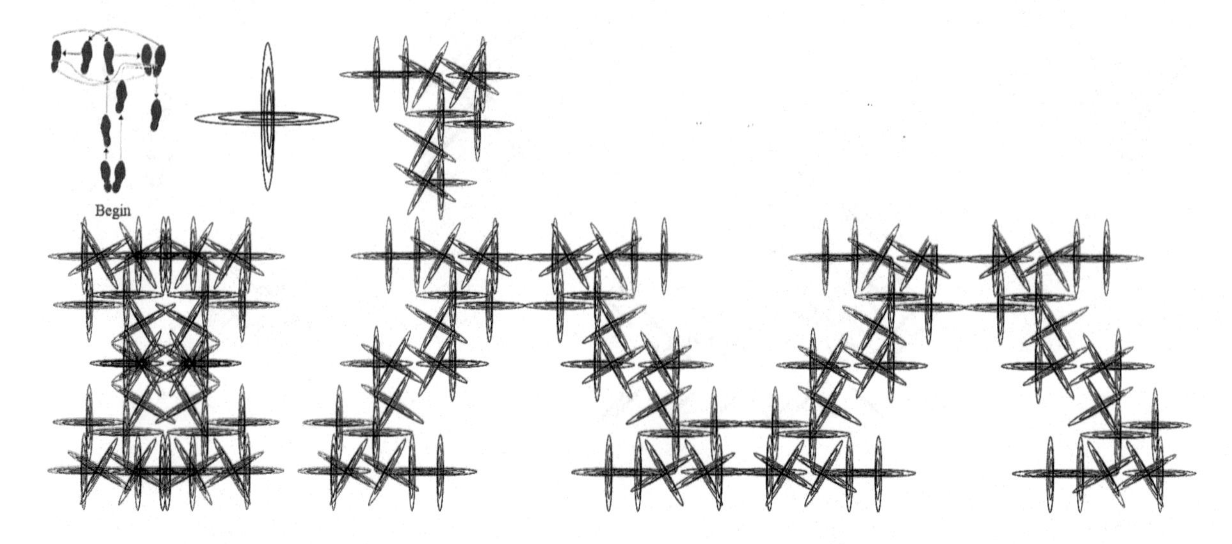

Figure 5. Josh Solomon, Pastoral
(© 2015, J. Solomon. Used with permission).

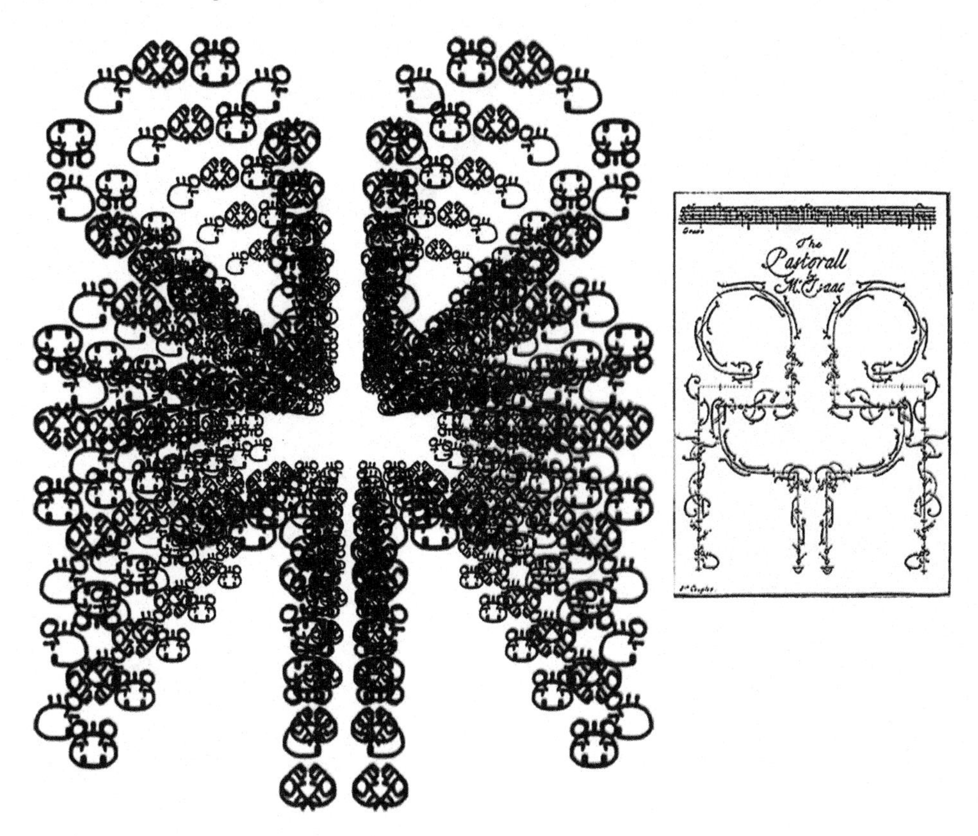

Figure 6. Josh Solomon, Swing
(© 2015, J. Solomon. Used with permission).

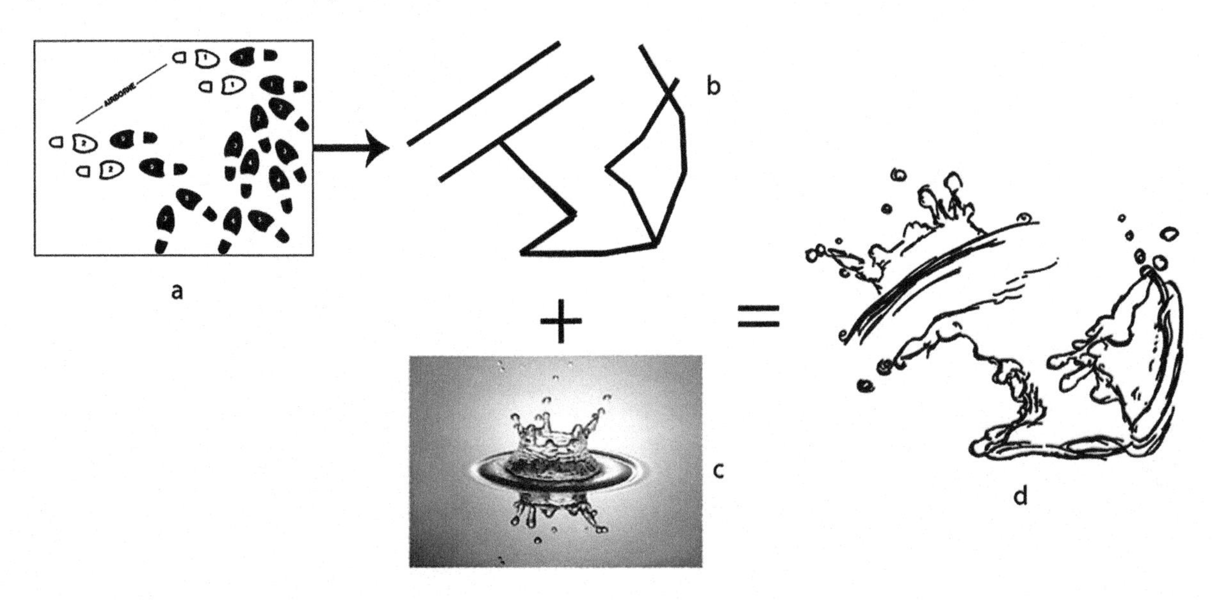

Figure 7. Josh Solomon, Circular pattern
(© 2015, J. Solomon. Used with permission).

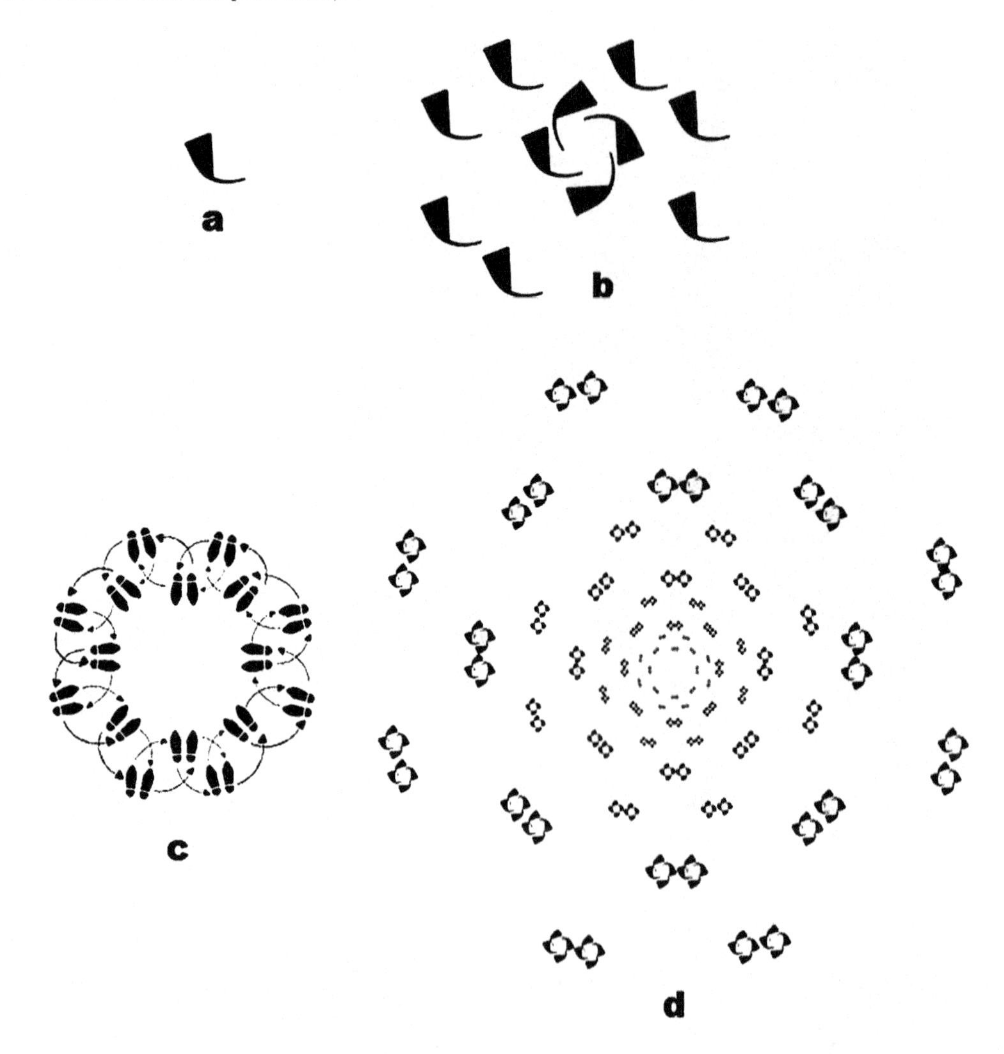

Jake Reuter, Drawing Program (Processing.org)

In order to start writing the codes for dancing, a student wrote, as a warm-up, codes for basic geometrical shapes.

Moving Ellipses

Shown in Figure 8.

```
void setup(){
  size(1000,800);
  background(0);
}
```

Figure 8. Jake Reuter
(© 2015, J. Reuter. Used with permission).

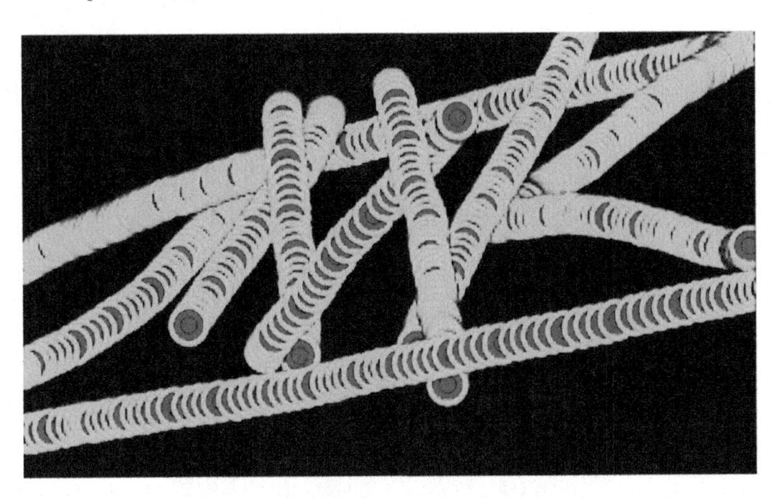

```
void draw() {
  if (mousePressed) {
    fill(0,255,0);
    ellipse(mouseX,mouseY, 50,50);
    fill(0,0,255);
    ellipse(mouseX,mouseY,35,35);
    fill(255,0,0);
    ellipse(mouseX,mouseY, 20,20);
  }
}
```

Moving a 3D Sphere

Shown in Figure 9.

```
void setup() {
  size(400, 400, P3D);
}
void draw() {
  background(200);
  stroke(255, 100);
  translate(200, 200, 0);
  rotateX(mouseY * 0.05);
  rotateY(mouseX * 0.05);
  fill(mouseX * 2, 0, 160);
  sphereDetail(mouseX / 4);
  sphere(150);
}
```

Figure 9. Jake Reuter. A 3D Sphere.
(© 2015, J. Reuter. Used with permission).

Line Drawing

Shown in Figure 10.

```
void setup() {
  size(800,800);
  stroke(255);
  background(0,200,0);
}
void draw() {
  if (mousePressed){
     fill(255,0,0);
     stroke(200,0,200);
     line(150,25,mouseX,mouseY);
  }
}
```

Ellipses 2

Shown in Figure 11.

```
void setup() {
       size(400, 400);
```

Figure 10. Jake Reuter. A Line Drawing.
(© 2015, J. Reuter. Used with permission).

Figure 11. Jake Reuter. Ellipses 2.
(© 2015, J. Reuter. Used with permission).

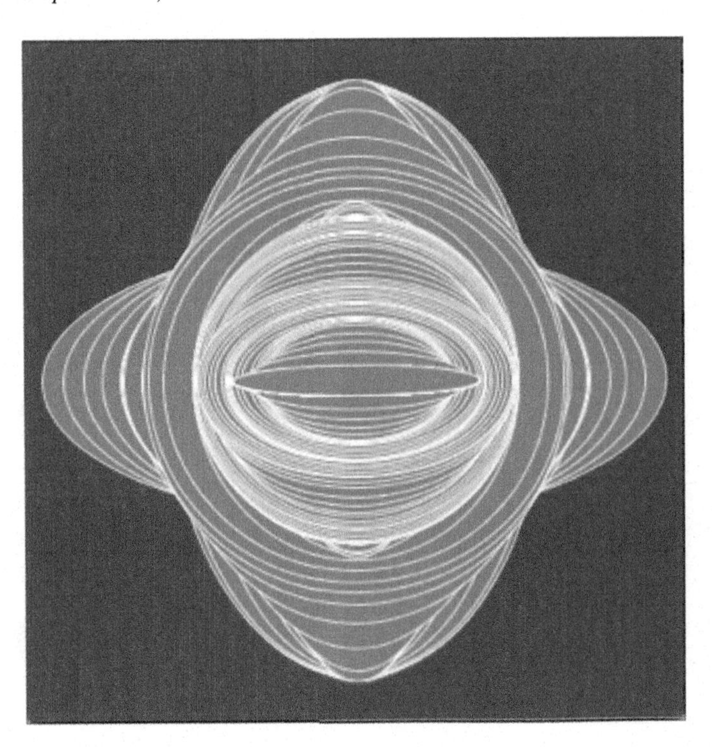

```
    stroke(255);
    background(192, 64, 0);
  }
  void draw() {
    ellipse(width/2, height/2, mouseX, mouseY);
    fill(0,150,255);
  }
```

Tango

Shown in Figure 12.

```
size(500,500);
background(0);
stroke(255,0,0);
line(30,40,40,110);
stroke(0,0,0);
fill(255,0,0);
ellipse(40,110,30,30);
stroke(150,50,50);
line(85,40,75,110);
stroke(0,0,0);
fill(150,50,50);
ellipse(75,110,30,30);
stroke(150,50,50);
line(230,40,220,110);
```

Figure 12. Jake Reuter. Tango.
(© 2015, J. Reuter. Used with permission).

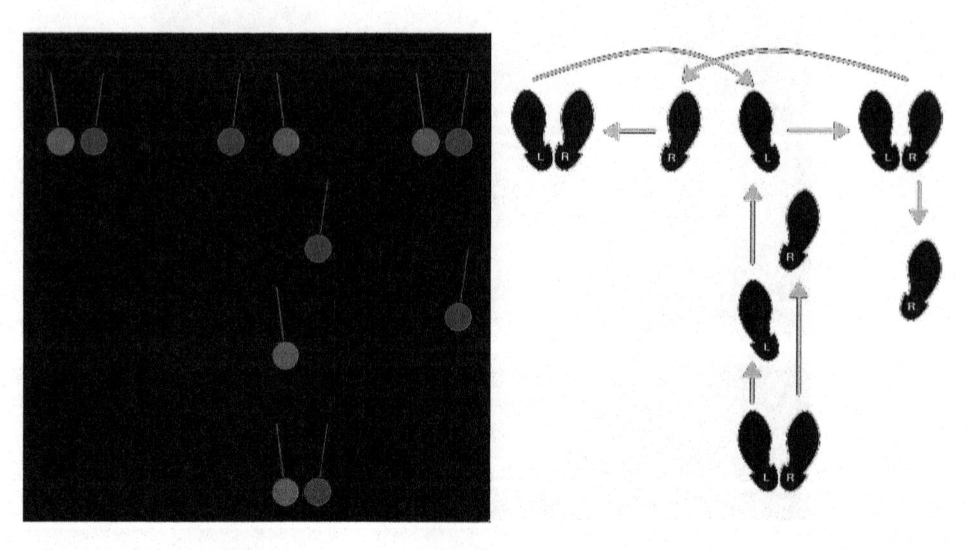

```
stroke(0,0,0);
fill(150,50,50);
ellipse(220,110,30,30);
stroke(255,0,0);
line(270,40,280,110);
stroke(0,0,0);
fill(255,0,0);
ellipse(280,110,30,30);
stroke(150,50,50);
line(325,150,315,220);
stroke(0,0,0);
fill(150,50,50);
ellipse(315,220,30,30);
stroke(255,0,0);
line(270,260,280,330);
stroke(0,0,0);
fill(255,0,0);
ellipse(280,330,30,30);
stroke(255,0,0);
line(270,400,280,470);
stroke(0,0,0);
fill(255,0,0);
ellipse(280,470,30,30);
stroke(150,50,50);
line(325,400,315,470);
stroke(0,0,0);
fill(150,50,50);
ellipse(315,470,30,30);
stroke(255,0,0);
line(420,40,430,110);
stroke(0,0,0);
fill(255,0,0);
ellipse(430,110,30,30);
stroke(150,50,50);
line(475,40,465,110);
stroke(0,0,0);
fill(150,50,50);
ellipse(465,110,30,30);
stroke(150,50,50);
line(475,220,465,290);
stroke(0,0,0);
fill(150,50,50);
ellipse(465,290,30,30);
```

Waltz

Shown in Figure 13.

```
size(500,500);
background(120);
stroke(255);
line(160,20,160,90);
stroke(250,190,0);
fill(250,190,0);
ellipse(160,20,30,30);
stroke(260,190,0);
line(195,20,195,90);
stroke(250,190,0);
fill(255);
ellipse(195,20,30,30);
stroke(255);
line(135,140,100,210);
stroke(250,190,0);
fill(250,190,0);
ellipse(135,140,30,30);
stroke(260,190,0);
line(15,295,85,295);
stroke(250,190,0);
fill(255);
ellipse(85,295,30,30);
stroke(255);
```

Figure 13. Jake Reuter, Waltz
(© 2015, J. Reuter. Used with permission).

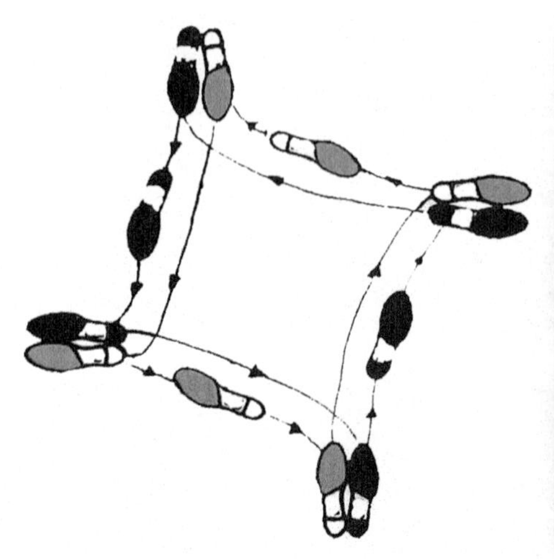

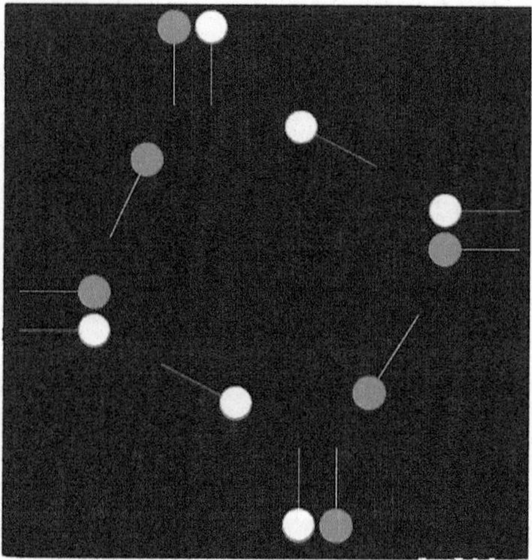

```
line(15,260,85,260);
stroke(250,190,0);
fill(250,190,0);
ellipse(85,260,30,30);
stroke(260,190,0);
line(150,325,220,360);
stroke(250,190,0);
fill(255);
ellipse(220,360,30,30);
stroke(260,190,0);
line(280,400,280,470);
stroke(250,190,0);
fill(255);
ellipse(280,470,30,30);
stroke(255);
line(315,400,315,470);
stroke(250,190,0);
fill(250,190,0);
ellipse(315,470,30,30);
stroke(255);
line(390,280,345,350);
stroke(250,190,0);
fill(250,190,0);
ellipse(345,350,30,30);
stroke(255);
line(415,220,485,220);
stroke(250,190,0);
fill(250,190,0);
ellipse(415,220,30,30);
stroke(260,190,0);
line(415,185,485,185);
stroke(250,190,0);
fill(255);
ellipse(415,185,30,30);
stroke(260,190,0);
line(280,110,350,145);
stroke(250,190,0);
fill(255);
ellipse(280,110,30,30);
```

Salsa

Shown in Figure 14.

Figure 14. Jake Reuter, Salsa
(© 2015, J. Reuter. Used with permission).

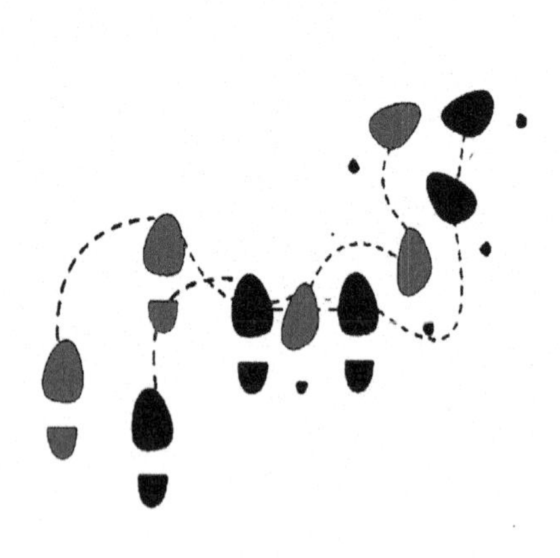

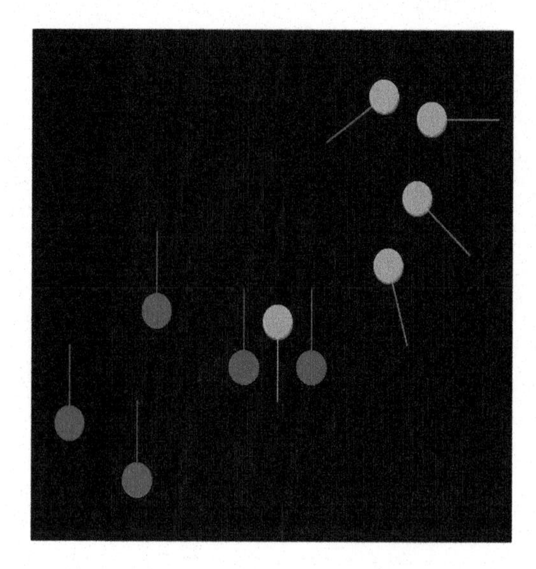

```
size(500,500);
background(0);
stroke(0,0,180);
line(40,300,40,370);
stroke(0,0,180);
fill(0,0,180);
ellipse(40,370,30,30);
stroke(0,0,180);
line(110,350,110,420);
stroke(0,0,180);
fill(0,0,180);
ellipse(110,420,30,30);
stroke(0,0,180);
line(130,200,130,270);
stroke(0,0,180);
fill(0,0,180);
ellipse(130,270,30,30);
stroke(0,0,180);
line(220,250,220,320);
stroke(0,0,180);
fill(0,0,180);
ellipse(220,320,30,30);
stroke(0,0,180);
line(290,250,290,320);
stroke(0,0,180);
fill(0,0,180);
```

```
ellipse(290,320,30,30);
stroke(100,100,255);
line(255,280,255,350);
stroke(100,100,255);
fill(100,100,255);
ellipse(255,280,30,30);
stroke(100,100,255);
line(370,230,390,300);
stroke(100,100,255);
fill(100,100,255);
ellipse(370,230,30,30);
stroke(100,100,255);
line(400,170,455,220);
stroke(100,100,255);
fill(100,100,255);
ellipse(400,170,30,30);
stroke(100,100,255);
line(415,100,485,100);
stroke(100,100,255);
fill(100,100,255);
ellipse(415,100,30,30);
stroke(100,100,255);
line(365,80,305,120);
stroke(100,100,255);
fill(100,100,255);
ellipse(365,80,30,30);
```

Jive

Shown in Figure 15.

Figure 15. Jake Reuter, Jive
(© 2015, J. Reuter. Used with permission).

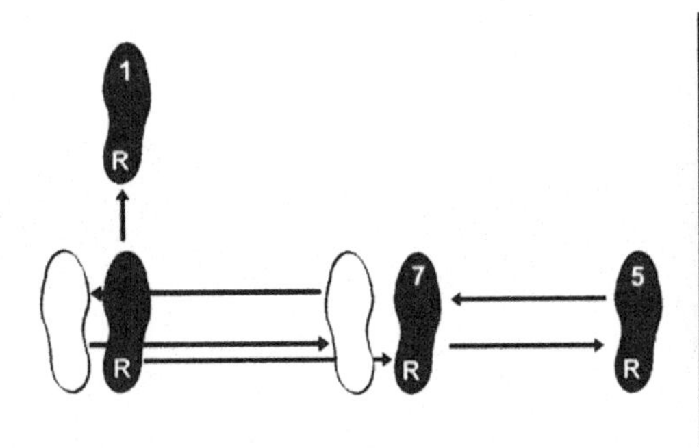

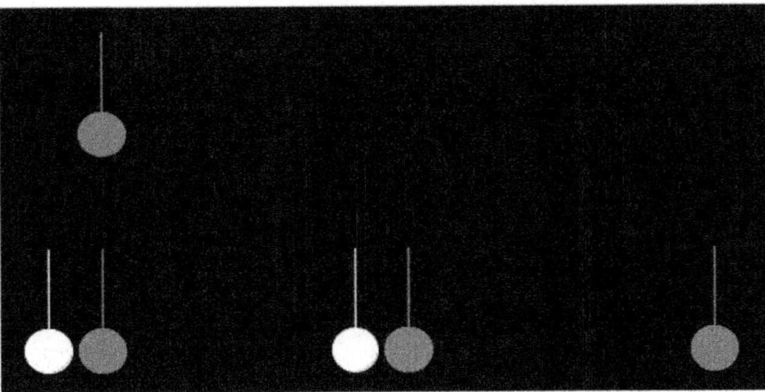

```
size(500,500);
background(0);
stroke(255);
line(230,400,230,470);
stroke(255);
fill(255);
ellipse(230,470,30,30);
stroke(155);
line(265,400,265,470);
stroke(155);
fill(155);
ellipse(265,470,30,30);
stroke(255);
line(30,400,30,470);
stroke(255);
fill(255);
ellipse(30,470,30,30);
stroke(155);
line(65,400,65,470);
stroke(155);
fill(155);
ellipse(65,470,30,30);
stroke(155);
line(465,400,465,470);
stroke(155);
fill(155);
ellipse(465,470,30,30);
stroke(155);
line(65,250,65,320);
stroke(155);
fill(155);
ellipse(65,320,30,30);
```

Irish Jig

Shown in Figure 16.

```
size(500,500);
background(0);
stroke(0,255,0);
line(230,370,230,450);
stroke(0,255,0);
fill(0,255,0);
ellipse(230,450,30,30);
```

Figure 16. Jake Reuter, Irish Jig
(© 2015, J. Reuter. Used with permission).

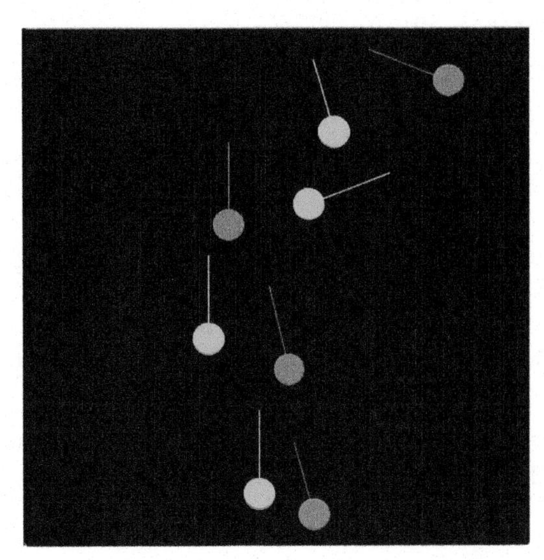

```
stroke(50,125,50);
line(265,400,285,470);
stroke(50,125,50);
fill(50,125,50);
ellipse(285,470,30,30);
stroke(50,125,50);
line(240,250,260,330);
stroke(50,125,50);
fill(50,125,50);
ellipse(260,330,30,30);
stroke(0,255,0);
line(180,220,180,300);
stroke(0,255,0);
fill(0,255,0);
ellipse(180,300,30,30);
stroke(50,125,50);
line(200,110,200,190);
stroke(50,125,50);
fill(50,125,50);
ellipse(200,190,30,30);
stroke(0,255,0);
line(360,140,280,170);
stroke(0,255,0);
fill(0,255,0);
ellipse(280,170,30,30);
```

```
stroke(50,125,50);
line(340,20,420,50);
stroke(50,125,50);
fill(50,125,50);
ellipse(420,50,30,30);
stroke(0,255,0);
line(285,30,305,100);
stroke(0,255,0);
fill(0,255,0);
ellipse(305,100,30,30);
```

Using Dance Steps as a Metaphor of a Program

In order to examine the pattern of the dancers' steps, student social dance was recorded; patterns made by the students' feet served as a metaphor used for explaining steps in programming (Figures 17 and 23).

Figure 17. Anna Ursyn. Social dance – waltz
(© 2015, A. Ursyn. Used with permission).

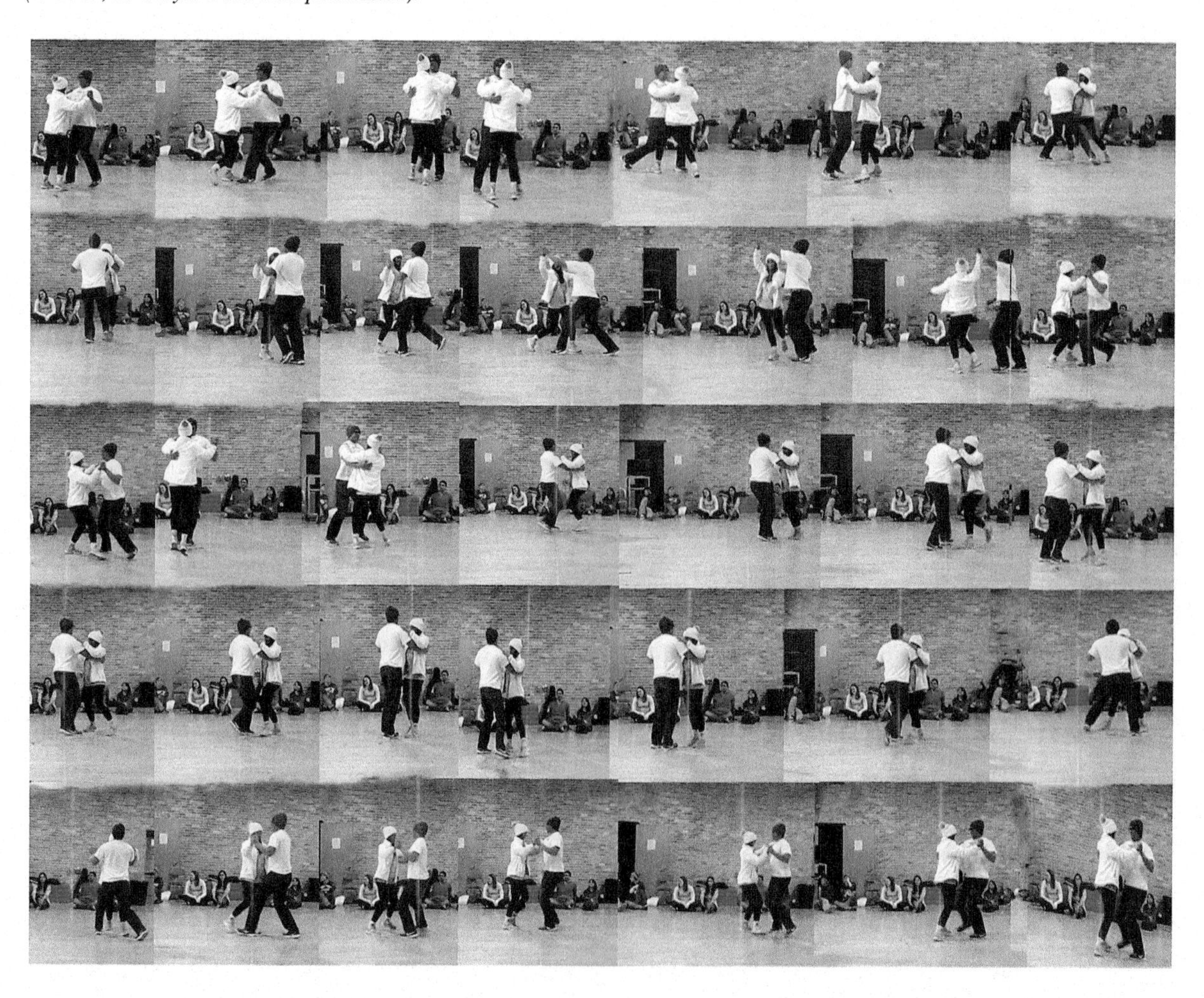

Students in a Social Dance class taught by Brittany Jacobs at the University of Northern Colorado are dancing tango. Since there were no restrictions for the attire at the physical education electives, one might spot a pom-pom hat worn for this highly formalized dance performed in a gym.

Mohammad Majid al-Rifaie

Tango

A Python code for basic Tango dance steps performed by a solo dancer is given below. This dance is repeated 100 times by 100 different solo dancers. Each time the dance starts at a different position on the dance floor. A sample visual output of the 100 cycles is illustrated in Figure 18.

Figure 18. Mohammad Majid al-Rifaie, Two turtles dance tango
(© 2015, M. Majid al-Rifaie. Used with permission).

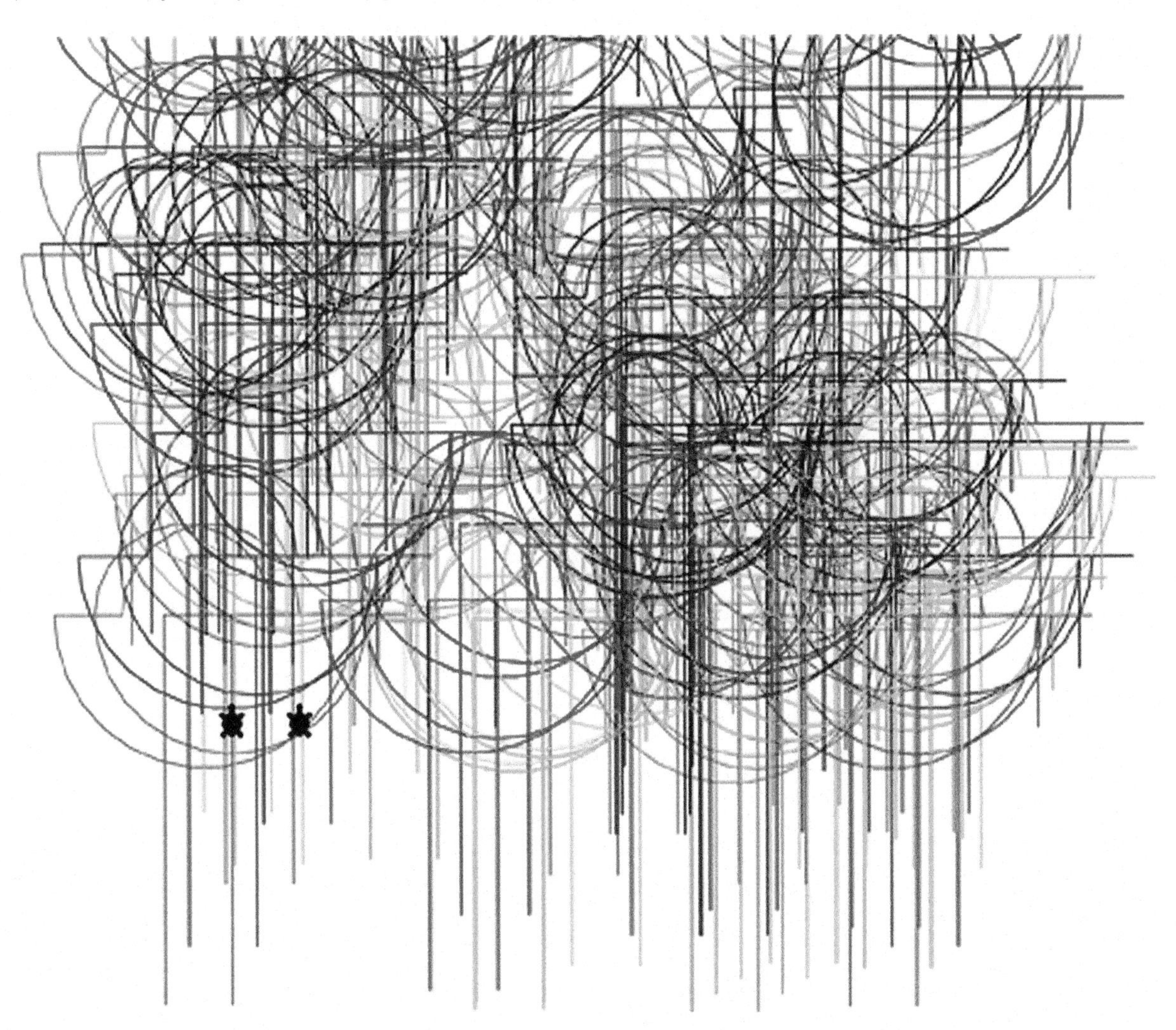

```
import turtle    # this line makes sure the turtle object can be used
import random    # this lines allows the program to generate random numbers
when needed
                 # e.g. in order to decide the step size
screen = turtle.Screen()
screen.screensize(300,300)
screen.colormode(255)
screen.bgcolor(0,0,0)
# the screen is made to be 300x300 with black background
l = turtle.Turtle() # l for left foot
r = turtle.Turtle() # r for right foot
# hiding the left and right foot until they are in their positions
l.hideturtle()
r.hideturtle()
l.penup()
r.penup()
# adjusting the speed and rhythom of the dance
speed = 50
l.speed(speed)
r.speed(speed)
# setting the shape of the feet to turtles
l.shape("turtle")
r.shape("turtle")
# initially the turtle is headed east,
# so they are rotated 90 degrees anti-clockwise to head upward (north)
# and go back 100 unit to be placed in the initial position
l.left(90); l.backward(100)
r.left(90); r.backward(100)
# here the right foot is given a distance from the left one
# (so they don't overlap)
r.right(90); r.forward(30); r.left(90)
# Now that each foot is in its place, they are made visible
l.showturtle()
r.showturtle()
# and the pen is down, so the user see the traces of their movements
l.pendown()
r.pendown()
# a stamp allows the trace of the feet to be printed whenever required
#l.stamp()
#r.stamp()
# Whenever this functin is called, one full basic tango movement is executed
def tango():
    # a random colour is generated for the traces of the feet
```

```
    red   = random.randint(0,255)
    green = random.randint(0,255)
    blue  = random.randint(0,255)
    l.pencolor(red,green,blue)
    r.pencolor(red,green,blue)

    # No 1 & 2
    l.forward(80)    # left moves forward 80 units
    r.forward(200)   # right moves forward 200 units
    #l.stamp()
    #r.stamp()
    # No 3
    l.forward(200)
    #l.stamp()
    # No 4
    r.forward(80)
    r.right(90)      # right foot is rotated to the right (clockwise) for 90
degrees
    r.forward(100)   # then it moves forward 100 units
    r.left(90)       # then rotated 90 degrees anti-clockwise
    #r.stamp()
    # No 5
    l.right(90)
    l.forward(100)
    l.left(90)
    #l.stamp()
    # No 6
    #r.left(90)
    #r.forward(150)
    #r.right(90)
    r.circle(65, 180)    # this lines allows a circle to be drawn showing the
arc made during the dance
    r.right(180)
    #r.stamp()
    # No 7
    #l.left(90)
    #l.forward(200)
    #l.right(90)
    l.right(180)
    l.circle(-100, 180)
    #l.stamp()
    # No 8
    r.left(90)
    r.forward(50)
```

```
    r.right(90)
    #r.stamp()
    # No 9
    #l.right(90)
    #l.forward(100)
    #l.left(90)
    l.circle(-65,180)
    l.left(180)
    #l.stamp()
    # No 10
    #r.right(90)
    #r.forward(200)
    #r.left(90)
    r.left(180)
    r.circle(110,180)
    #r.stamp()
    # No 11
    l.right(90)
    l.forward(100)
    l.left(90)
    #l.stamp()
    # No 12
    #r.right(180)
    #r.forward(80)
    #r.right(180)
    r.backward(80)
    #r.stamp()
# the FOR loop below makes sure that the lines are called n (n=100) times
for i in range(100):
    #l.stamp();     r.stamp()
    tango()     # this line calls all the steps of tango dance
    # once the first round is complete:
    # the pen is lifted
    l.penup(); r.penup()
    # a new dance starts at a random position
    # and feet left and right are places within the range of -300 to 300 which
    # is the within the dance floor
    randX = random.randint(-300,300)
    randY = random.randint(-300,300)
    l.goto(randX, randY)
    r.goto(randX-50, randY)
    # the pen is put down again to show the the movement of the feet
    l.pendown(); r.pendown()
```

```
# saving the final image as a pdf file (without the background colour)
screen.getcanvas().postscript(file="Tango.pdf")
```

Foxtrot

Python code for basic Foxtrot dance steps performed by a pair of dancers. This dance is repeated 100 times starting at a random position in the dance floor. The dance continues until the dancers leave the dance floor, at which point another pair of dancers come in the dance floor and start the dance again. The IF statement in the method "def checkLoc()" ensures that the dancers do not leave the dance floor. Also there is a minor offset introduced between the dancers, in order not to override the traces of the dancer with the fellow dancer. A sample visual output of the dance of 100 pair is given in Figure 19.

Figure 19. Mohammad Majid al-Rifaie, Turtles dance foxtrot
(© 2015, M. Majid al-Rifaie. Used with permission).

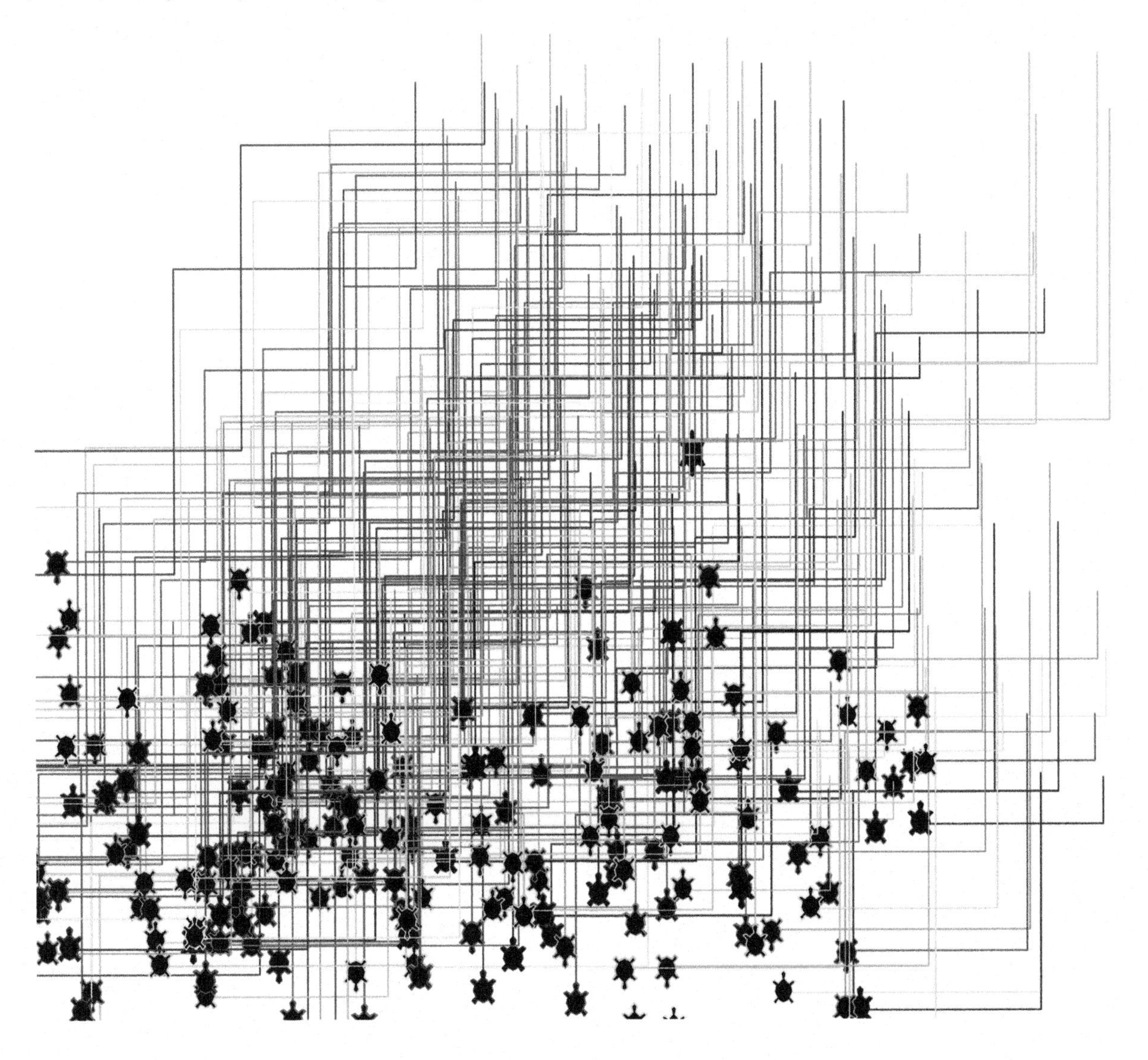

```
import turtle    # this line makes sure the turtle object can be used
import random    # this lines allows the program to generate random numbers
when needed
                 # e.g. in order to decide the step size
import time
screen = turtle.Screen()
screen.screensize(300,300)
screen.colormode(255)
screen.bgcolor(0,0,0)
sleepTime = 0.3
# a function that takes the name of the turtle, the angle at which
# it should be rotated and the distance that it should move
def move(turtle, angle, distance):
    turtle.left(angle)
    turtle.forward(distance)
# this function takes the turtle name as well as
# the x and y distances that the turtle should move
def moveXY(turtle, dx, dy):
    turtle.setpos(turtle.xcor()+dx,
                  turtle.ycor()+dy)
# four turtles are created
# lM for left foot of the man
# rM for right foot of the man
lM = turtle.Turtle()
rM = turtle.Turtle()
# lW for left foot of the woman
# rW for right foot of the woman
lW = turtle.Turtle()
rW = turtle.Turtle()
lM.hideturtle()
rM.hideturtle()
lW.hideturtle()
rW.hideturtle()
lM.shape("turtle")
rM.shape("turtle")
lW.shape("turtle")
rW.shape("turtle")
lM.showturtle()
rM.showturtle()
lW.showturtle()
rW.showturtle()
# adjust the direction of the foot to make sure that
# the man's feet are facing the woman's feet
```

```
move(lW, 90, 0)
move(rW, 90, 0)
move(lM, -90, 0)
move(rM, -90, 0)
screen.delay(10)
# this function takes two numbers (m and n) and generate a random integer num-
ber between m and n
def rand(m,n):
    random.randint(m,n)
# this function places the feet of the dances in the starting position
def startingPosition():
    # the initial position of the feet are marked and the pen is lifted
    lM.stamp();      lM.penup()
    rM.stamp();      rM.penup()
    lW.stamp();      lW.penup()
    rW.stamp();      rW.penup()
    # a random colour is generated for the man
    r = rand(0,255); b = rand(0,255); g = rand(0,255)
    lM.pencolor(r,g,b)#;      rM.fillcolor(r,g,b)
    rM.pencolor(r,g,b)#;      rM.fillcolor(r,g,b)
    # a random colour is generated for the the woman
    r = rand(0,255); b = rand(0,255); g = rand(0,255)
    lW.pencolor(r,g,b)#;      rM.fillcolor(r,g,b)
    rW.pencolor(r,g,b)#;      rM.fillcolor(r,g,b)
    # initial position of the feet is randomly selected on the dance floor
    initX = random.randint(-200,300)
    initY = random.randint(-200,300)
    # lowe and upper difference allows the feet to be slightly unsynched
    # rather than precisely opposite each other
    diffLower = 5
    diffUpper = 10
    lM.goto(initX,initY)
    rM.goto(initX-50,initY)
    rW.goto(initX+rand(diffLower,diffUpper),
             initY-50+rand(diffLower,diffUpper))

    lW.goto(initX-50+rand(diffLower,diffUpper),
             initY-50+rand(diffLower,diffUpper))
    # once the feet at their initial position, the pen of each foot is down
    # and ready to show the traces of their feet's movements
    lM.pendown()
    rM.pendown()
    lW.pendown()
    rW.pendown()
```

```
# setting the the speed at which the dance should take place
screen.delay(0.1)
# this function checks if the feet inside the dance floor
# if not, a new position is generated and the dance starts over
def checkLoc():
    if (rM.ycor() > 300 or rM.ycor() < -300 or
         lM.ycor() > 300 or lM.ycor() < -300 or
         rW.ycor() > 300 or rW.ycor() < -300 or
         lW.ycor() > 300 or lW.ycor() < -300):

        startingPosition()
def rand(a,b):
    return random.randint(a,b)
# this function performs one full cycle of the basic foxtrot basic dance steps
def foxtrot():

    stepSize = rand(100,200)    # a random step size is generated for each new
cycle

    # no. 1
    for i in range(stepSize):   # the FOR loop allows the relevant feet to
move together one unit at a time
        moveXY(rW, 0, -1)
        moveXY(lM, 0, -1)
    checkLoc()      # once done, the position of the feet are checked to see
that
                    # they are inside the dance floor, otherwise a new dance
starts
    time.sleep(sleepTime)   # this allows the dancers to stop for a little
while to show that once step is complete
    #move(lM, 0, 100)       # these line allows the fast movement of each foot
separately
    #move(rW, 0, -100)
    # no. 2
    for i in range(int(stepSize/2)):
        moveXY(rM, 0, -1)
        moveXY(lW, 0, -1)
    checkLoc()
    time.sleep(sleepTime)
    #move(rM, -45, 100)
    #move(lW, 180-45, 100)
    # no. 3
    for i in range(int(stepSize/4)):
        moveXY(lM, 0, -1)
```

```
        moveXY(rW, 0, -1)

    for i in range(stepSize):
        moveXY(lM, -1, 0)
        moveXY(rW, -1, 0)
    checkLoc()
    time.sleep(sleepTime)

    # no. 4
    for i in range(int(stepSize*3/4)):
        moveXY(rM, 0, -1)
        moveXY(lW, 0, -1)
    for i in range(int(stepSize)):
        moveXY(rM, -1, 0)
        moveXY(lW, -1, 0)
    checkLoc()
    time.sleep(sleepTime)

# ========================================
# initially the feet are located in an initial position
startingPosition()
# then the dance starts, in this case 100 cycle of the foxtrot basic step dance
for i in range(100):
    foxtrot()
# saving the final image as a pdf file (without the background colour)
screen.getcanvas().postscript(file="Foxtrot.pdf")
```

Rumba

Python code describes basic Rumba dance steps performed by two dancers. In this dance, dancers do not move simultaneously but rather one waits for one step to complete and then the other one moves the relevant foot. The purpose of this code is to present more simultaneous moves. Again there is an offset between the feet-location of the dancers in order to allow for the tracers not to be over-written. This dance is repeated 100 times and each time, the dancers take a step with a varying size yet staying loyal to the rules of the dance. A sample visual output of the 100 cycles is given in Figure 20.

```
import turtle    # this line makes sure the turtle object can be used
import random    # this lines allows the program to generate random numbers
when needed
                 # e.g. in order to decide the step size
screen = turtle.Screen()
screen.screensize(300,300)
```

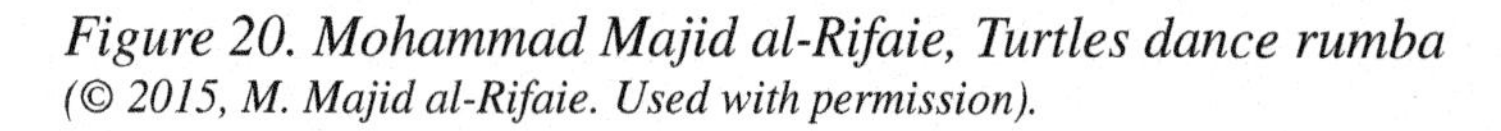

Figure 20. Mohammad Majid al-Rifaie, Turtles dance rumba
(© 2015, M. Majid al-Rifaie. Used with permission).

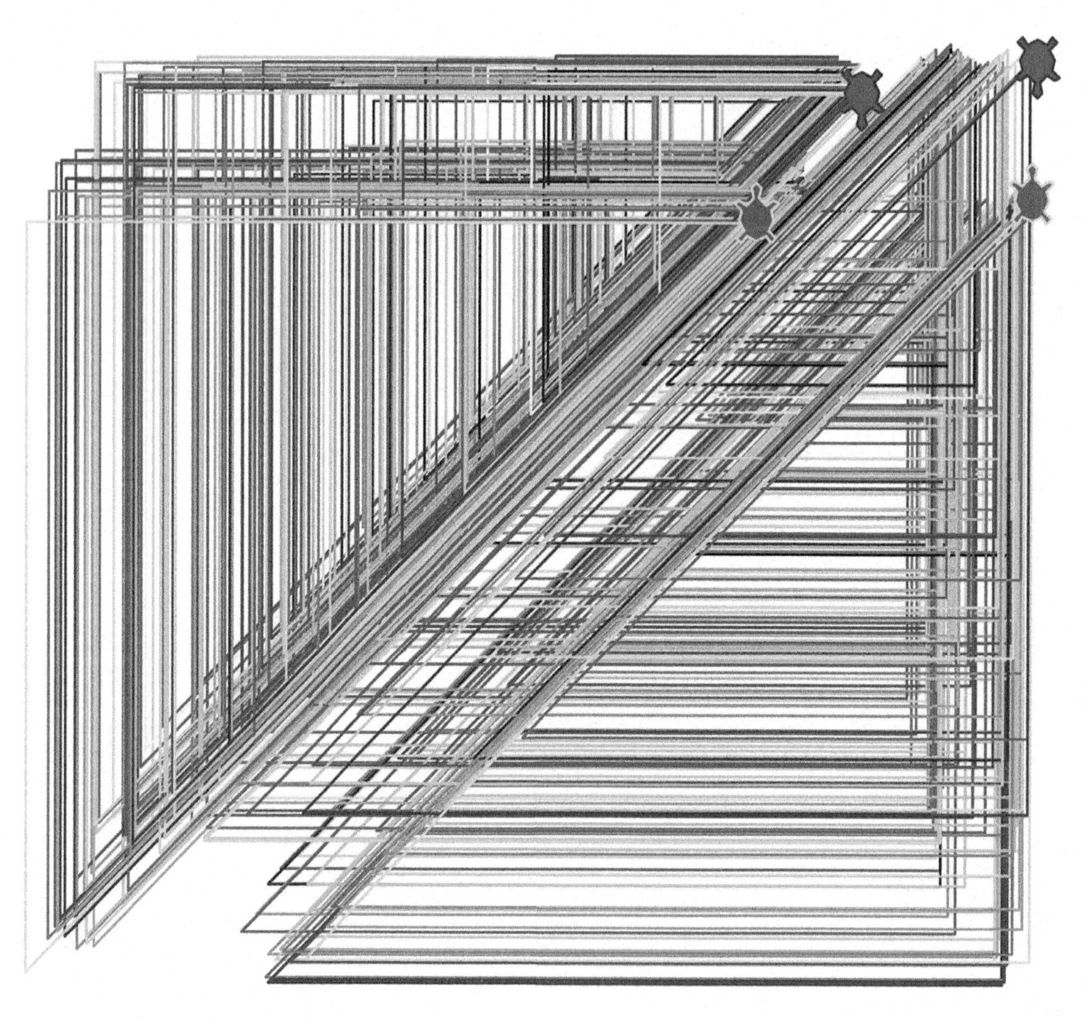

```
screen.colormode(255)
screen.bgcolor(0,0,0)
# a function that takes the name of the turtle, the angle at which
# it should be rotated and the distance that it should move
def move(turtle, angle, distance):
    turtle.left(angle)
    turtle.forward(distance)
# this function takes the turtle name as well as
# the x and y distances that the turtle should move
def moveXY(turtle, dx, dy):
    turtle.setpos(turtle.xcor()+dx,
                  turtle.ycor()+dy)
# four turtles are created
# lM for left foot of the man
```

```
# rM for right foot of the man
lM = turtle.Turtle()
rM = turtle.Turtle()
# lW for left foot of the woman
# rW for right foot of the woman
lW = turtle.Turtle()
rW = turtle.Turtle()
lM.hideturtle()
rM.hideturtle()
lW.hideturtle()
rW.hideturtle()
# the pens are lifted until the feet are in the starting position
lM.penup()
rM.penup()
lW.penup()
rW.penup()
# the man's feet are blue and woman's feet are red
lM.fillcolor("blue")
rM.fillcolor("blue")
lW.fillcolor("red")
rW.fillcolor("red")
lM.pencolor("blue")
rM.pencolor("blue")
lW.pencolor("red")
rW.pencolor("red")
lM.shape("turtle")
rM.shape("turtle")
lW.shape("turtle")
rW.shape("turtle")
lM.showturtle()
rM.showturtle()
lW.showturtle()
rW.showturtle()
screen.delay(10)
# adjust the direction of the foot to make sure that
# the man's feet are facing the woman's feet
lM.setpos(200,200)
rM.setpos(150,200)
move(lM, -90, 0)
move(rM, -90, 0)
rW.setpos(200,150)
lW.setpos(150,150)
move(lW, 90, 0)
move(rW, 90, 0)
```

```
stepSize = 200
# once the feet at their initial position, the pen of each foot is down
# and ready to show the traces of their feet's movements
lM.pendown()
rM.pendown()
lW.pendown()
rW.pendown()
screen.delay(1)
# this function takes two numbers (m and n) and generate a random integer num-
ber between m and n
def rand(a,b):
    return random.randint(a,b)
# Whenever this functin is called, one full basic rumba movement is executed
def rumba():
    # a random colour is generated for the man
    r = rand(0,255); b = rand(0,255); g = rand(0,255)
    lM.pencolor(r,g,b)#;      rM.fillcolor(r,g,b)
    rM.pencolor(r,g,b)#;      rM.fillcolor(r,g,b)
    # a random colour is generated for the the woman
    r = rand(0,255); b = rand(0,255); g = rand(0,255)
    lW.pencolor(r,g,b)#;      rM.fillcolor(r,g,b)
    rW.pencolor(r,g,b)#;      rM.fillcolor(r,g,b)

    stepSize = rand(50,300)     # a random step size is generated for each new
cycle
    r = 2;                      # this number offsets the location of the feet
byt a small distance
    # no. 0
    move(rW, 720, 0)            # turn woman's right foot 720 degrees anti-
clockwise
                                # this is used as an indicated showing that
the dance is about to start
    # no. 1
    # here woman's right foot and then man's left foot are moved along the y
axis
    moveXY(rW, 0, -stepSize+rand(-r,r))
    moveXY(lM, 0, -stepSize+rand(-r,r))
    # no. 2
    moveXY(rM, -stepSize+rand(-r,r), -stepSize+rand(-r,r))
    moveXY(lW, -stepSize+rand(-r,r), -stepSize+rand(-r,r))
    #move(rM, -45, 100)
    #move(lW, 180-45, 100)
    # no. 3
    moveXY(lM, -stepSize+rand(-r,r), 0)
```

```
    moveXY(rW, -stepSize+rand(-r,r), 0)
    # no. 4
    moveXY(rM, 0, stepSize+rand(-r,r))
    moveXY(lW, 0, stepSize+rand(-r,r))

    # no. 5
    moveXY(lM, stepSize+rand(-r,r), stepSize+rand(-r,r))
    moveXY(rW, stepSize+rand(-r,r), stepSize+rand(-r,r))
    # no. 6
    moveXY(rM, stepSize+rand(-r,r), 0)
    moveXY(lW, stepSize+rand(-r,r), 0)
    #rM.stamp(); lM.stamp(); rW.stamp(); lW.stamp()
# MAIN
# Here, 100 cycle of the rumba basic step dance is executed
for i in range(100):
    rumba()
# saving the final image as a pdf file (without the background colour)
screen.getcanvas().postscript(file="Rumba_couple.pdf")
```

Rumba Simultaneous

This dance is the same as the previous dance but the dancers move simultaneously as if trying to imitate the real dance steps. There are however no offsets between the feet-location of the dancers, therefore one step overwrites the trace of the other dancer's steps. A sample visual output of the 100 cycles is given in the Figure 21.

```
import turtle    # this line makes sure the turtle object can be used
import random    # this lines allows the program to generate random numbers
when needed
                 # e.g. in order to decide the step size
screen = turtle.Screen()
screen.screensize(300,300)
screen.colormode(255)
screen.bgcolor(0,0,0)
# a function that takes the name of the turtle, the angle at which
# it should be rotated and the distance that it should move
def move(turtle, angle, distance):
    turtle.left(angle)
    turtle.forward(distance)
# this function takes the turtle name as well as
# the x and y distances that the turtle should move
def moveXY(turtle, dx, dy):
```

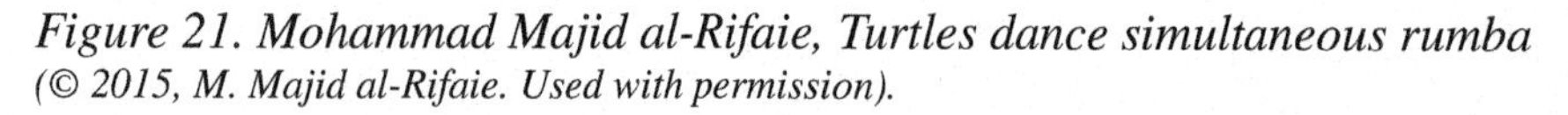

Figure 21. Mohammad Majid al-Rifaie, Turtles dance simultaneous rumba (© 2015, M. Majid al-Rifaie. Used with permission).

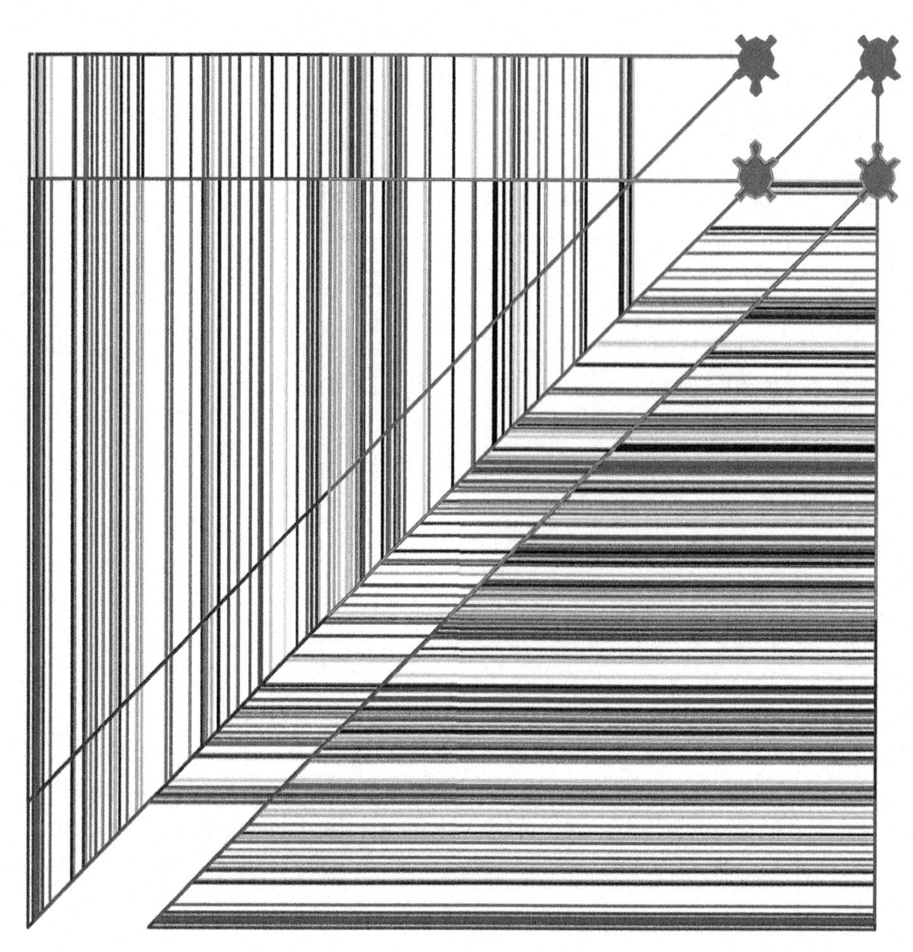

```
    turtle.setpos(turtle.xcor()+dx,
                  turtle.ycor()+dy)
# four turtles are created
# lM for left foot of the man
# rM for right foot of the man
lM = turtle.Turtle()
rM = turtle.Turtle()
# lW for left foot of the woman
# rW for right foot of the woman
lW = turtle.Turtle()
rW = turtle.Turtle()
lM.hideturtle()
rM.hideturtle()
lW.hideturtle()
rW.hideturtle()
```

```
# the pens are lifted until the feet are in the starting position
lM.penup()
rM.penup()
lW.penup()
rW.penup()
# the man's feet are blue and woman's feet are red
lM.fillcolor("blue")
rM.fillcolor("blue")
lW.fillcolor("red")
rW.fillcolor("red")
lM.pencolor("blue")
rM.pencolor("blue")
lW.pencolor("red")
rW.pencolor("red")
lM.shape("turtle")
rM.shape("turtle")
lW.shape("turtle")
rW.shape("turtle")
lM.showturtle()
rM.showturtle()
lW.showturtle()
rW.showturtle()
screen.delay(10)
# adjust the direction of the foot to make sure that
# the man's feet are facing the woman's feet
lM.setpos(200,200)
rM.setpos(150,200)
move(lM, -90, 0)
move(rM, -90, 0)
rW.setpos(200,150)
lW.setpos(150,150)
move(lW, 90, 0)
move(rW, 90, 0)
stepSize = 200
# once the feet at their initial position, the pen of each foot is down
# and ready to show the traces of their feet's movements
lM.pendown()
rM.pendown()
lW.pendown()
rW.pendown()
screen.delay(0.00001)
# this function takes two numbers (m and n) and generate a random integer num-
ber between m and n
def rand(a,b):
```

```
    return random.randint(a,b)
# Whenever this function is called, one full basic rumba movement is executed
def rumba():

    # a random colour is generated for the man
    r = rand(0,255); b = rand(0,255); g = rand(0,255)
    lM.pencolor(r,g,b)
    rM.pencolor(r,g,b)
    # a random colour is generated for the the woman
    r = rand(0,255); b = rand(0,255); g = rand(0,255)
    lW.pencolor(r,g,b)
    rW.pencolor(r,g,b)

    stepSize = rand(50,300)          # a random step size is generated for each
new cycle

    # no. 0
    move(rW, 720, 0)                  # turn woman's right foot 720 degrees
anti-clockwise
                                      # this is used as an indicated showing
that the dance is about to start

    # no. 1
    for i in range(stepSize):         # the FOR loop allows the relevant feet to
move together one unit at a time
        moveXY(rW, 0, -1)
        moveXY(lM, 0, -1)
    # no. 2
    for i in range(stepSize):
        moveXY(rM, -1, -1)
        moveXY(lW, -1, -1)
    # no. 3
    for i in range(stepSize):
        moveXY(lM, -1, 0)
        moveXY(rW, -1, 0)
    # no. 4
    for i in range(stepSize):
        moveXY(rM, 0, 1)
        moveXY(lW, 0, 1)

    # no. 5
    for i in range(stepSize):
        moveXY(lM, 1, 1)
        moveXY(rW, 1, 1)
```

```
    # no. 6
    for i in range(stepSize):
        moveXY(rM, 1, 0)
        moveXY(lW, 1, 0)
    #rM.stamp(); lM.stamp(); rW.stamp(); lW.stamp()
# MAIN
# Here, 100 cycle of the rumba basic step dance is executed
for i in range(100):
    rumba()
# saving the final image as a pdf file (without the background colour)
screen.getcanvas().postscript(file="Rumba_Couple_simultaneous.pdf")
```

Students taking a Social Dance class taught by Brittany Jacobs at the University of Northern Colorado are practicing tango while applying their own approach to the rules of this dance (Figure 22).

Figure 22. Anna Ursyn, Social dance – tango
(© 2015, A. Ursyn. Used with permission).

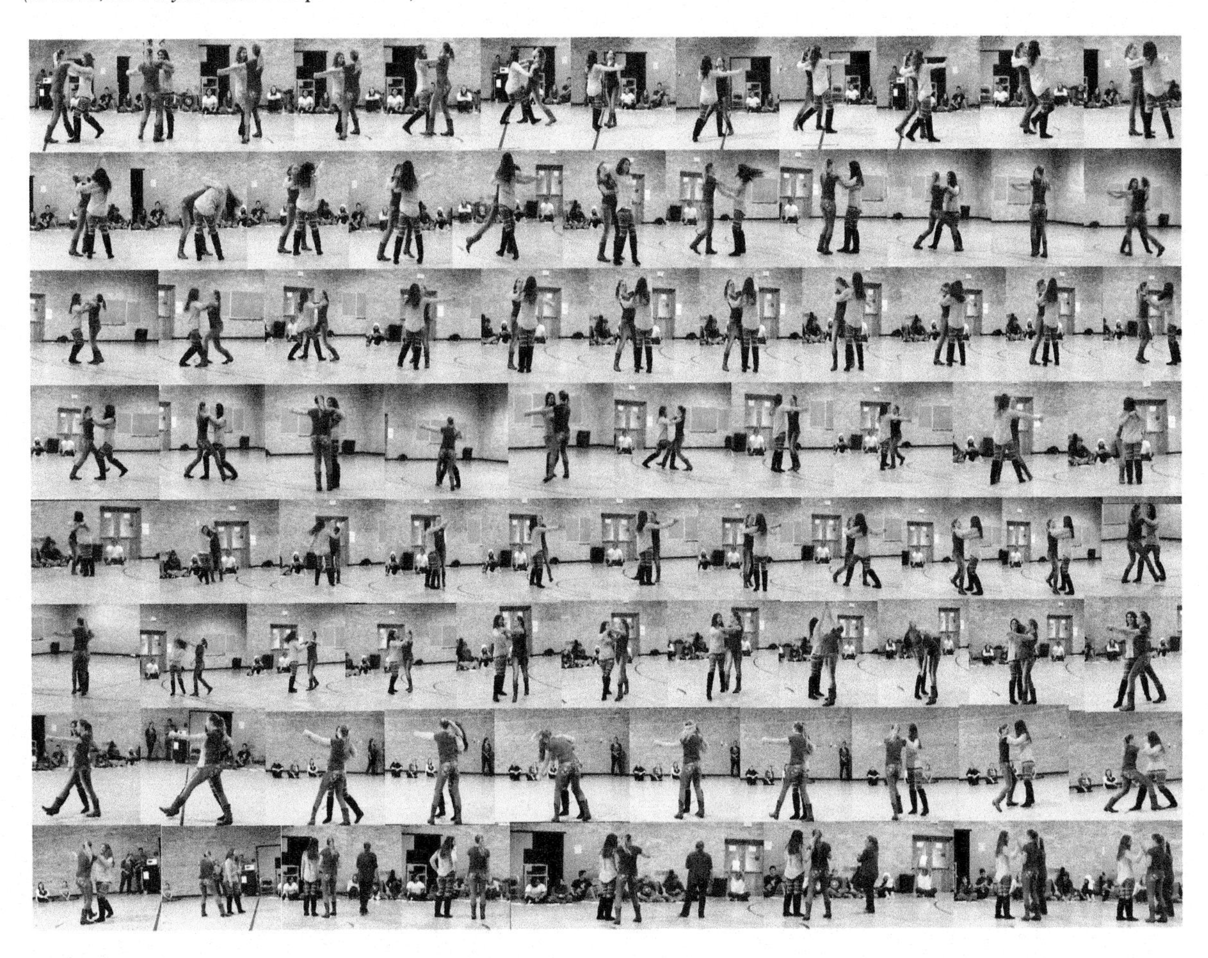

Md. Fahimul Islam

C# Portion for Dance as a Metaphor of Coding

Start development in C# in Microsoft Visual Studio 2010. First, create a new project and then add a timer to the form from the toolbox. Resize the form size to 600x600. Change the back color of the form to black. Double Click timer1 and it will open the code window; create timer1 (Tick method). Scroll the window up and the code window will appear in the middle. Window can be toggled between the Code and Design View window.

Now type code from the code section below and it will change the graphics according to the dance step sequence.

Code:

```
using System;
using System.Collections.Generic;
using System.ComponentModel;
using System.Data;                         Default class library
using System.Drawing;
using System.Linq;
using System.Text;
using System.Windows.Forms;
using System.Reflection;    // for brushes property retrieval


namespace DanceGraphics
{
    public partial class Form1: Form
    {
        int L = 0;                    //For left foot's coordinate
        int R = 1;                    //For right foot's coordinate
        int x = 0;                    //For x coordinate
        int y = 1;                    //For y coordinate
        int Step = 0;            //Step to retrieve 13 steps


    int[, ,] xy = new int[,,]{{{250,400},{310,400}},
                          {{250,274},{310,400}},
                          {{250,274},{310,190}},
                          {{250,76},{310,190}},
                          {{250,76},{450,84}},
                          {{400,76},{450,84}},
                          {{400,76},{200,76}},          //Coordinates
                          {{40,66},{200,76}},
                          {{40,66},{100,76}},
                          {{250,76},{100,76}},
                          {{250,76},{450,84}},
                          {{400,76},{450,84}},
                          {{400,76},{450,220}}};
```

```
        Random rnd = new Random();                        //Random variable rnd for
random color and x
        int xrnd = 0;                                //xrnd use to shift dance along
x axis
        Brush LBrush;                                //Left foot brush color variable
        Brush RBrush;                                //Right foot brush color vari-
able

        public Form1()
        {
            InitializeComponent();                        //Default initialize method
            this.Text = "Tango Dance";               //Set Title of the form
            LBrush = RandomBrush();               //Set random color to Left foot
            RBrush = RandomBrush();               //Set random color to Right foot
        }
        private void Form1_Load(object sender, EventArgs e)
        {
            this.DoubleBuffered = true;                         //Enable double buffer
to render smoothly
            this.Paint += new PaintEventHandler(Form1_Paint); //New paint
event add
        }
        private void Form1_Paint(object sender, System.Windows.Forms.Paint-
EventArgs e)
        {
                 //The following FillRectangle methods are used to draw left
and right
                 //rectangles represent the feet and lines attached to those.
         e.Graphics.FillRectangle(LBrush, xy[Step, L, x] + xrnd, xy[Step, L,
y], 11, 84);
         e.Graphics.FillRectangle(LBrush, xy[Step, L, x] + xrnd+5, 0, 1, 600);
         e.Graphics.FillRectangle(RBrush, xy[Step, R, x] + xrnd, xy[Step, R,
y], 11, 84);
         e.Graphics.FillRectangle(RBrush, xy[Step, R, x] + xrnd+5, 0, 1, 600);
            Step++;          // Next Step
            if (Step == 13) //Reset a new dance position and feet color
            {
                Step = 0;
                xrnd = rnd.Next(-300,300);
                do
                {
                    LBrush = RandomBrush();
                    RBrush = RandomBrush();
                } while (LBrush == RBrush); //Make sure that random color does
```

```
not match
                    }
            }
            Brush RandomBrush()           // method to return random Brush color
            {
                PropertyInfo[] brushInfo = typeof(Brushes).GetProperties();
                Brush[] brushList = new Brush[brushInfo.Length];
                for (int i = 0; i < brushInfo.Length; i++)
                {
                    brushList[i] = (Brush)brushInfo[i].GetValue(null, null);
                }
                Random randomNumber = new Random(DateTime.Now.Millisecond);
                return brushList[randomNumber.Next(1, brushList.Length)];
            }
            private void timer1_Tick(object sender, EventArgs e)
            {
                this.Refresh(); //Every 250 ms this method automatically refresh
the
            }                            //Form1 which results to invoke the Form1_
Load call
        }                                    //again and again
}
```

Figure 23 presents the output of this project.

In terms of Python, the easiest way is to use IDLE and run code by pressing F5.

In all the codes, each cycle is repeated 100 times and each time is different that the other cycle, in terms of step-size, color, etc. So in a way there is a limited set of improvisation involved within the constraints given.

RECOMMENDATIONS FOR FURTHER WORK

The follow up task would be to expand behind ready image and go into the time based explorations. Another interesting issue is to code for music related creations, and to explain scientific factors behind it. Expanding this work would aim at explaining physics behind a color image and sound, and show programming as an important means to communicate notions.

SUMMARY AND CONCLUSION

This is an untypical approach to teaching and learning programming, supported with a brief overview in various programming languages. It presents general information about basic concepts, rules, and mechanics behind programming introduced as a metaphor. Dances possess rules and definitions that need

Figure 23. Md. Fahimul Islam, Tango – output
(© 2015, F. Islam. Used with permission).

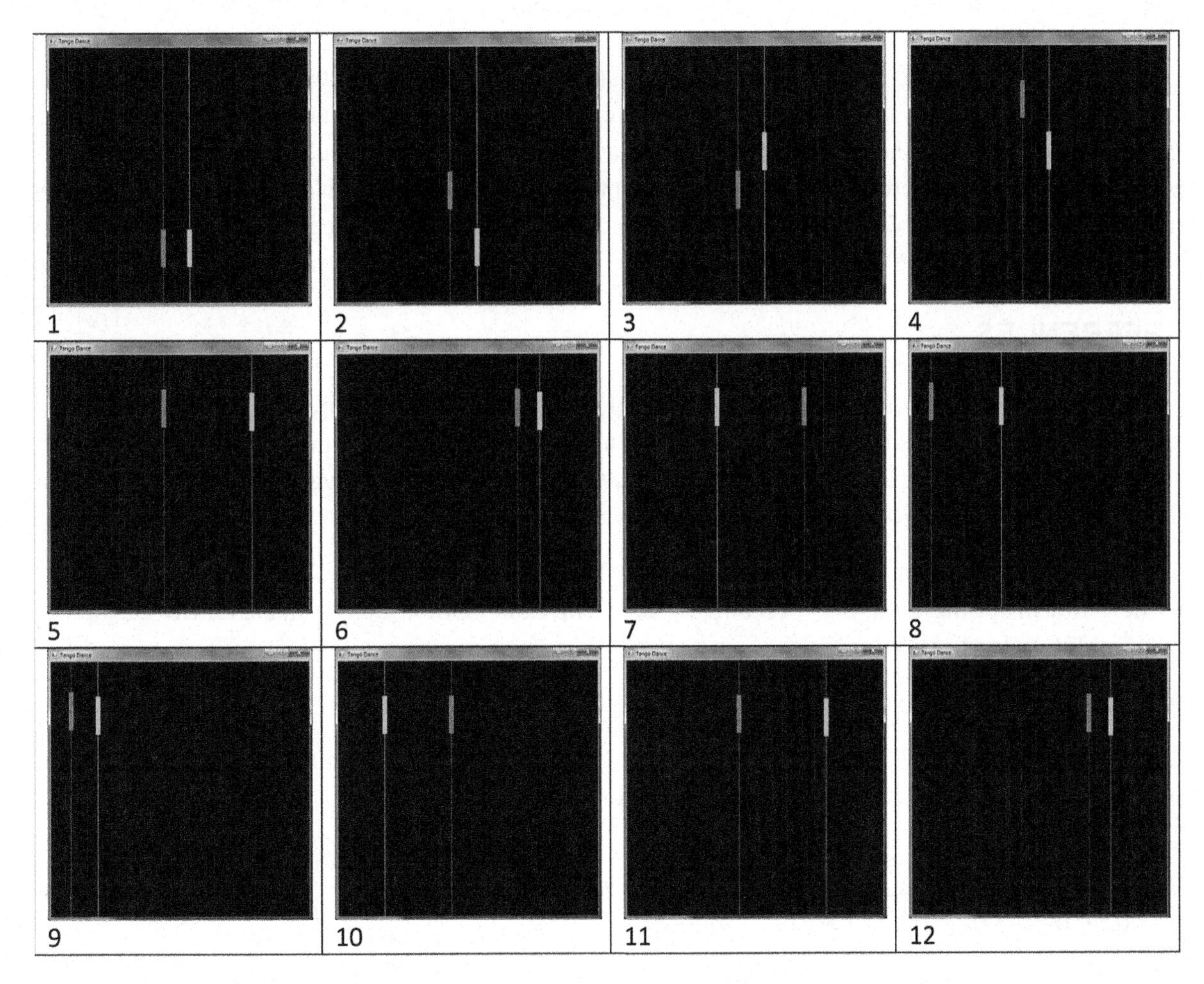

to be learned and practiced in order for the dancer to master them, thus being able to enjoy dancing. At the same time, programming is no different. We need first to grasp a general idea, syntax, conditions, and the concepts behind, and then apply them with a specific goal in mind. We also discussed the idea of randomness in coding, where the computer selects random numbers when dealing with a code. This is being compared to a dancer who is improvising, thus creating their own choreography of a newly invented dance, following the rhythm, tempo, and style of a music played. We all know that learning how to drive a car is not only about learning the rules and the meaning behind the road signs; a mileage driven makes a good driver. The more we drive the better connections our brain makes, building attention span, speed of reaction, readiness to react, and even some insight and intuition in relation to other drivers sharing a road.

We hope this chapter aid an inexperienced person feels more comfortable with the concept of coding and make them ready for experimentation; it would remove some possible fear of being not ready for programming due to some lack of training, background, or experience. At the same time, we hope it could become quite entertaining to try something beyond one's everyday routine, scope, or thinking.

More and more, programming becomes crucial for solving problems, connecting, experimenting, and creating a new content, ideas, and solutions. Programming is not just a set of tasks used to reach a goal. It aids us where the software stops. At the same time current workspaces within companies become extensively collaborative, so the tasks are shared and extended. The generalists work with specialists together, and it becomes more helpful when they can share the same language, lingo, or understanding. Thus, we hope that the readers who are non-savvy in programming would experience this type of environment, become more apt to discuss the in depth solutions to some problems or tasks, and become more ready, knowledgeable, and confident when requesting particular services from a coder.

REFERENCES

al-Rifaie, M. M., Aber, A., & Bishop, M. J. (2012). Cooperation of nature and physiologically inspired mechanisms in visualisation. Blood Flow Shapes the Drawings of Birds and Ants. In A. Ursyn (Ed.), *Biologically-Inspired Computing for the Arts: Scientific Data through Graphics*. IGI Global. doi:10.4018/978-1-4666-0942-6.ch003

al-Rifaie, M. M., Bishop, M., & Blackwell, T. (2011). An investigation into the diffusion search and particle swarm optimisation. In *Proceedings of the Gene Computation Conference GECCO'11* (pp. 37-44). Dublin, Ireland: ACM.

Alighieri, D. (2005). *De vulgari eloquentia.* (S. Botterill, Trans.). Cambridge University Press.

Bio. (2015). *Edmund Hillary Biography*. Retrieved August 2, 2015, from http://www.biography.com/people/edmund-hillary-9339111#death-and-legacy

Boden, M. (2016). Skills and the Appreciation of Computer Art. In al-Rifaie, M. M. (Ed.), Computational Creativity, Measurement and Evaluation, Connection Science. Taylor & Francis (submitted). doi:10.1080/09540091.2015.1130023

Eco, U. (1989). *The Open Work* (A. Cancogni, Trans.). Cambridge, MA: Harvard University Press.

Eco, U. (1990). *The Limits of Interpretation*. Bloomington, IN: Indiana University Press.

Make your Mark. (2015). Ringling College of Art + Design.

Western Union Tuning Out Singing Telegram. (1972, July 29). *The New York Times*, p. 27.

ADDITIONAL READING

Bousquet, M. (2015). *Physics for Animators.* Focal Press.

Cerny Milton, S. (2007). *Choreography: A Basic Approach Using Improvisation* (3rd ed.). Human Kinetics.

Dale, N., & Lewis, J. (2014). *Computer Science Illuminated* (6th ed.). Jones & Bartlett Learning.

Farrell, J. (2014). *Programming Logic and Design, Introductory* (8th ed.). Course Technology.

Lefevre, C. (2012). *The Dance Bible: The Complete Resource for Aspiring Dancers*. Barron's Educational Series; SPI edition.

KEY TERMS AND DEFINITIONS

Coding: Means designing, writing, testing, debugging, troubleshooting, and maintaining the source code of computer programs.

Dances: Are forms of art using a sequence of choreographed movements designed for human body toward a particular set of musical compositions with aesthetics in mind. Improvisation: in dance is used as choreography for dance composition when no particular preset specific rules are observed, and is based on experimentation and explorations between shape, space, and time. Conditions: and variables are solutions to control the flow of a command for particular rules, actions, exclusions, and preventive actions.

Metaphor: Describes content as being the same as something unrelated for the rhetorical effect, thus highlighting the similarities between them. Metaphors can be verbal or visual (thus offering semiotics and semantic comparisons).

Programming Languages: Help one to develop a code to control the behavior of a computer or to develop an algorithm.

Randomness: Refers to some lack of a specified pattern order, symbols, or predictability based on a concept of chance.

Syntax: Is a set of rules, procedures, and processes addressed for a particular language in order to guide its structure of particular ways of communication.

Section 4
Educational Applications and Cognitive Learning

The authors of this section provide theoretical and practical materials supporting teaching and learning science.

Chapter 11
Optimizing Students' Information Processing in Science Learning:
A Knowledge Visualization Approach

Robert Zheng
University of Utah, USA

Yiqing Wang
Shanghai Normal University, China

ABSTRACT

This chapter discusses a theoretical framework for designing effective visual learning in science education. The framework is based on several theories related to cognitive visual information processes and empirical evidence from authors' previous research in visual-based science learning. Emphasis has been made on the structural part of the framework that allows dynamic linking to critical factors in visual learning. The framework provides a new perspective by identifying the variables in visual learning and the instructional strategies aiming at the improvement of visual performance. The discussion of the theoretical and practical significance of the framework is made, followed by suggestions for future research.

INTRODUCTION

There have been genuine concerns about the decreasing of the numbers of students opting to study science in schools (Kennedy, 2014; Scogin & Stuessy, 2015). Studies have shown that the diminishing student population in science can be largely attributed to the following factors: (1) the difficulty of the subject characterized by abstract concepts and complex computation, (2) the dated instructional pedagogy that often lands in students' information overload, and (3) the disconnection between the science subject and the real world. (Savasci Acikalin, 2014; Sokolowska, de Meyere & Folmer, 2014).

Over the last several decades efforts have been made to improve student performance including strategies to promote cognitive and motivational aspects in science learning (Azevedo, 2015). Of particular

DOI: 10.4018/978-1-5225-0480-1.ch011

interest to researchers is the use of visual tools to facilitate cognitive information processes in science. For example, Chen and Yang (2014) investigated the relationship between visual tools and students' problem solving skills in science, and found a close correlation between the two. Similar findings were obtained by Liben, Kastens, & Christensen (2011) who studied students' learning in geology. Several learner-centered strategies have been proposed including inquiry-based learning (Gillies, Nichols, & Burgh, 2011) and self-regulated learning (Tang & Neber, 2008; Zimmerman, 1998, 2001) where learners are nurtured to develop high level thinking skills such as making inference and transferring knowledge to a new domain in visual-based science learning. Nonetheless, students may still experience frustration or fail to accomplish the learning goals if the design of above learning does not take into consideration the impact of various cognitive load in visual learning. That is, failing to consider the constraints of cognitive load in visual-based science learning may have serious consequences in learners' performance, especially when visuals cause redundancy or split attention in learning (Mayer, 2001). The purpose of the current chapter is to examine the functional role of visual representations by focusing on (1) the effects of cognitive load on visual learning and (2) instructional strategies targeting at reducing irrelevant cognitive load and increasing relevant cognitive load. By reading this chapter, the readers will be able to:

1. Identify challenges of cognitive loads in visual information processing.
2. Understand pedagogical strategies to reduce cognitive load in learning.
3. Be familiar with the visual learning framework in science learning.

THEORETICAL BACKGROUND

Over the last several decades, educators, psychologists and cognitive scientists have been interested in understanding the cognitive resources in working memory and their relations with visual learning (Paivio, 1986; Sweller, van Merrienboer, & Paas, 1998; Um, Plass, & Hayward, 2012). According to Mayer (2001), cognitive resources play an important role in successful performance in visual learning. Complex cognitive processes in learning such as association across domains, information retrieval from long-term memory, and engagement in deep learning, are closely related to cognitive resources in working memory (Johnson & Mayer, 2009; Paivio, 1986; Zheng, 2007). Several theories have contributed to the understanding of the role and function of cognitive resources in visual learning. They include working memory theory, dual-coding theory, cognitive theory of multimedia learning and cognitive load theory. A discussion of each theory in relation to cognitive resources in visual learning follows.

Working Memory Theory

When discussing cognitive resources, one would invariantly associate it with the epic theory of working memory by Baddeley and Hitch (1994). Based on their empirical findings pertaining to human information processes, Baddeley and colleagues (see Baddeley, 1986, 2000; Baddeley & Hitch, 1994) proposed a working memory model that includes a central executive system with three parts: phonological loop, visuospatial sketchpad, and episodic buffer. The phonological loop stores verbal content, whereas the visuospatial sketchpad caters to visuospatial information. The episodic buffer is a mechanism in working memory that dedicates to linking information across domains to form integrated units of visual, spatial and verbal information with time sequencing such as the memory of a story or event. Overloading any of the above three subsystems could have significant consequences in learning. For example, when phonological loop is overloaded, the learner will have little cognitive resources to process verbal information. The same is true with visuospatial sketchpad, in which the learner will not be able to process visual information such as calculating the arc of a circle in its visual format.

Baddeley's working memory model delineates, for the first time, the subsystems within the working memory and differentiates three kinds of cognitive resources associated with the subsystems. His model points out the distinction between verbal and visual information, which has a significant influence on other theories of visual learning including Paivo's (1986) dual-coding theory, Mayer's (2001) cognitive theory of multimedia learning and Sweller's (1988; Sweller & Chandler, 1991) cognitive load theory.

Dual-Coding Theory

Paivo (1986) posits that the efficacy of learning can be attained by expanding on learned material via verbal associations and visual imagery, which he defines as dual-coding processing. According to dual-coding theory, visual and verbal information is processed separately. That is, the visual information is processed through visual channel whereas the verbal information is processed through the verbal channel. When both information is processed in the working memory, they become associated with each other to form a mental representation of the incoming information. For example, when learning the concept of a car, the teacher presents to students a card with the word "car" on it and an image of a car. As the learner learns the word of car, he/she will associate the word with the unique characteristics of a car from the image such as four wheels, engine, and so forth. The word "car" thus becomes associated its image in working memory to form a mental representation of the car: a transportation vehicle for passengers or freight. The dual-coding theory postulates that when the incoming information is processed through verbal and visual channels, the learner is able to build a robust mental representation of the new information, which consequently enhances the learner's comprehension and knowledge transfer in learning. The significance of the dual-coding theory lies in its seminal approach to visual and verbal learning, which lays the groundwork for multimedia learning.

Cognitive Theory of Multimedia Learning

Mayer's cognitive theory of multimedia learning (Mayer, 2001) identifies seven principles of multimedia design. They include multimedia principle, spatial contiguity principle, temporal contiguity principle, coherence principle, modality principle, redundancy principle, and individual differences principle. Drawing from cognitive theory of information process, especially Baddeley's working memory theory, Mayer posits that the design of media including multimedia needs to take into consideration the cognitive resources in visual learning. His principles of multimedia learning examine the relations between visual and verbal information and the limitations of working memory. For example, his redundancy principle indicates that too many visuals can adversely affect learners' information processing in knowledge acquisition (Mayer, Heiser, & Lonn, 2001). In their study Mayer (2001) and colleagues found that narration with animation (i.e., audio + visual) was superior to narration with animation plus onscreen caption (i.e., audio + visual + visual). The latter creates a redundancy effect that can take significant portion of cognitive resources in working memory resulting in poor performance in visual learning. In a separate study Mayer and colleagues examined the relations between visual objects (e.g., words and pictures) and their spatial position (Mayer, Steinhoff, Bower, & Mars, 1995; Moreno & Mayer, 1999). They found that when corresponding words and pictures were near each other on the same page or computer screen, learners spent fewer cognitive resources in processing the information on the page. In fact, they were more likely to hold both word and image information in working memory while solving the problems.

Cognitive Load Theory

Cognitive load theory directly addresses the relationship between cognitive resources and information process in learning. Sweller and colleagues propose that there are three types of cognitive load: intrinsic cognitive load, extraneous cognitive load and germane cognitive load (Sweller, 1988; Sweller & Chandler, 1991; Sweller, van Merrienboer, & Paas, 1998). Each of them impact differently on learners' performance in visual learning.

Intrinsic cognitive load is the mental effort required by a learner to process a task. Sweller, van Merrienboër, and Paas (1998) note that element interactivity is the source of intrinsic load. As the level of element interactivity increases, the required mental effort to process the task increases as well, which causes a high demand on cognitive resources. Tasks with low element interactivity can be learned in isolation and therefore, consume less cognitive resources. Thus, low element interactivity tasks generally only become cognitively demanding when the sheer number of tasks to complete is surfeit. On the other hand, tasks that cannot be learned in isolation and that interact with other elements would impose a higher cognitive load (Ayres, 2006). For example, when learning the alphabet understanding the letter "A" does not require learning the letter "E". The level of element interactivity is very low because they can be learned independently. However, as one learns the word "apple", he/she must understand how to pronounce combinations of letters. This requires the understanding of not only the individual letters, but the interaction between them. A learner would have to learn that the letter "A" is pronounced in a different manner depending on the letters that follow it (e.g. difference between apple and ape). The increased element interactivity results in additional information that needs to be assimilated, and therefore imposes a greater cognitive load. So far, multiple efforts have been made to develop instructional strategies to lower the element interactivity in order to reduce intrinsic cognitive load in learning. They include pre-training, concept map and schema-induced analogical reasoning.

Extraneous cognitive load is caused by the format and manner in which information is presented (Brünken, Plass, & Leutner, 2003). For example, teachers may unwittingly require students to mentally integrate mutually referring, disparate sources of information that exact the limited cognitive resources in working memory, resulting in an increased cognitive load in learning. Chandler & Sweller (1991) examined the relationship between extraneous cognitive load and split-attention effect by studying the diagrams that were integrated with the text and the ones that were not. The results indicate that participants in the unintegrated condition performed poorly on the tasks and spent more time due to higher extraneous cognitive load in learning, whereas those in the integrated condition outperformed their counterparts in both measures. Consider the task presented in Figure 1. The learner would have to expend extra cognitive resources coordinating the information between the text and the image, which may overtax the working memory resources allocated to the visual information processing. Instructions like this would be likely to induce a high level of extraneous cognitive load.

Germane cognitive load was initially introduced to the cognitive load theory to separate useful, relevant demands on working memory from irrelevant and wasteful forms of cognitive processes like intrinsic

Figure 1. An example of instruction that may induce high extraneous load
(Adapted from Zheng, 2007. Used with permission).

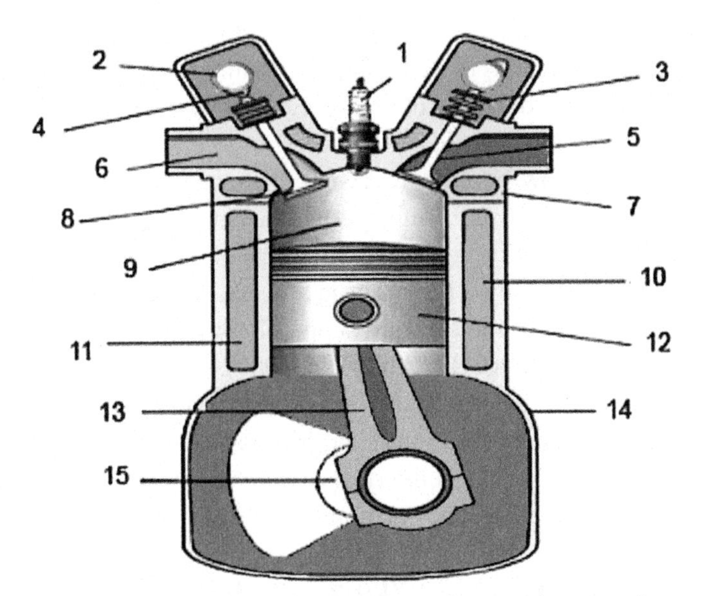

1. spark plug 2. camshaft 3. valve spring 4. cam
5. exhaust valve 6. mixture in 7. cylinder head
8. intake valve 9. combustion chamber 10. cooling water
11. cynlinder block 12. piston 13. connecting rod
14. crankcase 15. crankshaft

and extraneous cognitive loads (Sweller et al., 1998). According to Sweller et al., germane cognitive load is the mental effort a learner applies toward schema construction in learning. It is associated with motivation-, attitude-mediated cognitive resources directed towards achieving learning objectives. When a learner attends to the learning elements, attempts to establish connections between them and constructs a coherent mental representation in working memory, he/she invests germane mental effort. Researchers in germane cognitive load extends the research to motivational learning including self-efficacy, self-regulation and motivation (Um, Plass, & Hayward, 2012).

As it has been discussed above, intrinsic and extraneous cognitive loads can be detrimental to visual learning. Students who are overwhelmed with intrinsic cognitive load often fail to perform at the expected level in learning, whereas those who are affected by the extraneous cognitive load find themselves spread thin with the cognitive resources in information processing (Cook, Zheng, & Blaz, 2009; Zheng, 2007). Challenges thus remain in regard to how to optimize learners' visual learning by reducing both intrinsic and extraneous cognitive loads in their information process. The following section introduces approaches that have been empirically proven to be effective in minimizing intrinsic and extraneous cognitive loads in order to promote germane load in visual learning.

OPTIMIZE VISUAL LEARNING IN SCIENCE

Efforts have been made to optimize visual learning by alleviating the cognitive load in learning. Research indicates that pedagogical methods like two-step pre-training, concept map, and schema-induced analogical reasoning are effective in reducing intrinsic cognitive load in visual-based science learning (Miller, Geng, Zheng, & Dewald, 2012; Pollock, Chandler, & Sweller, 2002; Zheng, Yang, Garcia, & McCadden, 2008). Studies also show optimizing design elements by avoiding redundancy and split effects, for example, proves to be effective for minimizing the detrimental effects of extraneous cognitive load (Pociask & Morrison, 2008; Zheng, Miller, Snelbecker, & Cohen, 2006). Regarding germane cognitive load, researchers found that lowering intrinsic and extraneous cognitive loads would release the held-up cognitive resources, which would in turn support knowledge construction in visual learning (Zheng, McAlack, Wilmes, Kohler-Evans, & Williamson, 2009).

There are multiple approaches to managing the cognitive load in learning. Understanding the differences among the approaches has both theoretical and practical significances. In the next section our discussion will revolve around our previous empirical research in visual learning to demonstrate the effective approaches we took to reduce irrelevant cognitive load and promote relevant cognitive load.

Reducing Intrinsic Cognitive Load in Visual Learning

Oftentimes, learners are overwhelmed by the concepts, facts, and principles including formulas when learning the new content in science. This is partly due to a lack of schema in the domain area and partly due to a lack of cognitive scaffolds to support learners' acquisition of their schemata. Zheng, Udita, & Dewald (2012) proposed a structural approach to supporting learners' schema acquisition (Figure 2). The structure delineates the relationship among the complexity of the content, cognitive load and learning outcomes. The complexity of content could impose a high intrinsic cognitive load on learners, which would adversely affect their performance in visual learning. Zheng et al. (2012) point out that the key to

Figure 2. A structural approach to reducing the intrinsic cognitive load in visual learning

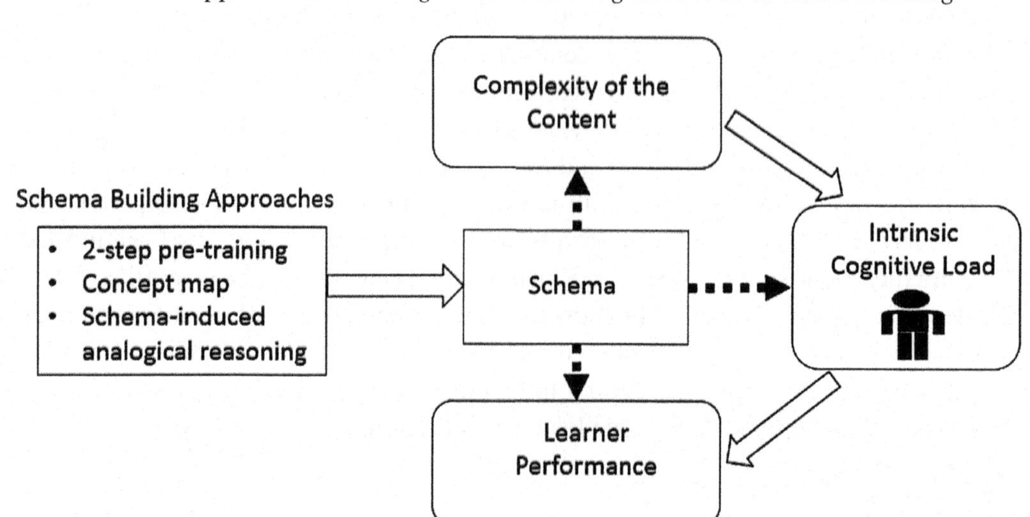

reverse the situation is to develop learners' schemata in the domain area. According to Zheng et al., schema can be a mediator that mediates the content difficult level, cognitive load, and learning outcomes. With a robust schema the learner will find the content not as difficult as it was before, which means a lower intrinsic cognitive load and more cognitive resources in working memory. The availability of cognitive resources and a robust schema in the domain area will lead to a better performance in learning. Among myriad approaches, two-step pre-training, concept map, and schema-induced analogical reasoning have been widely recognized as effective strategies to improve learners' schemata in learning.

Two-Step Pre-Training

Researchers (e.g., Kalyuga, 2005; McNamara & Kintsch, 1996) argue that prior knowledge can significantly influence learners' behavior and their consequent learning performance. They find that meaningful learning, especially learning at a deeper level, is closely related to learners' existing structure of prior knowledge often called schema. One of the widely accepted approaches to developing learners' schema is pre-training, which has been an important instructional strategy in helping learners learn complex materials (Lee, Plass, & Homer, 2006; Mehta & Russell, 2009). Pollock et al. (2002) propose a two-step pre-training model. In Pollock et al.'s model learners start with learning the isolated elements at a low element interactivity level (i.e., low intrinsic cognitive load), followed by a full content exposure with high element interactivity. Based on Pollock et al.'s model, we conducted a study to replicate the two-step process in learners' knowledge acquisition (see Zheng et al., 2012). Forty college students were recruited and were randomly assigned to two conditions: experimental and control. Participants in the experimental condition received the 2-step pre-training on the subject of diabetes with isolated concepts and facts presented first followed by the full content that was integrated with the concepts and facts from the isolated element step (Figure 3). We hypothesized that the two-step pre-training would provide necessary cognitive scaffolds for the learners to understand and make associations among different concepts and facts, which are essential for complex learning. In the control condition participants received the pre-training without the content being broken-down. Both conditions had the visual representations available for the learners to learn the content. Results showed that the experimental group outperformed the control group as measured by comprehension and knowledge transfer. The finding suggests that with its lower element interactivity (i.e., isolated concepts and facts), learners were able to develop their schemata without inflicting a high intrinsic cognitive load on themselves. As the learner built the schema in the domain area, he/she was able to learn more complex content without getting overloaded. Moreover, the transition from isolated element status to full content interactivity provided cognitive scaffolds that allowed the learner to understand how concepts and facts were meaningfully related to each other. In contrast, participants in the control group became cognitively overloaded due to a lack of schema and therefore showed a poor performance in learning. Our study provides the empirical evidence that visual learning in science can benefit from the 2-step pre-training by lowering the learners' intrinsic cognitive load and allowing them to access cognitive resources in complex visual learning.

Concept Map

Concept map is a visual tool that represents pattern of relationships among facts and concepts. It helps students (1) clarify their thinking and (2) facilitate their processing, organizing and prioritizing ideas (Wang & Dywer, 2006; Zheng & Dahl, 2010). According to Ausubel (1960), the most important issue

Figure 3. A sample of two-phase, isolated-interactive element pre-training
(Adapted from Zheng, Udita, & Dewald, 2012. Used with permission).

1. Glucose means sugar in blood.	**1. Glucose means sugar in blood.** *High or low level of glucose in blood than normal causes diabetes to occur.*
2. Insulin is the hormone in body.	**2. Insulin is the hormone in body.** *Insulin carries the sugar (glucose) into cells which use glucose as their fuel to survive. Lack of insulin causes high levels of blood glucose, interruption of metabolic processes, and physical problems.*
3. Balance means proper balance between insulin and food intakes.	**3. Balance means proper balance between insulin and food intakes.** *When blood glucose levels are low, food must be eaten. In contrast, when blood glucose is high or when food is ingested insulin must be injected to increase blood glucose.*
4. A glucometer is a monitoring device used to monitor blood sugar levels.	**4. A glucometer is a monitoring device used to monitor blood sugar levels.** *A glucometer uses individual test strips for each level check. Each test strip can only be used once, and blood must be drawn, through pricking, and blood input into the strip every time glucose levels are checked.*

influencing learning is what the learner already knows that can be associated with new learning. Concept map is a cognitive tool that scaffolds learners to bridge the new material with their prior knowledge by presenting an idea map that delineates the relationship among facts, concepts, and principles. Kinchin, Hay, and Adams (2000) point out that concept map resembles human knowledge structure in their schemata. This resemblance activates meaningful association through recall and mapping of concepts between the concept map and the stored information in learners' schemata. Figure 4 presents an example of concept map that shows the relationship among the concepts pertaining to plants and their utility functions.

There are various approaches to concept map application: presenting the concept map before or after the content; exposing students to the concept prepared by the teachers or asking the students to generate their own concept map. However, there is a lack of knowledge about how they differ from each other in terms of cognitive load and performance. To answer this question an experimental study was conducted at a large research I university in the Western United States (see Zheng and Dahl, 2010). Participants (N=27) were recruited and randomly assigned into three conditions: concept map provided before learning the science text (before concept map), concept map provided after learning the science text (after concept map), and the control condition. Participants were asked to study a text on diabetes. Depending on the condition, participants were asked to (1) first study the concept map that described the relationship between diabetes, sugar level, insulin, and medication followed by a text on diabetes, or (2) first study a text on diabetes followed by the concept map, or (3) text only without concept map. The study aimed to compare the differences between (1) concept map and no concept map, and (2) before and after concept map. The dependent measure included a knowledge test on diabetes.

Figure 4. An example of concept map for learning

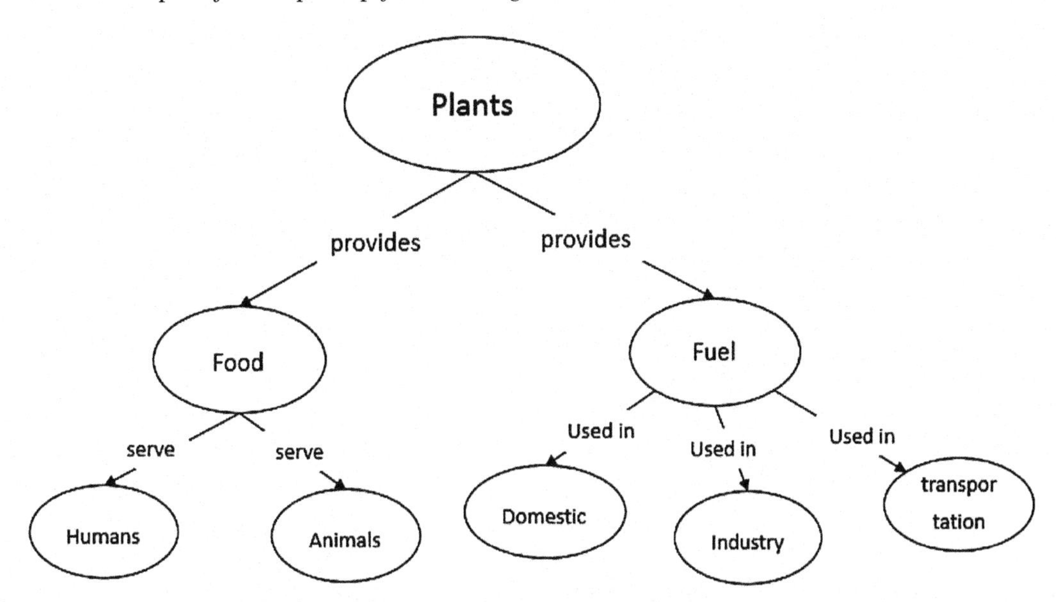

Results show that there was a significant difference between concept maps and non-concept maps. Participants who received the concept map demonstrated higher mean scores in the achievement test than those who did not, indicating that the concept map may help facilitate the construction of the prior knowledge and thus lead to better learning outcomes than those without concept maps. Additionally, differences were found between before concept map and after concept map conditions. The findings indicate that learners in the before-concept-map condition outperformed those in the after-concept-map condition. The results suggest that when the concept map is provided before learning, it functions as a tool for prior knowledge construction. However, when the concept map is provided after learning, it essentially functions as a summary of what has been learned. Although summary can be a powerful tool in reinforcing what the learner has learned, it may become less functional if the learner hasn't had a schema in the first place.

Schema-Induced Analogical Reasoning

Analogies are tools for understanding concepts. Differing from other schema construction approaches, the schema-induced analogical reasoning is to understand the new concept by matching it with a concept previously acquired in the schema. For example, when teaching a new science subject like atom, the new concept may take root in the learner's mind through an activation of prior knowledge by telling him/her that the atom resembles a miniature solar system. Most analogical reasoning takes the form of abstracting thinking via verbal presentation. However, we were interested in understanding how visual form of analogical reasoning would impact learners' schema activation in science learning (see Zheng et al., 2008). Eighty-nine fourth grade elementary students were recruited from an elementary school in the north-east of the United States. Students were randomly assigned to interactive multimedia with analogy, interactive multimedia without analogy and the control conditions. In the interactive multimedia with analogy condition, participants were presented with two multimedia learning objects: water cycle system and electrical circuits, with the former serving as the base domain of the analogy. The water cycle system includes a water-driven turbine, the water duct with controllable bars that can go up and down in the duct to affect the speed of water flow, and a water pump that pumps the water into the duct from one end to another. It simulates the electrical circulation with the pump as the battery, the water duct as the electric wires, the water bar as the electric resistance, and the water turbine as the light bulb. When the water pump increases its power and the water bars are down, the water flows fast which in turn makes the water turbine move fast. This process is similar to electric circuit in which the increased battery power with low electric resistance will increase the brightness of the light bulb (See Figure 5). In the interactive multimedia without analogy condition, only the electrical circuit multimedia object was presented. There was no water cycle system analogy based on which the participants could infer the principle of electricity. The students in each condition learned the electrical circuit in a visual format followed by the recall and transfer tests on electric circuit. Significant differences were obtained between analogy and non-analogy and between multimedia and non-multimedia. Our results indicate that schema-induced analogical reasoning in a multimedia environment can significantly reduce learners' cognitive load and facilitate their knowledge acquisition in science learning.

Reducing Extraneous Cognitive Load in Visual Learning

In science education, there is a trend to integrate visuals including visual arts to teach the content (Brown, Losoff, & Hollis, 2014; Dhanapal, Kanapathy, & Mastan, 2014). Brown et al. employ active

Figure 5. Electric circuit with water analogy
(Adapted from Zheng, Yang, Garcia, & McCadden, 2008. Used with permission).

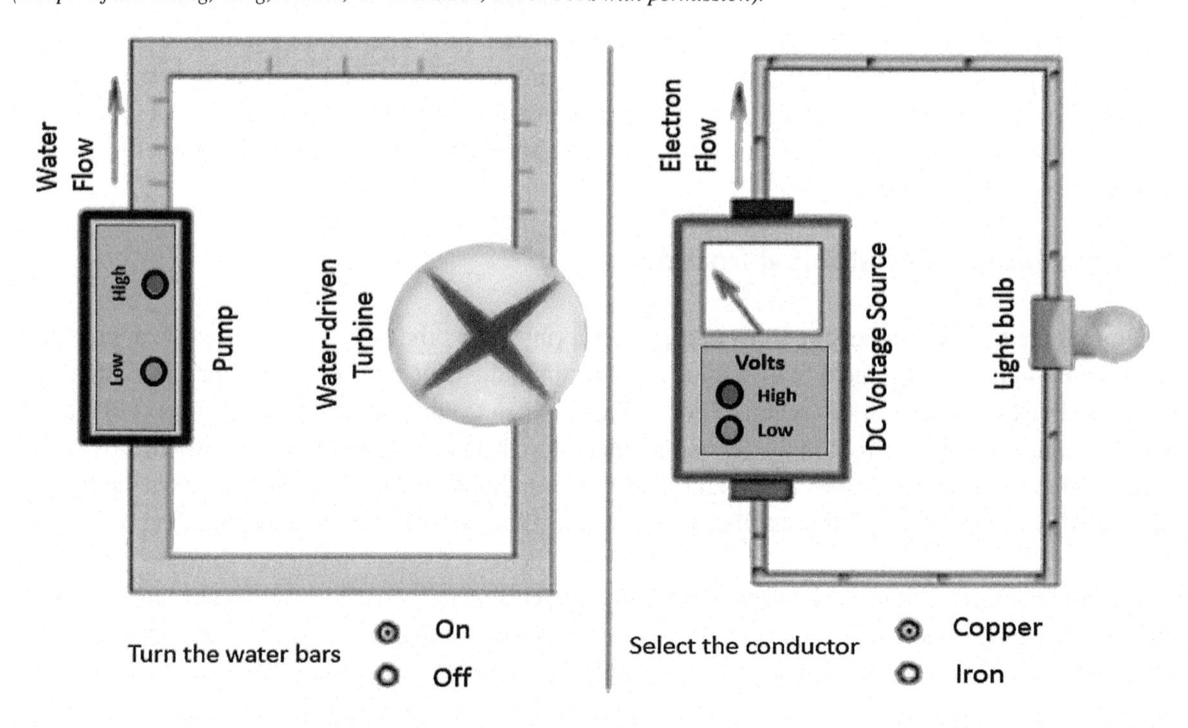

learning techniques that emphasize visual imagery to improve the quality of undergraduate learning in the sciences. This innovative approach, as Brown et al. point out, creates a space for student-driven, collaborative learning using historic and visual scientific materials. Bergey, Cromley and Kirchgessner (2015) compared the spaced restudy between visual representations and traditional text format. They found that students in visual condition consistently showed improvement in biology knowledge, biology diagram comprehension (near transfer), and geology diagram comprehension (far transfer) as compared to the students in the traditional text format. Chen and Gladding (2014) point out that a good visual representation activates perceptual symbols that are essential for the construction of the represented concept, whereas a bad representation does the opposite. Cook, Wiebe and Carter (2011) argue that effective design of multimedia based on cognitive principles enables learners to focus more on important information. They explain that optimally designed multimedia makes cognitive resources available in working memory by minimizing the extraneous cognitive load in learning. In the following section we provide three cases illuminating the relationship between visual design and extraneous cognitive load.

Visual vs. Non-Visual Learning

The graphic has been widely introduced to science education as an effective visual learning strategy. Its theoretical underlying assumption derives from Mayer's (2001) multimedia learning principle in which the author claims that presenting the content with graphics can significantly improve learners' recall and knowledge transfer. This is because the multimedia, which presents information through both visual and auditory forms offload the cognitive load in working memory. That is, information that previously went through one channel (e.g., visual) is now shared between two channels such as visual and auditory. Based on Mayer's multimedia learning principle, we compared the presence of graphics versus non-graphics in complex problem solving (Zheng & Cook, 2012). Forty-eight participants were recruited from a large research university in the western United States. They were asked to solve a series of multiple rule-based complex problems. Consider the multiple rule-based complex problem in Figure 6, which contains a set of conditions that restrict the order and parking positions of airplanes in a small regional airport. In solving the problem, the learner has to consider all the restrictions in parking decisions and then decide which flight would park at which gate without violating the conditions. For example, if the green plane parks at Gate 5, based on the condition restrictions (the red and blue planes must be separate, the blue and purple planes cannot be next to each other), the blue plane has to park at Gate 1, the red plane then parks at Gate 3, the purple plane at Gate 4 and the yellow plane at Gate 2. Learners often become overwhelmed in solving this type of problems, because they involve maintaining and manipulating several pieces of information simultaneously in the working memory while searching for the solutions. Our study was to examine the impact of extraneous cognitive load in graphic present and graphic non-present conditions. Using an eye-tracker to measure learners' eye movements as reflecting the cognitive load involved in learning (Beatty, 1982; Beatty & Lucero-Wagner, 2000), the results showed a main effect for graphic and non-graphic learning conditions indicating that graphic presentations aided in reducing extraneous cognitive load in complex problem solving. The finding is consistent with literature in cognitive load and multimedia learning research where effective visuals are shown to support learners' complex cognitive information process by reducing their extraneous cognitive load and making cognitive resources more available for learning (Kalyuga, 2014; Mayer, 2001; Mayer & Moreno, 2003; Sweller et al., 1998).

Figure 6. An example of multiple rule-based problem solving
(Adapted from Zheng, McAlack, Wilmes, Kohler-Evans, & Williamson, 2009. Used with permission).

Air Traffic Control

Five planes are going to land on a regional airport. The airport traffic controller will direct each plane to its gate based on the following conditions:

- The red plane and the blue plane must be separated by a gate between them.
- The purple plane must park at a gate next to Gate 3.
- The green plane can park either at the Gate 1 or Gate 5.
- The blue plane and the purple plane can't park next to each other.

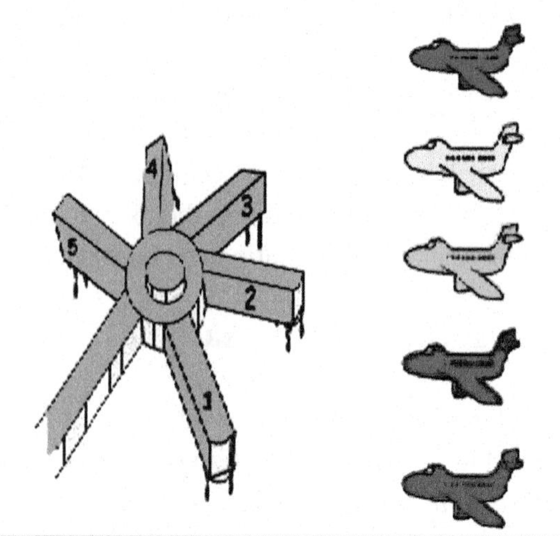

---

1. If the green plane parks at the Gate 5, which one of the following must be true?

    a) The blue plane is next to the green plane
    b) The red plane is at the Gate 2
    c) The purple plane is at the Gate 1
    d) The yellow plane is at Gate 2
    e) The red plane is next to the green plane

Interaction vs. Non-Interaction

When examining the relationship between multimedia and cognitive load, Zheng and Cook (2012) point out that different graphic representations may have different effects on learners' cognitive information processes dependent on the type of problem solving tasks involved and that interactive multimedia can be efficient in visualizing the complex processes in problem solving like multiple rule-based problems. In an earlier study Zheng and colleagues (Zheng et al., 2006) compared the effects between interactive multimedia and non-interactive multimedia in complex problem solving. In that study participants (N=114) were randomly assigned to interactive and non-interactive multimedia conditions. They were asked to solve six complex multiple rule-based problems, followed by a measure on recall and transfer test. It was hypothesized that manipulating the learning objects around to simulate different solutions could help learners visualize the problem solutions, hence improving their performance in complex learning. Results showed a main effect for multimedia presentation (i.e., interactive vs. non-interactive) at a statistically significant level with $p < .01$. More importantly, interactive presentation was significantly correlated with learners' spatial ability showing that high special ability learners learned better with interactive multimedia than those with low spatial ability. The former also proved to have a lower extraneous cognitive load than the latter in visual learning. The above study not only demonstrated the positive effect of interactive multimedia on extraneous cognitive load reduction but also called attention to the mediating factor of individual ability like spatial ability in visual learning.

Redundancy vs. Non-Redundancy

In his seminal article Sweller (1988) listed sources that cause extraneous cognitive load in learning, which include redundancy, split attention, and so forth. Mayer and his colleagues (2001) also identify redundancy as a source of extraneous cognitive load detrimental to learning. In their study, they note that multiple visuals (e.g., text, animation) presented at the same time can significantly hamper learners' information processing. The extra visual, which Mayer defines as redundancy, requires additional cognitive resources in working memory, thus affects the learner's performance. To further validate the deleterious effects of redundancy on learners' performance and its causal factor to extraneous cognitive load, Zheng and his colleagues (in preparation) conducted a study examining the relationship between extraneous cognitive load and redundancy. Forty-eight college students were recruited and were randomly assigned to redundancy and non-redundancy conditions. In the redundancy condition, the participants watched a video on health care with a caption shown simultaneously on the screen. In the non-redundancy condition, the video was shown without captions. It was hypothesized that learners in the redundancy condition would have to divide their attention between the content and the caption on the screen. That is, they have to coordinate two separate sources of information, which may induce a high extraneous cognitive load. Meanwhile, due to the coordination between caption and content, the learner would use extra cognitive resources in working memory, which in turn affected their ability to effectively process information during the learning. The initial results confirmed our hypothesis showing a redundancy effect with students in the redundancy condition having a higher extraneous cognitive load than those in the non-redundancy condition. Our study once again proved that redundancy can be detrimental to learners' performance due to its overtaxing of cognitive resources. In addition, our study has added to the literature that redundancy can cause high extraneous cognitive load. This finding explicates from the cognitive information process perspective the importance of the design in visual learning.

Promoting Germane Cognitive Load in Visual Learning

As it was mentioned earlier, the purpose of managing cognitive load, that is, reducing intrinsic and extraneous cognitive load, is to promote germane cognitive load where the learner is motivated to spend efforts in knowledge construction (Kalyuga, 2011). Cognitive efforts for generating positive affective states include optimizing the relation between the learning tasks, the schema, and the instructional support. One of the much studied topics in affective states is self-efficacy. According to Lodewyk and Winne (2005), self-efficacy is correlated with students' achievements in learning (Bandura, 1993; Schunk, 1994). They found that learners' confidence in and their perceptions about learning can significantly influence their exit behavior in learning. Their study showed that learners who demonstrated a high level of self-efficacy performed better in learning than those who did not. In a similar study Bleicher and Lindgren (2005) discovered a significant impact of self-efficacy on learners' hands-on and minds-on learning in science. Nonetheless, most research on self-efficacy and visual learning has been limited to the relationship between self-efficacy and achievement. More research is needed to understand the relationship among self-efficacy, achievements and cognitive load in visual learning. In other words, rather than treating self-efficacy as an independent variable, research should answer how self-efficacy mediates between cognitive load and achievements. Zheng et al. (2009) conducted a study to examine the relationship among multimedia, self-efficacy and achievements. By varying the design of multimedia (e.g., interactive multimedia vs. non-interactive multimedia), the authors were able to manipulate the cognitive load in visual learning. It was predicted that reduced cognitive load would lead to high self-efficacy and vice versa. Participants (N=222) were recruited from three universities in the United States and randomly assigned to interactive and non-interactive multimedia conditions. In the interactive condition the participants were able to manipulate the learning objects to find the solutions whereas in the non-interactive condition the participants studied the static pictures similar to the interactive multimedia condition but were not able to manipulate the learning objects. Learners were first given a set of complex problems, followed by an achievement test that consisted of recall and transfer questions. A pre-and posttests on self-efficacy was given to the participants to measure the change of self-efficacy in multimedia learning. The path analysis was employed to analyze the mediating role of self-efficacy between the multimedia and achievements. Results revealed a significant main effect for multimedia with an interaction between multimedia and cognitive load. Participants in the interactive multimedia showed a low cognitive load whereas those in the non-interactive multimedia had high cognitive load in visual learning. There was a significant change of self-efficacy in pre-and posttests. A path analysis was conducted with multimedia as causal factor and achievements as the result. Self-efficacy was treated as a mediator that mediated between multimedia and achievements (see Figure 7). The finding suggests that visual learning, when optimally designed, can significantly reduce cognitive load, which in turn leads to an enhanced self-efficacy. It was found that multimedia and self-efficacy were positively correlated, so were self-efficacy and achievements. Interestingly, multimedia was negatively correlated with achievements, which indicates that multimedia, when optimally designed, can support learners' learning. In

Figure 7. Path analysis model for interactive multimedia, self-efficacy and achievements
(Adapted from Zheng, McAlack, Wilmes, Kohler-Evans, & Williamson, 2009. Used with permission).

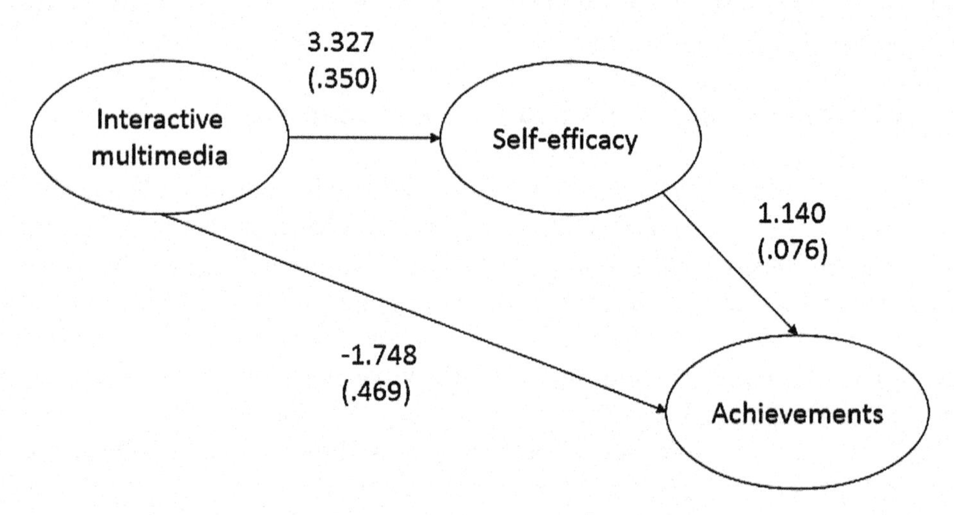

contrast, poorly designed multimedia can hamper learners' learning. In short, effective visual learning can enhance self-efficacy, which leads to an increased germane cognitive load that further contributes to the improvement of performance in learning.

A FRAMEWORK FOR OPTIMIZING EFFECTIVE VISUALS IN SCIENCE LEARNING

Successful science learning depends on the interaction of affect, cognition, and application of ideas. Simply inculcating students with science content does not make them into scientists, technologist, engineers or mathematicians (STEM), nor will they seek out STEM related courses or STEM based careers (Lamb, Akmal, & Petrie, 2015). Lamb et al. note a strong correlation among self-efficacy, science interest, visualization, and mental effort in science learning. Our studies have supported the causal relationship between cognitive load and learning outcomes, showing that cognitive load can exert significant impacts on learning outcomes. This finding is further supported by the mediating role of variables like schema, design, self-efficacy in visual learning. Given the strong correlations among the factors in visual learning, a model reflecting the causal relations of the factors is proposed, in which cognitive load is the direct cause of learning outcomes and the latent variables like schema, design and self-efficacy are the latent variables that impact cognitive load (Figure 8).

Figure 8. The causal model of cognitive load and outcomes in visual learning

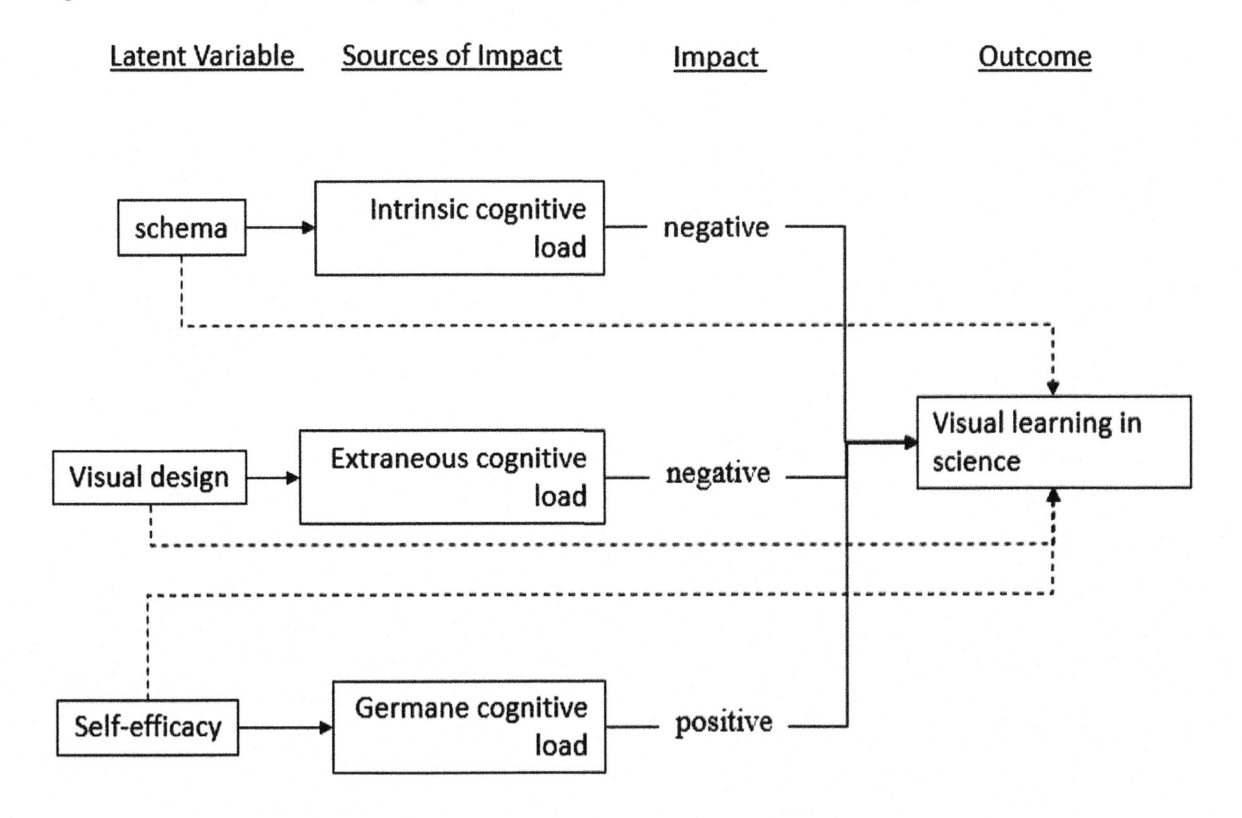

The causal model describes four relationships in visual learning. That is, the relationships between (1) cognitive load and the outcome, (2) the impact, the cognitive load and the outcome, (3) the latent variable (e.g., schema) and the cognitive load, and (4) the latent variable and the outcome. In this model, the outcome is directly influenced by cognitive load and indirectly influenced by latent variables. The relationship between cognitive load and outcome can be negative or positive dependent on the type of cognitive load involved.

Since latent variables play such a critical role in visual learning with its direct impact on cognitive load and its indirect impact on learning outcomes, we suggest that instructional intervention be added to the model focusing on the relationship between the latent variables and instructional strategies. Our previous studies have shown that instructional strategies like pre-training, concept map, etc. can improve learners' schemata, which leads to a significant improvement in learning outcomes (Zheng, 2007, 2010; Zheng & Cook, 2012; Zheng & Dahl, 2010; Zheng et al., 2006, 2008, 2009, 2012). Based on the empirical evidence obtained from our previous research, we modified our initial causal model to propose a framework for effective use of visual representations in cognitive and affective learning (Figure 9).

The proposed framework differs from previous research paradigm in visual learning where the focus was on the relationship between cognitive load and outcomes (Ayres, 2006; Niederhauser, Reynolds, & Salmen, 2000). We argue that the research on cognitive load should consider the latent variables that influence the direction and the status of cognitive load in learning. Our proposed framework defines the causal relationships between outcomes, cognitive load and its relevant latent variables, and explicates where the instructional strategies should be initiated, what impact they may have and how. Therefore, our proposed framework fills the gap in the literature in terms of operationalizing the instructional strategies in visual learning. For the framework to be operational, a heuristic guideline for the application of the framework is proposed with four steps for implementation.

1. Identify the relationship between sources of impact and outcomes.
2. Determine the direction of impact between sources of impact and outcomes, that is, negative or positive.

Figure 9. A framework for effective use of visual representations

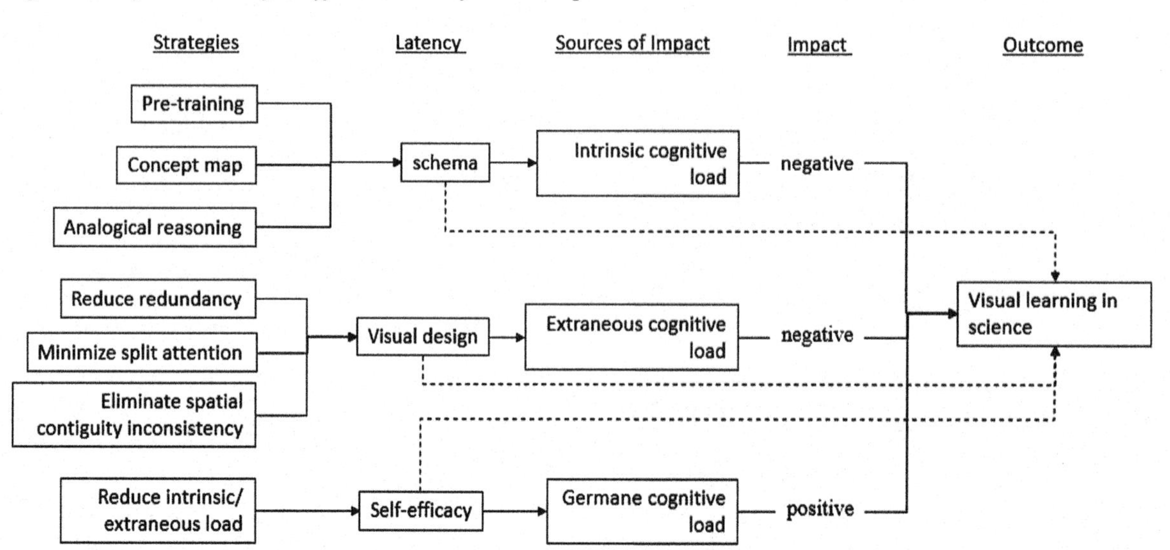

3. Identify the latent variable(s) that potentially mediate the sources of impact and outcomes, that is, its role in mitigating or improving the influence of sources of impact and outcomes.
4. Select instructional strategies that will reinforce the functional role of the mediator, which improves the relationship between sources of impact and outcomes.

Case Scenario for Implementing the Framework

Jackson Fullerton who is a fictional character in this case is an instructional designer at the *Visual Learning For Science,* whose role is to provide instructional support to schools in terms of integrating visual representations in science learning. After interviewing school administrators, teachers and students, Mr. Fullerton found that the content was developmentally challenging to the learners, which may impose a high intrinsic cognitive load on them. Once the sources of impact is identified, Mr. Fullerton began to conduct several pre-tests to determine the students' prior knowledge (i.e., schema) in the cognate area. He noticed that the learners had a low level of prior knowledge, which could negatively impact their learning outcomes. Since the learners lacked adequate schemata in the domain area, he decided to conduct a short pre-training to cover the basic concepts in the content area. Other instructional strategies like concept map and analogical reasoning will be used to further reduce their intrinsic cognitive load once the learners have developed a good schema in learning. It is expected that the learners may become more engaged and motivated once they have acquired an adequate schema in the domain area.

FUTURE RESEARCH

More studies are needed to investigate other instructional strategies such as worked examples in terms of their relevance to the proposed framework. Future research should be directed toward testing the generalizability of the framework that can be applied to various learning situations across all curricula and subject areas. Further emphasis should be made on the validation of the critical factors of the framework and their applicability in K-16 education and beyond.

DISCUSSION

One of the challenges in visual design pertinent to science education is to identify the types of cognitive load and its impact on learning. The current chapter has identified three types of cognitive load that impact learners' learning negatively or positively. These three types of cognitive loads include intrinsic, extraneous and germane cognitive loads. Of three cognitive load types, germane cognitive load is considered relevant to learning, thus needs to be promoted whereas other two types of cognitive load (i.e., intrinsic and extraneous cognitive loads) are detrimental to learning and therefore should be minimized. The chapter introduces instructional strategies that are empirically proven to be effective in reducing cognitive load. An important feature of the chapter is the identification of the relationship among instructional strategies, latent factors, cognitive load and learning outcomes. Our framework takes into consideration simultaneously multiple factors in visual learning and the strategies to improve visual performance in learning.

Therefore, our proposed framework reflects the operational relationship among the variables involved in visual learning. It is significant at both theoretical and practical levels. Theoretically, the framework has contributed to the understanding of the critical factors involved in visual learning, their relationship and the roles they play in supporting learners' cognitive information processing. It identifies the latent variables in relation to cognitive load and learning outcomes. It also raises the awareness of the critical role of latent variables and their relations with instructional strategies in visual learning. At the practical level, the framework serves as a guideline to teachers, instructional designers and other related professional in designing effective visual-based learning, particularly in science learning. It enables teachers to identify the types of cognitive load and their impact on visual-based learning, and determines the instructional strategies by identifying the variables that will (a) directly impact the cognitive load and (b) indirectly mediate the learning outcomes. The framework proposed in this chapter can help teaching professionals design and develop an effective visual learning for learners at all levels. However, the framework cannot be expected to address all aspects of the issues in visual learning. Future research is needed to test the model and its design principles in variety of online and offline learning environments.

CONCLUSION

This chapter discusses a theoretical framework for designing effective visual learning in science education. The framework is based on several theories related to cognitive visual information processes and empirical evidence from authors' previous research in visual-based science learning. Emphasis has been made on the structural part of the framework that allows dynamic linking to critical factors in visual learning. The framework provides a new perspective by identifying the variables in visual learning and the instructional strategies aiming at the improvement of visual performance. Given the increasing presence of visuals in science teaching and learning, the proposed framework has both theoretical and practical significances in terms of designing effective visual representations for STEM related subjects, and promoting deep learning in science education as measured by comprehension and knowledge transfer.

REFERENCES

Ausubel, D. P. (1960). The use of advance organizers in the learning and retention of meaningful verbal material. *Journal of Educational Psychology*, *51*(5), 267–272. doi:10.1037/h0046669

Ayres, P. (2006). Impact of reducing intrinsic cognitive load on learning in a mathematical domain. *Applied Cognitive Psychology*, *20*(3), 287–298. doi:10.1002/acp.1245

Azevedo, R. (2015). Defining and measuring engagement in learning in science: Conceptual, theoretical, methological, and analytical issues. *Educational Psychologist*, *50*(1), 84–94. doi:10.1080/00461520.2015.1004069

Baddeley, A. (2000). The episodic buffer: A new component of working memory? *Trends in Cognitive Sciences*, *4*(11), 417–423. doi:10.1016/S1364-6613(00)01538-2 PMID:11058819

Baddeley, A. D. (1986). *Working memory*. Oxford, UK: Oxford University Press.

Baddeley, J., & Hitch, G. (1994). Working memory. In G. H. Bower (Ed.), *The psychology of learning and motivation: Advances in research and theory* (Vol. 8, pp. 47–89). New York: Academic Press.

Bandura, A. (1993). Perceived self-efficacy in cognitive development and functioning. *Educational Psychologist, 28*(2), 117–148. doi:10.1207/s15326985ep2802_3

Beatty, J. (1982). Task-evoked pupillary responses, processing load, and the structure of processing resources. *Psychological Bulletin, 91*(2), 276–292. doi:10.1037/0033-2909.91.2.276 PMID:7071262

Beatty, J., & Lucero-Wagner, B. (2000). The pupillary system. In J. T., Cacioppo, L. G. Tassinary, & G. G. Berntson (Eds.), Handbook of psychophysiology (2nd ed.; pp. 142-162). Cambridge, UK: Cambridge University Press.

Bergey, B. W., Cromley, J. G., Kirchgessner, M. L., & Newcombe, N. S. (2015). Using diagrams versus text for spaced restudy: Effects on learning in 10th grade biology classes. *The British Journal of Educational Psychology, 85*(1), 59–74. doi:10.1111/bjep.12062 PMID:25529502

Bleicher, R. E., & Lindgren, J. (2005). Success in science learning and preservice science teaching self-efficacy. *Journal of Science Teacher Education, 16*(3), 205–225. doi:10.1007/s10972-005-4861-1

Brown, A. H., Losoff, B., & Hollis, D. R. (2014). Science instruction through the visual arts in special collections. *Libraries and the Academy, 14*(2), 197–216. doi:10.1353/pla.2014.0002

Brünken, R., Plass, J. L., & Leutner, D. (2003). Direct measurement of cognitive load in multimedia learning. *Educational Psychologist, 38*(1), 53–61. doi:10.1207/S15326985EP3801_7

Chandler, P., & Sweller, J. (1991). Cognitive load theory and the format of instruction. *Cognition and Instruction, 8*(4), 293–332. doi:10.1207/s1532690xci0804_2

Chen, Y.-C., & Yang, F.-Y. (2014). Probing the relationship between process of spatial problems solving and science learning: An eye tracking approach. *International Journal of Science and Mathematics Education, 12*(3), 579–603. doi:10.1007/s10763-013-9504-y

Chen, Z., & Gladding, G. (2014). How to Make a Good Animation: A grounded cognition model of how visual representation design affects the construction of abstract physics knowledge. *Physical Review Special Topics - Physics. Education Research, 10*(1), 010111-1–010111-24.

Cook, A., Zheng, R., & Blaz, J. W. (2009). Measurement of cognitive load during multimedia learning activities. In R. Zheng (Ed.), *Cognitive effectives of multimedia learning* (pp. 34–50). Hershey, PA: Information Science Reference/IGI Global Publishing. doi:10.4018/978-1-60566-158-2.ch003

Cook, M., Wiebe, E., & Carter, G. (2011). Comparing visual representations of DNA in two multimedia presentations. *Journal of Educational Multimedia and Hypermedia, 20*(1), 21–42.

Dhanapal, S., Kanapathy, R., & Mastan, J. (2014). A study to understand the role of visual arts in the teaching and learning of science. *Asia-Pacific Forum on Science Learning and Teaching, 15*(2), Article 12.

Gillies, R. M., Nichols, K., & Burgh, G. (2011). Promoting problem-solving and reasoning during cooperative inquiry science. *Teaching Education, 22*(4), 427–443. doi:10.1080/10476210.2011.610448

Johnson, C. I., & Mayer, R. E. (2009). A testing effect with multimedia learning. *Journal of Educational Psychology*, *101*(3), 621–629. doi:10.1037/a0015183

Kalyuga, S. (2005). Prior knowledge principle in multimedia learning. In R. E. Mayer (Ed.), *Cambridge handbook of multimedia learning* (pp. 325–337). Cambridge, UK: Cambridge University Press. doi:10.1017/CBO9780511816819.022

Kalyuga, S. (2011). Cognitive load theory: Implications for affective computing.*Proceedings of the Twenty-Fourth International Florida Intelligence Research Society Conference* (pp. 105-110).

Kalyuga, S. (2014). Managing cognitive load when teaching and learning e-skills.*Proceedings of the e-Skills for Knowledge Production and Innovation Conference 2014.*

Kennedy, D. (2014). The role of investigations in promoting inquiry-based science education in Ireland. *Science Education International*, *24*(3), 282–305.

Kinchin, I. M., Hay, D. B., & Adams, A. (2000). How a qualitative approach to concept map analysis can be used to aid learning by illustrating patterns of conceptual development. *Educational Research*, *42*(1), 43–57. doi:10.1080/001318800363908

Lamb, R., Akmal, T., & Petrie, K. (2015). Development of a Cognition-Priming Model Describing Learning in a STEM Classroom. *Journal of Research in Science Teaching*, *52*(3), 410–437. doi:10.1002/tea.21200

Lee, H., Plass, J. L., & Homer, B. D. (2006). Optimizing cognitive load for learning from computer based science simulations. *Journal of Educational Psychology*, *98*(4), 902–913. doi:10.1037/0022-0663.98.4.902

Liben, L. S., Kastens, K. A., & Christensen, A. E. (2011). Spatial foundations of science education: The illustrative case of instruction on introductory geological concepts. *Cognition and Instruction*, *29*(1), 45–87. doi:10.1080/07370008.2010.533596

Lodewyk, K. R., & Winne, P. H. (2005). Relations among the structure of learning tasks, achievement, and changes in self-efficacy in secondary students. *Journal of Educational Psychology*, *97*(1), 3–12. doi:10.1037/0022-0663.97.1.3

Mayer, R. (2001). *Multimedia learning*. Cambridge, UK: Cambridge Press. doi:10.1017/CBO9781139164603

Mayer, R., Heiser, H., & Lonn, S. (2001). Cognitive constraints on multimedia learning: When presenting more material results in less understanding. *Journal of Educational Psychology*, *93*(1), 187–198. doi:10.1037/0022-0663.93.1.187

Mayer, R., & Moreno, R. (2003). Nine ways to reduce cognitive load in multimedia learning. *Educational Psychologist*, *38*(1), 43–52. doi:10.1207/S15326985EP3801_6

Mayer, R. E., Steinhoff, K., Bower, G., & Mars, R. (1995). A generative theory of textbook design: Using annotated illustrations to foster meaningful learning of science text. *Educational Technology Research and Development*, *43*(1), 31–43. doi:10.1007/BF02300480

McNamara, D. S., & Kintsch, W. (1996). Learning from texts: Effects of prior knowledge and text coherence. *Discourse Processes*, *22*(3), 247–288. doi:10.1080/01638539609544975

Mehta, R., & Russell, E. (2009). Effects of pretraining on acquisition of novel configural discriminations in human predictive learning. *Learning & Behavior*, *37*(4), 311–324. doi:10.3758/LB.37.4.311 PMID:19815928

Miller, S., Geng, Y., Zheng, R., & Dewald, A. (2012). Presentation of complex medical information: Interaction between concept maps and spatial ability on deep learning. *International Journal of Cyber Behavior, Psychology and Learning*, *2*(1), 42–53. doi:10.4018/ijcbpl.2012010104

Moreno, R., & Mayer, R. E. (1999). Cognitive principles of multimedia learning: The role of modality and contiguity. *Journal of Educational Psychology*, *91*(2), 358–368. doi:10.1037/0022-0663.91.2.358

Niederhauser, D. S., Reynolds, R. E., & Salmen, D. J. (2000). The Influence of Cognitive Load on Learning from Hypertext. *Journal of Educational Computing Research*, *23*(3), 237–255. doi:10.2190/81BG-RPDJ-9FA0-Q7PA

Paivio, A. (1986). *Mental representations: A dual coding approach*. Oxford, UK: Oxford University Press.

Pociask, F. D., & Morrison, G. R. (2008). Controlling split attention and redundancy in physical therapy instruction. *Educational Technology Research and Development*, *56*(4), 379–399. doi:10.1007/s11423-007-9062-5

Pollock, E., Chandler, P., & Sweller, J. (2002). Assimilating complex information. *Learning and Instruction*, *12*(1), 61–86. doi:10.1016/S0959-4752(01)00016-0

Savasci Acikalin, F. (2014). Use of instructional technologies in science classrooms: Teachers' perspectives. *Turkish Online Journal of Educational Technology*, *13*(2), 197–201.

Schunk, D. H. (1994). Self-regulation of self-efficacy and attributions in academic settings. In D. H. Schunk & B. J. Zimmerman (Eds.), *Self-regulation of learning and performance: Issues and educational applications* (pp. 75–100). Hillsdale, NJ: Erlbaum.

Scogin, S. C., & Stuessy, C. L. (2015). Encouraging greater student inquiry engagement in science through motivational support by online scientist-mentors. *Science Education*, *99*(2), 312–349. doi:10.1002/sce.21145

Sokolowska, D., de Meyere, J., & Folmer, E. (2014). Balancing the needs between training for future scientists and broader societal needs – secure project research on mathematics, science and technology curricula and their implementation. *Science Education International*, *25*(1), 40–51.

Sweller, J. (1988). Cognitive load during problem solving: Effects on learning. *Cognitive Science*, *12*(2), 257–285. doi:10.1207/s15516709cog1202_4

Sweller, J., & Chandler, P. (1991). Evidence for cognitive load theory. *Cognition and Instruction*, *8*(4), 351–362. doi:10.1207/s1532690xci0804_5

Sweller, J., van Merrienboer, J. J. G., & Paas, F. (1998). Cognitive architecture and instructional design. *Educational Psychology Review*, *10*(3), 251–296. doi:10.1023/A:1022193728205

Tang, M., & Neber, H. (2008). Motivation and self-regulated science learning in high-achieving students: Differences related to nation, gender, and grade-level. *High Ability Studies*, *19*(2), 103–116. doi:10.1080/13598130802503959

Um, E., Plass, J. L., Hayward, E. O., & Homer, B. D. (2012). Emotional Design in Multimedia Learning. *Journal of Educational Psychology*, *104*(2), 485–498. doi:10.1037/a0026609

Wang, C. X., & Dwyer, F. M. (2006). Instructional effects of three concept mapping strategies in facilitating student achievement. *International Journal of Instructional Media*, *33*, 135–151.

Zheng, R. (2007). Cognitive functionality of multimedia in problem solving. In T. Kidd & H. Song (Eds.), *Handbook of Research on Instructional Systems and Technology* (pp. 230–246). Hershey, PA: Information Science. doi:10.4018/978-1-59904-865-9.ch017

Zheng, R. (2010). Effects of situated learning on students' knowledge gain: An individual differences perspective. *Journal of Educational Computing Research*, *43*(4), 463–483. doi:10.2190/EC.43.4.c

Zheng, R., & Cook, A. (2012). Solving complex problems: A convergent approach to cognitive load measurement. *British Journal of Educational Technology*, *43*(2), 233–246. doi:10.1111/j.1467-8535.2010.01169.x

Zheng, R., & Dahl, L. (2010). Using concept maps to enhance students' prior knowledge in complex learning. In H. Song & T. Kidd (Eds.), *Handbook of research on human performance and instructional technology* (pp. 163–181). Hershey, PA: IGI Global. doi:10.4018/978-1-60566-782-9.ch010

Zheng, R., McAlack, M., Wilmes, B., Kohler-Evans, P., & Williamson, J. (2009). Effects of multimedia on cognitive load, self-efficacy, and multiple rule-based problem solving. *British Journal of Educational Technology*, *40*(5), 790–803. doi:10.1111/j.1467-8535.2008.00859.x

Zheng, R., Miller, S., Snelbecker, G., & Cohen, I. (2006). Use of multimedia for problem-solving tasks. *Journal of Technology, Instruction. Cognition and Learning*, *3*(1-2), 135–143.

Zheng, R., Smith, D., Hill, J., Luptak, M., Hill, R., & Rupper, R. (in preparation). Exploring the roles of modality and crystalized knowledge in older adults' information processing. *University of Utah.*

Zheng, R., Udita, G., & Dewald, A. (2012). Does the format of pretraining matter? A study on the effects of different pretraining approaches on prior knowledge construction in an online environment. *International Journal of Cyber Behavior, Psychology and Learning*, *2*(2), 35–47. doi:10.4018/ijcbpl.2012040103

Zheng, R., Yang, W., Garcia, D., & McCadden, B. P. (2008). Effects of multimedia on schema induced analogical reasoning in science learning. *Journal of Computer Assisted Learning*, *24*(6), 474–482. doi:10.1111/j.1365-2729.2008.00282.x

Zimmerman, B. J. (1998). Academic studying and the development of personal skill: A self-regulatory perspective. *Educational Psychology*, *33*(2-3), 73–86. doi:10.1080/00461520.1998.9653292

Zimmerman, B. J. (2001). Theories of self-regulated learning and academic achievement: An overview and analysis. In B. J. Zimmerman & D. H. Schunk (Eds.), *Self-regulated learning and academic achievement: Theoretical perspectives* (2nd ed.; pp. 1–38). Mahwah, NJ: Lawrence Erlbaum.

KEY TERMS AND DEFINITIONS

Cognitive Theory of Multimedia Learning: Is proposed by Richard Mayer who has identified seven principles of multimedia design. They include multimedia principle, spatial contiguity principle, temporal contiguity principle, coherence principle, modality principle, redundancy principle, and individual differences principle. Drawing from cognitive theory of information process, especially Baddeley's working memory theory, Mayer posits that the design of media including multimedia needs to take into consideration the cognitive resources in visual learning. His principles of multimedia learning examine the relations between visual and verbal information and the limitations of working memory.

Concept Map: Is a visual tool that represents pattern of relationships among facts and concepts. It helps students (1) clarify their thinking and (2) facilitate their processing, organizing and prioritizing ideas. Concept map is a cognitive tool that scaffolds learners to bridge the new material with their prior knowledge by presenting an idea map that delineates the relationship among facts, concepts, and principles.

Dual-Coding Theory: Posits that the efficacy of learning can be attained by expanding on learned material via verbal associations and visual imagery, which is known as dual-coding processing. According to dual-coding theory, visual and verbal information is processed separately. That is, the visual information is processed through visual channel whereas the verbal information is processed through the verbal channel. When both types of information are processed in the working memory, they become associated with each other to form a mental representation of the incoming information.

Extraneous Cognitive Load: Is caused by the format and manner in which information is presented. For example, teachers may unwittingly require students to mentally integrate mutually referring, disparate sources of information that exact the limited cognitive resources in working memory, resulting in an increased cognitive load in learning.

Germane Cognitive Load: Is initially introduced to the cognitive load theory to separate useful, relevant demands on working memory from irrelevant and wasteful forms of cognitive processes like intrinsic and extraneous cognitive loads. Germane cognitive load is the mental effort a learner applies toward schema construction in learning. It is associated with motivation-, attitude-mediated cognitive resources directed towards achieving learning objectives.

Intrinsic Cognitive Load: Is the mental effort required by a learner to process a task. It germinates from the element interactivity within the content. As the level of element interactivity increases, the required mental effort to process the task increases as well, which causes a high demand on cognitive resources. Tasks with low element interactivity can be learned in isolation and therefore, consume less cognitive resources. Thus, low element interactivity tasks generally only become cognitively demanding when the sheer number of tasks to complete is surfeit. On the other hand, tasks that cannot be learned in isolation and that interact with other elements would impose a higher cognitive load.

Working Memory Theory: Was proposed by Baddeley and his colleagues. The working memory model includes a central executive system with three parts: phonological loop, visuospatial sketchpad, and episodic buffer. The phonological loop stores verbal content, whereas the visuospatial sketchpad caters to visuospatial information. The episodic buffer is a mechanism in working memory that dedicates to linking information across domains to form integrated units of visual, spatial and verbal information with time sequencing such as the memory of a story or event. Overloading any of the above three subsystems could have significant consequences in learning.

Chapter 12
Integrative Visual Projects for Cognitive Learning

Anna Ursyn
University of Northern Colorado, USA

ABSTRACT

This chapter comprises integrative studies on selected processes, events, and related technologies associated with several science categories. Learning projects are designed around themes drawn from events existing in everyday life, yet they familiarize the readers with complex disciplines and their applications. Complexity of apparently simple topics is presented in projects about familiar objects or actions. They are aimed at broadening the readers' general knowledge and experience rather than the technical or professional training. Topics and projects present nature- and science-related themes in terms of concept visualization including selected subjects pertaining to the basic sciences such physics, chemistry, biology, geography, or biology-inspired computing and modeling. The reader is encouraged to approach learning holistically and present concepts by creating technology based projects about visual presentation of information.

INTRODUCTION

When looking at nature we experience many concepts coming from various areas categorized as particular sciences. When examining particular action, process, or a product present in Nature, we may notice presence of descriptions belonging to various disciplines that are actually intertwined. This chapter focuses on few general concepts that have in common the merging of disciplines. Learning projects represent some multi-function tasks, possibilities, and operations, so they involve many scientific disciplines and describe and their role in everyday experiences. Tail characteristics and functions and related biological and physical concepts described in the following text may result in creating a learning project "The tail." Egg features and related concepts may support a creating learning project "An egg as a hotel." Next, this chapter offers a project "The beach and the ocean" followed by a project "Nanostructures and bioimaging." This chapter is illustrated with works of a group of students from the Computer Graphics area run by the author at the University of Northern Colorado.

DOI: 10.4018/978-1-5225-0480-1.ch012

1. THE TAIL STORY

Some comments about the role of a tail in animal life introduce the reader to this learning experience. They are followed by basic information about biological and physical factors that may determine the use of a tail and the tail's presence in a verbal and visual culture. A learning project is aimed at motivating the reader to create a story involving a tailed creature, and develop its representation. A vehicle for this story may be presented in 2d, 3d, or 4d in order to convey a positive, constructive motif or message.

Language of a Tail Communicating Mental and Emotional Situation

An animal tail may convey information and emotions, such as people used to verbalize. One may say that for a lack of tails, people lost a useful means of communication. Beyond verbal communication, they still can use the facial and body expression of their emotions, feelings, or passions, as well as they may convey an idea or meaning through the posture, or head and hand gestures. As they say, music started with hands and legs of a person. For example, while watching a TV program set to silent mode one can guess who is talking by viewing who vividly moves one's hands. A game could involve adding words and text to the person's gestures and expressions. Position or movement of a tail tells about the animal's physical and emotional state. In many cases tail movement may help to scare the opponent, support flight, landing, and balancing on a twig or a wire.

Animal tails often serve for social signaling. Distinct species of animals convey information with their tail in different ways; they announce possible dangers, feelings and emotions, or show superiority and dominance. For example, a rattlesnake uses its tail when it tries to threaten and deter its many predators such as hawks, weasels, or king snakes by shaking the noisemaker (rattle) placed at the end of its tail. A tail of a horse, if it's not braided, serves for brushing flies and other insects away.

A dog wags its tail to express love, joy, the sense of belonging, a need for attention, but also aggression, anxiety, or fear, for example of vaccination. A Japanese breed of a dog Akita wags a curved tail when is happy, but puts it straight when is feeling sick. Many of hunting dogs keep their whip-like tails straight when they follow the trail with a scent. An initial exploratory wagging of the two dogs' tails may later change either into play or into a dogfight.

A cat does it in a somewhat different way, still using its tail including its tip, which may twitch slightly during the hunting or playing, yet move stronger signaling aggression. Cat's vertically held tail signals good feelings, confidence, and happiness, for example when greeting its owner. In psychological terms, a cat may send a threatening signal by ruffling the hairs on its tail and back to increase its apparent size, along with arching its tail and back. In biochemical terms, experiments conducted long ago revealed the pilomotor action on the hairs of the cat's tail of adrenaline, acetylcholine (von Briicke, 1935, confirmed by Coon & Rothman, 1940), and nicotine (Burn, Leach, Rand, & Thompson, 1959) originating from within an organism. Pilomotor reaction means hair erection caused by activation of muscles placed at the base of each hair and controlled in this case by the sympathetic nervous system. Adrenaline is one of substances involved in transmission of messages that travel through this system. Ruffling of the cat's tail could be observed both after substance injection, as a response to cold – by creating a layer of insulation, or in response to the perceived stimuli causing anger or fear. For this reason the angle of the cat tail's hair has been considered a measure of the level of adrenaline in the blood.

A Tail in Biological and Physical Terms

We may ponder how time is included into the tail story (Carroll, 2012). Human embryos have developmental tails that have been retained through evolution but are absorbed in the later growth of a fetus. Some animals, e.g., earthworms or salamanders can regenerate their lost tails and even legs; however, it cannot happen in vertebrates where caudal vertebrae make a continuation of the backbone.

Fish and some other marine animals including snakes use their tails to support their locomotion. A vertical caudal fin called a tailfin redirects the flow of water past the fish, just reducing the retarding force of hydrodynamic drag of water. One may say the action of a rudder can be somehow compared to a fish tail, when a rudder in a ship or an aircraft serves to control yaw – a twisting, rotation, and oscillation around the vertical axis. Also, a tail of a swimming dog serves for this purpose.

By moving their tailfins from side to side, fish get a thrust by setting water in motion, which pushes them forward in water. Some species such as whales, dolphins, and porpoises – small toothed whales get thrust using their pectoral fins and moving horizontal tail fins up and down. Movements of a tail stabilize motion in water; they produce the longitudinal and/or transverse waves passing through a medium – water, which contribute to the longitudinal motion of a fish and its perpendicular movements. This medium carries energy generated by tails.

Waves can be described by their frequency (f) – number of vibrations per second, depending on the fish tail movements; period (T) – time of every cycle of a water wave; crests and troughs – high and low points of each wave. Frequency and period relate to each other: $f = 1/T$. Frequency is also related to wavelength (λ) – distance between two points in which a wave repeats itself. The speed of waves (v) would depend on these parameters: $v = \lambda/T$. Waves have amplitude – the maximum displacement from the rest or the equilibrium position; waves with a larger amplitude have more energy, and the rate of this energy is proportional to the square of the amplitude.

In bats, a tail may support thrust generation during takeoffs and flying. Several species of vespertilionid bats have a capacity to flap the tail-membrane (uropatagium) in order to generate thrust and lift during takeoffs and minimal-speed flight slower than one meter per a second (Adams, Snode, & Shaw, 2012). The tail flapping motion, termed "Tail-Assisted-Flight-Thrust" is similar to the way that dolphins and manatees use their flukes to thrust through water (Adams, 2014; UNC media channel, 2012).

Tails may also support a gliding flight, which is the birds' capacity to stay in the air without wing flapping. It is shared with other animals such as fish, lizards, snakes, squirrels, opossums, and even fossilized extinct species. The act of gliding requires energy to withstand air resistance or drag on the body, along with a gravity force. To keep flying, birds build a lift force by deflecting their wings and tails downward, thus pushing down the air. This happens according to the Newton's third law of motion telling that forces always act in pairs: actions cause equal but opposite reactions. Air pushed downward by the birds' wings pushes back on the bird upward and forward. When producing a lift force, birds' muscles convert gravitational potential energy into kinetic energy. Humans have been mimicking this technique for designing gliders; they use the power of wind without using motors.

Birds use their tail muscles to stabilize their flight, steer, and maneuver while flying (Gatesy & Dial, 1993). Birds' tails are especially important in high speed turning with a small radius, which is crucial for swallows or other insectivorous birds, which prey on insects. Researchers examined the efficiency of maneuvering by measuring how turning radius depends on the wings' span, area, and width (Thomas, 1993; Warrick, Bundle, & Dial, 2002). The lift from the tail was included in calculations of total available lift and resulting turning radius. The authors' calculations and models demonstrated that both maneuvering performance and the prey horizon increase with velocity (or acceleration performance), and explained why swallows foraging near the ground fly at high speed. By changing their tail angle, pigeons adjust thrust and drag to accelerate during takeoff and decelerate during landing. Tail angles shift from more horizontal orientations during takeoff to near-vertical orientations during landing, thereby reducing drag during takeoff and increasing drag during landing (Berg & Biewener, 2010).

In many animals, e.g., in cats, apes, or kangaroos a tail helps them to maintain physical stability and balance, so they do not topple over while making sharp turns or walking a narrow board. A tail may serve as an additional support. In lemurs a tail takes part in vertical clinging, leaping, and in a suspensory locomotion. Tails of monkeys and rats can grasp or hold objects; it's a possibility that would be useful for partygoers who would be able to shake hands while holding a glass and a treat.

Tail in a Verbal and Visual Culture

A notion of a tail is present in everyday spoken language. There is a saying, 'the tail wags the dog' used when a single factor dominates a whole, and another one, 'with one's tail between one's legs' about someone who feels the defeat and humiliation. Some would say someone acts like a peacock, as peacocks act for psychological reasons when they erect and expand their tails with very long feathers with the eyelike markings, and then fan their tails for courtship displays or to defeat the rival.

Costume design is often inspired by the animal world, with function, cause and effect in mind; for example, this may pertain to the masquerade balls in Venice, Italy. Trappers from the gold rush period were proudly wearing the raccoons' tails on their trapper hats. Human hairstyles and coiffures often draw from the shape of a horse tail, while trails attached to the back of the formal royal robes and the wedding gowns trail along in the way animal tails do.

Myths and legends in many cultures tell about dragons having the serpentine or reptilian features including the strong, armored tail, many times used for fighting. For example, Godzilla, a giant monster originating from Japan is frequently presented that way in many video games, novels, comic books, and movies. Tailed dragons, monsters, and demons are present in art and literature. They are usually considered evil creatures representing satanic forces, but sometimes act as guardians responsible for people or treasures. Literary examples may include Leviathan from the "Book of Job" of the 5th century BC, Beowulf from the 8th-11th century Old English epic, Jabberwocky from the Lewis Carroll's book "Through the Looking Glass" (1871), and magical dragons present in the J. R. R. Tolkien's "The Fellowship of the Ring" (1954).

Dragons, monsters, and demons with tails can be seen in paintings, for example created by Giotto di Bondone (1266/7-1337), Hans Memling (1430-1494), Hieronymus Bosch (1450-1516), Pieter Bruegel the Elder (1525-1569), or Salvador Dali (1904-1989). A woodcut created by Albrecht Dürer in 1498 shows "Saint Michael Fighting the Dragon." Most of these works of art are available online, e.g., the work of Hieronymus Bosch, (1450-1516) can be retrieved from http://www.ibiblio.org/wm/paint/auth/bosch/.

Many of these artists took inspiration for their images from the studies of supernatural beings that included demonology.

Materials

Hairs taken from animal tails serve for production of artistic materials, especially paintbrushes for oil, watercolor, acrylic, or ink painting. They include watercolor brushes with nibs made of sable, soft hair brushes having nibs made of ox, squirrel, sable, pony, goat, mongoose, or badger hair, along with stiff hog bristle, among many other kinds of natural and synthetic brushes. According to this tradition software tools for painting are called brushes. Brushes made of animal hairs serve also for exercising calligraphy, the art of lettering in a selected style in an expressive, harmonious, and skillful manner with the use of a broad instrument or a brush (Mediaville, 1996). Even for calligraphy made with a pen or a quill made of a wing feather, brushes served in the medieval times for illumination of the first letter of the each book chapter. Ink brushes, with nibs made from badger, goat, weasel, buffalo, wolf, rabbit, and also human hair, are used mostly in Chinese calligraphy (Ong, 2005) and Japanese ink painting also known as sumi-e (Okamoto, 1996).

Learning Project: The Tail – a Tailed Avatar

1. Imagine and then design a character with a tail, which can serve you for many purposes. Write a storyboard for a game, which involve the characters' tails. Your storytelling supporting the game events may be based on mythological tales and involve fantastic or legendary creatures; however, don't make a yet another dragon game that would look similar to many existing ones. How would you design a costume?
2. Invent a game where a tail would be your asset: a tool, a weapon, and means of communication. First, define an objective of you game and a goal you want to attain. Determine ways of communication with the use of your tail, both friendly and threatening.
3. Design an avatar for a game. Draw and describe a character with a tail, which would serve as your avatar in the game. Draw possible opponents, competitors, and also obstacles to achieving the goal of the game. Devise the ways your avatar could succeed in dealing with the opponents and how would this avatar overcome the obstructions and stumbling blocks using its marvelous tail. Think about the ways of using the tail to attain a faster movement when running on a surface, vaulting over barriers, and jumping into the air. You may also make your avatar suspending, thus resembling the way primates and sloths hang from tree branches. It may be applied to use locomotion in trees, swing on rock piles, feed by hanging on a tail while using both hands, or apply hanging or suspension of the body to rest and regenerate while feeling safe.

Remember to apply the laws of physics while you work on the game characters' movements. For example, when designing a swinging movement, you will examine the periodic motion of your character, which repeats back and forth movements around the central position in equal intervals of time (HyperPhysics, 2014). This pertains to a concept of the simple harmonic motion that means periodic motion where the force (F) is proportional to the displacement from the equilibrium position. Simple harmonic motion can be described by period (T) of the swinging that means the time the swinging character makes a full move back and forth, and amplitude (A) which means the maximum distance your avatar will move from the equilibrium position (in equilibrium position, there is no movement and only gravity counts). Your character will swing on its tail resembling a pendulum and obeying the rules of the simple harmonic motion. If you want to determine the period (T), you have first to determine the length (L) of the tail your character is swinging on. Then you may apply the equation $T=2\pi(L/g)^{1/2}$ where g is the acceleration of gravity, 9.8 m/s^2. The force of gravity is proportional to the distance from the equilibrium position, and 2π refers to the description of sinusoidal oscillation. Thus, the period (T) depends on the length of the character's tail and the acceleration due to gravity, which results from applying work by you or your avatar (work would mean force multiplied by distance). Applying work is needed to start swinging by building gravitational potential energy mgh (your character's weight m, times gravity g, times distance from the ground h). Now gravitational potential energy, which pushes one towards the earth, converts to kinetic energy $1/2\ mv^2$ (where m is mass and v is velocity – distance per time measured as m/s), so your character moves upward. Now you can design your character's movements based on a reliable source of information.

4. Design an educational material that conveys information using a 'tail' metaphor. It may be a handout, a graph, a newsletter, a pamphlet, a brochure, a video clip, or a poster. You may choose to describe the laws of physics that shape the use of a tail along with the speed and agility of its owner, recount changes in time due to evolutionary processes, or tell about the environmentally determined features of selected running, flying, or swimming animals with tails living on land or in water.

When working in a group, a shared folder may become useful to create a scene with an action involving all tailed characters. It would serve for the good of one's own composition and allow learning about a collaborative environment. Utilizing classmates' characters, and sharing own creations of all participants can be used to build a collaborative environment, so important in the working environments of the most media based companies. The following step is to create a 3D tailed creature to be executed as a printed sculpture.

Add 3D sculpture of a tailed creature here.

Figure 1 presents a work of Computer Graphics student at the University of Northern Colorado Lindsey Foy entitled "A Tail" along with her description of her tailed character. "For the tail project I layered a photo of a girl with a tail over top and reduced the opacity so the transformation could be seen. After taking other's tail projects, I made an imaginary land with all of the creatures and an island hanging in between the sky and the sea just floating in midair. I used the tail projects to show sort of an evolutionary stand from the bottom tadpole looking creatures, to the evolved human with a tail."

2. THE EGG STORY

A theme 'egg' may have many meanings and connotations, because it is one of the essential biological structures. Geometry behind its construction makes it sturdy and hard to break, while the shell supports an access for the air, and thus permits the natural growth of an embryo. The following learning experience is concerned with the significance of this ovoid or round object in human and animal life. It tells first about biology of an egg in several animal groups, ponders on a chicken or the egg causality dilemma, tells about egg culture related to traditions, language, food, and applications, and then invites the reader to design a learning project "An egg as a hotel."

Biology of an Egg in Animal Groups

Discussion in biological terms will pertain to the egg structure, its mechanical, chemical, immunological (and antibacterial), thermal features, along with its reproductive functions and biological developmental processes. Eggs are present in majority of animal species including many invertebrates (which have no

Figure 1. Lindsey Foy, Tail
(© 2015, L. Foy. Used with permission, along with permission of the authors of other tails collected from her classmates).

spine) such as insects, spiders (egg sacs of a house spider can be often seen on webs), mollusks, and crustaceans. The first eggs were laid, fertilized, and hatched in the ocean about 250 million years ago. The eggs of birds and terrestrial animals appeared about 100 million years later (USDA, 2013). The oldest eggshells, dated for 60,000 years ago were found in South Africa at Diepkloof Rock Shelter (Texier, Porraz, Parkington, Rigaud, Poggenpoel, Miller, Tribolo, Cartwright, Coudenneau, Klein, Steele, & Verna, 2010).

Generally, the reptiles, birds, and mammals, usually classified as amniotes lay their eggs on land or keep the fertilized egg in the mother's body, while anamniotes such as fishes and amphibians lay their eggs mostly in water. Female octopuses guard the hundreds or even thousands of their eggs against predators till their offspring hatch, even if they starve and must eat their own arms to survive (Mulcahy, 2012). When we compare and inspect online the images of several kinds of eggs: one that has been laid by a bird, other ones of a snake, ostrich, a cuckoo, a dinosaur egg with a fossilized dinosaur inside, and yet another one belonging to a turtle, we can discern major differences in their structure and in the part played in the offspring development.

In organisms where reproduction means the union of mobile male and immobile female gametes, the egg cell (in Latin called ovum, plural ova) denotes the female reproductive cell with genetic material placed in a central nucleus. In plants, especially the land plants and also in algae, egg cells are produced in female gametophytes, which have single sets of unpaired chromosomes (we can say a cell or a nucleus is haploid). After the egg is fertilized, the resulting zygote – the first cell of a new organism – contains two sets of chromosomes, one from each parent (so a cell or a nucleus forms a diploid generation). Birds, most fish, amphibians (frogs, toads, newts, and salamanders), reptiles (snakes, lizards, crocodiles, turtles, and tortoises), and also most insects, mollusks, and arachnids are laying eggs, so the embryo develops outside of the mother body, after external (like in fish and most of frogs) or internal (like in birds) fertilization. Animals living on land may lay soft-shelled eggs (like some lizards) or eggs with a protective eggshell (like birds and reptiles). Fish and amphibian eggs are jellylike. Embryos of all these animals develop on nutrients contained in eggs. In primates (where human species belongs) and other mammals, eggs contain only a small amount of substances essential for growth, so the development of the embryo depends on the female body.

The biggest animal egg ever recorded was a whale shark's egg measuring12 inches (30 cm) long (No author, SeaWorld Parks & Entertainment, 2015). Shark eggs are placed in a thin capsule made of collagen (No author, 2015, Whale Shark). Ostrich eggs measure about 7 inches (18 cm), and are the largest egg of living birds (Khanna & Yadav, 2005). Eggs of dinosaurs and the extinct birds such as the elephant bird were even larger. Eggs of dinosaurs differ in size: the biggest dinosaur eggs (which belong to segnosaurus) measure 19 inches (Lessem & Sovak, 2003).

Hundreds of thousands of dinosaur eggs have been found in different environments such as beach sands, floodplains, and sand dunes in various countries, mostly in northeastern Spain, northern China, and Mongolia (Carpenter, 1999). Eggs of bee hummingbirds, a decreasing population endemic to forests in Cuba, are the smallest known bird eggs weighting about 0.2 oz (half of a gram) (BirdLife International, 2012).

Shells can be rough or shiny, waterproofed by oily substances, or having pores. More than 7,000 pores in a chicken egg allow the embryo to breathe (no author, Scientific American, 2012). However, the number of pores in an eggshell decreases with altitude, preventing a loss of water caused by reduction of barometric pressure and an increase in water vapor permeability and diffusion at high altitude (Rahn, Carey, Balmas & Paganelli, 1977). One may ask about conditions when an eggshell is unnecessary, and attempt to find an analogy concerning humans. For example, some female turtles dig in holes in wet sand with their hind legs, to pack each egg in a hole, cover the eggs with sand, and then return to the ocean (Ernst & Lovich, 2009; a video at nothing2222229, 2008). Generally, animals living in water or in wet environment do not need to provide hard, impermeable shells on their eggs.

A chicken eggshell has an asymmetric shape close to an oval. In geometry, an oval is a two dimensional closed curve that may have one or two axes of symmetry. The three dimensional surface is called ovoid; it can be made by rotating an oval curve around one of its axes of symmetry. With two axes of symmetry, the figure looks similar from both ends, while with an only one axis of symmetry it resembles a chicken eggshell, having one end somehow pointed. An egg shaped in plaster serves as an assignment in sculpting at art departments aimed to develop in students capability to attain a demanded shape, in accordance with Michelangelo's saying that a sculpture is hidden in a block of a marble and an artist must extract it from the inside of it.

Colors of the bird eggs may vary from white, which is the color of calcium carbonate, to various colors caused by the shell pigments: all kinds of ivory, brown, pink, bluish, or greenish. They are often specked, which may play a part in ensuring the eggs' camouflage necessary in the ground-nesting. Different egg markings may also help females of seabirds belonging to the auk family to identify their own eggs in large groups of nests placed on cliff ledges (Birkhead, 1976).

Apart from dissimilarity in size, shape, color, or the presence and strength of a shell, eggs of distinct animal groups have dissimilar yolk distribution (Colbert & Morales, 2001). Small eggs of most worms, some mollusks, and many simple marine animals have small amount of evenly distributed yolk. Birds, reptiles, fish, and some primitive mammals have uneven distribution of yolk in the cytoplasm of the egg cell (ovum, with a central nucleus containing genetic material), and the fetus develops on the top of the yolk. Eggs of placental mammals are almost void of yolk; however, the ovum is one of the largest cells in human body. Research and laboratory production of human ova is a subject of legal debates. For example, in January 2015 the European Court of Justice ruled that the parthenogenetically activated non-fertilized human ova are not excluded from patentability (Hustinx, Kuipers, Morée, & Douma, 2015).

In non-mammalian vertebrates (such as birds, reptiles, amphibians, and fishes) the development of the offspring depends on nutrients stored in the egg yolk. Thus, the food source for the developing chicken embryo is the yolk of an egg, which nourishes the embryo as it grows. On the other hand one may ask whether nutrition from the yolk is always necessary. The mammals nourish their developing offspring during pregnancy through placenta, and later on due to the lactation. The eggs of mammals develop after fertilization from the immature oocytes; they show evidence of a loss of yolk-dependent nourishment. By means of comparative analyses concerning evolution of genomes and simulation of evolutionary processes, Brawand, Wahli & Kaessmann (2008) evidenced a loss during mammalian evolution (30-70 million years ago) of three ancestral genes encoding vitellogenin, the protein that is an egg yolk precursor. The milk resource genes appeared in the common mammalian ancestors about 200-310 million years ago. However, the preserved yolk sac in mammals is crucial for the early embryonic development and survival. It transfers to the embryo nutrients and other substances that control its differentiation and functions. Errors in yolk sac development and function could contribute to embryonic malformation, miscarriage, and growth diseases (Freyer & Renfree, 2009).

When we look at an avian egg from inside, we may find matter of several kinds and various structures:

- An eggshell, containing mostly calcium carbonate, may have different colors depending on chicken species and breeds. A cuticle, an outer membrane on the eggshell protects an egg from bacteria.

- Outer and inner membranes inside the shell also form barriers to bacteria. Eggshell membranes are clear films composed from structural fibrous proteins such as collagen.
- An air sac (or air cell) is small when an egg is laid and then increases in size as air penetrates the eggshell.
- Albumen, an egg white, which is thicker in the layer surrounding the yolk, contains mostly water (90%) and some proteins: albumins, which are the water-soluble globular proteins (proteins are globular, fibrous, and membrane proteins); mucoproteins, mostly mucopolysaccharides, which are present in many other organs throughout the body; and the insoluble in water globulins. Egg white protects the yolk and the embryo against stress and shocks. The Haugh unit measures the egg protein quality. It is obtained by weighting an egg and then measuring the height of the thick albumen (egg white) of an egg that was broken onto a flat surface (Eisen, Bohren, & McKean, 1962).
- A yolk feeds the developing embryo. It contains proteins, all the egg's fat, cholesterol, and provides vitamins: A (retinol), B_2 (riboflavin), B_6, folic acid (vitamin B_9), B_{12}, D, E, K, and also choline, iron, calcium, potassium, phosphorus, zinc, thiamine, and also lecithin, often used as an emulsifier, the essential fatty acids, and coenzyme Q10. The yolk contains an antibody called antiglobulin (IgY) that protects the embryo and hatchling from the microorganism invasion.
- A membranous yolk sac is attached to the embryos of birds and reptiles.
- A vitelline membrane separates yolk from the white.
- Chalazae are structures resembling ropes that keep the yolk centered (Ellis, 2012).
- The amniotic sac in mammalians, birds, and reptiles is a pair of membranes enclosing a developing embryo and later fetus immersed in amniotic fluid.

A Chicken or the Egg Causality Dilemma

It happened to be a difficult choice, which came first, the chicken or the egg (an adult chicken able to lay eggs is also called a hen). Ancient philosophers such as Aristotle (384-332 BC), Plato 429-c. 347 BC), Plutarch (46-126) and present-day philosophers, mathematicians, and physicists have worked on finding solutions for this dilemma. The 'chicken or the egg' problem has also many references in popular culture. The ancient philosophical references pertained to the questions about the beginning of life and the whole universe. Philosophers' investigations in terms of logic pointed at the futility of attempts to identifying the first case of a circular cause and consequence. Physicists and engineers called a circular reference a situation where one factor is necessary to calculate the factor itself. The van der Waals equation of the state of a fluid based on the ideal gas laws is often cited as an example of the circular reference.

The chicken or the egg problem could be seen as a circular reference: does the egg come first (which couldn't come without a chicken) or the chicken (that comes from the egg)? In a biological frame of reference, one may ask whether egg-laying species could pre-date the existence of chicken. The ''egg or chicken' dilemma is therefore considered a pointless question by a great number of scientists. For example, Freeman, Harding, Quigley, & Rodger (2010) stated that their research result may "actually further underlines that it's a fun but pointless question."

The proponents of an opinion that the egg must have come first (for example, Langdan, 2005) emphasize that all cells in an animal contain the same DNA, which comes from a zygote cell, the earliest developmental stage of the embryo, which is formed in an egg when egg and sperm fuse. Mutations resulting in new species occur in the reproductive rather than somatic DNA. The first chicken could be produced by mutations to the DNA that produced the zygote in non-chickens' eggs, so the egg must have been first; it contains a chicken but was not laid by a chicken.

A number of researchers would opt for the solution favoring the precedence of the eggs. The embryo in its first morphogenetic stage consists of a cluster of relatively equal-sized diploid cells that have two copies of each chromosome carrying genetic information in its genes. Cell reorganization in a cytoplasm, which follows fertilization, is often based on the self-organizing dynamic patterns that decide on the specialization of the offspring cells. According to Newman (2011) these primitive animal body plans might precede the appearance of eggs. Eggs might represent a set of independent evolutionary innovations: they have coevolved in their specific properties. Therefore, the author answers the question, 'which came first: egg or chicken' in favor of these chordate animals. Kalinka & Tomancak (2012) the answer the old question of which came first, the chicken or the egg in accordance with Newman, "viewing the early multicellular animal as a metaphor for the chicken, we see that the egg must always be considered an evolutionary novelty, emerging as it does from a functional adult organism."

EGG CULTURE

Egg Related Cultural Traditions

Eggs, young chicken, hens, and roosters make important icons in our culture. We may often see a figure of a rooster placed in front of a house or on a roof, when used as a wind gauge. Also a cock crowing early morning adds vivid coloring to a farmhouse.

By applying an origami technique, Faye Goldman (2014, 2016) created a mathematically inspired art work, *Brown and Green Egg*. The artist rearranged twelve pentagons made of polypropylene ribbon in a semi-regular pattern to obtain a Buckyball in an egg form. A Buckyball is a polyhedron made of pentagons and hexagons with every vertex meeting three edges.

Images of eggs often serve as pretenders that show metaphorical likeness to another forms and 'pretend' to be something else. Pretenders represent a special kind of design, as they carry hidden messages. For example, the timer serving to assist in the cooking of eggs is shown in Figure 2. It takes a form of an egg, and thus signals its function. We can see a pretender as a kind of a visual metaphor that may support visual communication through product design. We could even perceive this object to be an informer.

Decoration of eggshells or hard-boiled eggs is popular in many parts of the world; eggs are decorated with dyeing, painting, or spray-painting. Boiling eggs wrapped in onion skins gives the eggshells a mottled gold appearance. Egg painting is an old tradition held in many countries in different forms. For example, this tradition is popular mostly in Central and Eastern Europe; it is also cultivated in countries influenced by the culture of Persia, especially during the spring equinox, which is the Persian New Year.

Decorated eggs serve also for an egg hunt, an indoor or an outdoor game, where eggs (hard-boiled or the emptied eggshells filled with confetti or sweets) are hidden in places difficult for children to find. Easter egg hunting may satisfy the hunting instinct and the desire for sweets; at the same time it may support developing perceptivity, agility, and nimbleness. Egg tapping is a game of breaking the other player's hard-boiled egg without breaking one's own. Other games are the egg rolling while competing who could roll the egg the furthest, and egg tossing without breaking it. Dancing egg, a Catalonian (Spain) tradition of placing an emptied and waxed egg on a church fountain to make it turn without

Figure 2. The egg timer
(© 2015, Photo by A. Ursyn. Used with permission).

falling (de Mena, 1992, p. 91). Playing with eggs may also involve domestic pets: dogs, cats, weasels, parrots, or even tamed hedgehogs or raccoons; some of them love to play with a hardboiled egg like with a ball before eating it.

By decorating the hollow shell one may make an imitation of Fabergé eggs, jeweled eggs created in 1885-1917 by Peter Carl Fabergé (Faber, 2008), or other craft projects such as mosaic art projects made of eggshell pieces. Patterns with ovoid shapes can be found in the Fin-de-Siècle Art, Art Deco, and also Surrealism.

Eggs decorated with the Indian ink drawings by Anna Ursyn are presented in Figure 3. The specific shape of an egg serves for picturing on its surface some basic shapes related to the solar system, images of the Earth exterior as seen day and night, and human-made constructions along with related concept maps.

Egg in Language

The word 'egg' exists in English language in many frames of reference and contextual relationships. 'To walk on eggshells' means a need to act carefully and avoid upsetting someone while handling sensitive matters. The 'egghead' has been previously meaning an intellectual and a thinker, but in American slang it is a pejorative, anti-intellectual epithet, which for example was used by Richard Nixon against his political opponent. A saying, 'he is a good egg' means that this person has a special quality. There is a Yiddish wit, 'The eggs think they're smarter than the chickens (Yiddish Wit.com. 2015). The word 'egg' can be found in product names such as 'a chicken wire.'

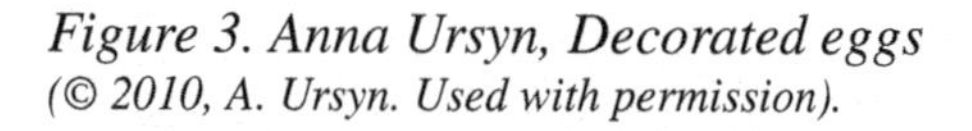

Figure 3. Anna Ursyn, Decorated eggs
(© 2010, A. Ursyn. Used with permission).

Food

Egg yolks and egg whites are important ingredients for all kinds of dishes, both when used separately and together. Some people favor omelets when they travel abroad; they believe the omelet is made and tastes much the same in all cultures, so they prefer this familiar course to the unknown, exotic dishes. Egg yolk is used as an ingredient in many dishes (e.g., crème brûlée or devil's eggs) and pastries, for meat and fruit pies, as well as in the cosmetic and pharmaceutical applications. Yolks are used sometimes as a shampoo substitute. It is also a component of some liqueurs such as eggnog, rompope, or Advocaat. Egg whites may serve for making frosting, meringues or Fedora cake; they are also used as glue that consolidates dough, sauces, and other dishes.

Caviar consists of salt-cured eggs of about 25 species of the freshwater and marine fish. The beluga (sturgeon) black caviar roe (egg mass in fish ovaries) and the red caviar salmon roe are a source of omega-3 fatty acids (Weatherby, 2010). In many countries, for example France and Japan sea urchin, crab, shrimp, and prawn roe are also considered a delicacy.

The most popular are the free-range and organic eggs, which are considered superior and are sold according to legal standards different in particular countries. Eggs are often sold pasteurized to reduce the risk of food-borne illness; mostly due to the possibility of Salmonella bacteria may be present. They are sold as the whole or liquid eggs. All egg products sold in the U.S must be pasteurized (USDA, 2013).

Egg as Glue

Egg yolk is used to make traditional egg-tempera for painting. The egg tempera paints, which are still made and used, contain egg yolk mixed with the pigments. The egg base colors don't decay like the water-based paints; they can be still seen in museums. Egg whites were used to make glair and for adhering gold leaf. Egg whites can be used for gluing paper or light cardboard because they are sticky as they dry, but they shouldn't be used on a flexible base such as canvas. After mixing with flour, water, sugar, and some alum egg whites are used to glue the papier-mâché projects. Eggs as glue consolidate also foods such as meat pies. Swallowing raw eggs has been used to alleviate the throat problems of opera singers, and also to relieve the hangover effects.

Environmental Concerns

Environmental concerns about the eggshell removal result from the fact that the US food industry generates hundreds of thousands tons of shell waste a year. After separation of eggshells (which is made up mostly of calcium carbonate) from the eggshell membrane (containing proteins), waste eggshells are used in many ways by pharmaceutical and food industries, as soil conditioner and supplement, animal feed supplement, coating pigments for inkjet printing (Yoo, Kokoszka, Zou, & Hsieh, 2009), low-cost solid catalyst for biodiesel production (Wei, Li, & Xu, 2009), and for adsorbing carcinogenic dyes in wastewater (Ee, 2013).

Several ways of the eggshell reuse are advised (Yeager, 2015) such as adding crushed eggshells to coffee grounds to make them less bitter and then composting them and thus adding calcium to the soil; using crushed eggshells as a nontoxic pest control, as a drain cleaner; using eggshell halves as seedling starters, and again, when transplanting tomatoes putting crushed eggshells into the hole to prevent the plant ends rot; using eggshell halves for making jello and chocolate molds, along with other possibilities. Egg cartons, dimpled forms for accommodating single eggs are designed to isolate eggs, protect them against stress, and absorb shocks. They are made of polystyrene foam or recycled paper molded through with the use of papier-mâché technology. Packing eggs competition serves sometimes as an entertainment and a challenge for designers; an egg is packed well when one can throw a package containing a raw egg over the building without breaking the eggshell. The dimpled forms are also used for producing camping mattresses and soundproofing acoustic foams.

Learning Project: An Egg as a Hotel

Translate information gathered about an egg into a concept of a hotel. Mentally inspect a chicken egg from the inside. You will find there the matter of several kinds as well as there are various structures; you can find many reasons why their properties have developed in a specific way. During the evolutionary processes, egg laying has evolved from live-bearing ancestors into the maternal input to embryos through many evolutionary transitions occurring in this process. Many independent origins of live-bearing may occur in particular groups of species such as sharks and rays (Dulvy & Reynolds, 1997). Examine the role of yolk in eggs of different types of animals, and how a yolk in human eggs is a remainder of the earlier evolutionary processes going in the phylogenetically earlier types of animals. Phylogenesis is a process of evolutionary development of organisms. Phylogenetics studies this process through the molecular sequencing data and by collecting morphological data about the forms of species, populations of organisms, and relations between their structures.

Decide what analogy you would like to create in order to compare an egg to a hotel on the basis of their common makeup, functions, and processes going on in both these structures. This analogy should be helpful in understanding the underlying biological, physical, and chemical concepts you would picture in both cases. Consider the subjects – individual organisms that inhabit an egg contrasted with the guests of a hotel. Characterize the properties of objects serving to host these subjects. Figure 4 and Figure 5 present two Computer Graphics at the University of Northern Colorado student works, Lindsey Foy, “Egg”

Figure 4. Lindsey Foy, Egg
(© 2015, L. Foy. Used with permission).

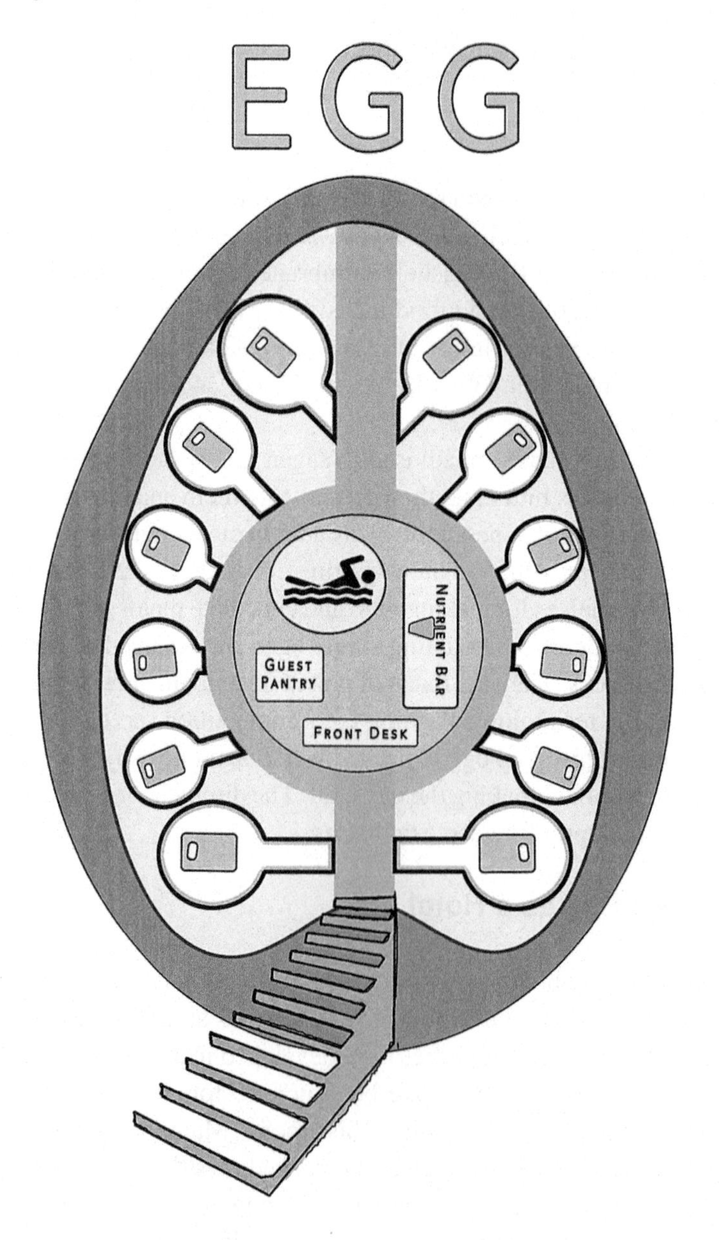

and Heather Southern, "Hotel." Lindsey Foy used the structure of an egg to relate to a hotel. Each of the side pods are rooms for guests, the center nucleus contains the main functions of the hotel: the swimming pool, bar, front desk, and guest pantry. The outside walls of the egg are the insulation of the hotel.

In case of humans, the design of a hotel may depend on climate related conditions, with a need of thermal insulation and (possibly solar) heating in cold regions and a lot of open spaces with pools for hotel guests in tropical areas. Protection from cold weather, water, or danger provided by an eggshell can be thus compared to accommodation offered to the hotel guests in this learning project.

Figure 5. Heather Southern, Hotel
(©2015, H. Southern. Used with permission).

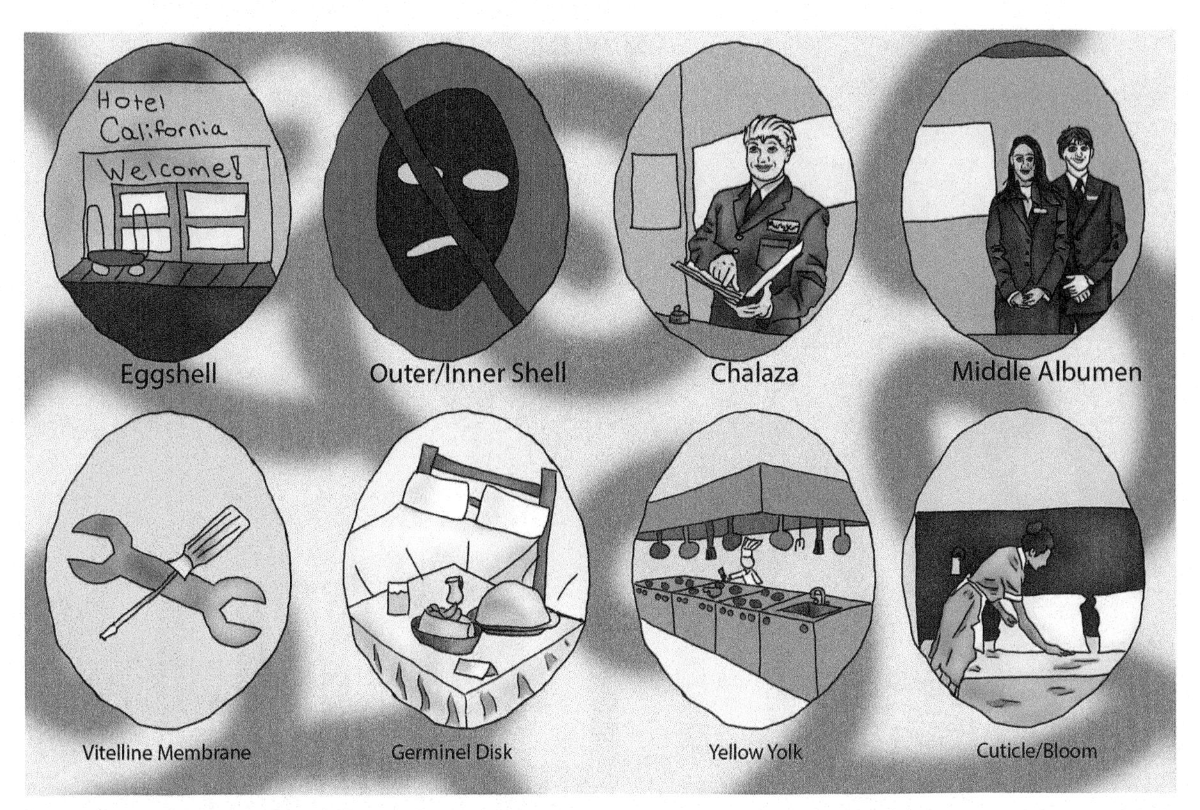

Quite different factors may determine the development of an offspring – the subjects of your project. At the same time, contrasting the mammalian and non-mammalian eggs could serve for creating a metaphor in your project. The hotel analogy may be suitable for a chicken egg, while it is the mother's organism that hosts the embryo in mammals. While creating a hotel metaphor for chicken development, find and set up the features common for both these objects. Both the external and the internal defense may be provided for the embryo by an eggshell.

You may want to compare and picture the strength of an eggshell (Gutierrez, 1987) and that of the hotel's roof, if it has a shape resembling the dome that covers the Renaissance Cathedral of Florence, Italy created by Filippo Brunelleschi. In spite of cracks, the Dome remains one of the largest brick domes ever built in the world (Borri, Betti, & Bartoli, 2010; Gibson, n.d.). The shape of a half of an eggshell may also resemble a geodesic dome, a structure that had been studied and described by numerous scientists and artists. About the year 1500, Leonardo da Vinci visualized the regular truncated icosahedron (Dresselhaus, Dresselhaus, & Eklund, 1996). Also about 1500, Albrecht Durer, created a drawing of the same construction by folding up a sheet of cardboard (Dresselhaus, Dresselhaus, & Eklund, 1996; Hart, 1999). Buckminsterfulerene (also called a buckyball), a spherical fullerene molecule of C_{60} in the form of a truncated icosahedron was discovered by Kroto, Health, O'Brien, Curl, & Smalley (1985) and awarded the Nobel Prize in chemistry. In a C_{60} fullerene molecule atoms of carbon are placed on the corners of the regular truncated icosahedron. Geodesic dome structures designed by Buckminster Fuller have a form of buckyballs. Figure 6 shows a 'CyberEgg," an image programmed in Fortran.

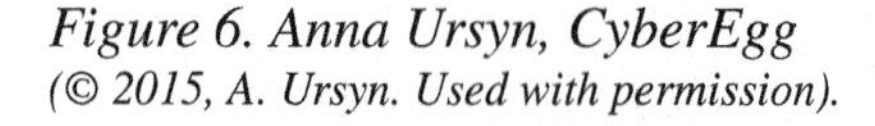

Figure 6. Anna Ursyn, CyberEgg
(© 2015, A. Ursyn. Used with permission).

As for the internal defense, the immune functions of an egg may secure the embryo's resistance to infection or toxins. Birds must defend their eggs against predators such as raccoons, weasels, foxes, skunks, mink, otters, gulls, and some other birds such as crows. The egg-eating snake has a distensible mouth and reduced teeth to accommodate large eggs; it also has spines on the ventral side of its neck vertebrae, which crush the eggshell. Hedgehogs seem to be privileged in a competition for eggs because of their spiny coat: no other animal would dare to robe a hedgehog from its booty. In a hotel, several procedures and measures are taken to ensure safety of the visitors' possessions, and provide the hotel guests with a peace of mind and feeling of safety. Metaphors may also include blankets and comforters compared to fat and fatty acids contained in yolks; one may choose to think about entertainment provided on the hotel's premises as a metaphor for yolk vitamins. Defense against bird predators carrying cameras as their heads has been pictured in a work by Cody DeVries entitled "Biomimicry" (Figure 7).

During the evolutionary processes, the egg laying has evolved from live-bearing ancestors into the maternal input to embryos through many evolutionary transitions occurring in this process (Dulvy & Reynolds, 1997). An analogy may be created between the nutrients in the egg yolk for the chicken embryos and the meals served on the hotel premises. Looking at the egg's internal structure one may design a hotel consisting of a large central area set up as a restaurant, which is surrounded by quiet, soundproof rooms providing safe environment for resting. Elevators connecting various areas of the hotel could be

Figure 7. Cody DeVries, Biomimicry
(©2015, C. DeVries. Used with permission).

compared to the egg's chalazae. Computer Graphics student at the University of Northern Colorado Brandon Malaty solved his project as an EggShip, which serves the vacationers using plasma ion propulsion and offering cryogenic sleeping tubes (Figure 8).

3. THE BEACH

A beach is a landform that can be a part of a shore by an ocean, seashore, lakeshore, or a riverside. A beach by the ocean or the sea is usually covered with shingle, smooth and round pebbles, sand, or gravel; it is located between the high- and low-water marks. Many coastal areas have dunes located inland from the beach.

The theme of a beach may be approached in many ways depending on one's frame of reference and interest. Wild, undeveloped beaches have their enthusiasts because of their natural beauty and preserved ecosystems. Travel sections in newspapers always grip our attention, especially when they involve a journey across the ocean (Figure 9).

Figure 8. Brandon Malaty, EggShip
(©2015, B. Malaty. Used with permission).

Meteorology may become a focus in another approach to the theme. Beach is a landform where many kinds of strong weather forces shape the conditions of human and animal life. Tsunami, long high waves (up to tens of meters high) in a body of an ocean are caused by earthquakes, volcanic eruptions, meteorite impacts, submarine slides or other events examined by specialists in dynamic geology, but also by human-made underwater nuclear detonations. Tsunami may have great destructive power for a whole coastal area, killing people and animals, causing landslides, and demolishing beach infrastructure. Computer models help predict tsunami arrival for the coastal areas.

Seaweeds are mostly red, green and brown algae that live near the seabed but can be found thrown on a beach by the tides. They are used as food and fertilizer, but also for filtration, the industrial, and medical uses, for example, as antiviral agents or in wound dressings. Agar made from seaweed is used in gastronomy and as culture medium for maintaining cells, bacteria, etc.

Figure 9. Anna Ursyn, New Worlds
(© 2006, A. Ursyn. Used with permission).

Architecture, human-made beach infrastructure and related cultural anthropology issues may also be of interest for many. Special features of seaside architecture typical of resorts, hotels, restaurants, and camps, harbor constructions, along with the beach related recreational infrastructure such as pools, changing rooms, showers, lifeguard posts, and beach furniture with umbrellas and folding deck chairs – all may be considered worth a study. On popular recreational beaches one may explore social issues related to this theme. Time spent on a beach can be seen as a special kind of a social encounter examined within a scope of sociology. Figures 10 and 11 show the beach related experiences and encounters.

Beyond this point you may encounter nude sunbathers.

Beaches are usually neighboring with the inland infrastructure, along with the ports and docks where ships arrive from the oceans. Sometimes we can fear the sight of a beach hedged with piers, docks, and harbors where ships unload, custom officers are stationed, and nobody knows what's really going on there (Figure 11).

Many beaches and marinas are awarded the Blue Flag certification by the Foundation for Environmental Education, when they meet the high environmental and quality standards. The Blue Flag criteria include standards for water quality, safety, environmental education and information, the provision of services and general environmental management criteria (When, 2015). Many times residents of big cities develop artificial beaches and design marinas equipped with a harbor and moorings for yachts and other boats.

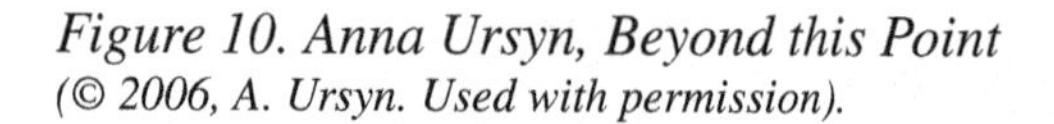

Figure 10. Anna Ursyn, Beyond this Point
(© 2006, A. Ursyn. Used with permission).

Figure 11. Anna Ursyn, Straight to the Beach
(© 2007, A. Ursyn. Used with permission).

Water sports differ according to water conditions: waterskiing, windsurfing, pool swimming, open water swimming, scuba diving and cave diving, snorkeling with a mask, fins, and a snorkel tube, sailing, canoeing and kayaking, boating and rowing, playing a beach ball, playing water polo, fishing, and many other sport activities. Underwater archeology and treasure hunting gatherers, both professionals and enthusiasts collect items from the bottom of the sea and on the beaches. More than 150 confirmed wrecks of fighter planes, landing crafts, and bombers lie beneath the waters around the Hawaiian islands (Borel, 2014). Underwater photography and videography may involve many of these activities. Natural objects and human-made materials gathered at the beach serve artists for their sculptures, collages, and installations.

Our planet holds mostly water (Figure 12). A good many things happens in water. Even some of our passions may often involve water. However, ecosystems characteristic of the ocean coast, seashore, or a riverside are suffering from some human actions, many of them polluting seawater. Oceans deposit on a beach thick tar and oil, along with pieces of wood. Tar and oil cause death of many birds and animals living in water. Procedures established in the use of oil and oil products are often devastating for the beach habitat. Oily storm water drainage collecting rain and ground water from cities, factories, and farms is often contaminated because of chemicals contained in urban runoff, gasoline, motor oil, heavy metals, trash and other pollutants from roadways and parking lots, as well as fertilizers and pesticides from lawns. According to the Knowledge Encyclopedia (2015), roads and parking lots are major sources of nickel, copper, zinc, cadmium, lead and polycyclic aromatic hydrocarbons, which are created as com-

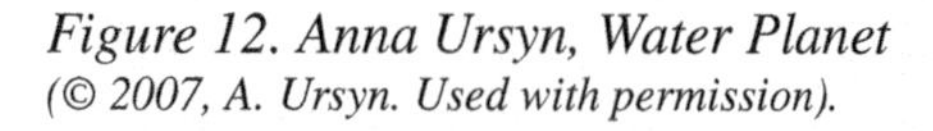

Figure 12. Anna Ursyn, Water Planet
(© 2007, A. Ursyn. Used with permission).

bustion byproducts of gasoline and other fossil fuels. Roof runoff contributes high levels of synthetic organic compounds and zinc (from galvanized gutters). Fertilizer use on residential lawns, parks and golf courses is a significant source of nitrates and phosphorus. Over half the ocean's waste oil comes from land-based sources and from unregulated recreational boating Knowledge Encyclopedia (2015). Oil spills cause harm to deep-ocean and coastal fishing and fisheries, as well as damage to wildlife and recreation. Figure 13 presents a work of Cody DeVries entitled "Food Chain."

Oceans serve as a source of energy. Renewable energy from water may come from geothermal energy available near tectonic plate boundaries (e.g., from hot springs and geothermal wells), hydropower, ocean current energy, offshore wind, salinity gradient energy, and wave power, among other solutions. Water covers about 70% of the earth's surface, so the wave power provides excellent source of clean, renewable energy. Electricity is produced by harnessing energy of ocean waves with the use of wave energy converters placed on the surface of water (Knowledge Encyclopedia, 2015). The energy provided is most often used in water desalination plants, electricity power plants, and the pumping of water into reservoirs (AENews (2015). Energy output is determined by wave height, wave speed, wavelength, and water density. Several countries study the ways of harnessing energy from water, including osmotic power plants, tidal energy, currents energy, and bio-power systems. Figure 14 shows a work of Lindsey Foy entitled "The Reef."

Figure 13. Cody DeVries, Food Chain
(© 2015, C. DeVries. Used with permission).

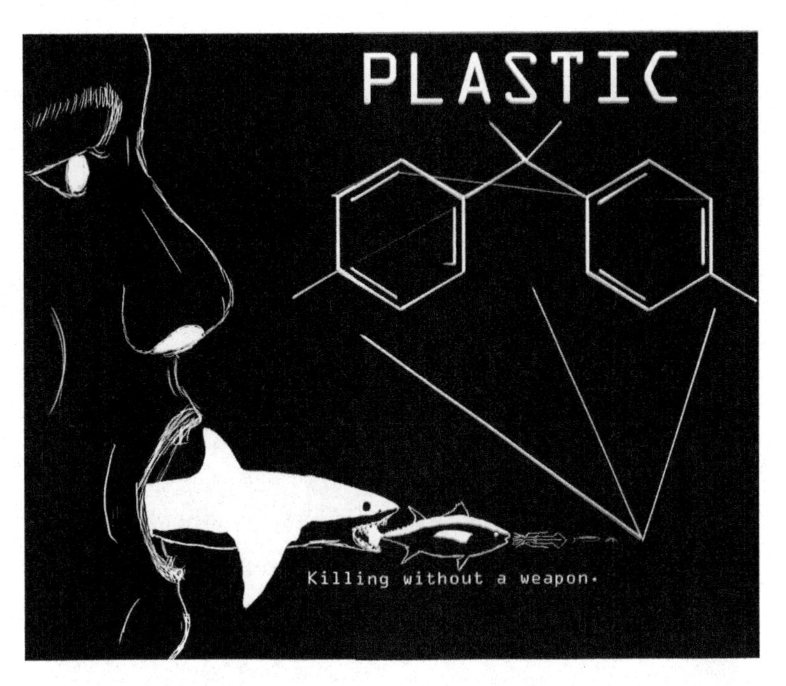

Learning Project: A New Experience

Quite unexpectedly you have found yourself at the seashore in a far away place. On your right there is a beach, a shingle, a dune, and a forest in the background. On your left a harbor and its mooring can be seen far away, with a city visible on a horizon. You find this place attractive and worth staying for a longer time. You got acquainted with somebody who suffers total amnesia due to some recent dramatic events. This person is unable to retrieve information that was acquired before the accident, but retains his intellectual, linguistic, artistic, and social skills. You have also encountered somebody staying under the witness protection program; this person has to forget about all past relationships. Your new friends are open to new actions, experiences, and projects. Together, you decided you would be able to consolidate some seed money, and would be inclined to initiate a startup company (Blank, & Dorf, 2012) that would fill a niche in the existing infrastructure of this place.

Now it's time to visualize this new enterprise and then draw a concept map – a diagram that shows relationships between concepts. Depict the nature of your project, whether you are going to create new objects, devices, materials, games, or maybe ideas about the making of new things a reality and a remedy for personal and social problems. Your concept map should present the crucial elements of this undertaking such as its necessary parts, links, resources, and its technological basis. Your concept map should show a structure of your project, would it be a company, an agency, a studio, a plant, or a workshop.

Probably, you will need to get new expertise in some areas to carry out this new task. Consider what kind of information you need to search for and learn, in addition to the existing skills possessed by you and your partners. To make your project satisfying, profitable, entertaining, and enjoyable, you

Figure 14. Lindsey Foy, The Reef
(© 2015, L. Foy. Used with permission).

and your partners may want to acquire new knowledge, use this knowledge to play with it as motivation for creating the project, work on it purposefully, use the acquired skills to resolve evolving problems, and then invent the winning shape of your startup venture using your new knowledge, skills, products, and leadership. This sequence of actions, described by Ivan Kokcharov (2015) as a hierarchy of skills, has five levels: know, play, work, solve, and invent. It has been based on the hierarchy of human needs delineated by Abraham Maslow (1943) in his paper "A Theory of Human Motivation."

According to your business concept map, you may now want to write a business plan for your startup (Entrepreneur, 2015). Describe the idea behind your project; characterize the important features of objects or concepts you want to develop. Present yourself and your partners as investors and then describe the financial variables, and estimate costs you will need to cover. If needed, think about the ways of raising money or looking for a bank loan. Define the novelty you intend to introduce, and make the possible competitor analysis. Think about the lifestyle you would like to create in your startup project; consider whether you want to attract talented employees. Examine who could be interested in your project's result; maybe it would have the health-oriented applications. Think whom you would like to cooperate with – artists, programmers, computing scientists, and/or management specialists. Describe the environmental, cultural, social, or educational impacts you may create by the successful implementing of you project.

Now draw the objects of your enterprise, thinking about their design and aesthetic value, as well as about future marketing of your products. If you are thinking about immaterial developments, picture a metaphor for your idea or concept.

4. NANOTECHNOLOGY AND BIOIMAGING

Data and information provided in this subchapter may be useful in creating a learning project, which convey the unity of processes and event occurring at quite different levels in natural domain, and give grounds for understanding the extensive use of biologically inspired solutions in computing, engineering, and materials science.

Nanostructures

Despite of the fact that we can see only the macroscopic three-dimensional world, we strive to learn about micro- and nano-structures in order to understand nature better and to make further progress in biology-inspired research, computing, and engineering. Nanotechnology is a dynamically growing discipline bringing about new solutions and products. For example, researchers demonstrated a new material involving glass and plastics in the form of a foldable glass, which can be reversible and repeatedly foldable without any mechanical failure (Nanowerk, 2015b).

Nanostructures can be naturally formed such as biomolecules and material particles, or they are human-made. Nanotechnology provides means to explore and produce these tiny structures. The range of structures under discussion is about 1-100 nm, with exceptions pertaining to some materials and devices such as several pharmaceutical substances. A nanometer is one-billionth of a meter; it means one thousand millionth of a meter.

Materials made nanotechnology display size-dependent properties resulting from quantum effects. These properties are distinct from those of materials controlled by laws of classical physics; hence their physical, chemical, and biological characteristics may be different at the macroscopic scale (Nanowerk, 2015a). For example, visualizing butterfly wing scales look not the same when seen at different magnifications: a wing that is dark blue when looked at without magnification becomes yellow at 220x magnification, purple at 5000x magnification, and green at 20000x magnification (PennState modules, 2011). The U.S. National Nanotechnology Initiative (NNI) provides the following definition: "Nanotechnology is the understanding and control of matter at dimensions between approximately 1 and 100 nanometers, where unique phenomena enable novel applications. Encompassing nanoscale science, engineering, and technology, nanotechnology involves imaging, measuring, modeling, and manipulating matter at this length scale" (Nanowerk, 2015a).

Biology-Inspired Applications and Materials

We learn about animal life and translate this knowledge into ideas enabling creating new, biology-inspired solutions. Swarm intelligence is a term describing behavior of natural or artificial systems displaying decentralized, self-organized activity of animals or agents. Natural systems displaying swarm intelligence may include fish schooling, bird flocking, ant and some other insect colonies, animal herding, and microbial intelligence. Swarm intelligence provides inspiration for creating algorithms, models, of artificial systems, and swarm robotics.

The collective behavior of fish may serve as an example of swarm intelligence. Advantages from fish schooling behavior may include better hydrodynamic efficiency, defense against predators, enhanced foraging success, and easiness in finding a mate. Researchers describe fish school structure and spatial order (e.g., Hemelrijk & Hildebrandt, 2011): its size, density, fish movement, directions, and positions. Then they create mathematical models, evolutionary models, examine, and simulate fish collective behavior using imaging technologies such as acoustic imaging (Ward, Krause, & Sumpter, 2012).

Biology-Inspired Computing

Natural structures inspire also specialists in computing. Natural computation includes biologically inspired computation, computationally motivated biology, and computing with biology. Biologically inspired computation relates to biomimicry and biomimetics; it is aimed at devising mathematical and engineering tools to generate solutions to computation problems. Computationally motivated biology involves investigating biology with computers, and is focused on simulating natural phenomena. Examples are artificial life, fractal geometry (L-systems, iterative function systems, particle systems, Brownian motion) and cellular automata. Computation with biology means investigation of substrates other than silicon. Examples are molecular or DNA computing and quantum computing.

Computational intelligence is a sub-field of AI using sub-symbolic techniques: messy, scruffy that prefers empiricism to formalism, and soft. Sub-disciplines include adaptive and intelligence systems (Browniee, 2012):

- Evolutionary computation (e.g., genetic algorithms, evolution strategy, genetic and evolutionary programming, and differential evolution).
- Swarm intelligence, using adaptive strategies: interaction and cooperation of large numbers of lesser intelligent agents. Examples are particle swarm optimization – probabilistic algorithms inspired by the flocking and foraging behavior of birds and fish, and ant colony optimization inspired by the foraging behavior of ants.
- Fuzzy intelligence means investigation of fuzzy logic, which is not constrained to true and false, and accepts approximate truth or degree of truth. It is applied for expert systems and control system domains.
- Artificial immune systems are inspired by the acquired immune systems of vertebrates; they refer to clonal selection, the dendritic cell algorithm, immune network algorithm, and are used for optimization and pattern recognition domains.
- Artificial neural networks: architecture and learning strategies inspired by modeling of neurons in the brain. Neural network learning is considered adaptive learning.

- Metaheuristics. Heuristic is an algorithm that locates good-enough solutions (not necessarily proven, correct, or optimal). Metaheuristics combines heuristic procedures using a higher-level strategy algorithm.

Figure 15 presents “A School of Fish” by Anna Ursyn.

According to the National Research Council of the National Academies, strategies for creation of new materials and systems may be characterized as bio-mimicry, bio-inspiration, and bio-derivation. Bioinspired strategies for working in new directions may go toward applying bioinspiration by creating systems that perform as natural systems. By applying bio-mimicry people create materials that mimic living systems, design structures that function in the same way, and create synthetic materials that respond to external stimuli. Bioderivation means incorporating biomaterials into human-made structures, that means using biomaterials to create a hybrid with artificial material (No author, Inspired by Biology, 2008).

Bioinspired materials, may for example use biological molecules as the active element in sensors, and materials inspired by biology such as layered, hierarchical, abalone shell-like composites may serve as lightweight, tough armor (Bioinspired and Bioderived Materials, 2015). Genetic algorithms may serve as another example of bio-inspired systems. In accordance with the theory of evolution, genetic algorithms are applied to solve optimization tasks by mimicking the evolutionary behavior of living organisms. Evolution of a random population of possible solutions is evolving due to application of the selection, mutation, and crossover operators inspired by natural evolution (Rivero, Dorado, Fernandez-Blanco, & Pazas, 2009).

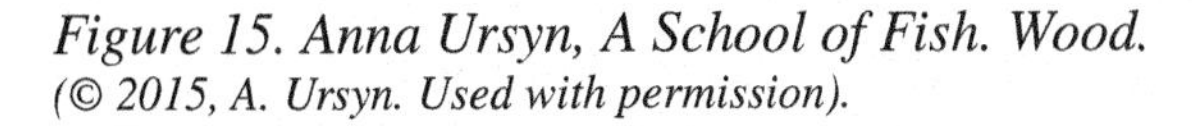

Figure 15. Anna Ursyn, A School of Fish. Wood.
(© 2015, A. Ursyn. Used with permission).

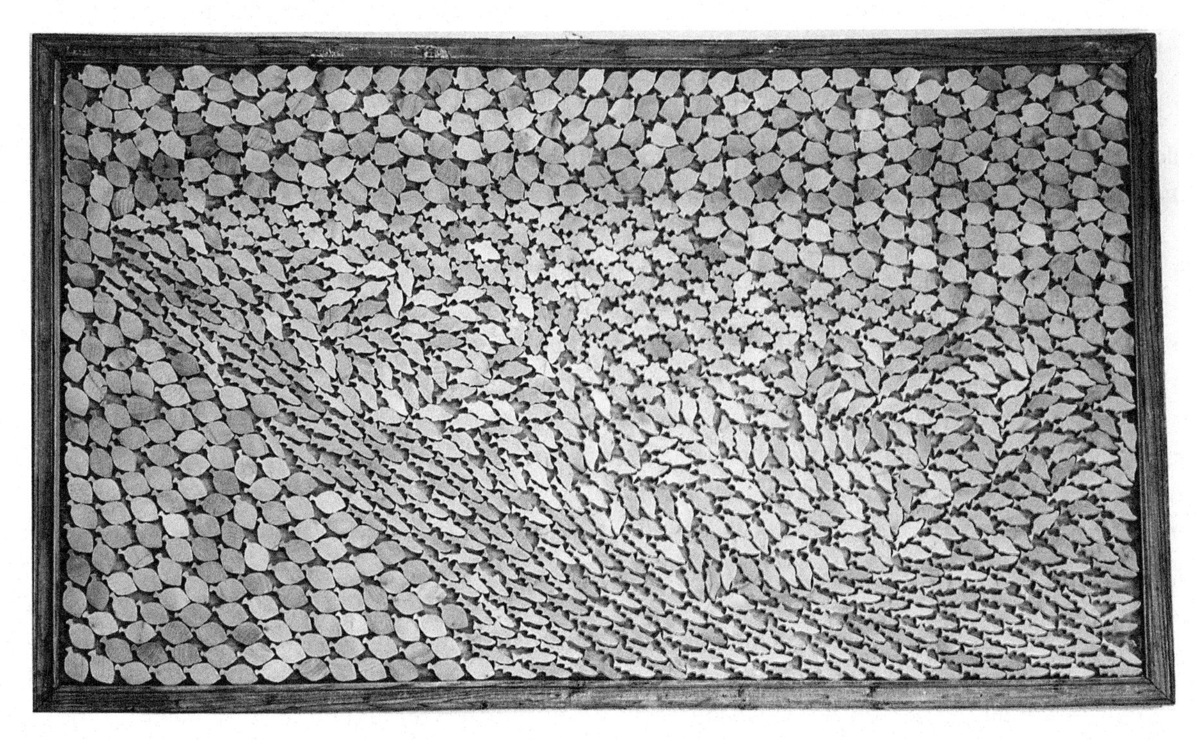

Biomimicry can be exemplified as sensors detecting minute quantities of molecules with the sensitivity and accuracy of the immune system, or advanced materials that adapt their properties or self-heal disruptions. Other examples are the flight techniques mimicking birds and bats, Velcro with multiple hooks resembling burrs and other prickly seeds, climbing boots resembling gecko's feet, and hundreds of other applications.

Bioderivation enables creating hybrid materials, for example using biological enzymes to convert organic matter into usable fuels (No author, Inspired by Biology, 2008). Molecules, structures, systems, and natural fabrication processes could serve as the basis for synthetic materials with enhanced properties. Biological molecules or cells become the active components of a device; the challenge is not only incorporation of the sensing entity but also the preservation of biological function in a non-biological environment (Bioinspired and Bioderived Materials, 2015). Researchers from the University of California at Irvine have developed an adhesive film with a protein derived from a squid skin. When stretched, it reflects near-infrared light. This technology could serve soldiers to avoid detection by enemy thermal cameras (Squid skin, 2015).

Figure 16 shows a work by Heather Southern, which draws analogies between the animal and human world by showing human survival tactics (biomimicry), protection of a body with armor (bioinspiration), and creating a drug vaccine, or serum obtained from animals (bioderivation).

Materials science is an interdisciplinary area that examines the relationship between the structure, properties, and performance of the natural and human-made materials. Researchers are constructing new biology-inspired soft materials based on nanotechnologies (soft matter can be found in everyday things including food, soap, ink, paint, cosmetics, putty, and gels). A nanomembrane made out of the graphene produced by the researchers at ETH Zurich is extremely light and breathable, so it may serve for making waterproof clothing (Celebi, Buchheim, Wyss, Droudian, Gasser, Shorubalko, Kye, Lee, &, 2014). Also, it may allow the ultra-rapid water filtration and separation of chemical mixtures, as it is thinner than a nanometer, 100,000 times thinner than the diameter of a human hair.

Adaptive materials and structures can sense external stimuli and react by changing their organization following a feedback loop (Tsukruk, 2013). The designing of adaptive structural materials can be modeled on forms that exist in nature (Lahann, 2013). Synthetic nanostructures can be based on biological materials. Tsukruk (2013) provides examples of the natural adaptive materials and systems:

- Adaptive colors in butterflies and octopuses providing photonic, sensing, camouflaging abilities.
- Dynamic adhesion in gecko feet – climbing, holding.
- Self-healing biological parts – self-healing materials.
- Reptilian locomotion – movement on complex terrains.
- Dogs and other canines – remote trace chemical sensing.
- Silk materials – tough lightweight nanocomposites.
- Night vision in some species – thermal sensors.
- Wave tracking in seals and fish – underwater monitoring.
- Spider hair air flow receptors – mechanical sensors.

Synthetic nanostructures with a membrane arrangement exhibit temperature sensitivity: temperature changes deform membranes. Similar approach was used to mimic the ability of spiders to detect sound waves using unique hair structures, and to mimic the ability of fish to sense fluid velocity Tsukruk (2013).

Figure 16. Heather Southern, Biomimicry
(© 2015, H. Southern. Used with permission).

Figure 17 presents a work by Lindsey Foy entitled "Xroads." created for the international competition along this theme. In order to visualize the intersections across art, science and technology, this student took a picture of a leaf and used all of the different connection points as roads. She highlighted one path in particular, and inserted an electrical spark connecting the paths to signify technology's use in helping us make connections. Lindsey Fox won the first place in the SPACE – Student Poster and Animation Competition and Exhibition at the ACM/Siggraph 2015 Computer Graphics and Interactive Techniques Conference 'Xroads of Discovery' (ACM is the Association for Computing Machinery; SIGGRAPH is the ACM Special Interest Group on Computer Graphics and Interactive Techniques).

Figure 17. Lindsey Foy, Xroads
(© 2015, L. Foy. Used with permission).

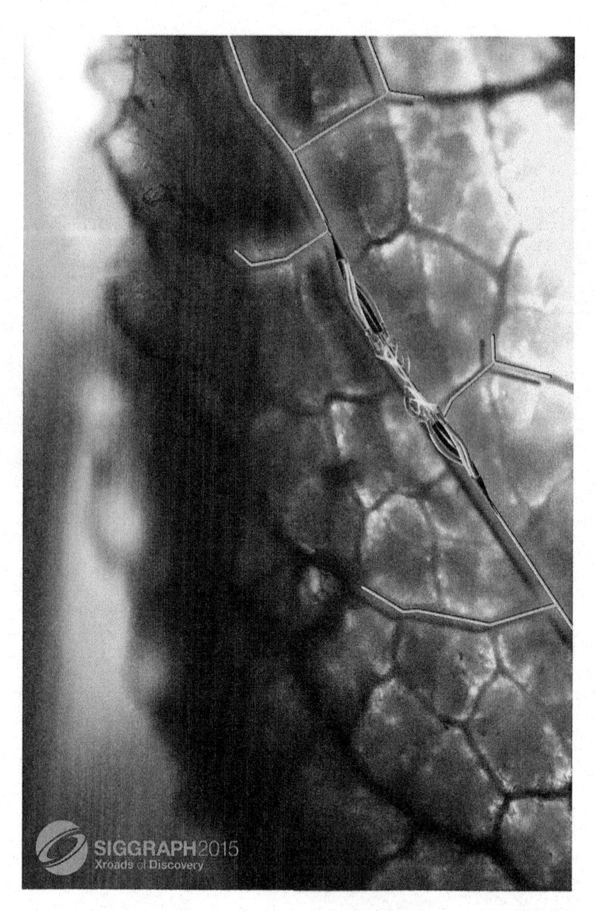

Biophotonics

Photonics is a science that studies the generation, detection, and application of light, mostly the visible and near-infrared light. Biophotonics combines biology and photonics for studying photons, quantum units of light. Biological photonics study photonic micro- and nanostructures in insects, birds, plants, and also fossil organisms. Examples of the photonic systems in nature include color-changing chameleons and cephalopods (such as squids and octopuses). Bioinspired photonics comprises the design, fabrication, and utilization of photonic materials that have been inspired by nature. Bioinspired photonic systems combine biological inspiration with nanoscience; they result in developing improvements in materials and devices. Bioinspired photonic systems involve several fields of study including photonics, nanofabrication, neuroscience, optical system study, and several others fields of study (Greanya, 2015). For example, t is possible now to record, identify, spectra of individual bacteria and differentiate them at the strain level (Sandros & Adar, 2015).

Bioimaging

The discipline of bioimaging explores biological structures and functions to create information visualizations in two, three, or four dimensions. Bioimaging researchers gather and analyze information from sources such as light waves, nuclear magnetic resonance, x-rays, or ultrasounds. Current methods of bioimaging may comprise optical light microscopy, several medical imaging techniques such as magnetic resonance imaging (MRI), nuclear magnetic resonance imaging (NMRI), magnetic resonance tomography (MRT), functional magnetic resonance imaging (fMRI), and many other advanced approaches involving microspectroscopy to investigate structures at the scale of micrometers, for example Raman spectroscopy, photoacoustic imaging techniques, bioimaging with fluorescent, magnetic supernanoparticles, bioimaging with laser near-infrared (NIR) fluorescence imaging with multifunctional nanostructures, among a number of other technologies. Researchers are developing the IR camouflage tape inspired by the squid skin properties (No author, Squid skin, 2015).

Optical Microscopy

The quality of imaging with optical microscopy depends on the right selection of a microscope according to the imaging depth needed (de Grand & Bonfig, 2015). This is important because features such as light scatter, absorption, background signals, achieving a sufficient collection of photons at the detector, and differences in refractive index become more challenging in the deeper specimens. Super resolved fluorescence microscopy brought optical microscopy into the nano dimension and won the 2014 Nobel Prize (Chang, 2014). Below, several examples are provided, of the developments in optical microscopy.

The total internal reflection fluorescence system serves for imaging just below the surface of a specimen; it allows imaging adhesion, hormone binding, neurotransmitter secretion, and membrane dynamics. This kind of imaging is limited to about 200 micrometers in the Z-axis.

The wide-field fluorescence emission system serves for imaging 0-700 micrometers beneath the specimen surface, and serves for imaging nuclei, stem cells, certain brain slices, neuron signaling, DNA, and chromosomes.

Confocal imaging serves for viewing thick and multilayered tissue samples at the depth of 0-1000 micrometers. Confocal microscopes allow three-dimensional fluorescence imaging in real time; they emit excitation light to generate fluorescence in the sample. Thin optical sectioning by building stacks of images is possible, because they remove out-of-focus light from outside of the focal plane.

Multiphoton excitation microscopy provides 3D resolution and allows imaging even deeper: 0-2000 micrometers below the specimen surface, using pairs of photons with a long wavelengths of light, which are absorbed by a single fluorophore and stimulate fluorescence only at the plane of focus. Multiphoton microscopy allows viewing thick brain slices, eye tissue, developing embryos, and other living tissues over hours or even weeks. The rapid pulsing of lasers enables researchers to study in real time the fast changing processes such as Ca^{++} signaling along neurons in response to stimuli. Recently, imaging up to 8000 micrometers (8 mm) into a fixed tissue is possible due to developments in clearing techniques with the use of clearing agents, and ultra-long working distance optics in multiphoton microscopes. These systems do not require slicing and thus damaging tissues, and thus can be used to provide neurobiological and other information about deep, intact structures. Ultrafast, dual-wavelength lasers with ultrafast (femtosecond) pulses provide simultaneous activation of multiple neurons (Arrigoni & Gallaher, 2015).

Structured illumination microscopy methods are now being developed, to improve the super resolution of a microscope, both in the lateral (the X and Y axes) and the axial (the depth in the Z axis) resolution. These methods require hardware for photo-switching, special laser systems, and mathematical models, computational or with software based options.

Medical Imaging Techniques at Micro- and Nano-Scale

Medical imaging techniques used in radiology: magnetic resonance imaging (MRI), nuclear magnetic resonance imaging (NMRI), functional magnetic resonance tomography (fMRI) or magnetic resonance tomography (MRT) apply strong magnetic fields and radio waves to form images of a body (Irvin & Koenning, 2015). This type of imaging serves for diagnosis, estimation of the disease stages, and tissue responses to ionizing radiation. It serves in cardiovascular, musculoskeletal, gastrointestinal, and neuroimaging examinations.

Magnetic resonance imaging (MRI) is a form of nuclear magnetic resonance spectroscopy. This medical imaging technique was awarded the 2003 Nobel Prize. In the MRI procedures, hydrogen atoms excited by an oscillating magnetic field emit a radio frequency signal that is measured by the receiver. MRI gives physiological information by showing images of regions with contrasting water content, such as different areas of the brain. It now uses contrast enhancement by the use of spin-exchange optical pumping to produce hyperpolarized noble gas, which enables new applications, especially in pulmonary imaging, for diagnosis and other applications. To achieve hyperpolarized gas, photons are absorbed by alkali atoms in vapor (Irvin & Koenning, 2015).

Nuclear magnetic resonance spectroscopy (NMR) offers a non-invasive technique for determining the structure of organic compounds with 1.0 mm anatomical resolution without ionizing radiation and with no known adverse biological effects. Data not available by other techniques may be obtained from samples weighing less than a milligram (Reusch, 2013; Partain et al., 1984).

Functional magnetic resonance imaging (fMRI) allows monitoring in real time responses of the brain structures to external stimuli through measuring the hemodynamic response to transient neural activity resulting from changes in the local blood oxygenation level: the ratio of oxyhemoglobin to deoxyhemoglobin.

Microspectroscopy for Real-Time Imaging of Singlet Oxygen

Oxygen is important in maintaining life; however, the lowest excited state of oxygen – singlet oxygen (dioxygen O_2) has unique reactivity, which can result in polymer degradation or the death of biological cells. For that reason it serves as an intermediate in photodynamic therapy aimed at cancer treatments, where light is utilized as a medical tool. With a spectral imaging method developed at the Charles U., Prague, near-infrared NIR luminescence microspectroscopy allows researchers to direct, real-time imaging of singlet oxygen (its very weak near-infrared phosphorescence) (Scholz & Shah, 2015).

Raman Spectroscopy for Monitoring Tumor Surgery

Raman spectroscopy allows observation of low-frequency vibrations and rotations of a system, and is used in chemistry to identify molecules. It relies on Raman inelastic scattering of monochromatic, near-infrared or near-ultraviolet light, which interacts with molecular vibrations (DoITPoMS, 2015).

Raman spectroscopy has been enhanced by a possibility of choosing the multiple laser wavelengths and software for coupling multivariate algorithms to 2D and 3D imaging. This allows applying tip-enhanced Raman spectroscopy for nanoscale resolution. It is now possible to identify in real time tumor margins during surgery. Stimulated Raman scattering microscopy (SRS) is faster than the spontaneous Raman microscopy; coherent antiStokes Raman spectroscopy (CARS) gives even higher signal intensity and better 3D spatial resolution. Raman measurements are done through an endoscope for studies of atherosclerosis, cancer, and bone diseases (Sandros & Adar, 2015).

Intravascular Photoacoustic Imaging Technique

Photoacoustics combined with ultrasound produces real-time images of blood vessels, organs, and body parts using a hand-held probe with a laser diode stack emitting 130-ns, 0.56-mJ pulses at 805 nm, with a rate of up to 10 kHz. As reported by Khalid Daoudi (2014), ultrasound detection can be synchronized with laser pulsing.

Ultrasound signals are generated by a fast-pulsing (2,000 near-infrared pulses per second) Raman laser. Photoacoustic imaging produces 3D images and reveals the presence of carbon-hydrogen bonds building lipid molecules in arterial atherosclerotic plaques. According to Ji-Xin Cheng (Wang, Ma, Slipchenko, Liang, Hui, Shung, Roy, Sturek, Zhou, Chen, & Cheng, 2014), the 2-kHz barium nitrite Raman laser, mounted in an endoscope and put into blood vessels of the Ossabaw miniature pig, generates pressure waves at the ultrasound frequency. A transducer device picks the waves. It heats and expands the tissue but not damages it.

Bioimaging with Fluorescent, Magnetic Supernanoparticles

Fluorescent, magnetic nanoparticles is an emergent class of materials, which enable bioimaging and tracking in the body. They can be manipulated with magnetic fields, so they are not lost in a jumble of molecules within a cell. Magnetic nanoparticles cluster when coated uniformly with fluorescent quantum dots, and also with silica to bind to particular molecules, for example as markers for tumor cells. Imaging is done with multiphoton microscopy techniques. Supernanoparticles can be used in magnetic resonance imaging (MRI) to probe biological functions in cells. Chen, Riedemann, Etoc, Herrmann, Coppey, Barch, Farrar, Zhao, Bruns, Wei, Guo, Cui, Jensen, Chen, Harris, Cordero, Wang, Jasanoff, Fukumura, Reimer, et al (2014) developed a method of co-assembling magnetic nanoparticles with fluorescent quantum dots, to form colloidal magneto-fluorescent supernanoparticles. The resulting superstructure consists of a core of closed-packed nanoparticles surrounded by a shell of fluorescent quantum dots, additionally covered with silica coating. Magneto-fluorescent supernanoparticles can be magnetically manipulated inside living cells and optically tracked. They can also serve as *in vivo* imaging probes, as multi-photon and magnetic resonance dual-modal probes (Chen et al, 2014).

The RGB (red-blue-green) marking of individual cells enables following the temporal and spatial distribution of single neurons in the adult brain. To achieve this, scientists (Gomez-Nicola, Riecken, Fehse, & Perry, 2014) used different viral envelopes of the y-retroviral and lentiviral vector sets. They tracked the generation of neurons vs. glial cells, as well as new astrocytes. In another *in vivo* experiments, researchers (Packer, Russell, Dalgleish, & Häusser, 2015) simultaneously observed and controlled neural circuit activity with cellular resolution, which gave the 'read-write' access to the brain: interacting with and understanding brain activity. The researchers recorded images of holographic targeting of laser beams to individual neurons in the brain cortex.

Multifunctional Nanostructures and Laser Near-Infrared Fluorescence Imaging

Multifunctional nanostructures developed in University of California at Riverside serve as vehicles for laser near-infrared fluorescence imaging and photothermal removal of cancer cells in vitro, as well as imaging of the over-expression of human epidermal growth factor receptor-2 (HER-2). Imaging and molecular targeting by multifunctional nanostructures were possible with the use of low cytometry and fluorescence microscopy, and continuous NIR laser irradiation at 808 nm. (Image cytometry enables statistical imaging a large number of cells with the use of optical microscopy). Nanoconstructs used a NIR dye, indocyanine green as theranostic materials (theranostics means combination of diagnostics and therapy) to merge the drug therapy and diagnostics, and to advance personalized medicine. It was encapsulated within a polymer, which was functionalized with monoclonal antibodies. Functionalized nanoconstructs are more effective in targeting the HER2 receptor. They can be applied to intraoperative detection, imaging, and phototherapy of ovarian cancer nodules (Bahmani, Guerrero, Bacon, Kundr, Vullev, & Anvari, 2014).

Examples of the light beans application in service of medicine and photopharmacology may include a phosphorescent bandage and light-controlled medication techniques. Researchers generate a transparent liquid bandage, which displays a quantitative, oxygenation-sensitive color map of the underlying tissue. Phosphorescent fluid paint-on bandage (Zongxi Li et al., 2014) can indicate oxygen concentration in damaged tissue. Phosphor glows longer and more brightly when the oxygen concentration is reduced. Oxygen-sensitive phosphor and a green oxygen-insensitive reference dye show changes in tissue oxygenation. Electronic flash units, with 400/70-nm bandpass filters provide pulses of excitation light, and a NIR CMOS camera collects the light that is emitted by the bandage. A violet-blue LED may control (through skin) a release of a diabetes medication JB253, which releases insulin from pancreatic cells (Broichhagen et al., 2014). The photo switchable drug changes shape when exposed to a violet-blue light.

Learning Project: Creating a Story about a Bioinspired, Biomimicking, or Hybrid Creature

1. Imagine yourself as an ornithologist, a beekeeper, a researcher of the ocean life and processes, or any other specialist involved in animal life. Think about special features or abilities displayed by animals under your attention that you would like to possess. Try to understand the nature of these capacities and ponder about their micro- and nano-structures in order to understand them better. Consider which bioimaging (or other) technologies would you apply to examine structures enabling these animals to surpass human abilities in particular areas. Then, think about selecting technologies that could be helpful in devising bio-inspired applications based on these special features, creating biomimicry based materials, or create a hybrid of the animal structures with artificial materials.
2. Draw an avatar on a computer, with a pencil, or any other artist's tool. It doesn't have to look like you; it may be quite fantastic. Make it a bioinspired, biomimicking, or hybrid creature that would be able to exceed your own capabilities in an area you choose.

3. Create animation, a cartoon, or an illustrated story where your avatar would be presented as super clever and proficient because of having abilities characteristic of the animals you are working on or with. In a background of your image, present the animals that inspired you in developing applications or materials that make you (and your avatar) exceptional. Design an action where your abilities become crucial to solve harmful problems to be dealt with, win, or survive.
4. Record your facial expression using software such as PhotoBooth, or some smart phone apps. You may apply your faces to your creatures.

CONCLUSION

This chapter provides background information related to selected themes, and then offers integrative learning projects designed in the spirit of the STEAM education. By adding the art related components to the science based concepts one is provided with a task of depicting interconnections of processes, events, rules, products, and intersections so often present in science. Therefore, the process of analyzing those functions becomes open to comparing, contrasting, and drawing conclusions. At the same time applying the notions of beauty or, as some could see it, the dilemmas behind scientific and nature based concepts as an inspiration for artistic creation, creates an opportunity to develop personal environment, often unknown to others, where new ideas and solutions can be created. The task requires evolving and applying one's abstract thinking, and drawing notations and connotations from facts, research, and cross-sections between various disciplines. This type of task also invites the idea of sharing knowledge and dwelling on new ideas through exchanging them with someone from another discipline. Learning about different points of view and gaining experience in specific areas or disciplines can be compared and understood as a factor leading toward more global view at selected processes, concepts, and ideas. There is a need both for specialists and generalists, but wider and broader understanding of various concepts can build insight, intuition, and better understanding that would replace rote learning by memorizing.

Visual projects support teaching science with an integrative approach and make learning topics closer to future professional tasks. Creating individual solutions may build up motivation and interest in learning. Teaching with the use of current applications and devices may arise motivation and support students' wish for achievement.

REFERENCES

Adams, R. A., Snode, E. R., & Shaw, J. B. (2012). Flapping Tail Membrane in Bats Produces Potentially Important Thrust during Horizontal Takeoffs and Very Slow Flight. *PLoS ONE*, *7*(2), e32074. doi:10.1371/journal.pone.0032074 PMID:22393378

AENews. (2015). *Wave Power*. Retrieved August 13, 2015, from http://www.alternative-energy-news.info/technology/hydro/wave-power/

Arrigoni, M., & Gallaher, N. (2015, February-March). Diverse Applications Drive Lasers for Multiphoton Microscopy. *BioPhotonics*, 32-36.

Bahmani, B., Guerrero, Y., Bacon, D., Kundr, V., Vullev, V., & Anvari, B. (2014). Functionalized polymeric nanoparticles loaded with indocyanine green as theranostic materials for targeted molecular near infrared fluorescence imaging and photothermal destruction of ovarian cancer cells. *Lasers in Surgery and Medicine*, *46*(7), 582–592. doi:10.1002/lsm.22269 PMID:24961210

Berg, A. M., & Biewener, A. A. (2010). *Wing and body kinematics of takeoff and landing flight in the pigeon (Columba livia)*. doi: .10.1242/jeb.038109

Bioinspired and Bioderived Materials. (2015). *Materials Research to Meet 21st Century Defense Needs*. The National Academies Press. Retrieved August 12, 2015, from http://www.nap.edu/openbook.php?record_id=10631&page=181

Bird Life International. (2012). *Mellisuga helenae*. The IUCN Red List of Threatened Species. Version 2014.3. Retrieved March 1, 2015, from http://www.iucnredlist.org/details/22688214/0

Birkhead. (1978). Behavioural adaptations to high density nesting in the common guillemot *Uria aalge*. *Animal Behaviour, 26*(2), 321-324.

Blank, S., & Dorf, B. (2012). *The Startup Owner's Manual: The Step-By-Step Guide for Building a Great Company*. K & S Ranch.

Borel, B. (2014). Map Maritime treasures at the Bottom of the Sea. *Popular Science*. Retrieved August 14, 2015, from http://www.popsci.com/article/science/map-maritime-treasures-bottom-sea

Borri, C., Betti, M., & Bartoli, G. (2010). Brunelleschi's dome in Florence: The masterpiece of a genius. London: Taylor & Francis Group.

Brawand, D., Wahli, W., & Kaessmann, H. (2008). Loss of Egg Yolk Genes in Mammals and the Origin of Lactation and Placentation. *PLoS Biology*, *6*(3), e63. doi:10.1371/journal.pbio.0060063 PMID:18351802

Broichhagen, B., Schönberger, M., Cork, S. C., Frank, J. A., Marchetti, P., Bugliani, M., & Trauner, D. et al. (2014). Optical control of insulin release using a photoswitchable sulfonylurea. *Nature Communications*, *5*, 5116. doi:10.1038/ncomms6116 PMID:25311795

Browniee, J. (2012). Clever Algorithms: Nature-Inspired Programming Recipes. lulu.com.

Burn, J. H., Leach, E. H., Rand, M. J., & Thompson, J. W. (1959). Peripheral Effects of Nicotine and Acetylcholine Resembling those of Sympathetic Stimulation. *The Journal of Physiology*, *148*(2), 332–352. doi:10.1113/jphysiol.1959.sp006291 PMID:13806185

Carpenter, K. (2000). *Eggs, Nests, and Baby Dinosaurs: A Look at Dinosaur Reproduction*. Indiana University Press.

Carroll, S. M. (2012). From Eternity to Here: The Quest for the Ultimate Theory of Time. ONEWorld Publ.

Celebi, K., Buchheim, J., Wyss, R. M., Droudian, A., Gasser, P., Shorubalko, I., & Park, H. G. et al. (2014). Ultimate permeation across atomically thin porous graphene. *Science*, *344*(6181), 289–344. doi:10.1126/science.1249097 PMID:24744372

Chang. (2014, October 14). Nobel Laureates Pushed Limits of Microscopes. *The New York Times*. Retrieved February 8, 2015, from http://www.nytimes.com/2014/10/09/science/nobel-prize-chemistry.html?_r=0

Chen, O., Riedemann, L., Etoc, F., Herrmann, H., Coppey, M., Barch, M., ... Reimer, R. (2014). Magneto-fluorescent core-shell supernanoparticles. *Nature Communications*. doi: .10.1038/ncomms609

Colbert, E. H., & Morales, M. (2001). *Colbert's Evolution of the Vertebrates: A History of the Backboned Animals Through Time* (5th ed.). New York: Wiley-Liss.

Coon, J. M., & Rothman, S. (1940). Nature of pilomotor response to acetylcholine; some observations on pharmaco-dynamics of skin. *Journal de Pharmacologie*, *68*, 301–311.

Daoudi, K., van den Berg, P. J., Rabot, O., Kohl, A., Tisserand, S., Brands, P., & Steenbergen, W. (2014). Handheld probe integrating laser diode and ultrasound transducer array for ultrasound/photoacoustic dual modality imaging. *Optics Express*, *22*(21), 26365–26374. doi:10.1364/OE.22.026365 PMID:25401669

De Grand, A., & Bonfig, S. (2015, January). Selecting a Microscope Based on Imaging Depth. *Bio-Photonics*, 26-29.

De Mena, J. M. (1992). Curiosidades y leyendas de Barcelona. Plaza & Janes Editores, S.A.

DoITPoMS. (2015). *Raman Spectroscopy*. University of Cambridge. Retrieved August 10, 2015, from http://www.doitpoms.ac.uk/tlplib/raman/index.php

Dresselhaus, M. S., Dresselhaus, G., & Eklund, P. C. (1996). *Science of fullerenes and carbon nanotubes*. San Diego, CA: Academic Press.

Dulvy, N. K., & Reynolds, J. D. (1997). Evolutionary transitions among egg-laying, live-bearing and maternal inputs in sharks and rays. *Proceedings. Biological Sciences*, *264*(1386), 1309–1315. doi:10.1098/rspb.1997.0181

Ee, C. C. (2013). *Removal of Dye by Adsorption of Eggshell Powder*. Malaysia: Technical University.

Eisen, E. J., Bohren, B. B., & McKean, H. E. (1962). The Haugh Unit as a Measure of Egg Albumen Quality. *Oxford Journals Science & Mathematics Poultry Science*, *41*(5), 1461–1468. doi:10.3382/ps.0411461

Ellis, D. (2012). *Eggs 101: The Anatomy of an Egg*. Retrieved February 23, 2015, from http://davidstable.com/2012/12/eggs-101-the-basics/

Encyclopedia, K. (2015). *Waves*. Retrieved August 13, 2015, from http://www.waterencyclopedia.com/Tw-Z/Waves.html

Entrepreneur. (2015). *Plan your business plan*. Retrieved August 13, 2015, from http://www.entrepreneur.com/article/38292#ixzz2OreoXTKd

Ernst, C. H., & Lovich, J. E. (2009). *Turtles of the United States and Canada* (2nd ed.). Johns Hopkins University Press.

Faber, T. (2008). *Faberge's Eggs: The Extraordinary Story of the Masterpieces That Outlived an Empire. Random House*. Retrieved from http://en.wikipedia.org/wiki/Faberge_egg

Freeman, C. L., Harding, J. H., Quigley, D., & Rodger, M. (2010). Structural Control of Crystal Nuclei by an Eggshell Protein. *Angewandte Chemie International Edition, 49*(30), 5135–5137. doi:10.1002/anie.201000679 PMID:20540126

Freyer, C., & Renfree, M. B. (2009, September15). The mammalian yolk sac in placenta. *Journal of Experimental Zoology. Part B, Molecular and Developmental Evolution, 312*(6), 545–554. doi:10.1002/jez.b.21239 PMID:18985616

Gatesy, S. M., & Dial, K. P. (1993). Tail muscle activity patterns in walking and flying pigeons (Columba livia). *The Journal of Experimental Biology, 176*, 55–76.

Gibson, C. (no date). *Florence Cathedral's Dome, Italy*. Retrieved February 22, 2015, from http://www.brunelleschisdome.com/

Goldman, F. (2014). *Geometric Origami*. Thunder Bay Press.

Goldman, F. (2016). Artist statement. In 2016 Joint Mathematics Meetings Exhibition of Mathematical Art. Tesselation Publishing.

Gomez-Nicola, D., Riecken, K., Fehse, B., & Perry, V. H. (2014). *In vivo* RGB marking and muliticolour single-cell tracking in the adult brain. *Scientific Reports, 4*, 7520. doi:10.1038/srep07520 PMID:25531807

Gutierrez, A. (1987). Dynamic Geometry: Euclidean Egg with 8 Arcs, Step-by-Step Construction. HTML5 Animation for Tablets (iPad, Nexus). *Geometry for Kids, School, Mathematics Education*. Retrieved February 23, 2015, from http://www.gogeometry.com/geogebra/euclidean-egg-8-arcs-step-construction-html5-animation-ipad-tablet.html

Hart, G. H. (1999). *Virtual Polyhedra*. Retrieved March 1, 2015, from http://www.georgehart.com/virtual-polyhedra/durer.html

Hemelrijk, C. K., & Hildenbrandt, H. (2011). Some causes of the variable shape of flocks of birds. *PLoS ONE, 6*(8), e22479. doi:10.1371/journal.pone.0022479 PMID:21829627

Hemelrijk, C. K., & Hildenbrandt, H. (2011). Some causes of the variable shape of flocks of birds. *PLoS ONE, 6*(8), e22479. doi:10.1371/journal.pone.0022479 PMID:21829627

Hustinx, J. P., Kuipers, G., Morée, I., & Douma, T. (2015). *Netherlands: ECJ: Non-Fertilised Human Ovum Activated By Parthenogenesis Could Be Patentable*. De Brauw Blackstone Westbroek N.V. Retrieved March 3, 2015, from http://www.debrauw.com/newsletter/ecj-non-fertilised-human-ovum-activated-parthenogenesis-patentable/

HyperPhysics. (2014). *C.R. Nave, Georgia State University*. Retrieved January 21, 2015, from http://hyperphysics.phy-astr.gsu.edu/hbase/pend.html#c1

Irvin, D., & Koenning, T. (2015, January). Line-Narrowed Laser Module Enables Spin-Exchange Optical Pumping. *BioPhotonics*, 36-39.

Kalinka, A. T., & Tomancak, P. (2012). The evolution of early animal embryos: Conservation or divergence? *Trends in Ecology & Evolution, 27*(7), 385–393. doi:10.1016/j.tree.2012.03.007 PMID:22520868

Khanna, D. R., & Yadav, P. R. (2005). *Biology of Birds*. Discovery Publishing House Pvt.Ltd.

Kokcharov, I. (2015). *Hierarchy of Skills*. Retrieved August 13, 2015, from http://www.slideshare.net/igorkokcharov/kokcharov-skillpyramid2015

Kroto, H. W., Health, J. R., O'Brien, S. C., Curl, R. F. & Smalley, R. E. (1985). C60: Buckminsterfullerene. *Nature, 318*(6042), 162. doi:10.1038/318162a0

Lahann, J. (2013). *Adaptive soft and biological materials. In Adaptive materials and structures: A workshop report*. Washington, DC: The National Academies Press. Retrieved May 12, 2014, from http://www.nap.edu/catalog. php?record_id=18296

Langdan, C. M. (2005). *Cognitive theoretic Model of the Universe*. Retrieved February 27, 2015, from http://www.megafoundation.org/CTMU/Q&A/Archive.html#Chicken

Lessem, D., & Sovak, J. (Illustrator) (2003). Scholastic Dinosaur A To Z. Scholastic Reference.

Li, Z., Roussakis, E., Koolen, P. G. L., Ibrahim, A. M. S., Kim, K., Rose, L. F., & Evans, C. L. et al. (2014). Non-invasive transdermal two-dimensional mapping of cutaneous oxygenation with a rapid-drying liquid bandage. *Biomedical Optics Express*, *5*(11), 11. doi:10.1364/BOE.5.003748 PMID:25426308

Maslow, A. H. (1943). A theory of human motivation. *Psychological Review*, *50*(4), 370–396. doi:10.1037/h0054346

Mediaville, C. (1996). *Calligraphy: From Calligraphy to Abstract Painting*. Belgium: Scirpus Publications.

Mulcahy, K. (2012). *Top 10 Fascinating Eggs*. Listverse. Retrieved March 3, 2015, from Lehttp://listverse.com/2012/03/01/top-10-fascinating-eggs/

Nanowerk. (2015a). *Introduction to nanotechnology*. Retrieved August 10, 2015, from http://www.nanowerk.com/nanotechnology/introduction/introduction_to_nanotechnology_1.php#ixzz3iXtNEYyL

Nanowerk. (2015b). *Nantechnology Spotlights*. Retrieved August 11, 2015, from http://www.nanowerk.com/#ixzz3iZH91GZ7

Newman SA. 2011. Animal egg as evolutionary innovation: a solution to the "embryonic hourglass" puzzle. *J. Exp. Zool. (Mol. Dev. Evol.), 316*(7), 467–483. Doi: 10.1002/jez.b.21417

No author. (2008). Inspired by Biology: From Molecules to Materials to Machines (Free Executive Summary). *Committee on Biomolecular Materials and Processes, National Research Council.* Retrieved August 10, 2015, from http://www.nap.edu/catalog/12159.html

No author. (2012). *Porous Science: How Does a Developing Chick Breathe Inside Its Egg Shell?* Scientific American, Science Buddies. Retrieved March 1, 2015, from http://www.scientificamerican.com/article/bring-science-home-chick-breathe-inside-shell/

No author. (2015, May-June). Squid skin. *BioPhotonics,* 16.

No author. (2015). *SeaWorld Parks & Entertainment. Whale Shark, Cartilaginous Fish.* Retrieved February 27, 2015, from http://seaworld.org/animal-info/animal-bytes/cartilaginous-fish/whale-shark/

No author. (2015). Whale Shark – Cartillaginous Fish. *SeaWorld Parks & Entertainment.* Retrieved January 28, 2015, from http://seaworld.org/en/animal-info/animal-bytes/cartilaginous-fish/whale-shark/

nothing2222229. (2008). *Red-Eared Slider Turtle Digging a Hole.* A video retrieved February 23, 2015, from https://www.youtube.com/watch?v=BCl8wXltv7E

Okamoto, N. (1996). *Japanese Ink Painting: The Art of Sumi-e.* Sterling.

Ong, S. C. (2005). *China condensed: 5000 years of history & culture.* Singapore: Marshall Cavendish.

Packer, A. M., Russell, L. E., Dalgleish, H. P. W., & Häusser, M. (2015). Simultaneous all-optical manipulation and recording of neural circuit activity with cellular resolution *in vivo. Nature Methods, 12*(2), 140–146. doi:10.1038/nmeth.3217 PMID:25532138

Partain, C. L., Price, R. R., Patton, J. A., Stephens, W. H., Price, A., Runge, V. M., . . . James, A. E., Jr. (1984). Nuclear Magnetic Resonance Imaging. The Radiological Society of North America. *RadioGraphics, 4*, 5-25. Retrieved August 12, 2015, from issuehttp://pubs.rsna.org/doi/pdf/10.1148/radiographics.4.1.5

PennState Modules. (2011). *Nano4Me.org. NACK educational resources.* The Pennsylvania State University. Retrieved May 30, 2014, from http://nano4me.live.subhub.com/categories/modules

Rahn, H., Carey, C., Balmas, K., & Paganelli, C. (1977). Reduction of pore area of the avian eggshell as an adaptation to altitude (water vapor permeability). *Proceedings of the National Academy of Sciences of the United States of America, 74*(7), 3095–3098. doi:10.1073/pnas.74.7.3095 PMID:16592423

Reusch, W. (2013). *Nuclear Magnetic Resonance Spectroscopy.* Retrieved August 11, 2015, from https://www2.chemistry.msu.edu/faculty/reusch/VirtTxtJml/Spectrpy/nmr/nmr1.htm#nmr1

Rivero, D., Dorado, J., Fernandez-Blanco, & Pazas, A. (2009). A Genetic Algorithm for ANN Design, Training and Simplification. In Bio-Inspired Systems: Computational and Ambient Intelligence: 10th International Work-Conference on Artificial Neural Networks, (pp. 391-398). Academic Press.

Sandros, M. G., & Adar, F. (2015, January). Raman Spectroscopy and Microscopy: Solving Outstanding Problems in the Life Sciences. *BioPhotonics*, 40-43.

Scholz, M., & Shah, M. (2015, January). Microspectoscopy Enables Real-Time Imaging of Singet Oxygen. *BioPhotonics*, 44-46.

Texier, P. J., Porraz, G., Parkington, J., Rigaud, J. P., Poggenpoel, C., Miller, C., . . . Verna, C. (2010). A Howiesons Poort tradition of engraving ostrich eggshell containers dated to 60,000 years ago at Diepkloof Rock Shelter, South Africa. *Proceedings of the National Acadademy of Science USA.* Retrieved January 28, 2015, from http://www.ncbi.nlm.nih.gov/pubmed/20194764?dopt

The Tail Story Adams. R. A. (2014). *University ofNorthern Colorado Bat Research Lab Portal.* Retrieved January 25, 2015, from http://www.researchgate.net/profile/Rick_Adams2/publications

Thomas, A. L. R. (1993). On the aerodynamics of bird tails. *Philosophical Transactions of the Royal Society of London. Series B, Biological Sciences, 340*(1294), 361–380. doi:10.1098/rstb.1993.0079

Tsukruk, V. (2013). Learning from nature: Bioinspired materials and structures. In *Adaptive materials and structures: A workshop report*. Washington, DC: The National Academies Press. Retrieved May 12, 2014, from http://www.nap. edu/catalog.php?record_id=18296

UNC Media Channel. (2012). Retrieved January 25, 2015, from http://www.unco.edu/news

USDA. United States Department of Agriculture. (2013). *Food Safety and Inspection Service*. Retrieved March 1, 2015, from http://www.fsis.usda.gov/wps/portal/fsis/topics/food-safety-education/get-answers/food-safety-fact-sheets/egg-products-preparation/shell-eggs-from-farm-to-table/CT_Index

Von Brücke, F. (1935). Über die Wirkung von Acetylcholine auf die Pilomotoren. *Klinische Wochenschrift, 14*(1), 7–9. doi:10.1007/BF01778952

Wang, P., Ma, T., Slipchenko, M. N., Liang, S., Hui, J., Shung, K. K., & Cheng, J.-X. et al. (2014). High-speed Intravascular Photoacoustic Imaging of Lipid-laden Atherosclerotic Plaque Enabled by a 2-kHz Barium Nitrite Raman Laser. *Scientific Reports, 4*, 6889. doi:10.1038/srep06889 PMID:25366991

Ward, A. J., Krause, J., & Sumpter, D. J. (2012). Quorum decision-making in foraging fish shoals. *PLoS ONE, 7*(3), e32411. doi:10.1371/journal.pone.0032411 PMID:22412869

Ward, A. J., Krause, J., & Sumpter, D. J. (2012). Quorum decision-making in foraging fish shoals. *PLoS ONE, 7*(3), e32411. doi:10.1371/journal.pone.0032411 PMID:22412869

Warrick, D. R., Bundle, M. W., & Dial, K. P. (2002). Bird Maneuvering Flight: Blurred Bodies, Clear Heads1. *Integrative and Comparative Biology, 42*(1), 141–148. doi:10.1093/icb/42.1.141 PMID:21708703

Weatherby, C. (2010). Study ranks salmon eggs as one of the three roes richest in omega-3s. *Vital Choice, Wild Seafood and Organics*. Retrieved March 1, 2015, from https://www.vitalchoice.com/shop/pc/articlesView.asp?id=948

Wei, Z., Li, B., & Xu, C. (2009). Application of waste eggshell as low-cost solid catalyst for biodiesel production. *Bioresource Technology, 100*(11), 2883–2885. doi:10.1016/j.biortech.2008.12.039 PMID:19201602

When. (2015). *Beach*. Retrieved August 13, 2015, from http://us.when.com/wiki/Beach?s_chn=1&s_pt=aolsem&type=content&v_t=content

Yeager, J. (2015). *12 Eggscellent Things You Can Do with Eggshells*. Retrieved February 23, 2015, from http://www.goodhousekeeping.com/home/green-living/reuse-eggshells-460809

yiddishwit. com. (2015). Retrieved March 3, 2015, from http://www.yiddishwit.com/gallery/eggs-smart.html

Yoo, S., Kokoszka, J., Zou, P., & Hsieh, J. S. (2009). Utilization of calcium carbonate particles from eggshell waste as coating pigments for ink-jet printing paper. *Bioresource Technology, 100*(24), 6416–6421. doi:10.1016/j.biortech.2009.06.112 PMID:19665373

ADDITIONAL READING

Brennan, A. B., & Kirschner, C. M. (2014). *Bio-inspired Materials for Biomedical Engineering (Wiley-Society for Biomaterials)*. Wiley. doi:10.1002/9781118843499

Browniee, J. (2012). Clever Algorithms: Nature-Inspired Programming Recipes. lulu.com. ISBN 1446785068.

Dunn, J. L., & Alderfer, J. (2011). *Field Guide to the Birds of North America* (6th ed.). National Geographic.

Feher, J. J. (2012). *Quantitative Human Physiology: An Introduction (Academic Press Series in Biomedical Engineering)*. Academic Press.

Floreano, D. (2008). *Bio-Inspired Artificial Intelligence: Theories, Methods, and Technologies (Intelligent Robotics and Autonomous Agents series)*. The MIT Press.

Greanya, V. (2015). *Bioinspired Photonics: Optical Structures and Systems Inspired by Nature*. CRC Press.

Jenkins, S., & Page, R. (2015). *Egg: Nature's Perfect Package*. HMH Books for Young Readers.

Levere, T. H. (2001). *Transforming Matter: A History of Chemistry from Alchemy to the Buckyball (Johns Hopkins Introductory Studies in the History of Science)*. Johns Hopkins University Press.

KEY TERMS AND DEFINITIONS

Cytometry: The measurement of the characteristics of cells: cell size, cell count, cell shape and structure, cell cycle phase, DNA content, and the presence of specific proteins (https://www.google.com/webhp?sourceid=chrome-instant&ion=1&espv=2&ie=UTF-8#q=cytometry).

Gravitational Potential Energy: Energy of an object resulting from its position in a gravitational field. Gravitational acceleration of an object near the Earth is assumed constant and equal of about 9.8 m/s^2. Gravitational potential energy GPE = mgh where m is mass (kg), g is acceleration due to gravity (around 9.8 on Earth), and h is the height of the object above the ground (m).

Hydrodynamic Drag: Means the resistance to the motion of a body by water of other fluid around it around it or resistance to the motion of a fluid in pipes, channels, etc. The hydrodynamic drag depends on velocity, characteristic area of the body, the shape of the body, and the direction of motion.

Pilomotor Reflex: Causes the rising of body's hair in animals including humans, feathers in birds, or quills in porcupines. It results in developing goose bumps on human skin, as a part of a response to strong emotions such as fear of an attack, sadness, joy, euphoria, or sexual arousal. Hair rising depend on activation of muscles at the base of each hair, which in turn is stimulated by the nerve endings of the sympathetic nervous system.

Raman Spectroscopy: Serves to examine low frequency vibrational and rotational energy states of objects under research such as molecules. Photons are scattered between the direction of motion. n electrons in an atom or a molecule. Several methods relay on inelastic Raman scattering of monochromatic (using only one color) light provided by laser in the visible, near-infrared (NIR), or near-ultraviolet range. Illuminated spots emit electromagnetic radiation, which enables recording processes occurring in objects.

Resonance: Occurs when an oscillating object is driven by another vibrating system. As a result, at a specific resonant frequency the system stores vibrational energy, and the amplitude of object's oscillation increases.

Suspensory: Locomotion involves hanging or suspension of a body exhibited by sloths and primates living on trees such as small gibbons or lemurs.

Sympathetic Nervous System: A part of the autonomic nervous system, along with the parasympathetic nervous system. It conveys information both ways, as the sensory and efferent (outward) signals. The autonomic nervous system regulates the unconscious actions of the body organs such as eyes, heart, lungs, blood vessels, sweat glands, digestive track, kidneys, pancreas, and penis.

Chapter 13
The Difference between Evaluating and Understanding Students' Visual Representations of Scientists and Engineers

Donna Farland-Smith
The Ohio State University, USA

Kevin D. Finson
Bradley University, USA

ABSTRACT

This chapter is a discussion of multiple tools for analyzing children's representations of scientists and engineers. Draw-A-Scientist and Draw-An-Engineer protocols have been utilized by science education researchers to investigate learners' perceptions of scientists and engineers. The chapter discusses the methods for analyzing students' perceptions of scientists and engineers how aspects of analysis lead to deeper understanding of the visual data. The discussion presented here is framed in the context in which refined protocols and rubrics are tools that uncover ranges of conceptions, and sometimes visual data are best examined by simple evaluation methods and sometimes by a qualitative rubric. The overarching question of this section is how can researchers use analysis of visual data to further what they already know about conceptions of scientists and engineers.

INTRODUCTION

We live in a world today where we are overloaded with visual data. Now more than ever, we have masses of resources and information at our fingertips. But how much of the visual data do we really understand? As teachers, we lean towards thinking of visual data of something that is external, in which our students absorb, or experience in textbooks or learning environments versus representations that our students

DOI: 10.4018/978-1-5225-0480-1.ch013

create for us to learn more about them. As researchers who have spent the last ten years intrigued by what students were telling educators about their interest in science careers through their representations or visualizations, we have sought a deeper understanding of what these images mean to us, as classroom educators. While analyzing and evaluating these pictures of scientists and engineers is interesting and often enjoyable, the work we do is often haunted by one nagging question: Do these representations of those in science fields really mean anything? How can this visual data be viewed from multiple perspectives and provide deeper meaning? As more visual data become available, rubrics appear to be changing the accepted understanding of the visual data students provide in terms of science and engineering fields.

Science and engineering are two disciplines that are commonly referred to as being comparable in nature. Over the last fifteen years, engineering has made its way into science curriculum at all levels, elementary, middle and high school. The most notable example is in the science standards when the National Research Council created the *Next Generation of Science Standards* (NRC, 2012) to include engineering practices. The inclusion of engineering practices expands the previous *National Science Standards* from 1996 to include relevant practices germane to both science and engineering in addition to disciplinary core ideas in the science content areas and crosscutting concepts. Key ideas in science, like, asking questions are not only reinforced as guiding principles of scientific investigations but also are inclusive from the engineering perspective as in asking questions to solve problems and to elicit ideas that lead to the constraints and specifications for its solution. "STEM" programs are becoming increasingly popular to address these engineering perspectives. And so it seems appropriate that we examine visual data of scientists to include the specific field of engineering. However, there seems to be a newer contemporary understanding of *how* to understand this visual data versus past studies in which only evaluated the what of visual data.

EVALUATING VISUAL DATA FROM STUDENTS ABOUT SCIENTISTS AND ENGINEERS

Well before children are able to express and verbalize which careers may be interesting to them, they are forming opinions and impressions from the world in which they live. Children collect and store ideas about scientists and in some cases engineers as well based on their experiences. For this reason, asking children to complete what science educators frequently term the Draw-A-Scientist Test (DAST) became a popular method for providing insight into how children represent and identify with those in the science fields. In its many forms, visual information and visual data within these illustrations have an important role to play with respect to learning and career choice. These assumptions might exhibit themselves in multiple ways, such as influencing someone's conceptions or even serving as tools that enable educators to investigate, collect information, and utilize that information to improve their instruction and student learning. The focus of this portion of the chapter is to discuss the analysis of this form of visual data and provide multiple means of evaluation as a way to understand these visual representations.

Students' images have been the focus of research for decades (i.e., Mead & Metraux, 1957; Chambers, 1983; Finson, 2002) with the consistent consensus of an emerging stereotype including scientists as white men, in lab coats, working alone in laboratories. A number of studies that asked students to "draw a picture of a scientist" contributed to what is commonly referred to as the stereotypical image of scientists (Chambers, 1983; Chiang & Guo, 1996; Fung, 2002; Maoldomhnaigh & Hunt, 1988; Newton & Newton, 1992; 1998; Song, Pak, & Jang; 1992; She, 1998). These seminal studies led researchers to begin to label and discuss this visual data in terms of stereotypical (white men in laboratories working independently often from society) and non-stereotypical (woman, outside doing science in collaboration). Analyzing these images frequently involved labeling the representations of scientists as stereotypical, or alternative, based on the appearance of individuals in the drawing, not to mention objects in the drawing, including things like lab coats, glasses or beakers. While the majority of DAST research concentrated on students' stereotypical images and their perceptions of scientists, the manner in which data derived from these studies has been analyzed has often been restricted to such analyses as the reporting of frequencies and the computation of simple t-tests. This level of analysis has provided a starting point for the investigation of students' perceptions of scientists and interesting discussion regarding potential for future investigations. The result from analyzing images of scientists was the labeling of visual data as stereotypical or non-stereotypical. This may be due to the multifaceted complexity involved with such investigations.

A scoring mechanism called the Draw-A-Scientist-Checklist (DAST-C) allowed researchers to focus on something else besides the initial "stereotypical image of the scientist" as expectations portrayed in students' drawings (Finson, Beaver, & Crammond, 1995). The checklist was created from the common aspects or features found in illustrations from previous studies and were based initially on the scientists, but not explicitly about their appearance, location, and activity. The DAST-C was an initial attempt to understand what students drew, the elements were derived from characteristics of drawings and interviews from children as reported in prior research. The checklist was designed to allow researchers to check off those items that had appeared most commonly (hence more stereotypical) in prior research while notations were made for other items such as the magnifying glasses, etc. so that later analysis could account for those drawing components.

Recent research studies conducted on students' perceptions of engineers (Yap, Ebert, & Lyons, 2003; Knight & Cunninham, 2004; Lyons & Thompson, 2006) are very similar to previous studies on perceptions of scientists because they share a common purpose. The purpose of the Draw-An-Engineer Test is to have students describe their knowledge about engineers and engineering through drawing and sometimes written responses. These illustrations have often been analyzed for stereotypical features much like the illustrations of scientists. Karatas, Micklos, & Bodner (2011) in their study of sixth-grade students' views of the nature of engineering and images of engineers suggested that the students' concepts of engineers and engineering was fragile, unstable, and likely to change within the time frame of the interviews they conducted. Students associate engineering with fixing, building, and working on things, and when asked to draw engineers, students portrayed engineers as physical laborers, or working on cars (Oware, Capobianco, & Diefes-Dux, 2007, Cunningham, Lachapelle, & Lindgren-Streicher, 2005). In some cases, students believed that in order to for engineers to perform their work, engineers needed materials, or objects, such as blueprints, computers, and safety gear (change). One conclusion to be drawn from all this is that children have a tendency to view engineers in stereotypical ways (mechanics of laborers fixing things) much in the same way they view scientists in stereotypical ways (chemists in laboratories surrounded by bubbling liquids). Hence, what we glean from these studies is that the initial conceptions of engineers and scientists held by children are somewhat limited and are in want of expansion.

PERCEPTIONS TO CONCEPTIONS

One of the challenges facing science educators is how we can equip teachers so they are enabled to broaden their students' views of scientists. Such a broadening, by necessity, requires we become cognizant of students' perceptions about what scientists might look like and their generalized ideas of what work scientists actually do – the basic "perceptions" -- which is essentially the earlier starting point for assessments such as the Draw-A-Scientist Test. Frequently, terms used with "perceptions of scientists" research include "images," "perceptions," "visions of," "views about" or "views of," and "look like" (Finson & Farland-Smith, 2013). But once we have a better idea about students' perceptions, we need to dig more deeply to reveal other ideas students may have regarding scientists. These ideas include what scientists actually do and where they do it. It includes how scientists engage in their work. In short, it includes determining what students know about the scientific enterprise and science itself. Science educators clearly recognize the importance of this, as evidenced by its inclusion in the national standards, beginning back with the National Science Education Standards (NRC, 1996). This digging deeper then moves us past simple perceptions and into "conceptions", which are more deeply held, more complex thinking structures. Then, we are dealing with "conceptions of scientists", or CoS.

The construct of CoS, however, does not operate alone in the student's schema. There are other things that impact what a student thinks when thinking about a scientist. Finson & Farland-Smith (2013) identified two significant things that work in concert with CoS: conceptual change theory and personal science identity (see Figure 1). Understanding how these three are related and interact with each other becomes important in effective science teaching and design.

Conceptual change theory tells us how a student progresses through learning and moves from simple stereotypical perceptions of scientists toward a more robust understanding about scientists and the scientific enterprise. Such theories are complex in and of themselves, and they serve to inform teachers of the effective and efficient ways to help their students through that progression. For example, one theory, posited by Posner, Strike, Hewson and Gerzog in 1982, presented the learning progression in essentially a linear sequence. Overly simplified, their theory was that a learner could not progress to the "next step" until he/she had learned the previous step. Misconceptions that occurred in learning had to be addressed by presenting the learner with the more scientifically "correct" concept and then giving students experiences with it. Once done, the misconception was erased and replaced by the newer and more appropriate conception. To enable the learner to progress, these end-points were to be targeted by the teacher. A later model proposed by Vosniadou (2007) – again overly simplified – viewed learning as a three-dimensional networking in which multiple components influenced each other simultaneously. Misconceptions are seen as ideas appropriate for the learner at his/her given developmental level, and never really go away – but more developed ideas can be learned and then utilized by the student in appropriate contexts. Further, these ideas are strongly influenced by background experiences and social factors. However, to enable the learner to progress, the end-point is not the target of instruction, but the presuppositions upon which those end-points are based are targeted. Addressing the presuppositions will result in changes in the end-points of learning. Whichever conceptual change theory one wishes to employ, the structure of the theory dictates the level of material and the sequencing of material that is presented to students. Hence, understanding something about conceptual change and how it occurs is important for the design and delivery of effective science instruction.

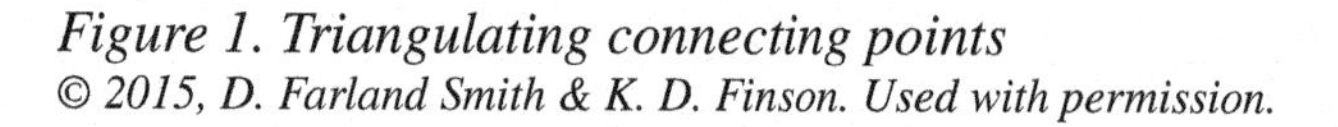

Figure 1. Triangulating connecting points
© 2015, D. Farland Smith & K. D. Finson. Used with permission.

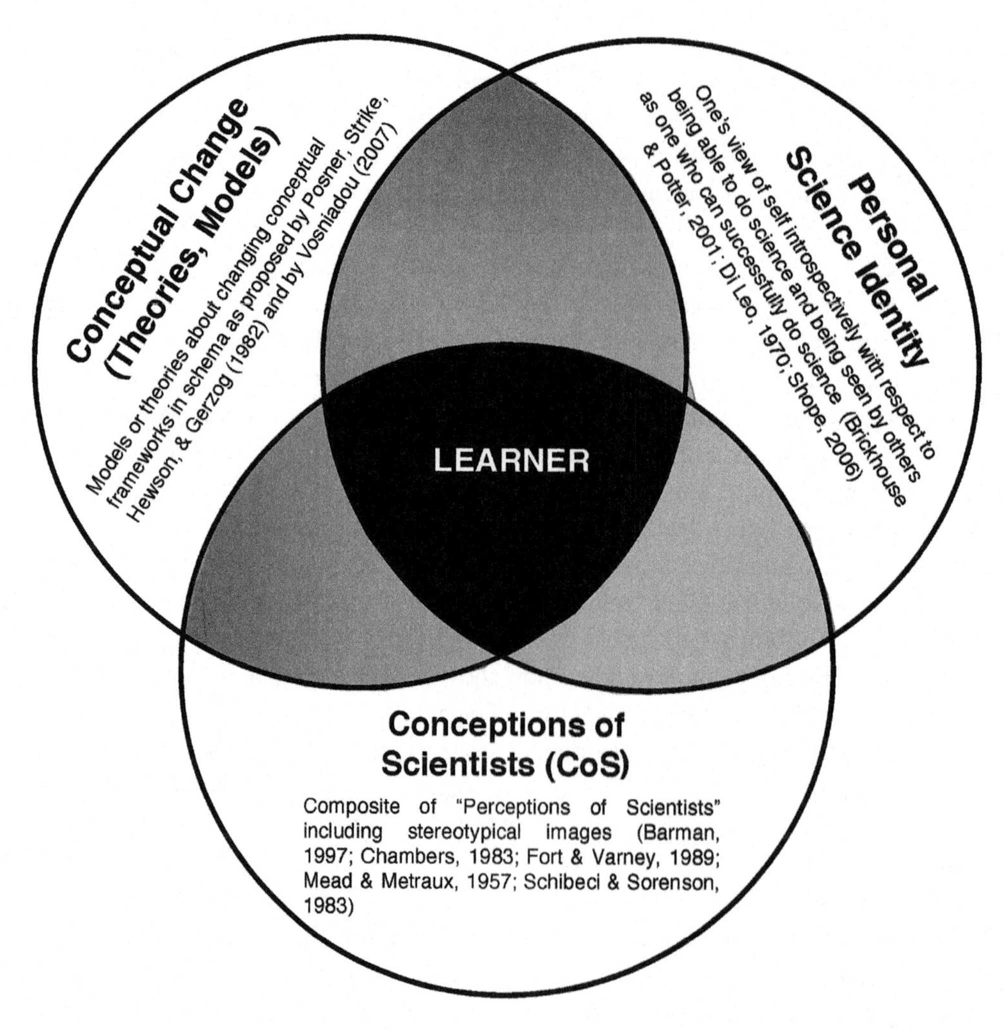

The third part is personal science identity. Essentially, this is the introspective nature of how one sees himself/herself as one who can do science and is viewed by others as being one who does science and/or can do science. One's science identity is generated through attitudes, feelings, and intellect that are embedded within a particular context. Typically, that context is a community of practice, such as a classroom, and cannot be isolated as a single construct. For example, science identity is related to self-efficacy. A student's self-efficacy with respect to our discussion means how the learner sees himself/herself as being able to be a scientist and do the kinds of investigations scientists do, and valuing that perception in a positive way. One's attitudes toward science certainly contribute to one's self-efficacy in this respect. Further, the degree of the leaner's scientific literacy can have a significant influence on his/her ability to understand and deal with science, and consequently is an influence on his/her science identity (Finson & Farland-Smith, 2013).

To explain some of the Figure 1 components' interactions a little further, consider the difference between science concepts, CoS, and science identity. Science concepts can be such things as change, entropy, gradient, organism, etc. Although science concepts are part of what constitutes scientific literacy, they in and of themselves do not result in a particular CoS, but they do inform the learner of what scientists may be investigating about. This can contribute, in turn, to development of a learner's CoS. If the learner comes to believe he/she can understand something about a concept and how to investigate it, his/her self-efficacy (with respect to science) can increase. In turn, this might be recognized by peers in the social context as a strength of that learner, thus contributing to his/her science identity. How a student understands any particular phenomenon will strongly influence how he/she experiences it as well as influencing the approaches he/she takes to interact with it – something that can be guided by the teacher's utilization of an appropriate conceptual change theory. A key take-away from all this is that nonidentity-related conceptions (entropy, change, etc.) only involve the concept itself, whereas identity-related conceptions involve the student's concept of self.

In an operational sense, personal science identity can be exhibited through children's pictures (drawings) of scientists. Their drawings are in a real sense collective reflections of their attitudes, experiences and perceptions. Because of this, utilizing students' drawings to assess their perceptions of scientists, CoS, and some about their science identity can yield information the teacher can find helpful in guiding the child's further science learning. But these tidbits are not always easily derived by just taking a quick look at a drawing. The drawings need to be examined in a deliberate manner. This is where the development and utilization of a rubric enters the picture. Rubrics can be innovative, and can clearly provide an accurate way to derive quantitative scores for a child's drawings of scientists and engineers. Those quantified scores can then be subjected to statistical testing (t-tests, etc.). The rubric, then, becomes a lens through which science educators and teachers can analyze students' drawings with some degree of precision and glean deeper understandings of how students are being socialized around scientific knowledge and science-based science identities.

MOVING BEYOND EVALUATION TO UNDERSTANDING

The combination of the rubric approach for analysis and protocol refinement enables clarities to emerge and subsequently increased detail to what one could ascertain from students about their expressed images of scientists, engineers and each field respectively. For example, insights into how engineers use math science in their work were explored along with the idea that engineers are their own subset under the broader umbrella of scientists. Mathematics and computational tools are critical to both science and engineering. Though these elements have yet to be studied in children representations. In order to this two specific criterion needed to be addressed; 1) the directions, and 2) the use of multiple pictures.

Protocols

One of the important components to examining this type of visual data was specific directions. In 2003 Farland modified the Draw-A-Scientist Test (DAST) directions from a simple "draw a scientist" to ensure three aspects appearance, location and activity for evaluation with the DAST Rubric. As a result, the prompt was modified from simply "draw a scientist" to

Imagine that tomorrow you are going on a trip (anywhere) to visit a scientist in a place where the scientist is working right now. Draw the scientist busy with the work this scientist does. Add a caption, which tells what this scientist might be saying to you about the work you are watching the scientist do. Do not draw yourself or your teacher.

Farland (2006) used information collected from these specific directions to create a rubric based on the variations collected, observed and analyzed to create a comprehensive picture of students' perceptions of scientists. Hundreds of drawings revealed, for instance, (1) who is doing science, (Caucasian, African American, or male, female,) and the overall appearance of scientists (crazy, mad scientist, normal-looking); (2) what location the scientist is in (basement, laboratory, etc); (3) what activity is being done (mixing chemicals, studying rocks, finding fossils), and what tools are being used (from explosives to more commonly used tools like magnifying glasses). Using a rubric to analyze children's illustrations of scientists is a contemporary approach and appropriate because it allows for individual creativity as there is no right or wrong answer for illustrations as there often is with science content. Thomas et al (2016) applied the same approach to the Draw-An-Engineering Test (DAET), thus creating the modified directions, which included the drawings of engineers to be working specifically. Thus leading to a new acronym (DAEWT) Draw-An-Engineer-At-Work Test as opposed to the previous acronym (DAET) Draw-An-Engineer Test. The directions are as follows:

Draw an engineer at work. In this space you will create an illustration of an engineer. There are no wrong answers – so think about what you know about engineers' work and draw the ideas and details that you remember. This is not a test of your artistic skills – so don't worry about your drawing ability. You will notice that there are a few more opportunities for you to describe your engineer (and explain your drawing too). So, now, let's begin. Please draw an engineer-at-work in the box provided.

Up until these modified directions in 2105, the evaluation of the visual data of engineers included drawings with simple directions to "draw-an-engineer" and follow-up interviews (Capobianco, Dieffes-Dux, Mena & Weller, 2011). Analysis of the interpretive nature of the Draw-An-Engineer-Test (DAET) and interview were both inductive processes; that is instead of searching for patterns which were pre-determined, themes were allowed to emerge first, then, researchers constructed meaning from students' responses (Capobianco, Dieffes-Dux, Mena & Weller, 2011). In this study, two themes emerged which would be the focus of future studies, artifacts and actions. In discussing this type of visual data, these two words become a framework for which researchers and educators can begin to discuss and understand the context of the pictures, just as in the same way as Finson's, Beaver's, & Cramond's DAST-C (1995) allowed for discussion about the content included in illustrations of scientists. The action portrayed in the picture was coded, for example, many pictures of engineers included the action of fixing or repairing something. The artifact is the object used by the engineer, i.e., air conditioner, toaster, lightbulb (all labeled electronics), vehicles, or buildings.

It wasn't until a scoring mechanism called the Draw-A-Scientist-Checklist (DAST-C) was developed that researchers were able to focus on something else besides the initial "stereotypical image of the scientist" as expectations portrayed in students' drawings. The checklist was created from the common aspects or features found in illustrations from previous studies and were based initially on the scientists, but not explicitly about their appearance, location, and activity. The DAST-C was an initial attempt to understand what students drew, the elements were derived from characteristics of drawings and interviews from children as reported in prior research. The checklist was designed to allow researchers to check off those items that had appeared most commonly (hence more stereotypical) in prior research while notations were made for other items such as the magnifying glasses, etc. so that later analysis could account for those drawing components.

Engineers followed a similar protocol in arriving at what components to include in the Draw an Engineer at Work Test. In their study, Capobianco, et al 2011 revealed four distinct characteristics of engineers emerged: (1) an engineer is a mechanic who fixes engines or drives cars and trucks; (2) an engineer is a laborer who fixes engines or drives cars and trucks; (3) An engineer is a technician who fixes electronics and computers; and (4) an engineer is someone who designs.

Later, the modified DAST was developed by Farland-Smith and McComas (2007). The major difference between the DAST-C and the modified DAST and DAST Rubric combination was that the DAST-C simply limited scoring to whether or not a particular drawing element was present in the drawing, whereas the modified DAST and DAST Rubric takes the assessment a couple steps further, actually on a continuum, as the drawing's are labeled as "Sensationalized," "Traditional," "Broader than Traditional". A description of scoring ranges from 0-3 of the DAST Rubric in each of these three categories: 1) the APPEARANCE of scientists, 2) the LOCATION of scientists and 3) the ACTIVITY of scientists.

In the same way as the development and application of a single instrument and protocol has provided for countless research studies about scientists, we anticipate, probing debates and lively discussion amongst researchers across fields of engineering with a specified protocol. Through the examination and evaluation of students' pictures with specific directions a wealth of information was attained while looking broadly at one's conceptions over multiple pictures. The next section will discuss the purpose of using three images to understand students' visual data of scientists and engineers.

MULTIPLE REPRESENTATIONS OF SCIENTISTS AND ENGINEERS

For decades, researchers had been convinced that one stereotypic image of scientists existed among children worldwide (Chambers, 1983; Chiang & Guo, 1996; Fung, 2002; Maoldomhnaigh & Hunt, 1988; Newton & Newton, 1992; 1998; Song, Pak, & Jang; 1992; She, 1998). This stereotypic image includes, a white male in a laboratory setting with beakers and/or chemicals. While most people find this amusing and even can recall meeting or seeing a scientist in media or real life to reinforce this stereotype is does bode well for encouraging students to enter the field of science.

The purpose of having students create multiple drawings of scientists and engineers is an attempt to expose and possibly exhaust the conceptions students' hold about people who work in these fields. It is reasonable to assume that in multiple pictures students would have a sufficient opportunity to represent their ideas about anything, in this case scientists and engineers. For example, if they hold a view of the work of science and the true nature of who can be a scientist, it will be evident across the three pictures. On the other hand, if a student draws the same image consecutively there is good reason to believe it is the student's only view (Farland-Smith & McComas, 2007).

This idea that young children hold a range of perceptions versus the commonly accepted singular, insular image was something gained from evaluation with a rubric. It is during this process that competing and/or conflicting images can be uncovered. For this reason, the multiple drawing task to gain a wider vide of what students really know about scientists rather than limiting students to one opportunity to represent their view. Most of us, if asked to draw only *one* image of a scientist (or anything else for that matter), are likely to draw a stereotypical image to get the point across even though we may appreciate the limitations of that single drawing. With multiple opportunities, if a student draws scientists of differing ethnicities, of both genders doing work in a variety of settings it is reasonable to assume that they have a sufficiently broad view of the work of science and engineering. Furthermore, combining a contemporary perspective and creative method of analyzing student perceptions may be helpful in developing a theoretical understanding of how students interpret scientists, engineers and their work. It also provides a language for researchers to use in discussing these illustrations and frames the context, (i.e., appearance, location, activity) for the visual representations of scientists to ease comparison.

For example, data from a preliminary study evaluating the use of multiple pictures showed that if we were to make assumptions about what students' think of scientists with respect to activity from a single drawing, we would be correct only 24% of the time. In 76% of cases, the students' first drawing is *not* their only conception of a scientist. The following pictures demonstrate the range of perceptions students may hold about scientists and is included for evidence that the first picture a child draws may not contain all their feelings and/or conceptions of scientists (Farland-Smith & McComas, 2007). What follows is a discussion of how using rubrics and multiple pictures moves beyond evaluation to an actual understanding of students' visual representations of scientists. A discussion of how value is assigned when scoring three pictures of scientists will be described in the next section. A discussion of how to value three pictures of engineers follow the discussion of scientist pictures.

Limited Visual Representations of Scientist

If the cumulative modified DAST score of three pictures is between 3 and 4.5 than it is labeled limited, and it means the visual data represented a narrow representation of the category (i.e., appearance, location, and activity). In terms of appearance, the visual data might suggest the child drew very similar appearances of all three scientists. Most likely, the child drew very stereotypical appearances of scientists on each opportunity. In terms of location, the visual data might suggest the child drew very similar location in all three pictures. Most likely, the child drew a very stereotypical location, like a basements, or laboratory on each opportunity. In terms of activities, it may suggest the visual data may suggest the child drew very similar activities of all three scientists. Most likely, the child drew stereotypical representations of the activities done by scientists on each opportunity.

Competing Visual Representations of Scientists

If a child's cumulative modified DAST score of the three pictures is between 4.5 and 7.5 than it is labeled competing, and it means that the visual data represents very different conceptions of what scientists look like and compete within their own understanding of the appearance of scientists. Most likely, the set of three pictures include a stereotypical representation of the appearance of a scientist and a non-stereotypical representation of the appearance of a scientist. In terms of location, a label of competing on the cumulative scores indicates that the child drew very different conceptions of the places where scientists work. Most likely, the set of three pictures include a stereotypical representation of the location (basement) of a scientist and a non-stereotypical representation of the location (outdoors) of a scientist, or just had three traditional pictures of a scientist working in a laboratory setting. In terms of activity, if the sum of the three pictures is between 4.5 and 7.5 and can suggest that the child drew very different holds very different perceptions of the activities done by scientists, which compete within their understanding of what do. Most likely, the set of three pictures include a stereotypical representation of the activity of a scientist and a non-stereotypical representation of the activity of a scientist.

Expansive Visual Representations of Scientists

If a child's cumulative modified DAST appearance category is labeled expansive, it means that the cumulative score of all three pictures was between 7.5 and 9 and suggests the child holds very different conceptions about the appearance of scientists. Most likely, they included women in their pictures, or multiple scientists in more than one picture. If a child's cumulative location category is labeled expansive it means that the cumulative score of all three pictures was between 7.5 and 9 and suggests the child holds broad conceptions about the location of scientists. Most likely, they included the possibilities of outdoors into their perception of where scientists work, and carried this through to more than one picture. In the category of activity, the label expansive also means that the cumulative score of all three pictures was between 7.5 and 9 and suggests the child holds broad conceptions about the activity of scientists. Most likely, they included several different fields of science in more than one picture.

SCORING VISUAL DATA IN ENGINEERING ILLUSTRATIONS

While the DAST Rubric assigns numerical scores (0-3) for three categories (appearance, location & activity), the Draw-An-Engineer-At-Work Test (DAEWT) Rubric assigns a numerical score in four separate and specific categories: 1) the use of math in engineering (0-2); 2) the use of science in engineering (0-2); 3) gender stereotypes (0-3); and 4) the work of an engineer (0-3). The DAEWT Rubric assigns a score from 0 to 2 for math and science categories and assigns a score from 0 to 3 for gender and work of an engineer categories within each picture. Details of the scoring of the *Use of Mathematics* and the *Use of Science* will now be discussed briefly.

- **No Conception:** Illustrations that score a "0" in both the science and math categories are labeled No Conception and can be referred to as vague or nonsensical and provide no answer or specifically indicate that math or science was not used. These drawings may contain responses like, thinking, no, yes, I am not sure or I don't know.
- **Basic Conception:** Illustrations that score a "1" in science and/or math can be referred to as Basic Conceptions and identify a science/math skill an engineer might use. These drawings may contain responses like; adding, subtracting, electricity, testing, fixing.
- **Advanced Conception:** Illustrations that score a "3" in science and/or math can be referred to as Advanced Conceptions and identify a science/math skill an engineer might use and give the reason why an engineer apply math or science. These drawings may contain responses like; trying to figure out where it hit the building so it won't fall on him, or what kind of plant to plant.
- **Gender and the Work of an Engineer:** Have four components and are scored with numbers assigned 0-3. In gender the categories are *No Conception*, *Traditional Conception*, *Non-Traditional Conception*, or *Expanded Conception*.
- **No Conception in Gender:** Illustrations that score a "0" in gender are labeled so because the stereotype could not be determined or the conflicting gender information revealed a misalignment between the descriptions and the identified gender in the illustration.
- **Traditional Conception in Gender:** Illustrations which score a "1" in gender are labeled so because a single male was depicted.

- **Non-Traditional Conception in Gender:** Illustrations which score a "2" in gender are labeled so because a single female was depicted.
- **Expanded Conception in Gender:** Illustrations which score a "3" in gender depict a group of individuals working together as a team or in cases where the artist indicates that gender does not matter.

While Gender and the Work of an Engineer have four components and are scored with numbers assigned 0-3, there is a slight difference in the labeling of categories. In Work of an Engineer the categories are *No Conception, Naïve Conception, Basic Conception, Advanced Conception.*

- **No Conception in Work of an Engineer:** Illustrations that score a "0" may represent vague activity that cannot be determined, or no answer is provided. It could be that the illustration provides a mistaken conception of the work of an engineer, for example, driving a train, moving things or talking to people. Also included in this category are references to another career or profession such as teaching, physician, scientist, hairstylist, policeman or pilot.
- **Naïve Conception in Work of an Engineer:** Illustrations that score "1" identifies stereotypic notions about the work of an engineer, for example, children may draw a mechanic, skilled laborer, or technician. The overall picture suggests engineers are fixers, builders or doers, for example, they repair cars, replace wires, build houses, make TV's, help the air, clean the world. Naïve conceptions may also include engineering experiences that mirror classroom situations which portray child or teacher as an engineer. Some of these examples within the context of the classroom include making foil boats, building with blocks, or gum drop structures.
- **Basic Conceptions in Work of an Engineer:** Illustrations scoring "2" identifies a field of engineering like mechanical engineering, or environmental engineering but not the "why" a engineer is doing it. These pictures may also suggest that engineers create or design new things (i.e., phones, gum, houses) but does not reference a specific problem or need.
- **Advanced Conceptions in Work of an Engineer:** Illustrations scoring "3" identify the work of an engineer and references a specific problem or a need to be resolved. This is different from a 2, in which only the what (problem) is included and not the why. These pictures may also reference the way engineers design or "find a way" to solve a problem. The also may suggest that engineers work to improve things like, gas mileage, structural integrity, usable fuel, etc. The visual data represents overall that engineers design, test, experiment and plan, however, not necessarily in that order.

COMPARING STEREOTYPICAL ILLUSTRATIONS OF SCIENTISTS AND ENGINEERS

We are moving toward a stereotypical engineer in the same ways we have talked about stereotypical scientists. These engineers are the doers, not designers. Multiple ways of analysis offer a broader scope of possibilities with respect to investigating children's visual data. Rubrics, in this case, provide deeper analysis in particular areas, like appearance, location, gender, work of an engineer, and activities of scientists. The combination of the modified directions with particular rubrics enables clarities to emerge and subsequently increased detail to what one could ascertain from students about their expressed images of the scientific and engineering fields. Insights into how engineers use math and science in their work are new to the field. They could be aligned with how educators are inclusive in their ideas where engineers form their own subset under the broader umbrella of scientists. Through the examination and evaluation of visual data we, as researchers gain a wealth of information while looking broadly at one's perceptions or conceptions over multiple pictures. In the same way as the development and application of a single instrument and protocol has provided for countless research studies about scientists, we anticipate and probe debates and lively discussions amongst researchers across fields of engineering. Examining visual data under different lenses can be beneficial to us as educators, and may propel new analysis and ways of thinking about the pictures children make for us.

SOLUTIONS AND RECOMMENDATIONS

Rubrics for the DAST and DAEWT are by no means a solution, but can be a useful tool helping science educators to conduct well-designed investigations about students' conceptions of scientists and engineers. In the past, methods of evaluation of visual representations of scientists in particular have failed to gain credibility amongst many science education researchers. The development and use of rubrics can help address concerns that have been expressed with respect to this line of investigation.

FUTURE RESEARCH DIRECTIONS

Contemporary perspectives of analysis students' conceptions of scientists and engineers move research in new directions with each modification to protocols and multiple opportunities for children to express their conceptions. Each new insight brings new questions and considerations with possible connections to instructional strategies in the classroom. Future research might examine the methods of interventions for for both DAST and DAEWT. A task for educators and researchers is to challenge the question of why do students' images of scientists and engineers their work remain stereotypical? Researchers should continue to examine the relevance and influences holding a non-stereotypical conception of a scientist and/or engineer has on a child's success in school science education. A study examining one populations visual representations of engineers and scientists for comparison might prove helpful to new understandings of the power of these images.

In addition, future investigations with teachers and other educators about their conceptions of scientists and engineers might help researchers to understand how their epistemological beliefs shape their students and classroom. For example, how are elementary teacher's beliefs about how to teach science and how children learn science shaped by their own perceptions of scientists and engineers? Is there a relationship between teachers' perceptions of scientists/engineers and students' conceptions of scientists/engineers?

CONCLUSION

The focus of this chapter was the creative way researchers have examined visual data in the past and present and how newer methods of evaluation offer different perspectives. This is not to say that new methods of evaluation offer better analysis, just different perspectives. Both methods of analyzing visual data in this case of scientists and engineers broaden what can be learned from data collection. Multiple ways of analysis offer broader scopes of children's visual data. Rubrics, in this case, provide deeper analysis in particular areas, like appearance, location, gender, or work of an engineer and activities of scientists. The combination of the modified directions with particular rubric enables clarities to emerge and subsequently increased detail to what one could ascertain from students about their expressed images of scientists and the engineering field. Insights into how engineers use math science in their work are new to the field align with how educators are inclusive in their ideas that engineers are their own subset under the broader umbrella of scientists. As researchers we need to feel comfortable and embrace the unknown when we pose questions, such as why not analyze illustrations of scientists under the lens of science and math and how it's used, and analyze illustrations of engineers by appearance, location and activity? Why were the DAST Rubric and DEAWT Rubric created with their individual focus? And what more do we know now that we did not know as a result of this analysis? Through the examination and evaluation visual data we, as researchers gain a wealth of information while looking broadly at one's perceptions or conceptions over multiple pictures. In the same way as the development and application of a single instrument and protocol has provided for countless research studies about scientists, we anticipate, probing debates and lively discussion amongst researchers across fields of engineering. Examining visual data under different lenses can be beneficial to us as educators and propel new analysis and ways of thinking about the pictures children make for us.

REFERENCES

Barman, C. (1997). Students' views of scientists and science: Results from a national study. *Science and Children*, *35*(1), 18–23.

Brickhouse, N. W., & Potter, J. T. (2001). Young women's science identity formation in an urban context. *Journal of Research in Science Teaching*, *38*(8), 965–980. doi:10.1002/tea.1041

Capobianco, B. M., Diefes-dux, H. A., Mena, I., & Weller, J. (2011). What is an engineer? Implications of elementary school student conceptions for engineering education. *The Journal of Engineering Education*, *100*(2), 304–328. doi:10.1002/j.2168-9830.2011.tb00015.x

Chambers, D. (1983). Stereotypic images of the scientist: The draw-a-scientist test. *Science Education*, *67*(2), 255–265. doi:10.1002/sce.3730670213

Chiang, C., & Guo, C. (1996). *A study of the images of the scientist for elementary school children.* Paper presented at the National Association for Research in Science Teaching, St. Louis, MO.

Cunningham, C., Lachapelle, C., & Lindgren-Streicher, A. (2005). *Assessing elementary school students' conceptions of engineering and technology*. Paper presented at the annual American Society for Engineering Education Conference & Exposition, Portland, OR.

DiLeo, J. H. (1970). *Young children and their drawings*. New York, NY: Brunner/Mazel Publishers.

Farland, D. (2006). The effect of historical, nonfiction trade books on elementary students' perceptions of scientists. *Journal of Elementary Science Education*, *18*(2), 31–48. doi:10.1007/BF03174686

Farland-Smith, D., & McComas, W. F. (2007). *The enhanced (DAST): A more reliable, valid, efficient reliable & complete method of identifying students' perceptions of scientists.* Paper presented at the meeting of the National Association of Research in Teaching Conference, New Orleans, LA.

Finson, K., & Farland-Smith, D. (2013). Applying Vosniadou's conceptual change model to visualizations on conceptions of scientists. In K. Finson & J. Pederson (Eds.), *Visual Data and Their Use in Science Education* (pp. 47–76). Information Age Publishing Inc.

Finson, K. D. (2002). Drawing a scientist: What we do and do not know after fifty years of drawings. *School Science and Mathematics, 102*(7), 335–345. doi:10.1111/j.1949-8594.2002.tb18217.x

Finson, K. D., Beaver, J. B., & Cramond, B. L. (1995). Development and field test of a checklist for the draw-a-scientist test. *School Science and Mathematics, 95*(4), 195–205. doi:10.1111/j.1949-8594.1995.tb15762.x

Fort, D. C., & Varney, H. L. (1989). How students see scientists: Mostly male, mostly white, and mostly benevolent. *Science and Children, 26*, 8–13.

Fung, Y. (2002). A comparative study of primary and secondary school students' images of scientists. *Research in Science & Technological Education, 20*(2), 199–213. doi:10.1080/0263514022000030453

Karatas, F., Micklos, A., & Bodner, G. (2011). Sixth-grade students' views of the nature of engineering and images of engineers. *Journal of Science Education and Technology, 20*(2), 123–135. doi:10.1007/s10956-010-9239-2

Knight, M., & Cunninham, C. (2004). *Draw an engineer test (DAET): Development of a tool to investigate students' ideas about engineers and engineering*. Paper presented as the annual American Society for Engineering Education Conference & Exposition, Salt Lake City, UT.

Lyons, J., & Thompson, S. (2006). *Investigating the long-term impact of an engineering-base GK-12 program on students' perceptions of engineering*. Paper presented at the annual American Society for Engineering Education Conference and Exposition, Chicago, IL.

Maoldomhnaigh, M., & Hunt, A. (1988). Some factors affecting the image of the scientist drawn by older primary school pupils. *Research in Science & Technological Education, 6*(2), 159–166. doi:10.1080/0263514880060206

Mead, M., & Metraux, R. (1957). Image of the scientist among high school students. *Science, 126*(3270), 384–390. doi:10.1126/science.126.3270.384 PMID:17774477

National Research Council (NRC). (2012). *A Framework for K-12 science education. Practices, Crosscutting Concepts, and core ideas*. Washington, DC: National Academy Press.

Newton, D., & Newton, D. (1998). Primary children's perceptions of science and the scientist: Is the impact of a national curriculum breaking down the stereotype? *International Journal of Science Education, 20*(9), 1137–1149. doi:10.1080/0950069980200909

Newton, D., & Newton, L. (1992). Young children's perceptions of science and the scientist. *International Journal of Science Education, 14*(3), 331–348. doi:10.1080/0950069920140309

Oware, E., Capobianco, B., & Diefes-Dux, H. (2007). *Gifted students' perceptions of engineers? A study of students in a summer outreach program.* Paper presented as the annual American Society for Engineering Education Conference & Exposition, Honolulu, HI.

Posner, G. J., Strike, K. A., Hewson, P. W., & Gerzog, W. A. (1982). Accomodation of a scientific conception: Towards a theory of conceptual change. *Science Education, 66*(2), 211–227. doi:10.1002/sce.3730660207

Schibeci, R. A., & Sorenson, I. (1983). Elementary school children's perceptions of scientists. *School Science and Mathematics, 83*(1), 14–19. doi:10.1111/j.1949-8594.1983.tb10087.x

She, H. (1998). Gender and grade level differences in Taiwan students' stereotypes of science and scientists. *Research in Science & Technological Education, 16*(2), 125–135. doi:10.1080/0263514980160203

Shope, R. E., III. (2006). *The Ed3U science model: Teaching science for conceptual change.* Retrieved at http://theaste.org/publications/proceedings/2006proceedings/shope.html

Song, J., Pak, S., & Jang, K. (1992). Attitudes of boys and girls in elementary and secondary schools towards science lessons and scientists. *Journal of the Koran Association for Research in Science Education, 12*, 109–118.

Thomas, J., Colston, N., Ley, T., Ivey, T., Utley, J., DeVore-Wedding, B., & Hawley, L. (2016). *Developing a rubric to assess drawings of engineers at work.* Paper presented at the Annual Conference and Exposition of the American Society for Engineering Educators, New Orleans, LA.

Vosniadou, S. (2007). Conceptual change and education. *Human Development, 50*(1), 47–54. doi:10.1159/000097684

Yap, C., Ebert, C., & Lyons, J. (2003). *Assessing students' perceptions of the engineering profession.* Paper presented as the South Carolina educators for the practical use of research annual conference, Columbia, SC.

KEY TERMS AND DEFINITIONS

Evaluation: The action of analysis.
Illustrations: Artwork that represents an idea.
Learning: Process of acquiring new knowledge.
Perception: The representation of what is perceived.
Rubric: Specific protocol for a specific purpose.
Science: Nature and study of the World.
Scientists: Humans who have endeavored over time and participate in the study of science.

Compilation of References

10 cool optical illusions. (n.d.). Retrieved September 17, 2015 from http://psychology.about.com/od/sensationandperception/tp/cool-optical-illusions.htm

88 constellations for Wittgenstein (to be played with the left hand). (n.d.). In *Electronic Literature Collection Volume Two online.* Retrieved from http://collection.eliterature.org/2/works/clark_wittgenstein.html

Abbott, A. (2008). Swiss "dignity" law is threat to plant biology. *Nature, 452*(7190), 919. doi:10.1038/452919a PMID:18441543

Abbott, E. A. (1983). *Flatland: A Romance in Many Dimensions. Harper Perennial.*

Abelson, H., Ledeen, K., & Lewis, H. (2008). *Blown to Bits.* Boston, MA: Pearson.

Ace of the Fungal Kingdom. (2006). *Terme di Diocleziano.* Retrieved December 10, 2015, from http://www.tiedyedfreaks.org/ace/diocleziano/DSCN5314crop.jpg

Adams, R. A., Snode, E. R., & Shaw, J. B. (2012). Flapping Tail Membrane in Bats Produces Potentially Important Thrust during Horizontal Takeoffs and Very Slow Flight. *PLoS ONE, 7*(2), e32074. doi:10.1371/journal.pone.0032074 PMID:22393378

Adler, I. (1972). *The new mathematics.* New York: John Day Company.

AENews. (2015). *Wave Power.* Retrieved August 13, 2015, from http://www.alternative-energy-news.info/technology/hydro/wave-power/

Aids, S. (n.d.). *ScreenScope - Mirror Stereoscope.* Retrieved August 3, 2015, from http://www.stereoaids.com.au

Albers, J. (1975). *Interaction of Color* (Revised Edition). New Haven, CT: Yale University Press.

Alighieri, D. (2005). De vulgari eloquentia. (S. Botterill, Trans.). Cambridge University Press.

al-Rifaie, M. M., Aber, A., & Bishop, M. J. (2012). Cooperation of nature and physiologically inspired mechanisms in visualisation. Blood Flow Shapes the Drawings of Birds and Ants. In A. Ursyn (Ed.), *Biologically-Inspired Computing for the Arts: Scientific Data through Graphics.* IGI Global. doi:10.4018/978-1-4666-0942-6.ch003

al-Rifaie, M. M., Bishop, M., & Blackwell, T. (2011). An investigation into the diffusion search and particle swarm optimisation. In *Proceedings of the Gene Computation Conference GECCO'11* (pp. 37-44). Dublin, Ireland: ACM.

American Heritage Dictionary. (1969). Boston, PA: Houghton Mifflin.

American Museum of Natural History. (2013). *Students Use 3D Printing to Reconstruct Dinosaurs.* Retrieved August 8, 2015, from https://www.youtube.com/watch?v=_KBxG1_WO8k

AMS. (n.d.). *Conant prize*. Retrieved September 20, 2015, from http://www.ams.org/profession/prizes--awards/ams--prizes/conant--prize

Anderson, R. A. (1992). Diversity of eukaryotic algae. *Biodiversity and Conservation*, *1*(4), 267–292. doi:10.1007/BF00693765

Andrew, T. J., & Purves, D. (1997). Similarities in normal and binocular rivalrous viewing. *Proceedings of the National Academy of Sciences of the United States of America*, *94*(18), 9905–9908. doi:10.1073/pnas.94.18.9905 PMID:9275224

Animation, P. (n.d.). Retrieved September 17, 2015 from http://www.psy.vanderbilt.edu/faculty/blake/rivalry/BR.html

Aprile, I., Ferrarin, M., Padua, L., Di Sipio, E., Simbolotti, C., Petroni, S., & Dickmann, A. et al. (2014, July). Walking strategies in subjects with congenital or early onset strabismus. *Frontiers in Human Neuroscience*, *8*, 484. doi:10.3389/fnhum.2014.00484 PMID:25071514

Armstrong, C., & de Zegher, C. (2004). *Ocean flowers: Impressions from nature*. Princeton, NJ: Princeton University Press.

Arnheim, R. (1969/2004). Visual Thinking. University of California Press.

Arnheim, R. (1988). The Power of the Center, A Study of Composition in the Visual Arts. Univ. of California Press.

Arnheim, R. (1990). *Thoughts on art education* (Occasional Paper Series Vol 2). Oxford University Press.

Arnheim, R. (1974/1983). *Art and visual perception: A psychology of the creative eye* (2nd ed.). University of California Press.

Arrigoni, M., & Gallaher, N. (2015, February-March). Diverse Applications Drive Lasers for Multiphoton Microscopy. *BioPhotonics*, 32-36.

Art of Life. (2012). Retrieved August 27, 2015, from http://biodivlib.wikispaces.com/Art+of+Life

Asscher, S., & Widger, D. (2008). *Plato - The Republic*. Project Gutenberg. Retrieved September 20, 2015, from http://www.gutenberg.org/files/1497/1497-h/1497-h.htm#2H_4_0004>

Atkins, A. (1850). *Photographs of British Algae: Cyanotype impressions*. Academic Press.

Ausubel, D. P. (1960). The use of advance organizers in the learning and retention of meaningful verbal material. *Journal of Educational Psychology*, *51*(5), 267–272. doi:10.1037/h0046669

Ayres, P. (2006). Impact of reducing intrinsic cognitive load on learning in a mathematical domain. *Applied Cognitive Psychology*, *20*(3), 287–298. doi:10.1002/acp.1245

Azevedo, R. (2015). Defining and measuring engagement in learning in science: Conceptual, theoretical, methological, and analytical issues. *Educational Psychologist*, *50*(1), 84–94. doi:10.1080/00461520.2015.1004069

Backus, J. G. (1969). The Acoustical Foundations of Music – Musical Sound: a Lucid Account of its Properties, Production, Behavior, and Reproduction. W. W. Norton.

Baddeley, A. (1992). Working Memory. *Science. New Series*, *255*(5044), 556–559.

Baddeley, A. (1998). *Human memory*. Boston: Allyn & Bacon.

Baddeley, A. (2000). The episodic buffer: A new component of working memory? *Trends in Cognitive Sciences*, *4*(11), 417–423. doi:10.1016/S1364-6613(00)01538-2 PMID:11058819

Baddeley, A. (2003). Working memory: Looking back and looking forward. *Nature Reviews. Neuroscience*, *4*(10), 829–839. doi:10.1038/nrn1201 PMID:14523382

Baddeley, A. D. (1986). *Working memory*. Oxford, UK: Oxford University Press.

Baddeley, J., & Hitch, G. (1994). Working memory. In G. H. Bower (Ed.), *The psychology of learning and motivation: Advances in research and theory* (Vol. 8, pp. 47–89). New York: Academic Press.

Bahmani, B., Guerrero, Y., Bacon, D., Kundr, V., Vullev, V., & Anvari, B. (2014). Functionalized polymeric nanoparticles loaded with indocyanine green as theranostic materials for targeted molecular near infrared fluorescence imaging and photothermal destruction of ovarian cancer cells. *Lasers in Surgery and Medicine*, *46*(7), 582–592. doi:10.1002/lsm.22269 PMID:24961210

Banchoff, T. (1990). *Beyond the Third Dimension*. New York: Scientific American Library.

Bandura, A. (1993). Perceived self-efficacy in cognitive development and functioning. *Educational Psychologist*, *28*(2), 117–148. doi:10.1207/s15326985ep2802_3

Banff Centre New Media Institute. (1999). BNMI Co-Production Archives 'G'. *The Split-Brain Human Computer User Interface*. Retrieved July 10, 2015, from http://www.banffcentre.ca/bnmi/coproduction/archives/s.asp#thesplit

Banich, M. T. (2003). The divided visual field technique in laterality and interhemispheric integration. In K. Hughdahl (Ed.), *Experimental Methods in Neuropsychology* (pp. 47–63). New York: Kluwer. doi:10.1007/978-1-4615-1163-2_3

Banich, M. T., & Shenker, J. I. (1994). Investigations of interhemispheric processing: Methodological considerations. *Neuropsychology*, *8*(2), 263–277. doi:10.1037/0894-4105.8.2.263

Barman, C. (1997). Students' views of scientists and science: Results from a national study. *Science and Children*, *35*(1), 18–23.

Barnes, J. (Ed.). (1984). *Aristotle, De Interpretatione. In The Complete Works of Aristotle*. Princeton University Press.

Bay Observatory Gallery. (2015). *Observing Landscapes*. Retrieved August 27, 2015, from http://www.exploratorium.edu/visit/bay-observatory-gallery

Bazin, A. (2004). *What is cinema?* (2nd ed.; Vol. 1). Oakland, CA: University of California Press.

Beatty, J., & Lucero-Wagner, B. (2000). The pupillary system. In J. T., Cacioppo, L. G. Tassinary, & G. G. Berntson (Eds.), Handbook of psychophysiology (2nd ed.; pp. 142-162). Cambridge, UK: Cambridge University Press.

Beatty, J. (1982). Task-evoked pupillary responses, processing load, and the structure of processing resources. *Psychological Bulletin*, *91*(2), 276–292. doi:10.1037/0033-2909.91.2.276 PMID:7071262

Bederson, B., & Shneiderman, B. (2003). *The Craft of Information Visualization: Readings and Reflections*. San Francisco, CA: Morgan Kaufmann Publishers.

Behrens, R. (1997). Eyed awry: The ingenuity of Del Ames. *The North American Review*, *282*(2), 26–33.

Berg, A. M., & Biewener, A. A. (2010). *Wing and body kinematics of takeoff and landing flight in the pigeon (Columba livia)*. doi: .10.1242/jeb.038109

Bergey, B. W., Cromley, J. G., Kirchgessner, M. L., & Newcombe, N. S. (2015). Using diagrams versus text for spaced restudy: Effects on learning in 10th grade biology classes. *The British Journal of Educational Psychology*, *85*(1), 59–74. doi:10.1111/bjep.12062 PMID:25529502

Berkowitz, L. (1989). Frustration-aggression hypothesis: Examination and reformulation. *Psychological Bulletin, 106*(1), 59–73. doi:10.1037/0033-2909.106.1.59 PMID:2667009

Bertschi, S., Bresciani, S., Crawford, T., Goebel, R., Kienreich, W., & Lindner, M. et al.. (2011). What is Knowledge Visualization? Perspectives on an Emerging Discipline.*Proceedings of the Information Visualisation 15th International Conference* (pp. 329-336), London. doi:10.1109/IV.2011.58

Binocular rivalry bibliography. (n.d.). Retrieved October 2, 2015, from https://sites.google.com/site/oshearobertp/publications/binocular-rivalry-bibliography

Bio. (2015). *Edmund Hillary Biography*. Retrieved August 2, 2015, from http://www.biography.com/people/edmund-hillary-9339111#death-and-legacy

Biodiversity Heritage Library. (2015). Retrieved August 27, 2015, from http://biodiversitylibrary.org/

Bioinspired and Bioderived Materials. (2015). *Materials Research to Meet 21st Century Defense Needs*. The National Academies Press. Retrieved August 12, 2015, from http://www.nap.edu/openbook.php?record_id=10631&page=181

Bird Life International. (2012). *Mellisuga helenae*. The IUCN Red List of Threatened Species. Version 2014.3. Retrieved March 1, 2015, from http://www.iucnredlist.org/details/22688214/0

Birkhead. (1978). Behavioural adaptations to high density nesting in the common guillemot *Uria aalge*. *Animal Behaviour, 26*(2), 321-324.

Biro, M. (2013). *Anselm Kiefer*. New York, NY: Phaidon.

Blake, R. (1989). A neural theory of binocular rivalry. *Psychological Review, 96*(1), 145–167. doi:10.1037/0033-295X.96.1.145 PMID:2648445

Blake, R. (2001). A Primer on Binocular Rivalry, Including Current Controversies. *Brain and Mind, 2*(1), 5–38. doi:10.1023/A:1017925416289

Blake, R. (2005). Landmarks in the History of Binocular Vision. In D. Alais & R. Blake (Eds.), *Binocular Rivalry* (pp. 1–27). Cambridge, MA: MIT Press.

Blake, R., Brascamp, J., & Heeger, D. J. (2014). Can binocular rivalry reveal neural correlates of consciousness? *Philosophical Transactions of the Royal Society of London. Series B, Biological Sciences, 369*(1641). doi:10.1098/rstb.2013.0211 PMID:24639582

Blake, R., & Logothetis, N. K. (2002, January). Visual Competition. *Nature Reviews. Neuroscience, 1*(1), 13–21. doi:10.1038/nrn701 PMID:11823801

Blake, R., Westendorf, D., & Overton, R. (1979). What is suppressed during binocular rivalry? *Perception, 9*(2), 223–231. doi:10.1068/p090223 PMID:7375329

Blank, S., & Dorf, B. (2012). *The Startup Owner's Manual: The Step-By-Step Guide for Building a Great Company*. K & S Ranch.

Bleicher, R. E., & Lindgren, J. (2005). Success in science learning and preservice science teaching self-efficacy. *Journal of Science Teacher Education, 16*(3), 205–225. doi:10.1007/s10972-005-4861-1

Blind Spot. (n.d.). Retrieved September 17, 2015 from http://www.exploratorium.edu/snacks/blind_spot/index.html

Blunt, W. (1971). *The compleat naturalist: A life of Linnaeus*. New York, NY: Viking.

Boden, M. (2016). Skills and the Appreciation of Computer Art. In Computational Creativity, Measurement and Evaluation, Connection Science. Taylor & Francis. doi:10.1080/09540091.2015.1130023

Bogen, J. E. (1975, Spring). Some Educational Aspects of Hemispheric Specialization. *UCLA Educator*, *17*(2), 24–32.

Bogen, J. E., & Vogel, P. J. (1962). Cerebral commissurotomy in man. *Bulletin of the Los Angeles Neurological Society*, *27*, 169–172.

Bogue, R. (2003). *Deleuze on cinema*. New York, NY: Routledge.

Borel, B. (2014). Map Maritime treasures at the Bottom of the Sea. *Popular Science*. Retrieved August 14, 2015, from http://www.popsci.com/article/science/map-maritime-treasures-bottom-sea

Bork, A. M. (1964). *The Fourth Dimension in Nineteenth-Century Physics*. The University of Chicago Press on behalf of The History of Science Society. Retrieved September 20, 2015, from http://www.jstor.org/stable/228574

Bork, A., & Gunnarsdottir, S. (Eds.). (2012). Tutorial Distance Learning: Rebuilding Our Educational System. Springer Science + Business Media, LLC.

Borri, C., Betti, M., & Bartoli, G. (2010). Brunelleschi's dome in Florence: The masterpiece of a genius. London: Taylor & Francis Group.

Boster, F. J., Meyer, G. S., Roberto, A. J., Inge, C., & Strom, R. (2006). Some effects of video streaming on educational achievement. *Communication Education*, *55*(1), 46–62. doi:10.1080/03634520500343392

Boster, F. J., Meyer, G. S., Roberto, A. J., Lindsey, L., Smith, R., Inge, C., & Strom, R. (2007). The impact of video streaming on mathematics performance. *Communication Information*, *56*, 134–144.

Bourke, P. (1990). *Hyperspace User Manual*. Retrieved September 20, 2015, from http://paulbourke.net/geometry/hyperspace/

Bourne, V. J. (2006). The divided visual field paradigm: Methodological considerations. *Laterality*, *11*(4), 373–393. doi:10.1080/13576500600633982 PMID:16754238

Braun, M. (2010). *Eadweard Muybridge*. London: Reaktion Books.

Brawand, D., Wahli, W., & Kaessmann, H. (2008). Loss of Egg Yolk Genes in Mammals and the Origin of Lactation and Placentation. *PLoS Biology*, *6*(3), e63. doi:10.1371/journal.pbio.0060063 PMID:18351802

Bredl, K., Groß, A., Hünniger, J., & Fleischer, J. (2012). The Avatar as a Knowledge Worker? How Immersive 3D Virtual Environments may Foster Knowledge Acquisition. *Electronic Journal of Knowledge Management*, *10*(1), 15–25.

Breese, B. B. (1899). On inhibition. *Psychological Monographs*, *3*(1), 1–65. doi:10.1037/h0092990

Bregman, A. S. (1990). *Auditory scene analysis*. Cambridge, MA: MIT Press.

Brewster, D. (1856). *The Stereoscope; its History, Theory, and Construction, with its Application to the fine and useful Arts and to Education: With fifty wood Engravings*. London: John Murray.

Brickhouse, N. W., & Potter, J. T. (2001). Young women's science identity formation in an urban context. *Journal of Research in Science Teaching*, *38*(8), 965–980. doi:10.1002/tea.1041

Broadbent, D. E. (1954). The role of auditory localization in attention and memory span. *Journal of Experimental Psychology*, *44*, 51–55. doi:10.1037/h0056491 PMID:13152294

Broichhagen, B., Schönberger, M., Cork, S. C., Frank, J. A., Marchetti, P., Bugliani, M., & Trauner, D. et al. (2014). Optical control of insulin release using a photoswitchable sulfonylurea. *Nature Communications*, *5*, 5116. doi:10.1038/ncomms6116 PMID:25311795

Brooks, R. (1980, January). Hemispheric Differences in Memory: Implications for Education. *The Clearing House: A Journal of Educational Strategies, Issues and Ideas*, *53*(5), 248–250. doi:10.1080/00098655.1980.9959221

Broudy, H. S. (1987). *The Role of Imagery in Learning.* Occasional Paper 1, The Getty Center for Education in the Arts.

Brown, A. H., Losoff, B., & Hollis, D. R. (2014). Science instruction through the visual arts in special collections. *Libraries and the Academy*, *14*(2), 197–216. doi:10.1353/pla.2014.0002

Browniee, J. (2012). Clever Algorithms: Nature-Inspired Programming Recipes. lulu.com.

Brünken, R., Plass, J. L., & Leutner, D. (2003). Direct measurement of cognitive load in multimedia learning. *Educational Psychologist*, *38*(1), 53–61. doi:10.1207/S15326985EP3801_7

Buchberg, K. (Ed.). (2014). *Henri Matisse: The cut-outs*. New York, NY: MOMA.

Burkhard, R., Meier, M., Smis, J. M., Allemang, J., & Honish, J. (2005). Beyond Excel and PowerPoint:Knowledge Maps for the Transfer and Creation of Knowledge in Organizations. In *International Conference on Information Visualisation* (pp. 403-408). London, UK: IEEE Computer Society Press.

Burn, J. H., Leach, E. H., Rand, M. J., & Thompson, J. W. (1959). Peripheral Effects of Nicotine and Acetylcholine Resembling those of Sympathetic Stimulation. *The Journal of Physiology*, *148*(2), 332–352. doi:10.1113/jphysiol.1959.sp006291 PMID:13806185

Bush, V. (1945). As we may think. In From Memex to Hypertext. San Diego, CA: Academic Press.

Byrne, J. H. (Ed.). (1997–Present). *Chapter 14. Visual Processing: Eye and Retina: Figure 14.7.* Neuroscience Online. Department of Neurobiology and Anatomy. The University of Texas Medical School at Houston. Retrieved November 21, 2015, from http://neuroscience.uth.tmc.edu/s2/chapter14.html

Callahan, P. (1997). *The Enigma of the Towers.* Retrieved September 28, 2015, from http://whale.to/b/callahan.html

Callahan, P. (1992). *Nature's silent music: A rucksack naturalist's Ireland.* Kansas City, MO: Acres U.S.A.

Campbell, F. W., & Howell, E. R. (1972). Monocular alternation; a method for the investigation of pattern vision. *The Journal of Physiology*, *225*, 19–21. PMID:5074381

Capobianco, B. M., Diefes-dux, H. A., Mena, I., & Weller, J. (2011). What is an engineer? Implications of elementary school student conceptions for engineering education. *The Journal of Engineering Education*, *100*(2), 304–328. doi:10.1002/j.2168-9830.2011.tb00015.x

Carpenter, K. (2000). *Eggs, Nests, and Baby Dinosaurs: A Look at Dinosaur Reproduction.* Indiana University Press.

Carroll, S. M. (2012). From Eternity to Here: The Quest for the Ultimate Theory of Time. ONEWorld Publ.

Carroll, L. (1865). *Alice's Adventures in Wonderland.* London: MacMillan and Company.

Casey, J. (2007). *The First Six Books of the Elements of Euclid.* Project Gutenberg. Retrieved September 20, 2015, from http://www.gutenberg.org/ebooks/21076

Cavanna, A. E., & Nanni, A. (2014). *Consciousness: Theories in Neuroscience and Philosophy of Mind.* Springer. doi:10.1007/978-3-662-44088-9

Ceaparu, I., Lazar, J., Bessiere, K., Robinson, J., & Shneiderman, B. (2004). Determining Causes and Severity of End-User Frustration. *International Journal of Human-Computer Interaction*, *17*(3), 333–356. doi:10.1207/s15327590ijhc1703_3

Celebi, K., Buchheim, J., Wyss, R. M., Droudian, A., Gasser, P., Shorubalko, I., & Park, H. G. et al. (2014). Ultimate permeation across atomically thin porous graphene. *Science*, *344*(6181), 289–344. doi:10.1126/science.1249097 PMID:24744372

Chambers, D. (1983). Stereotypic images of the scientist: The draw-a-scientist test. *Science Education*, *67*(2), 255–265. doi:10.1002/sce.3730670213

Chandler, P., & Sweller, J. (1991). Cognitive load theory and the format of instruction. *Cognition and Instruction*, *8*(4), 293–332. doi:10.1207/s1532690xci0804_2

Chang. (2014, October 14). Nobel Laureates Pushed Limits of Microscopes. *The New York Times*. Retrieved February 8, 2015, from http://www.nytimes.com/2014/10/09/science/nobel-prize-chemistry.html?_r=0

Chen, O., Riedemann, L., Etoc, F., Herrmann, H., Coppey, M., Barch, M., ... Reimer, R. (2014). Magneto-fluorescent core-shell supernanoparticles. *Nature Communications*. doi: .10.1038/ncomms609

Chen, B., Vansteenkiste, M., Beyers, W., Boone, L., Deci, E. L., Duriez, B., & Verstuyf, J. et al. (2015). Basic psychological need satisfaction, need frustration, and need strength across four cultures. *Motivation and Emotion*, *39*(2), 216–236. doi:10.1007/s11031-014-9450-1

Chen, C. (2010). *Information Visualization: Beyond the Horizon* (2nd ed.). Springer.

Chen, C. H. (2011). *Emerging topics in computer vision and its applications*. World Scientific Publishing Company.

Chen, Y.-C., & Yang, F.-Y. (2014). Probing the relationship between process of spatial problems solving and science learning: An eye tracking approach. *International Journal of Science and Mathematics Education*, *12*(3), 579–603. doi:10.1007/s10763-013-9504-y

Chen, Z., & Gladding, G. (2014). How to Make a Good Animation: A grounded cognition model of how visual representation design affects the construction of abstract physics knowledge. *Physical Review Special Topics - Physics. Education Research*, *10*(1), 010111-1–010111-24.

Cherry, C. (1953). Some experiments on the recognition of speech, with one and with two ears. *The Journal of the Acoustical Society of America*, *25*(5), 975–979. doi:10.1121/1.1907229

Cheshire Cat Experiment. (n.d.). Retrieved September 17, 2015 from http://www.exploratorium.edu/snacks/cheshire_cat/

Chiang, C., & Guo, C. (1996). *A study of the images of the scientist for elementary school children*. Paper presented at the National Association for Research in Science Teaching, St. Louis, MO.

Chittka, L., & Brockmann, A. (2005). Perception Space–The Final Frontier. *PLoS Biology*, *3*(4), e137. doi:10.1371/journal.pbio.0030137 PMID:15819608

Christie, A. (2011). A taste for seaweed: William Kilburn's late eighteenth-century designs for printed cottons. *Journal of Design History*, *24*(4), 299–314. doi:10.1093/jdh/epr037

Clark, J. M., & Paivio, A. (1991). Dual coding theory and education. *Educational Psychology Review*, *3*(3), 149–210. doi:10.1007/BF01320076

Colbert, E. H., & Morales, M. (2001). *Colbert's Evolution of the Vertebrates: A History of the Backboned Animals Through Time* (5th ed.). New York: Wiley-Liss.

Colvin, M. K., Funnell, M. G., Hahn, B., & Gazzaniga, M. S. (2005). Identifying functional channels in the corpus callosum: Correlating interhemispheric transfer time with white matter organization. *Journal of Cognitive Neuroscience*, *139*(suppl. 5), 2409–2419.

Compton, R. J. (2002). Interhemispheric interaction facilitates face processing. *Neuropsychologia*, *40*(13), 2409–2419. doi:10.1016/S0028-3932(02)00078-7 PMID:12417469

Connolly, M. (2009). *The place of artists' cinema: Space, site and screen*. Bristol, UK: Intellect.

Constant, J. (2014). *Wasan geometry*. Retrieved from http://hermay.org/jconstant

Contrast Influences Predominance. (n.d.). Retrieved September 17, 2015 from http://www.psy.vanderbilt.edu/faculty/blake/rivalry/BR.html

Conway, M. A., & Loveday, C. (2015). Remembering, imagining, false memories & personal meanings. *Consciousness and Cognition*, *33*, 574–581. doi:10.1016/j.concog.2014.12.002 PMID:25592676

Cook, A., Zheng, R., & Blaz, J. W. (2009). Measurement of cognitive load during multimedia learning activities. In R. Zheng (Ed.), *Cognitive effectives of multimedia learning* (pp. 34–50). Hershey, PA: Information Science Reference/IGI Global Publishing. doi:10.4018/978-1-60566-158-2.ch003

Cook, M., Wiebe, E., & Carter, G. (2011). Comparing visual representations of DNA in two multimedia presentations. *Journal of Educational Multimedia and Hypermedia*, *20*(1), 21–42.

Coon, J. M., & Rothman, S. (1940). Nature of pilomotor response to acetylcholine; some observations on pharmacodynamics of skin. *Journal de Pharmacologie*, *68*, 301–311.

Cooper, M. (2015). *MaxCooper*. Retrieved September 20, 2015, from http://maxcooper.net

Cooper, L. (1945). *Louis Agassiz as a teacher; illustrative extracts on his method of instruction*. Ithaca, NY: Comstock.

Corballis, M. C. (1994). Split decisions: Problems in the interpretation of results from commissurotomized subjects. *Behavioural Brain Research*, *64*(1-2), 163–172. doi:10.1016/0166-4328(94)90128-7 PMID:7840883

Corballis, M. C., & Sergent, J. (1988). Imagery in a commissurotomized patient. *Neuropsychologia*, *26*(1), 13–26. doi:10.1016/0028-3932(88)90027-9 PMID:3362338

Corballis, P. M., Funnell, M. G., & Gazzaniga, M. S. (1999). A dissociation between spatial and identity matching in callosotomy patients. *Neuroreport*, *10*(10), 2183–2187. doi:10.1097/00001756-199907130-00033 PMID:10424695

Craft, B., & Cairns, P. (2005). Beyond guidelines: What can we learn from the visual informationseeking mantra? In *Proceedings of the 9th International Conference on Information Visualization* (pp. 110-118). IEEE. doi:10.1109/IV.2005.28

Craft, B., & Cairns, P. (2008). Directions for methodological research in information visualization. In *Proceedings of the 12th International Conference on Information Visualisation*. IEEE. doi:10.1109/IV.2008.88

Craig, A. D. (2003). Interoception: The sense of the physiological condition of the body. *Current Opinion in Neurobiology*, *13*(4), 500–505. doi:10.1016/S0959-4388(03)00090-4 PMID:12965300

Crick, F. C. & Koch. (2003 February). A Framework for Consciousness. *Nature Neuroscience*, *6*(2), 119-126. Retrieved November 2, 2015, from http://www.klab.caltech.edu/koch/crick-koch-03.pdf

Crick, F. C., & Koch, C. (1992, September). The Problem of Consciousness. *Scientific American*, *267*(3), 152–159. doi:10.1038/scientificamerican0992-152 PMID:1502517

C-SPAN. (1991). *Thomas Confirmation Hearings*. Retrieved September 6, 2015, from http://www.c-span.org/search/?searchtype=Videos&sort=Newest&seriesid[]=24

Cunningham, C., Lachapelle, C., & Lindgren-Streicher, A. (2005). *Assessing elementary school students' conceptions of engineering and technology*. Paper presented at the annual American Society for Engineering Education Conference & Exposition, Portland, OR.

D'Alembert, J. L. R. (1751). *Dictionnaire raisonné des sciences, des arts et métiers*. Briasson. Retrieved September 20, 2015, from http://www.lexilogos.com/encyclopedie_diderot_alembert.htm

Da Vinci, L. (2014). The Notebooks of Leonardo Da Vinci. CreateSpace Independent Publishing Platform.

Damasio, A. (2012). *Self Comes to Mind: Constructing the Conscious Brain*. Vintage.

Damasio, A. R. (1999). *The feeling of what happens: Body and emotion in the making of consciousness*. New York: Harcourt Brace.

Damasio, A., & Carvalho, G. B. (2013). The nature of feelings: Evolutionary and neurobiological origins. *Nature Reviews. Neuroscience*, *14*(2), 143–152. doi:10.1038/nrn3403 PMID:23329161

Daoudi, K., van den Berg, P. J., Rabot, O., Kohl, A., Tisserand, S., Brands, P., & Steenbergen, W. (2014). Handheld probe integrating laser diode and ultrasound transducer array for ultrasound/photoacoustic dual modality imaging. *Optics Express*, *22*(21), 26365–26374. doi:10.1364/OE.22.026365 PMID:25401669

Darwin, C. (2004). The Descent of Man. Penguin Classics.

De Grand, A., & Bonfig, S. (2015, January). Selecting a Microscope Based on Imaging Depth. *BioPhotonics*, 26-29.

De Mena, J. M. (1992). Curiosidades y leyendas de Barcelona. Plaza & Janes Editores, S.A.

de Villers-Sidani, E., Chang, E. F., Bao, S., & Merzenich, M. M. (2007). Critical period window for spectral tuning defined in the primary auditory cortex (A1) in the rat. *The Journal of Neuroscience*, *27*(1), 1809. doi:10.1523/JNEUROSCI.3227-06.2007 PMID:17202485

Deci, E. L., & Ryan, R. M. (1985). *Intrinsic motivation and self-determination in human behaviour*. New York: Plenum Press. doi:10.1007/978-1-4899-2271-7

Deci, E. L., & Ryan, R. M. (2000). The "what" and "why" of goal pursuits: Human needs and the self-determination of behaviour. *Psychological Inquiry*, *11*(4), 227–268. doi:10.1207/S15327965PLI1104_01

Deleuze, G. (2013). *Cinema 1 the movement-image*. New York, NY: Bloomsbury Academic.

Dennerlein, J. T., Becker, T., Johnson, P., Reynolds, C., & Picard, R. (2003). Frustrating computers users increases exposure to physical factors.*Proceedings of the 15th Congress of theInternational Ergonomics Association*.

Descartes, R. (1644). *Traité de l'homme*. Paris: Angot.

Deutsch, D. (2013). *The Psychology of Music* (3rd ed.). Elsevier Inc.

DeVry University. (2014). *Web Game Programming Degree Specialization*. Retrieved April 3, 2014 from http://www.devry.edu/degree-programs/college-media-arts-technology/web-game-programming-about.html

Dhanapal, S., Kanapathy, R., & Mastan, J. (2014). A study to understand the role of visual arts in the teaching and learning of science. *Asia-Pacific Forum on Science Learning and Teaching, 15*(2), Article 12.

Dickinson, E. (2006). *Emily Dickinson's herbarium*. Cambridge, MA: Harvard University Press.

DigiVol. (2015). Retrieved August 27, 2015, from http://volunteer.ala.org.au/

DiLeo, J. H. (1970). *Young children and their drawings*. New York, NY: Brunner/Mazel Publishers.

Dimond, S., & Beaumont, G. (1972). Processing in perceptual integration between the cerebral hemispheres. *British Journal of Psychology*, *63*(4), 509–514. doi:10.1111/j.2044-8295.1972.tb01300.x PMID:4661080

Dion, M. (2010). *Herbarium Perrine (Marine Algae)*. Retrieved December 16, 2010, from http://www.youtube.com/watch?v=F4voeIl-bXY

DoITPoMS. (2015). *Raman Spectroscopy*. University of Cambridge. Retrieved August 10, 2015, from http://www.doitpoms.ac.uk/tlplib/raman/index.php

Dollard, J., Doob, L., Miller, N., Mowrer, O., & Sears, R. (1939). *Frustration and Aggression*. New Haven, CT: Yale University Press. doi:10.1037/10022-000

Dresselhaus, M. S., Dresselhaus, G., & Eklund, P. C. (1996). *Science of fullerenes and carbon nanotubes*. San Diego, CA: Academic Press.

Drucker, J. (1995). *The century of artists' books*. New York, NY: Granary Books.

Ducker, S. C. (1988). *The contented botanist: Letters of W.H. Harvey about Australia and the Pacific*. Melbourne, Australia: Melbourne University Press.

Duke, G. (2015). *Smorball and Beanstalk: Games that aren't just fun to play but help science too*. Retrieved from http://blog.biodiversitylibrary.org/2015/08/smorball-and-beanstalk-games-that-arent.html

Dulvy, N. K., & Reynolds, J. D. (1997). Evolutionary transitions among egg-laying, live-bearing and maternal inputs in sharks and rays. *Proceedings. Biological Sciences*, *264*(1386), 1309–1315. doi:10.1098/rspb.1997.0181

Dunn, J. (2011). The Evolution of Classroom Technology. *Edudemic: Connecting Education and Technology*. Retrieved from http://www.edudemic.com/classroom-technology/

Durant, M. A. (2002). Some assembly required: Ten fragments toward a picture of collage. In T. Piché (Ed.), *Some assembly required: Collage culture in post-war America* (pp. 19–28). Syracuse, NY: Everson Museum of Art.

Dutour, É. F. (1760). Discussion d'une question d'optique. Mémoires de Mathématique et de Physique Présentés par Divers Savants. *L'Académie des Sciences, 3*, 514-530.

Eco, U. (1989). *The Open Work* (A. Cancogni, Trans.). Cambridge, MA: Harvard University Press.

Eco, U. (1990). *The Limits of Interpretation*. Bloomington, IN: Indiana University Press.

Edgerton, S. Y. (2008). The Renaissance Rediscovery of Linear Perspective. *ACLS Humanities*.

Edwards, B. (2012). *Drawing on the Right Side of the Brain: The Definitive* (4th ed.). New York: Tarcher/Penguin.

Ee, C. C. (2013). *Removal of Dye by Adsorption of Eggshell Powder*. Malaysia: Technical University.

Eggebrecht, A. T., White, B. R., Chen, C., Zhan, Y., Snyder, A. Z., Dehlgani, H., & Culver, J. P. (2012). A quantitative spatial comparison of high-density diffuse optical tomography and fMRI cortical mapping. *NeuroImage*, 2012. PMID:22330315

Einstein, A., Klein, M. J., Kox, A. J., Renn, J., & Schulman, R. (Eds.). (1997). The collected papers of Albert Einstein. Princeton University Press.

Eisen, E. J., Bohren, B. B., & McKean, H. E. (1962). The Haugh Unit as a Measure of Egg Albumen Quality. *Oxford Journals Science & Mathematics Poultry Science*, *41*(5), 1461–1468. doi:10.3382/ps.0411461

Eisenstein, S. (1949). *Film form; essays in film theory* (J. Leyda, Trans.). New York, NY: Harcourt, Brace.

Eisenstein, S. (2004). The dramaturgy of film form. In L. Braudy & M. Cohen (Eds.), *Film theory and criticism: Introductory readings* (6th ed.; pp. 23–40). New York, NY: Oxford University Press.

Elkins, J. (2003). *Visual studies: A skeptical introduction*. New York: Routledge.

Elkins, J. (2005). Letter to John Berger. In J. Savage (Ed.), *John Berger: Drawings of John Berger* (pp. 111–114). Cork, Ireland: Occasional Press.

Ellis, D. (2012). *Eggs 101: The Anatomy of an Egg*. Retrieved February 23, 2015, from http://davidstable.com/2012/12/eggs-101-the-basics/

Encyclopaedia Britannica. (1994). Depth Perception: Correspondence of Points. *Encyclopaedia Britannica*. Retrieved December 20, 2015, from http://www.britannica.com/topic/depth-perception

Encyclopedia Britannica. (2015). *Combination tone*. Retrieved October 27, 2015, from http://www.britannica.com/science/combination-tone

Encyclopedia of Life. (2015). Retrieved August 27, 2015, from http://eol.org/

Encyclopedia, K. (2015). *Waves*. Retrieved August 13, 2015, from http://www.waterencyclopedia.com/Tw-Z/Waves.html

Engel, A. A. K., Fries, P., Konig, P., Brecht, M., & Singer, W. (1999). Temporal binding, binocular rivalry and consciousness. *Consciousness and Cognition*, *8*(2), 128–151. doi:10.1006/ccog.1999.0389 PMID:10447995

Entrepreneur. (2015). *Plan your business plan*. Retrieved August 13, 2015, from http://www.entrepreneur.com/article/38292#ixzz2OreoXTKd

Ernst, C. H., & Lovich, J. E. (2009). *Turtles of the United States and Canada* (2nd ed.). Johns Hopkins University Press.

Exchange, E. (n.d.). Retrieved September 17, 2015 from http://www.psy.vanderbilt.edu/faculty/blake/rivalry/BR.html

Experiments. (n.d.). Retrieved September 17, 2015 from http://www.richardgregory.org/experiments/

Faber, T. (2008). *Faberge's Eggs: The Extraordinary Story of the Masterpieces That Outlived an Empire. Random House*. Retrieved from http://en.wikipedia.org/wiki/Faberge_egg

Farland, D. (2006). The effect of historical, nonfiction trade books on elementary students' perceptions of scientists. *Journal of Elementary Science Education*, *18*(2), 31–48. doi:10.1007/BF03174686

Farland-Smith, D., & McComas, W. F. (2007). *The enhanced (DAST): A more reliable, valid, efficient reliable & complete method of identifying students' perceptions of scientists*. Paper presented at the meeting of the National Association of Research in Teaching Conference, New Orleans, LA.

Farlex. (2012). *Otoacoustic emission*. Farlex Partner Medical Dictionary.

Farrugia, N., Jakubowski, K., Cusack, R., & Stewart, L. (2015). Tunes stuck in your brain: The frequency and affective evaluation of involuntary musical imagery correlate with cortical structure. *Consciousness and Cognition*, *35*, 66–77. doi:10.1016/j.concog.2015.04.020 PMID:25978461

Fathauer, R. (2016). Artist statement in Fathauer, R., & Selikoff, N. (Eds.), 2016 Joint Mathematics Meetings Exhibition of Mathematical Art. Tesselation Publishing.

Fergonzi, F. (2007). Episodes in the critical debate on collage, from Apollinaire to Greenberg. In M. M. Laberti & M. G. Messina (Eds.), *Collage/Collages from cubism to new dada* (pp. 322–335). Milan, Italy: Mondadori Electa.

Field Book Project. (2015). Retrieved August 27, 2015, from http://www.mnh.si.edu/rc/fieldbooks/

Fineman, M. (2012). *Faking it: Manipulated photography before Photoshop*. New York, NY: Metropolitan Museum of Art.

Finson, K. D. (2002). Drawing a scientist: What we do and do not know after fifty years of drawings. *School Science and Mathematics*, *102*(7), 335–345. doi:10.1111/j.1949-8594.2002.tb18217.x

Finson, K. D., Beaver, J. B., & Cramond, B. L. (1995). Development and field test of a checklist for the draw-a-scientist test. *School Science and Mathematics*, *95*(4), 195–205. doi:10.1111/j.1949-8594.1995.tb15762.x

Finson, K., & Farland-Smith, D. (2013). Applying Vosniadou's conceptual change model to visualizations on conceptions of scientists. In K. Finson & J. Pederson (Eds.), *Visual Data and Their Use in Science Education* (pp. 47–76). Information Age Publishing Inc.

Fischbein, E. (2002). *Intuition in Science and Mathematics*. New York: Kluwer.

Forster, B. A., Corballis, P. M., & Corballis, M. C. (2000). Effect of luminance on successiveness discrimination in the absence of the corpus callosum. *Neuropsychologia*, *38*(4), 441–450. doi:10.1016/S0028-3932(99)00087-1 PMID:10683394

Fort, D. C., & Varney, H. L. (1989). How students see scientists: Mostly male, mostly white, and mostly benevolent. *Science and Children*, *26*, 8–13.

Framing Game. (n.d.). Retrieved December 26, 2015 from http://www.vision3d.com/frame.html

Franz, E. A., Waldie, K. E., & Smith, M. J. (2000). The effect of callosotomy on novel versus familiar manual actions: A neural dissociation between controlled and automatic processes. *Psychological Science*, *11*(1), 82–85. doi:10.1111/1467-9280.00220 PMID:11228850

Freeman, C. L., Harding, J. H., Quigley, D., & Rodger, M. (2010). Structural Control of Crystal Nuclei by an Eggshell Protein. *Angewandte Chemie International Edition*, *49*(30), 5135–5137. doi:10.1002/anie.201000679 PMID:20540126

Freyer, C., & Renfree, M. B. (2009, September15). The mammalian yolk sac in placenta. *Journal of Experimental Zoology. Part B, Molecular and Developmental Evolution*, *312*(6), 545–554. doi:10.1002/jez.b.21239 PMID:18985616

Frøkjær, E., Hertzum, M., & Hornbæk, K. (2000). Measuring Usability: Are Effectiveness, Efficiency, and Satisfaction Really Correlated? In *CHI 2000 Conference Proceedings*, (pp. 345-352). ACM Press.

Fukagawa, H., & Rothman, T. (2008). *Sacred Mathematics: Japanese Temple Geometry*. Princeton, NJ: Princeton University Press.

Fung, Y. (2002). A comparative study of primary and secondary school students' images of scientists. *Research in Science & Technological Education*, *20*(2), 199–213. doi:10.1080/0263514022000030453

Funk, V. (2014). The erosion of collections-based science: Alarming trend or coincidence? Plant Press, 17(4), 13-14.

Funnell, M. G., Corballis, P. M., & Gazzaniga, M. S. (1999). A deficit in perceptual matching in the left hemisphere of a callosotomy patient. *Neuropsychologia*, *38*, 441–450. PMID:10509836

Gap. In (1969). *The American Heritage Dictionary of the English Language*. New York, NY: American Heritage Publishing.

Gardner, H. (1983/2011). *Frames of mind: The theory of multiple intelligences* (3rd ed.). Basic Books.

Gardner, H. (1993). *Art, Mind, and Brain: A Cognitive Approach to Creativity*. New York: Basic Books, A Division of Harper Collins Publishers.

Gardner, H. (1993/2006). *Multiple intelligences: New horizons in theory and practice*. Basic Books.

Gardner, H. (1997). *Extraordinary minds: Portraits of exceptional individuals and an examination of our extraordinariness*. Basic Books, Harper Collins Publishers.

Garoian, C., & Gaudelius, Y. (2008). *Spectacle pedagogy, art, politics and visual culture*. Albany, NY: State University of New York Press.

Garvey, G. P. (2011). The Wheatstone Stereoscopic Random Line Pair Generator with Fitness Function for Non-objective Art. *Proceedings of the 8th Conference on Creativity & Cognition*. Atlanta, GA: The High Museum of Art. doi:10.1145/2069618.2069728

Garvey, G. P. (2002). The Split Brain Human Computer User Interface. *Leonardo*, *35*(3), 319–325. doi:10.1162/002409402760105352

Garvey, G. P. (2005). *Homage to Square: Suprematist Composition*. Flash Animation. Artist Collection.

Gatesy, S. M., & Dial, K. P. (1993). Tail muscle activity patterns in walking and flying pigeons (Columba livia). *The Journal of Experimental Biology*, *176*, 55–76.

Gatty, M. (1872). British sea-weeds. London, UK: Bella dn Daldy.

Gazzaniga, M. S. (1998, June). The Split Brain Revisited. *Scientific American*. Retrieved July 20, 2015, from http://www.scientificamerican.com/article/the-split-brain-revisited

Gazzaniga, M. S. (1982). Split brain research: A personal history. *Cornell University Alumni Quarterly.*, *45*, 2–12.

Gazzaniga, M. S. (1985). *The Social Brain. Discovering the Networks of the Mind*. New York, NY: Basic Books, Inc.

Gazzaniga, M. S. (2005). Forty-five years of split-brain research and still going strong. *Nature Reviews. Neuroscience*, *6*(8), 653–659. doi:10.1038/nrn1723 PMID:16062172

Gazzaniga, M. S. (2008). Spheres of Influence. *Scientific American*, *19*(3), 32–39. Retrieved from http://issuu.com/samuelsantos52/docs/scientific_american_mind_vol_19__3_ PMID:18783046

Gazzaniga, M. S., Bogen, J. E., & Sperry, R. W. (1962). Some functional effects of sectioning the cerebral commissures in man. *Proceedings of the National Academy of Sciences of the United States of America*, *48*(10), 1765–1769. doi:10.1073/pnas.48.10.1765 PMID:13946939

Gelsthorpe, D. (2013). *Margaret Gatty's algal herbarium in St Andrews*. Retrieved from https://naturalsciencecollections.wordpress.com/2013/08/27/margaret-gattys-algal-herbarium-in-st-andrews/

Geschwind, D. H., & Crabtree, E. (2003). Handedness and Cerebral Laterality. In H. Whitaker (Ed.), *Concise Encyclopedia of Brain and Language* (pp. 221–230). Oxford, UK: Elsevier Science.

Gibson, C. (no date). *Florence Cathedral's Dome, Italy*. Retrieved February 22, 2015, from http://www.brunelleschisdome.com/

Gibson, S. (2015). *Animal, vegetable, mineral*. New York: Oxford University Press.

Gillies, R. M., Nichols, K., & Burgh, G. (2011). Promoting problem-solving and reasoning during cooperative inquiry science. *Teaching Education*, *22*(4), 427–443. doi:10.1080/10476210.2011.610448

Gilot, F., & Lake, C. (1964). *Life with Picasso*. New York, NY: McGraw-Hill.

Giocomo, L. M., Moser, M.-B., & Moser, E. I. (2011). Computational models of grid cells. *Neuron*, *71*(4), 589–603. doi:10.1016/j.neuron.2011.07.023 PMID:21867877

Glattke, T. J., & Kujawa, S. G. (1991). Otoacoustic Emissions. *American Journal of Audiology*, *1*(1), 29–40. doi:10.1044/1059-0889.0101.29 PMID:26659426

Gleizes, A. (1912). Du cubisme. Eugène Figuière, Éditeur. University of California Libraries.

Gödel, K. F. (1949). *A remark on the relationship between relativity theory and idealistic philosophy.*In P. Schilpp (Ed.), *Library of Living Philosophers* (Vol. 7, pp. 555–562). Open Court.

Goldman, F. (2016). Artist statement. In 2016 Joint Mathematics Meetings Exhibition of Mathematical Art. Tesselation Publishing.

Goldman, F. (2014). *Geometric Origami*. Thunder Bay Press.

Gomez-Nicola, D., Riecken, K., Fehse, B., & Perry, V. H. (2014). *In vivo* RGB marking and muliticolour single-cell tracking in the adult brain. *Scientific Reports*, *4*, 7520. doi:10.1038/srep07520 PMID:25531807

Gonzalez, H. B., & Kuenzi, J. J. (2012). *Science, Technology, Engineering, and Mathematics (STEM) Education: A Primer*. Congressional Research Service. Retrieved July 5, 2014, from http://fas.org/sgp/crs/misc/R42642.pdf

Goodenough, U. (1998). *The sacred depths of nature*. New York: Oxford University Press.

Gray, H. (1918). *Anatomy of the Human Body*. Philadelphia, PA: Lea and Febiger. Retrieved December 2, 2015, from https://commons.wikimedia.org/wiki/File:Gray722.png

Gray, L. (n.d.). Auditory System: Pathways and Reflexes. In *Neuroscience Online, the Open-Access Neuroscience Electronic Textbook*. The University of Texas Health Science Center at Houston (UTHealth). Retrieved September 6, 2015, from http://neuroscience.uth.tmc.edu/

Greville, R. K. (1830). *Algae Britannicae*. Edinburgh, UK: Maclachlan and Stewart.

Grujović, N., Radović, M., Kanjevac, V., Borota, J., Grujović, G., & Divac, D. (2011). 3D printing technology in education environment. *34th International Conference on Production Engineering* (pp. 29-30). Academic Press.

Guilford, J. P. (1950). Creativity. *The American Psychologist*, *5*(9), 444–454. doi:10.1037/h0063487 PMID:14771441

Guilford, J. P. (1959). Traits of creativity. In *Creativity and its cultivation*. New York: Harper and Row.

Guilford, J. P. (1967). *The nature of human intelligence*. McGraw-Hill.

Guilford, J. P. (1968). *Intelligence, creativity and their educational implications*. San Diego, CA: Robert Knapp, Publ.

Gunn, M., & Kraemer, E. W. (2011). The agile teaching library: Models for integrating information literacy in online learning experiences. In S. Kelsey & K. St. Amant (Eds.), Computer-mediated communication: Issues and approaches in education, (pp. 191-206). Hershey, PA: Information Science Reference.

Gunning, T. (2004). What's the point of an index? or, faking photographs. *Nordicom Review, 1-2,* 39-49.

Gurkewitz, R., & Arnstein, B. (2016). Artist statement in Fathauer, R., & Selikoff, N. (Editors*), 2016 Joint Mathematics Meetings Exhibition of Mathematical Art*. Tesselation Publishing.

Gutierrez, A. (1987). Dynamic Geometry: Euclidean Egg with 8 Arcs, Step-by-Step Construction. HTML5 Animation for Tablets (iPad, Nexus). *Geometry for Kids, School, Mathematics Education*. Retrieved February 23, 2015, from http://www.gogeometry.com/geogebra/euclidean-egg-8-arcs-step-construction-html5-animation-ipad-tablet.html

Habitat Earth. (2015). Retrieved August 27, 2015, from https://www.calacademy.org/exhibits/habitat-earth

Haeckel, E. (1904). Art forms in nature. New York: Dover.

Hall, D., & Fifer, S. J. (1990). Illuminating video: An essential guide to video art. New York, NY: Aperture in association with the Bay Area Video Coalition.

Harshbarger, E. (2010). Inception's Penrose Staircase. *Wired Magazine*. Retrieved September 30, 2015, from http://www.wired.com/2010/08/the-never-ending-stories-inceptions-penrose-staircase/

Hart, G. H. (1999). *Virtual Polyhedra*. Retrieved March 1, 2015, from http://www.georgehart.com/virtual-polyhedra/durer.html

Harvey, W. H. (1849). *A manual of British Algae*. London: van Voorst.

Hawkins, J. E. (n.d.). The Human Ear Anatomy. *Encyclopeadia Britannica*. Retrieved September 6, 2015, from http://www.britannica.com/science/ear

Hawkins, T. (1984). The Erlanger Programm of Felix Klein: Reflections on Its Place. In the History of Mathematics. *Historia Mathematica*, *11*(4), 442–470. doi:10.1016/0315-0860(84)90028-4

Heinlein, R. (1940). *And He Built A Crooked House*. Tor Books.

Heller, E. J. (2012). *Why You Hear What You Hear: An Experiential Approach to Sound, Music, and Psychoacoustics*. Princeton University Press.

Hemelrijk, C. K., & Hildenbrandt, H. (2011). Some causes of the variable shape of flocks of birds. *PLoS ONE*, *6*(8), e22479. doi:10.1371/journal.pone.0022479 PMID:21829627

Henderson, L. D. (1983). *The Fourth Dimension and Non-Euclidean Geometry in Modern Art*. The MIT Press.

Henry, L. H. (1981). The Economic Benefits of the Arts: A Neuropsychological Comment. *Journal of Cultural Economics*, *5*(1), 52–60. doi:10.1007/BF00189206

herbaria@home. (2015). Retrieved August 27, 2015, from http://herbariaunited.org/atHome/

Herbology Manchester. (2015). Retrieved August 27, 2015, from https://herbologymanchester.wordpress.com/

Herring, E. (1868). The Theory of Binocular Vision. Plenum.

Hervey, A. B. (1881). *Sea mosses: A collector's guide and an introduction to the study of marine Algae*. Boston: S.E. Cassino. Retrieved from http://www.biodiversitylibrary.org/item/111603

Higgins, D. (1996). A book. In J. Rothenberg & D. Guss (Eds.), *the book, spiritual instrument* (pp. 102–104). New York, NY: Granary Books.

Hinton, C. H. (1888). *A New Era of Thought. Selected Writings of Charles H. Hinton.* Dover Publications.

Hirstein, W. (2005). *Brain Fiction: Self-deception and the Riddle of Confabulation.* Cambridge, MA: MIT Press.

Hoelzer, M. (2012). *SUN Chloroplast E-book*. Retrieved August 27, 2015, from http://www.markhoelzer.com/SUN-chlorophyllEbookWorking/index.html

Höffler, T. N., & Leutner, D. (2007). Instructional animation versus static pictures: A meta-analysis. *Learning and Instruction, 17*(6), 722–738. doi:10.1016/j.learninstruc.2007.09.013

Hofstadter, D. (1985). *Metamagical Themas.* New York: Basic Books.

Hollasch, S. R. (1991). *Four-Space Visualization of 4D Objects.* (Master of Science thesis). Arizona State University. Retrieved September 30, 2015, from http://steve.hollasch.net/thesis/

Hollins, M., & Hudnell, K. (1980). Adaptation of the binocular rivalry mechanism. *Investigative Ophthalmology & Visual Science, 19,* 1117–1120. PMID:7410003

Holmes, O. W. (1859, June). The Stereoscope and the Stereograph. Boston: *The Atlantic Monthly.* Retrieved September 12, 2015, from http://www.theatlantic.com/magazine/archive/1859/06/the-stereoscope-and-the-stereograph/303361/

Honey, M., Pearson, G., & Schweingruber, H. (Eds.). (2014). *National Academy of Engineering and National Research Council. In STEM Integration in K-12 Education: Status, Prospects, and an Agenda for Research.* Washington, DC: The National Academies Press.

Honey, M., Pearson, G., & Schweingruber, H. (Eds.). (2014). *STEM Integration in K-12 Education: Status, Prospects, and an Agenda for Research. National Academy of Engineering and National Research Council.* Washington, DC: The National Academies Press.

Hopkins, R. L. (1984, Summer). Education and the Right Hemisphere of the Brain: Thinking in Patterns for a Computer World. *Journal of Thought, 19*(2), 104–114.

Howard, I. P., & Rogers, B. J. (1995). *Binocular Vision and Stereopsis.* Oxford, UK: Oxford University Press.

Hsu, J. (2007). Innovative Technologies for Education and Learning: Education and Knowledge-oriented applications of blogs, Wikis, podcasts, and more. *International Journal of Information and Communication Technology Education, 3*(3), 70–89. doi:10.4018/jicte.2007070107

Huelsenbeck, R. (1981). Collective dada manifesto. In R. Motherwell (Ed.), The dada painters and poets: An anthology (2nd ed.; pp. 242-245). Cambridge, MA: The Belknap Press of Harvard University Press. (Reprinted and translated from Dada almanach, 1920)

Huffman, K. R. (n.d.). *The Electronic Arts Community Goes Online: A Personal View.* Retrieved September 30, 2015, from http://www.lehman.cuny.edu/vpadvance/artgallery/gallery/talkback/issue3/centerpieces/huffman.html

Hughes, L. (2014). *Digital Collections as Research Infrastructure.* Retrieved June 24, 2014, from http://www.educause.edu/ero/article/digital-collections-research-infrastructure

Hupé, J., & Rubin, N. (2003, March). The Dynamics of Bi-stable Alternation in Ambiguous Motion Displays: A Fresh Look at Plaids. *Vision Research, 43*(5), 531–548. doi:10.1016/S0042-6989(02)00593-X PMID:12594999

Hustinx, J. P., Kuipers, G., Morée, I., & Douma, T. (2015). *Netherlands: ECJ: Non-Fertilised Human Ovum Activated By Parthenogenesis Could Be Patentable.* De Brauw Blackstone Westbroek N.V. Retrieved March 3, 2015, from http://www.debrauw.com/newsletter/ecj-non-fertilised-human-ovum-activated-parthenogenesis-patentable/

Huvent, G. (2008). *Sangaku. Le mystère des énigmes géométriques japonaises.* Paris: Dunod.

HyperPhysics. (2014). *C.R. Nave, Georgia State University.* Retrieved January 21, 2015, from http://hyperphysics.phy-astr.gsu.edu/hbase/pend.html#c1

Illusions, V. (n.d.). Retrieved November 2, 2015, from http://faculty.washington.edu/chudler/chvision.html

iNaturalist. (2015). Retrieved August 27, 2015, from http://www.inaturalist.org/

Inhelder, B. (1977). Genetic epistemology and developmental psychology. *Annals of the New York Academy of Sciences*, *291*(1), 332–341. doi:10.1111/j.1749-6632.1977.tb53084.x

Inhelder, B., & Piaget, J. (1958). *The Growth of Logical Thinking from Childhood to Adolescence*. London: Routledge and Kegan Paul. doi:10.1037/10034-000

Irvin, D., & Koenning, T. (2015, January). Line-Narrowed Laser Module Enables Spin-Exchange Optical Pumping. *BioPhotonics*, 36-39.

Irwin, J. L., Opplinger, D. E., Pearce, J. M., & Anzalone, G. (2015). Evaluation of RepRap 3D Printer Workshops in K-12 STEM (Program/Curriculum Evaluation).*122nd ASEE Annual Conference & Exposition*. doi:10.18260/p.24033

Ivins, W. (1953). *Prints and visual communication*. Cambridge, MA: Harvard University Press.

Janis, H., & Blesh, R. (1967). *Collage: Personalities, concepts, techniques* (Rev. ed.). Philadelphia, PA: Chilton Book.

Jaynes, J. (1976, 2000). The Origin of Consciousness in the Breakdown of the Bicameral Mind. New York: Houghton Mifflin/Mariner Books.

Jensen, A. R. (1972, October12). Genetics and Education. A second look. *New Scientist*, 96–99.

Jepson Herbarium. (2015). Retrieved September 22, 2015, from http://ucjeps.berkeley.edu/jeps/

Johnson, C. I., & Mayer, R. E. (2009). A testing effect with multimedia learning. *Journal of Educational Psychology*, *101*(3), 621–629. doi:10.1037/a0015183

Johnson, E. A. (2014). *Ask the beasts: Darwin and the god of love*. New York, NY: Bloomsbury.

Julesz, B. (1971). *Foundations of Cyclopean Perception*. Cambridge, MA: MIT Press.

Kaku, M. (1995). *Hyperspace: A Scientific Odyssey Through Parallel Universes*. Anchor publisher.

Kalinka, A. T., & Tomancak, P. (2012). The evolution of early animal embryos: Conservation or divergence? *Trends in Ecology & Evolution*, *27*(7), 385–393. doi:10.1016/j.tree.2012.03.007 PMID:22520868

Kalyuga, S. (2005). Prior knowledge principle in multimedia learning. In R. E. Mayer (Ed.), *Cambridge handbook of multimedia learning* (pp. 325–337). Cambridge, UK: Cambridge University Press. doi:10.1017/CBO9780511816819.022

Kalyuga, S. (2011). Cognitive load theory: Implications for affective computing.*Proceedings of the Twenty-Fourth International Florida Intelligence Research Society Conference* (pp. 105-110).

Kalyuga, S. (2014). Managing cognitive load when teaching and learning e-skills.*Proceedings of the e-Skills for Knowledge Production and Innovation Conference 2014*.

Kandel, E. R., Schwartz, J. H., Jessell, T. M., Siegelbaum, S. A., & Hudspeth, A. J. (2012). *Principles of Neural Science* (5th ed.). McGraw-Hill Education / Medical.

Kapoora, A., Burlesonc, W., & Picard, R. W. (2007). Automatic prediction of frustration. *International Journal of Human-Computer Studies*, *65*(8), 724–736. doi:10.1016/j.ijhcs.2007.02.003

Karatas, F., Micklos, A., & Bodner, G. (2011). Sixth-grade students' views of the nature of engineering and images of engineers. *Journal of Science Education and Technology*, *20*(2), 123–135. doi:10.1007/s10956-010-9239-2

Keefe, T. (2003). Enhancing a face-to-face course with online lectures: Instructional and pedagogical issues. *Proceedings of the eighth annualMid-South Instructional Technology Conference*.

Keeney, E. (1992). *The botanizers: Amateur scientists in nineteenth-century America*. Chapel Hill, NC: University of North Carolina Press.

Kelly, M. (2015). *Of Microscopes and Monsters*. Retrieved August 27, 2015, from https://microscopesandmonsters.wordpress.com/

Kelly, M. (2012). The semiotics of slime: Visual representation of phytobenthos as an aid to understanding ecological status. *Freshwater Reviews*, *5*(2), 105–199. doi:10.1608/FRJ-5.2.511

Kemp Smith, N. (1933). *Kant, I. Critique of Pure Reason*. The Macmillan Press Ltd. Retrieved September 30, 2015, from https://archive.org/details/immanuelkantscri032379mbp

Kennedy, D. (2014). The role of investigations in promoting inquiry-based science education in Ireland. *Science Education International*, *24*(3), 282–305.

Keverne, E. B. (2015). Genomic imprinting, action, and interaction of maternal and fetal genomes. *PNAS*, *112*(22), 6834–6840. Retrieved October 25, 2015, from www.pnas.org/cgi/doi/10.1073/pnas.1411253111

Keystone View Company. (1879-1930). *Keystone View Company*. Retrieved September 18, 2015, from http://ark.digitalcommonwealth.org/ark:/50959/sq87dd56f

Khanna, D. R., & Yadav, P. R. (2005). *Biology of Birds*. Discovery Publishing House Pvt.Ltd.

Kimura, D. (1961a). Cerebral dominance and the perception of verbal stimuli. *Canadian Journal of Psychology*, *15*(3), 166–171. doi:10.1037/h0083219

Kimura, D. (1961b). Some effects of temporal-lobe damage on auditory perception. *Canadian Journal of Psychology*, *15*(3), 156–165. doi:10.1037/h0083218 PMID:13756014

Kinchin, I. M., Hay, D. B., & Adams, A. (2000). How a qualitative approach to concept map analysis can be used to aid learning by illustrating patterns of conceptual development. *Educational Research*, *42*(1), 43–57. doi:10.1080/001318800363908

Kingdon, J. (2011). In the eye of the beholder. In M. R. Canfield (Ed.), *Field notes on science and nature* (pp. 129–160). Washington, DC: Smithsonian Institution Press. doi:10.4159/harvard.9780674060845.c8

Kitterle, F. L., Christman, S., & Hellige, J. B. (1990). Hemispheric differences are found in the identification, but not the detection, of low versus high spatial frequencies. *Perception & Psychophysics*, *48*(4), 297–306. doi:10.3758/BF03206680 PMID:2243753

Klarreich, E. (2014). A Tenacious Explorer of Abstract Surfaces. *Quanta Magazine*. Retrieved from https://www.quantamagazine.org/20140812-a-tenacious-explorer-of-abstract-surfaces/

Knight, M., & Cunninham, C. (2004). *Draw an engineer test (DAET): Development of a tool to investigate students' ideas about engineers and engineering*. Paper presented as the annual American Society for Engineering Education Conference & Exposition, Salt Lake City, UT.

Koerner, L. (1996). Carl Linnaeus in his time and place. In N. Jardine, J. A. Secord, & E. C. Spary (Eds.), *Cultures of natural history* (pp. 145–162). Cambridge, UK: Cambridge University Press.

Kohler, T. (2012). Manifesto collage: Collage in arts and sciences. In C. Salm (Ed.), *Manifesto collage: Defining collage in the twenty-first century* (pp. 9–10). Nürnberg, Germany: Verlag für Moderne Kunst.

Kokcharov, I. (2015). *Hierarchy of Skills*. Retrieved August 13, 2015, from http://www.slideshare.net/igorkokcharov/kokcharov-skillpyramid2015

Kosara, R. (2007). Visualization criticism – The missing link between information visualization and art, *Proceedings of the 11th International Conference on Information Visualisation (IV)*, (pp. 631–636). doi:10.1109/IV.2007.130

Kroto, H. W., Health, J. R., O'Brien, S. C., Curl, R. F. & Smalley, R. E. (1985). C60: Buckminsterfullerene. *Nature, 318*(6042), 162. doi:10.1038/318162a0

Kuhn, T. (1970). *The structure of scientific revolutions* (2nd ed.). Chicago, IL: University of Chicago Press.

Lagrange, J.-L. (1811). *Mécanique Analytique*. Courcier.

Lahann, J. (2013). *Adaptive soft and biological materials. In Adaptive materials and structures: A workshop report*. Washington, DC: The National Academies Press. Retrieved May 12, 2014, from http://www.nap.edu/catalog.php?record_id=18296

Lambert, A. J. (1991). Interhemispheric interaction in the splitbrain. *Neuropsychologia, 29*(10), 941–948. doi:10.1016/0028-3932(91)90058-G PMID:1762673

Lamb, R., Akmal, T., & Petrie, K. (2015). Development of a Cognition-Priming Model Describing Learning in a STEM Classroom. *Journal of Research in Science Teaching, 52*(3), 410–437. doi:10.1002/tea.21200

Landow, G. (1999). Hypertext as collage-writing. In P. Lunenfeld (Ed.), *The digital dialectic: New essays on new media* (pp. 150–171). Cambridge, MA: MIT Press.

Langdan, C. M. (2005). *Cognitive theoretic Model of the Universe*. Retrieved February 27, 2015, from http://www.megafoundation.org/CTMU/Q&A/Archive.html#Chicken

Latour, B. (1990). Drawing things together. In M. Lynch & S. Woolgar (Eds.), *Representation in scientific practice* (pp. 19–68). Cambridge, MA: MIT Press.

Lazar, J., & Norcio, A. (2000). System and Training Design for End-User Error. In S. Clarke & B. Lehaney (Eds.), Human-Centered Methods in Information Systems: Current Research and Practice, (pp. 76-90). Hershey, PA: Idea Group Publishing. doi:10.4018/978-1-878289-64-3.ch005

Le Clerc, S. (1679). Discours Touchant de Point de Veue, dans lequel il es prouvé que les chose qu'on voit distinctement, ne sont veues que d'un oeil. In Destined for Distinguished Oblivion: The Scientific Vision of William Charles Wells (1757-1817). New York: Springer-Science+Business Media.

LeCompte, N., & Rush, J. C. (1981). The Jack Sprat Syndrome: Can Split-Brain Theory Improve Education by Including the Arts? *The Journal of Education, 163*(4), 335–343.

Leefmans, B. M.-P. (1983). Das undbild: A metaphysics of collage. In J. P. Plottel (Ed.), *Collage* (pp. 189–228). New York, NY: New York Literary Forum.

Lee, H., Plass, J. L., & Homer, B. D. (2006). Optimizing cognitive load for learning from computer based science simulations. *Journal of Educational Psychology, 98*(4), 902–913. doi:10.1037/0022-0663.98.4.902

Lehning, H. (1995). Learning mathematics with CAS.*Proceedings of the World Conference on Computers in Education VI*. London: Chapman & Hall. doi:10.1007/978-0-387-34844-5_78

Leighten, P. (1985). Picasso's collages and the threat of war, 1912-13. *The Art Bulletin, 67*(4), 653–672. doi:10.1080/00043079.1985.10788297

Lengler, R. (2007). How to induce the beholder to persuade himself: Learning from advertising research for information visualization. In *Proceedings of the 11th International Conference on Information Visualization*, (pp. 382-392). IEEE Computer Society Press. doi:10.1109/IV.2007.66

Lessem, D., & Sovak, J. (Illustrator) (2003). Scholastic Dinosaur A To Z. Scholastic Reference.

Levelt, W. (1965). *On Binocular Rivalry*. Soesterberg, The Netherlands: Institute for Perception RVO-TNO.

Levi, D. (2013). Linking assumptions in amblyopia. *Visual Neuroscience*, *30*(5-6), 277–287. doi:10.1017/S0952523813000023 PMID:23879956

Levy, E. K. (2014). Sleuthing the Mind: Curator's Introduction. *Leonardo*, *47*(5), 427–428. doi:10.1162/LEON_a_00864

Lewis, D., & Larsen, M. J. (1927). The Cancellation, Reinforcement, and Measurement of Subjective Tones. *Proceedings of N.A.S.*, *23*(7), 415–421. doi:10.1073/pnas.23.7.415 PMID:16588176

Li, H.-L. (2016). Artist statement in Fathauer, R., & Selikoff, N. (Eds.), 2016 Joint Mathematics Meetings Exhibition of Mathematical Art. Tesselation Publishing.

Liben, L. S., Kastens, K. A., & Christensen, A. E. (2011). Spatial foundations of science education: The illustrative case of instruction on introductory geological concepts. *Cognition and Instruction*, *29*(1), 45–87. doi:10.1080/07370008.2010.533596

Lifemapper. (2015). Retrieved from http://lifemapper.org/

Lima, M. (2014). *The book of trees: Visualizing the branches of knowledge*. New York: Princeton Architectural Press.

Lippard, L. (1995). *The pink glass swan: Selected essays on feminist art*. New York: New Press.

Li, Z., Roussakis, E., Koolen, P. G. L., Ibrahim, A. M. S., Kim, K., Rose, L. F., & Evans, C. L. et al. (2014). Non-invasive transdermal two-dimensional mapping of cutaneous oxygenation with a rapid-drying liquid bandage. *Biomedical Optics Express*, *5*(11), 11. doi:10.1364/BOE.5.003748 PMID:25426308

Lodewyk, K. R., & Winne, P. H. (2005). Relations among the structure of learning tasks, achievement, and changes in self-efficacy in secondary students. *Journal of Educational Psychology*, *97*(1), 3–12. doi:10.1037/0022-0663.97.1.3

Logothetis, N. K. (1998). Single units and conscious vision. *Philosophical Transactions of the Royal Society of London. Series B, Biological Sciences*, *353*(1377), 1801–1818. doi:10.1098/rstb.1998.0333 PMID:9854253

Logothetis, N. K., Pauls, J., Augath, M., Trinath, T., & Oeltermann, A. (2001). Neurophysiological investigation of the basis of the fMRI signal. *Nature*, *412*(6843), 150–157. doi:10.1038/35084005 PMID:11449264

Loizides, A. (2012). *Andreas Loizides research home page*. Retrieved August 1, 2015, from http://www0.cs.ucl.ac.uk/staff/a.loizides/research.html

Loizides, A., & Slater, M. (2001). The empathic visualisation algorithm (EVA): Chernoff faces revisited. *Technical Sketch SIGGRAPH*, *2001*, 179.

Loizides, A., & Slater, M. (2002). The empathic visualisation algorithm (EVA) - An automatic mapping from abstract data to naturalistic visual structure.*Proceedings of 6th International Conference on Information Visualisation* (pp. 705-712). Los Alamitos, CA: IEEE. doi:10.1109/IV.2002.1028852

Lorenz, K. (1937). Imprinting. *The Auk*, *54*(3), 245–273. doi:10.2307/4078077

Lyons, J., & Thompson, S. (2006). *Investigating the long-term impact of an engineering-base GK-12 program on students' perceptions of engineering*. Paper presented at the annual American Society for Engineering Education Conference and Exposition, Chicago, IL.

MacLean, P. D. (1990). The Triune Brain in Evolution: Role in Paleocerebral Functions. Springer.

Macroalgal Herbarium Portal. (2015). Retrieved August 27, 2015, from http://macroalgae.org/portal/index.php

Maeda, J. (2012). *STEM to STEAM: Art in K-12 is Key to Building a Strong Economy. Edutopia*. Retrieved July 5, 2014, from http://www.edutopia.org/blog/stem-to-steam-strengthens-economy-john-maeda

Magic Eye Random Dot Stereograms. (n.d.). Retrieved September 17, 2015 from http://www.magiceye.com/

Majerich, D. M., & Schmuckler, J. S. (2008). *Compendium of science demonstration-related research from 1918 to 2008*. Xlibris.

Makela, M. (1996). *The Photomontages of Hannah Höch*. Minneapolis, MN: Walker Art Center.

Manovich, L. (2002). *The language of new media*. Cambridge, MA: MIT Press.

Maoldomhnaigh, M., & Hunt, A. (1988). Some factors affecting the image of the scientist drawn by older primary school pupils. *Research in Science & Technological Education*, *6*(2), 159–166. doi:10.1080/0263514880060206

Marr, D. (1982/2010). *Vision: A Computational Investigation into the Human Representation and Processing of Visual Information*. MIT Press.

Martinovic, D., McDougall, D., & Karadag, Z. (2012). *Technology in Mathematics Education: Contemporary Issues*. Informing Science Press.

Maslow, A. H. (1943). A theory of human motivation. *Psychological Review*, *50*(4), 370–396. doi:10.1037/h0054346

Mathews, M. M., & Moore, F. R. (1970). GROOVE – a program to compose, store, and edit functions of time. *Communications of the ACM*, *13*(12), 715–721. doi:10.1145/362814.362817

Maya Mathematical System . (2015). Yucatan's Maya calendar. Retrieved August 10, 2015, from http://www.mayacalendar.com/f-mayamath.html

Mayer, R. E. (1997). Multimedia learning: Are we asking the right questions? *Educational Psychologist*, *32*(1), 1–19. doi:10.1207/s15326985ep3201_1

Mayer, R. E. (1999). The promise of educational psychology: Vol. 1. *Learning in the content areas*. Upper Saddle River, NJ: Prentice Hall.

Mayer, R. E. (2001). *Multimedia learning*. New York: Cambridge University Press. doi:10.1017/CBO9781139164603

Mayer, R. E. (2002). The promise of educational psychology: Vol. 2. *Teaching for meaningful learning*. Upper Saddle River, NJ: Prentice Hall.

Mayer, R. E., & Moreno, R. (1998). A split-attention effect in multimedia learning: Evidence for dual processing systems in working memory. *Journal of Educational Psychology*, *90*(2), 312–320. doi:10.1037/0022-0663.90.2.312

Mayer, R. E., & Moreno, R. (2003). Nine ways to reduce cognitive load in multimedia learning. *Educational Psychologist*, *38*(1), 43–52. doi:10.1207/S15326985EP3801_6

Mayer, R. E., Moreno, R., Boire, M., & Vagge, S. (1999). Maximizing constructivist learning from multimedia communications by minimizing cognitive load. *Journal of Educational Psychology*, *91*(4), 638–643. doi:10.1037/0022-0663.91.4.638

Mayer, R. E., Steinhoff, K., Bower, G., & Mars, R. (1995). A generative theory of textbook design: Using annotated illustrations to foster meaningful learning of science text. *Educational Technology Research and Development*, *43*(1), 31–43. doi:10.1007/BF02300480

Mayer, R., Heiser, H., & Lonn, S. (2001). Cognitive constraints on multimedia learning: When presenting more material results in less understanding. *Journal of Educational Psychology*, *93*(1), 187–198. doi:10.1037/0022-0663.93.1.187

McCarthy, R. A., & Warrington, E. K. (1990). *Cognitive Neuropsychology: A Clinical Introduction.* New York: Academic Press.

McHale, B. (1991). *Postmodernist fiction.* London, UK: Routledge.

McLuhan, M. H. (1979). Figure and Grounds in Linguistic Criticism. Interpretation of Narrative by Mario J. Valdés, Owen J. Miller. *A Review of General Semantics, 36*(3), 289-294.

McNamara, D. S., & Kintsch, W. (1996). Learning from texts: Effects of prior knowledge and text coherence. *Discourse Processes*, *22*(3), 247–288. doi:10.1080/01638539609544975

McNeil. B.1., & Nelson, K.R. (1991). Meta-analysis of interactive video instruction: A 10 year review of achievement effects. *Journal of Computer-Based Instruction*, *18*(1), 1–6.

Mead, M., & Metraux, R. (1957). Image of the scientist among high school students. *Science*, *126*(3270), 384–390. doi:10.1126/science.126.3270.384 PMID:17774477

Mediaville, C. (1996). *Calligraphy: From Calligraphy to Abstract Painting.* Belgium: Scirpus Publications.

Meenes, M. (1930). A phenomenological description of retinal rivalry. *The American Journal of Psychology*, *42*(2), 260–269. doi:10.2307/1415275

Mehta, R., & Russell, E. (2009). Effects of pretraining on acquisition of novel configural discriminations in human predictive learning. *Learning & Behavior*, *37*(4), 311–324. doi:10.3758/LB.37.4.311 PMID:19815928

Meier, A. (2014). *From ocean to ornament, the Most Extraordinary Victorian Seaweed Scrapbook.* Retrieved November 11, 2014, from http://www.atlasobscura.com/articles/objects-of-intrigue-seaweed-scrapbook

Mercuri, R., & Meredith, K. (2014). An educational venture into 3D Printing.*Integrated STEM Education Conference (ISEC)*. IEEE.

Merlot. (2015). Retrieved August 27, 2015, from https://www.merlot.org/merlot/index.htm

Merlot, E., Couret, D., & Otten, W. (2008). Prenatal stress, fetal imprinting, and immunity. *Brain, Behavior, and Immunity*, *22*(1), 42–51. doi:10.1016/j.bbi.2007.05.007 PMID:17716859

Merriam-Webster Dictionary and Thesaurus. (2015). *An Encyclopedia Britannica Company.* Retrieved August 8, 2015, from http://www.merriam-webster.com/

Merrifield, M. P. (1860). *A sketch of the natural history of Brighton.* Brighton, UK: Pierce.

Miles, B. (2012). The future leaks out: A very magical and highly charged interlude. In C. Fallows & S. Genzmer (Eds.), *Cut-ups, cut-ins, cut-outs: The art of William S. Burroughs* (pp. 22–31). Nürnberg, Germany: Verlag für Moderne Kunst.

Miller, S., Geng, Y., Zheng, R., & Dewald, A. (2012). Presentation of complex medical information: Interaction between concept maps and spatial ability on deep learning. *International Journal of Cyber Behavior, Psychology and Learning*, *2*(1), 42–53. doi:10.4018/ijcbpl.2012010104

Mitry, J. (1997). *The aesthetics and psychology of the cinema.* Bloomington, IN: Indiana University Press.

Mohr, B., Pulvermuller, F., Rayman, J., & Zaidel, E. (1994). Interhemispheric cooperation during lexical processing is mediated by the corpus-callosum: Evidence from the split-brain. *Neuroscience Letters*, *181*(1-2), 17–21. doi:10.1016/0304-3940(94)90550-9 PMID:7898762

Moore, A., & Malinowski, P. (2009). Meditation, mindfulness and cognitive flexibility. *Consciousness and Cognition, 18*(1), 176–186. doi:10.1016/j.concog.2008.12.008 PMID:19181542

Moray, N. (1959). Attention in dichotic listening: Affective cues and the influence of instructions. *The Quarterly Journal of Experimental Psychology, 11*(1), 56–60. doi:10.1080/17470215908416289

Moreno, R., & Mayer, R. E. (1999). Cognitive principles of multimedia learning: The role of modality and contiguity. *Journal of Educational Psychology, 91*(2), 358–368. doi:10.1037/0022-0663.91.2.358

Moreno, R., & Mayer, R. E. (2000). A Learner Centred Approach to Multimedia Explanations: Deriving instructional design principles from cognitive theory. *Interactive Multimedia Electronic Journal of Computer-Enhanced Learning, 2*(2), 12–20.

Moser, E. I., Moser, M.-B., & Roudi, Y. (2014). Network mechanisms of grid cells. *Phil. Trans. R. Soc. B, 369*(1635), 20120511. doi:10.1098/rstb.2012.0511 PMID:24366126

Mulcahy, K. (2012). *Top 10 Fascinating Eggs*. Listverse. Retrieved March 3, 2015, from Lehttp://listverse.com/2012/03/01/top-10-fascinating-eggs/

Murr, L. E., & Williams, J. B. (1988). Half-Brained Ideas about Education: Thinking and Learning with Both the Left and Right Brain in a Visual Culture. *Leonardo, 21*(4), 413–419. doi:10.2307/1578704

Murzi, M. (2015). *Poincare, J. H.* Internet Encyclopedia of Philosophy. Retrieved September 20, 2015, from http://www.iep.utm.edu/poincare/

Nanowerk. (2015a). *Introduction to nanotechnology*. Retrieved August 10, 2015, from http://www.nanowerk.com/nanotechnology/introduction/introduction_to_nanotechnology_1.php#ixzz3iXtNEYyL

Nanowerk. (2015b). *Nantechnology Spotlights*. Retrieved August 11, 2015, from http://www.nanowerk.com/#ixzz3iZH91GZ7

Nash, K. S. (2015, October 14). The Next Security Frontier: The Human Body. *The Wall Street Journal*.

Nasim, O. W. (2013). *Observing by hand: Sketching the nebulae in the nineteenth century*. Chicago, IL: University of Chicago Press. doi:10.7208/chicago/9780226084404.001.0001

National Research Council (NRC). (2012). *A Framework for K-12 science education. Practices, Crosscutting Concepts, and core ideas*. Washington, DC: National Academy Press.

Nebes, R. (1972). Superiority of the minor hemisphere in commissurotomized man on a test of figural unification. *Brain, 95*(3), 633–638. doi:10.1093/brain/95.3.633 PMID:4655286

Nebes, R. (1973). Perception of spatial relationships by the right and left hemispheres of a commissurotomized man. *Neuropsychologia, 7*, 333–349. PMID:4792179

Necker, L. A. (1832). Observations on some remarkable optical phaenomena seen in Switzerland; and on an optical phaenomenon which occurs on viewing a figure of a crystal or geometrical solid. *London and Edinburgh Philosophical Magazine and Journal of Science, 1*(5), 329–337.

Neisser, U. (1966). Cognitive Psychology. New York: Appleton.

Newman SA. 2011. Animal egg as evolutionary innovation: a solution to the "embryonic hourglass" puzzle. *J. Exp. Zool. (Mol. Dev. Evol.), 316*(7), 467–483. Doi: 10.1002/jez.b.21417

Newton, D., & Newton, D. (1998). Primary children's perceptions of science and the scientist: Is the impact of a national curriculum breaking down the stereotype? *International Journal of Science Education, 20*(9), 1137–1149. doi:10.1080/0950069980200909

Newton, D., & Newton, L. (1992). Young children's perceptions of science and the scientist. *International Journal of Science Education, 14*(3), 331–348. doi:10.1080/0950069920140309

Nicolaïdis, S. (2008). Prenatal imprinting of postnatal specific appetites and feeding behavior. *Metabolism: Clinical and Experimental, 57*(Suppl 2), S22–S26. doi:10.1016/j.metabol.2008.07.004 PMID:18803961

Niederhauser, D. S., Reynolds, R. E., & Salmen, D. J. (2000). The Influence of Cognitive Load on Learning from Hypertext. *Journal of Educational Computing Research, 23*(3), 237–255. doi:10.2190/81BG-RPDJ-9FA0-Q7PA

Nielsen, J. A., Zielinski, B. A., Ferguson, M. A., Lainhart, J. E., & Anderson, J. S. (2013). An evaluation of the left-brain vs. right brain hypothesis with resting state functional connectivity magnetic resonance imaging. *PLOS One*. Retrieved September 17, 2015, from http://www.plosone.org/article/info%3Adoi%2F10.1371%2Fjournal.pone.0071275

No author. (1972, July 29). Western Union Tuning Out Singing Telegram. *The New York Times*, p. 27.

No author. (2008). Inspired by Biology: From Molecules to Materials to Machines (Free Executive Summary). *Committee on Biomolecular Materials and Processes, National Research Council.* Retrieved August 10, 2015, from http://www.nap.edu/catalog/12159.html

No author. (2012). *Porous Science: How Does a Developing Chick Breathe Inside Its Egg Shell?* Scientific American, Science Buddies. Retrieved March 1, 2015, from http://www.scientificamerican.com/article/bring-science-home-chick-breathe-inside-shell/

No author. (2015). *Make your Mark.* Ringling College of Art + Design.

No author. (2015). *SeaWorld Parks & Entertainment. Whale Shark, Cartilaginous Fish.* Retrieved February 27, 2015, from http://seaworld.org/animal-info/animal-bytes/cartilaginous-fish/whale-shark/

No author. (2015). Whale Shark – Cartillaginous Fish. *SeaWorld Parks & Entertainment.* Retrieved January 28, 2015, from http://seaworld.org/en/animal-info/animal-bytes/cartilaginous-fish/whale-shark/

No author. (2015, May-June). Squid skin. *BioPhotonics,* 16.

Nobuyuki Kayahara's Spinning Dancer Illusion. (n.d.). Retrieved September 19, 2015, from http://www.procreo.jp/labo/silhouette.swf

Norman, D. (1983). Design rules based on analyses of human error. *Communications of the ACM, 26*(4), 254–258. doi:10.1145/2163.358092

nothing2222229. (2008). *Red-Eared Slider Turtle Digging a Hole.* A video retrieved February 23, 2015, from https://www.youtube.com/watch?v=BCl8wXltv7E

Okamoto, N. (1996). *Japanese Ink Painting: The Art of Sumi-e.* Sterling.

Ong, S. C. (2005). *China condensed: 5000 years of history & culture.* Singapore: Marshall Cavendish.

Optical Illusion Pictures. (n.d.). Retrieved September 17, 2015 from http://brainden.com/optical-illusions.htm

Optical Illusions & Visual Phenomena 123 of them. (n.d.). Retrieved September 17, 2015 from http://michaelbach.de/ot/

O'Shea, R. P. (1999). *Translation of Dutour (1760).* Retrieved October 1, 2015, from https://sites.google.com/site/oshearobertp/publications/translations/dutour-1760

O'Shea, R. P. (2004, June). Psychophysics: Catching the Old Codger's Eye. *Current Biology*, *14*(12), R478–R479. doi:10.1016/j.cub.2004.06.014 PMID:15203021

Overduin, J. (2007). *Einstein's Space-time*. Retrieved September 20, 2015, from https://einstein.stanford.edu/SPACETIME/spacetime2.html

Oware, E., Capobianco, B., & Diefes-Dux, H. (2007). *Gifted students' perceptions of engineers? A study of students in a summer outreach program*. Paper presented as the annual American Society for Engineering Education Conference & Exposition, Honolulu, HI.

Packer, A. M., Russell, L. E., Dalgleish, H. P. W., & Häusser, M. (2015). Simultaneous all-optical manipulation and recording of neural circuit activity with cellular resolution *in vivo*. *Nature Methods*, *12*(2), 140–146. doi:10.1038/nmeth.3217 PMID:25532138

Paivio, A. (1986). *Mental representation: A dual coding approach*. Oxford, UK: Oxford University Press.

Paivio, A. (1986). *Mental representations: A dual coding approach*. Oxford, UK: Oxford University Press.

Palimpsest. In (1969). *The American Heritage Dictionary of the English Language*. New York, NY: American Heritage Publishing.

Panksepp, J., & Biven, L. (2012). The Archaeology of Mind: Neuroevolutionary Origins of Human Emotions (Norton Series on Interpersonal Neurobiology). W. W. Norton & Company.

Parker, M. (2014). *Things To Make and Do in the Fourth Dimension*. Farrar. *Straus and Giroux, LLC*.

Parker-Pope, T. (2008). *The Truth About the Spinning Dancer*. Retrieved October 1, 2015, from http://well.blogs.nytimes.com/2008/04/28/the-truth-about-the-spinning-dancer/?_r=0

Park, J. E. (2004). *Understanding 3D Animation Using Maya*. Springer.

Parncutt, R. (2011). *Harmony: A Psychoacoustical Approach*. Springer.

Parshall, N., & Millar, D. (2010). *The William Henry Harvey exsiccatae volumes*. Retrieved December 19, 2015, from http://www.aussiealgae.org/HarveyColl/manuscript.php

Partain, C. L., Price, R. R., Patton, J. A., Stephens, W. H., Price, A., Runge, V. M., . . . James, A. E., Jr. (1984). Nuclear Magnetic Resonance Imaging. The Radiological Society of North America. *RadioGraphics*, *4*, 5-25. Retrieved August 12, 2015, from issuehttp://pubs.rsna.org/doi/pdf/10.1148/radiographics.4.1.5

Paul, A. M. (2011). What babies learn before they are born. *TED Global 2011*. Retrieved October 25, 2015, from https://www.ted.com/talks/annie_murphy_paul_what_we_learn_before_we_re_born/transcript?language=en

PennState Modules. (2011). *Nano4Me.org. NACK educational resources*. The Pennsylvania State University. Retrieved May 30, 2014, from http://nano4me.live.subhub.com/categories/modules

Perkowitz, S. (2011). *Slow light invisibility, teleportation and other mysteries of light*. London, UK: Imperial College Press.

Phillpot, C. (1987). Some contemporary artists and their books. In J. Lyons (Ed.), *Artists' books: A critical anthology and sourcebook* (pp. 97–132). Rochester, NY: Visual Studies Workshop Press.

Phycological Research. (2015). Retrieved August 29, 2015, from http://ucjeps.berkeley.edu/CPD/algal_research.html

Piaget, J. (1970). *Science of education and the psychology of the child*. New York: Orion Press.

Piaget, J., & Inhelder, B. (1971). *Mental Imagery in the Child. A study of the development of imaginal representation* (P. A. Chilton, Trans.). New York: Basic Books Inc.

Pierce, J. R. (1970). *Telstar, A History*. SMEC Vintage Electrics.

Platoni, K. (2015). *We Have the Technology*. New York: Basic Books.

Pociask, F. D., & Morrison, G. R. (2008). Controlling split attention and redundancy in physical therapy instruction. *Educational Technology Research and Development*, *56*(4), 379–399. doi:10.1007/s11423-007-9062-5

Poffenberger, A. (1912). Reaction time to retinal stimulation with special reference to the time lost in conduction through nervous centers. *Archives de Psychologie*, *23*, 1–73.

Poggi, C. (1992). *In defiance of painting: Cubism, futurism, and the invention of collage*. New Haven, CT: Yale University Press.

Pogrebin, R. (2015, August 23). Sarah Sze aims for precise randomness in installing her gallery show. *The New York Times*, p. C1.

Polansky, A., Blake, R., Braun, J., & Heeger, D. (2000). Neuronal activity in human primary visual cortex correlates with perception during binocular rivalry. *Nature Neuroscience*, *3*(11), 1153–1159. doi:10.1038/80676 PMID:11036274

Pollock, E., Chandler, P., & Sweller, J. (2002). Assimilating complex information. *Learning and Instruction*, *12*(1), 61–86. doi:10.1016/S0959-4752(01)00016-0

Porta, J. B. (1593). *De Refractione. Optices Parte. Libri Novem.* Carlinum and Pacem, Naples Retrieved October 1, 2015, from http://testyourself.psychtests.com/testid/3178

Posner, G. J., Strike, K. A., Hewson, P. W., & Gerzog, W. A. (1982). Accomodation of a scientific conception: Towards a theory of conceptual change. *Science Education*, *66*(2), 211–227. doi:10.1002/sce.3730660207

Psychology Concepts. (n.d.). Retrieved October 1, 2015, from http://www.psychologyconcepts.com/depth-perception/

Quigg, C. (2003). *Envisioning Particles and Interactions*. Retrieved September 20, 2015, from http://boudin.fnal.gov/~quigg/JGV/EnvPFintro.html

Raaberg, G. (1998). Beyond fragmentation: Collage as feminist strategy in the arts. *Mosaic: A Journal for the Interdisciplinary Study of Literature, 31*(3), 153-171.

Rahn, H., Carey, C., Balmas, K., & Paganelli, C. (1977). Reduction of pore area of the avian eggshell as an adaptation to altitude (water vapor permeability). *Proceedings of the National Academy of Sciences of the United States of America*, *74*(7), 3095–3098. doi:10.1073/pnas.74.7.3095 PMID:16592423

Ramon y Cajal, S., & DeFelipe, J. (1988). Cajal on the Cerebral Cortex: An Annotated Translation of the Complete Writings. Oxford University Press.

Rapaczynski, W., & Ehrlichman, H. (1979). Opposite visual hemifield superiorities in face recognition as a function of cognitive style. *Neuropsychologia*, *17*(6), 645–652. doi:10.1016/0028-3932(79)90039-3 PMID:522978

Ravikumar, R., & Mohsin Khan, I. (2015). Design & Development of a 3D Printer. *Proceedings of 12th IRF International Conference*. Academic Press.

Reed, S. K. (2006). Cognitive architectures for multimedia learning. *Educational Psychologist*, *41*(2), 87–98. doi:10.1207/s15326985ep4102_2

Reitzel, E. (1984). *Le Cube Ouvert, structures and foundations*. International Conference on Tall Buildings. Singapore. World Heritage Encyclopedia. Retrieved September 20, 2015, from http://self.gutenberg.org/articles/grande_arche

Reusch, W. (2013). *Nuclear Magnetic Resonance Spectroscopy*. Retrieved August 11, 2015, from https://www2.chemistry.msu.edu/faculty/reusch/VirtTxtJml/Spectrpy/nmr/nmr1.htm#nmr1

Rivalry, F. (n.d.). Retrieved September 17, 2015 from http://www.psy.vanderbilt.edu/faculty/blake/rivalry/BR.html

Rivero, D., Dorado, J., Fernandez-Blanco, & Pazas, A. (2009). A Genetic Algorithm for ANN Design, Training and Simplification. In Bio-Inspired Systems: Computational and Ambient Intelligence: 10th International Work-Conference on Artificial Neural Networks, (pp. 391-398). Academic Press.

Robbe-Grillet, A. (1962). *Alain Robbe-Grillet's last year at Marienbad: a ciné-novel* (R. Howard, Trans.). London, UK: John Calder.

Rogers, M. (2013). *Researchers debunk myth of "right brain" and "left-brain" personality traits*. University of Utah, Office of Public Affairs. Retrieved September 17, 2015, from http://www.plosone.org/article/info%3Adoi%2F10.1371%2Fjournal.pone.0071275

Rohdie, S. (2006). *Montage*. Manchester, UK: Manchester University Press.

Rosenberg, H. (1989). Collage: Philosophy of put-togethers. In K. Hoffman (Ed.), *Collage: Critical views* (pp. 59–66). Ann Arbor, MI: UMI Research Press.

Roussel, R. (1995). *How I Wrote Certain of My Books* (T. Winkfield, Trans.). Boston, MA: Exact Change. (Original work published 1935)

Rubin, E. (2001). Readings in perception. In S. Yantis (Ed.), *Visual perception: Essentials readings*. Philadelphia: Psychology Press. (Original work published 1921)

Rubin, N. (2001). Figure and ground in the brain. *Nature Neuroscience*, *4*(9), 857–858. doi:10.1038/nn0901-857 PMID:11528408

Rucker, R. (2014). The Fourth Dimension: Toward a Geometry of Higher Reality. Mineola, NY: Dover Books. (Reprint, 1984 edition.)

Ryan, R. M., & Deci, E. L. (2002). An overview of self-determination theory: An organismic-dialectical perspective. In E. L. Deci & R. M. Ryan (Eds.), Handbook of self-determination research, (pp. 3-33). Rochester, NY: The University of Rochester Press.

Sandros, M. G., & Adar, F. (2015, January). Raman Spectroscopy and Microscopy: Solving Outstanding Problems in the Life Sciences. *BioPhotonics*, 40-43.

Savasci Acikalin, F. (2014). Use of instructional technologies in science classrooms: Teachers' perspectives. *Turkish Online Journal of Educational Technology*, *13*(2), 197–201.

Sax, L. (2009). *Boys adrift: the five factors driving the growing epidemic of unmotivated boys and underachieving young men*. Basic Books.

Schank, R., & Neaman, A. (2001). Motivation and failure in educational systems design. In K. Forbus & P. Feltovich (Eds.), *Smart Machines in Education*. Cambridge, MA: AAAI Press and MIT Press.

Schelly, C., Anzalone, G., Wijnen, B., & Pearce, J. M. (2015). Open-Source 3-D Printing Technologies for Education: Bringing Additive Manufacturing to the Classroom. *Journal of Visual Languages and Computing*, *28*, 226–237. doi:10.1016/j.jvlc.2015.01.004

Schibeci, R. A., & Sorenson, I. (1983). Elementary school children's perceptions of scientists. *School Science and Mathematics*, *83*(1), 14–19. doi:10.1111/j.1949-8594.1983.tb10087.x

Schnupp, J., Nelken, I., & King, A. (2011). *Auditory Neuroscience*. MIT Press.

Scholl, R. W. (2001). *Cognitive style and the Meyers-Briggs type inventory*. The University of Rhode Island, The Charles T. Schmidt, Jr. Labor Research Center. Retrieved August 9, 2015, from http://www.uri.edu/research/lrc/scholl/webnotes/Dispositions_Cognitive-Style.htm

Scholz, M., & Shah, M. (2015, January). Microspectoscopy Enables Real-Time Imaging of Singet Oxygen. *BioPhotonics*, 44-46.

Schouten, J. F. (1940). The residue and the mechanism of hearing. *Proceedings of the Koninklijke Akademie van Wetenschap*, *43*, 991–999.

Schouten, J. F., Ritsma, R. J., & Cardozo, B. L. (1962). Pitch of the residue. *The Journal of the Acoustical Society of America*, *34*(9B), 1418–1424. doi:10.1121/1.1918360

Schulz, I. (2010). Kurt Schwitters: Color and collage. In I. Schulz (Ed.), *Kurt Schwitters: Color and collage* (pp. 51–63). Houston, TX: Menil Collection.

Schunk, D. H. (1994). Self-regulation of self-efficacy and attributions in academic settings. In D. H. Schunk & B. J. Zimmerman (Eds.), *Self-regulation of learning and performance: Issues and educational applications* (pp. 75–100). Hillsdale, NJ: Erlbaum.

Schwebel, M., & Raph, J. (Eds.). (1973). Piaget in the Classroom. New York: Basic Books, Inc.

Scogin, S. C., & Stuessy, C. L. (2015). Encouraging greater student inquiry engagement in science through motivational support by online scientist-mentors. *Science Education*, *99*(2), 312–349. doi:10.1002/sce.21145

Seam. In (1969). *The American Heritage Dictionary of the English Language*. New York, NY: American Heritage Publishing.

Seaweed Collections Online. (2015). Retrieved August 29, 2015, from http://seaweeds.myspecies.info/

Secord, A. (2011). Pressed into service: specimens, space and seeing in botanical practice. In D. Livingstone & C. Withers (Eds.), *Geographies of nineteenth-century science* (pp. 283–310). Chicago, IL: University of Chicago Press. doi:10.7208/chicago/9780226487298.003.0012

Seitz, W. (1961). The art of assemblage: Catalogue of an exhibition held at the Museum of Modern Art, New York. New York, NY: Museum of Modern Art.

Severed Corpus Callosum. (2008, June 25). Scientific American Frontiers. Retrieved September 17, 2015, from https://www.youtube.com/watch?v=82tlVcq6E7A

Sewall, L. (1995). The skill of ecological perception. In T. Roszak, M. E. Gomes, & A. D. Kanner (Eds.), *Ecopsychology: Restoring the earth, healing the mind* (pp. 201–215). Berkeley, CA: Counterpoint Press.

Shattuck. (1992). The art of assemblage: A symposium (1961). In J. Elderfield (Ed.), *Studies in modern art, no. 2: Essays on assemblage* (pp. 118-159). New York, NY: Museum of Modern Art.

She, H. (1998). Gender and grade level differences in Taiwan students' stereotypes of science and scientists. *Research in Science & Technological Education*, *16*(2), 125–135. doi:10.1080/0263514980160203

Sheldon, K. M., & Gunz, A. (2009). Psychological needs as basic motives, not just experiential requirements. *Journal of Personality*, *77*(5), 1467–1492. doi:10.1111/j.1467-6494.2009.00589.x PMID:19678877

Shephard, K. (2003). Questioning, promoting and evaluating the use of streaming video to support student learning. *British Journal of Educational Technology*, *34*(3), 295–308. doi:10.1111/1467-8535.00328

Shneiderman, B. (2014). *Treemaps for space constrained visualization of hierarchies*. Retrieved August 10, 2015, from: http://www.cs.umd.edu/hcil/treemap-history/

Shneiderman, B. (1996). The Eyes Have It: A Task by Data Type Taxonomy for Information Visualizations. In *Proceedings of the IEEE Symposium on Visual Languages*. Washington, DC: IEEE Computer Society Press. doi:10.1109/VL.1996.545307

Shope, R. E., III. (2006). *The Ed3U science model: Teaching science for conceptual change*. Retrieved at http://theaste.org/publications/proceedings/2006proceedings/shope.html

Shorkey, C. T., & Crocker, S. B. (1981). Frustration theory: A source of unifying concepts for generalist practice. *Social Work*, *26*(5), 374–379.

Simanek, D. (n.d.). *How to View 3D Without Glasses*. Retrieved September 28, 2015, from http://www.lhup.edu/~dsimanek/3d/view3d.htm

Smith, E. E. (2015, July 27). One Head, Two Brains. *The Atlantic*. Retrieved August 24, 2015, from http://www.theatlantic.com/health/archive/2015/07/split-brain-research-sperry-gazzaniga/399290/

Snow, C. (1993). The two cultures (Canto ed.). London, UK: Cambridge University Press.

Sokolowska, D., de Meyere, J., & Folmer, E. (2014). Balancing the needs between training for future scientists and broader societal needs – secure project research on mathematics, science and technology curricula and their implementation. *Science Education International*, *25*(1), 40–51.

Song, J., Pak, S., & Jang, K. (1992). Attitudes of boys and girls in elementary and secondary schools towards science lessons and scientists. *Journal of the Koran Association for Research in Science Education*, *12*, 109–118.

Sorli, A., & Fiscaletti, D. (2012). Special theory of relativity in a three--dimensional Euclidean space. *Physics Essays*, *25*(1). Retrieved September 20, 2015, from http://physicsessays.org/browse-journal-2/category/28-issue-1-march-2012.html

Spary, E. C. (2004). Scientific symmetries. *History of Science*, *42*(1), 1–46. doi:10.1177/007327530404200101

Sperry, R. (1966). Brain bisection and the neurology of consciousness. In J. C. Eccles (Ed.), Brain and conscious experience, (pp. 298-313). New York: Springer-Verlag.

Sperry, R. (1968). Hemisphere Deconnection and Unity in Conscious Awareness. *The American Psychologist*, *23*(10), 723–733. doi:10.1037/h0026839 PMID:5682831

Spies, W. (1982). *Focus on art*. New York, NY: Rizzoli.

Spreading Waves of Dominance. (n.d.). Retrieved September 17, 2015 from http://www.psy.vanderbilt.edu/faculty/blake/rivalry/BR.html

STEM to STEAM. (2014). Retrieved July 5, 2014, from http://stemtosteam.org/

STEM to STEAM. (2014). Retrieved September 9, 2015, from http://stemtosteam.org/

Sternberg, R. (2011). *Cognitive Psychology* (6th ed.). Wadsworth Publishing.

Sternberg, R. J. (2007). *Wisdom, intelligence, and creativity synthesized.* New York: Cambridge University Press.

Sternberg, R. J., & Kaufman, S. B. (Eds.). (2011). *The Cambridge Handbook of Intelligence (Cambridge Handbooks in Psychology).* Cambridge University Press. doi:10.1017/CBO9780511977244

Stevens, S. S., & Warshofsky, F. (1981). *Sound and Hearing* (Revised Edition). Time Life Education.

Stewart, I. (2002). *The annotated Flatland. Basic Books.*

Stone, J. (2012). *Vision and Brain: How We Perceive the World.* Cambridge, MA: MIT Press.

Sweller, J. (1988). Cognitive Load during problem solving: Effects on learning. *Cognitive Science, 12*(2), 257–285. doi:10.1207/s15516709cog1202_4

Sweller, J., & Chandler, P. (1991). Evidence for cognitive load theory. *Cognition and Instruction, 8*(4), 351–362. doi:10.1207/s1532690xci0804_5

Sweller, J., & Chandler, P. (1994). Why some material is difficult to learn. *Cognition and Instruction, 12*(3), 185–233. doi:10.1207/s1532690xci1203_1

Sweller, J., van Merrienboer, J. J. G., & Paas, F. (1998). Cognitive architecture and instructional design. *Educational Psychology Review, 10*(3), 251–296. doi:10.1023/A:1022193728205

Tang, M., & Neber, H. (2008). Motivation and self-regulated science learning in high-achieving students: Differences related to nation, gender, and grade-level. *High Ability Studies, 19*(2), 103–116. doi:10.1080/13598130802503959

Taylor, B. (2004). *Collage: The making of modern art.* London, UK: Thames & Hudson.

Terhardt, E. (1974). Pitch, consonance, and harmony. *The Journal of the Acoustical Society of America, 55*(5), 1061–1069. doi:10.1121/1.1914648 PMID:4833699

Texier, P. J., Porraz, G., Parkington, J., Rigaud, J. P., Poggenpoel, C., Miller, C., . . . Verna, C. (2010). A Howiesons Poort tradition of engraving ostrich eggshell containers dated to 60,000 years ago at Diepkloof Rock Shelter, South Africa. *Proceedings of the National Acadademy of Science USA*. Retrieved January 28, 2015, from http://www.ncbi.nlm.nih.gov/pubmed/20194764?dopt

The blind spot. (n.d.). Retrieved September 17, 2015 from http://faculty.washington.edu/chudler/chvision.html

The Farlex Medical Dictionary. (2015). Retrieved August 20, 2015, from http://medical-dictionary.thefreedictionary.com/abstract+thinking

The S2 Eye Tracker. Eyeworks software. (n.d.). Retrieved September 25, 2015, from (http://www.mirametrix.com/products/?gclid=CN3Jl8Xz5sgCFZWRHwodaBEBdA)

The Tail Story Adams. R. A. (2014). *University of Northern Colorado Bat Research Lab Portal.* Retrieved January 25, 2015, from http://www.researchgate.net/profile/Rick_Adams2/publications

The Young Lady Versus Old Lady Optical Illusion. (n.d.). Retrieved September 22, 2015, from http://brainden.com/face-illusions.htm

Thomas, J., Colston, N., Ley, T., Ivey, T., Utley, J., DeVore-Wedding, B., & Hawley, L. (2016). *Developing a rubric to assess drawings of engineers at work.* Paper presented at the Annual Conference and Exposition of the American Society for Engineering Educators, New Orleans, LA.

Thomas, A. L. R. (1993). On the aerodynamics of bird tails. *Philosophical Transactions of the Royal Society of London. Series B, Biological Sciences, 340*(1294), 361–380. doi:10.1098/rstb.1993.0079

Thomsen, D. (1985). Shadow matter. *Science News*, *127*(19), 296. doi:10.2307/3969495

Todd, J. W. (1912). *Reaction to multiple stimuli*. New York: The Science Press. doi:10.1037/13053-000

Tong, F., Nakayama, K., Vaughan, J. T., & Kanwisher, N. (1998). Binocular rivalry and visual awareness in human extrastriate cortex. *Neuron*, *21*(4), 753–759. doi:10.1016/S0896-6273(00)80592-9 PMID:9808462

Treisman, A. (1964). Verbal cues, language and meaning in selective attention. *The American Journal of Psychology*, *77*(2), 206–209. doi:10.2307/1420127 PMID:14141474

Tri-Trophic Collection Network Portal. (2013). Retrieved August 29, 2015, from https://www.google.com/search?q=tritrophic+portal&ie=utf-8&oe=utf-8

Tsukruk, V. (2013). Learning from nature: Bioinspired materials and structures. In *Adaptive materials and structures: A workshop report*. Washington, DC: The National Academies Press. Retrieved May 12, 2014, from http://www.nap.edu/catalog.php?record_id=18296

Tufte, E. (1983). *The visual display of quantitative information*. Cheshire, CT: Graphics Press.

Tufte, E. (1990). *Envisioning Information*. Cheshire, CT: Graphics Press.

Tufte, E. R. (1990). *Envisioning information* (2nd ed.). Graphics Press.

Tufte, E. R. (1997). *Visual explanations: Images and quantities, evidence and narrative*. Graphics Press.

Ulmer, G. L. (1983). The object of post-criticism. In H. Foster (Ed.), *The Anti-aesthetic: Essays on postmodern culture* (pp. 83–110). Port Townsend, WA: Bay Press.

Um, E., Plass, J. L., Hayward, E. O., & Homer, B. D. (2012). Emotional Design in Multimedia Learning. *Journal of Educational Psychology*, *104*(2), 485–498. doi:10.1037/a0026609

UNC Media Channel. (2012). Retrieved January 25, 2015, from http://www.unco.edu/news

Unitary vs. Piecemeal Rivalry. (n.d.). Retrieved September 17, 2015 from http://www.psy.vanderbilt.edu/faculty/blake/rivalry/BR.html

United States. Congress. Senate. Committee on the Judiciary. (1993). Nomination of Judge Clarence Thomas to be Associate Justice of the Supreme Court of the United States: hearings before the Committee on the Judiciary, United States Senate, first session ... Washington: U.S. G.P.O. For sale by the U.S. G.P.O., Supt. of Docs. Unknown. *The Holmes stereoscope, with the inventions and improvements added by Joseph L. Bates*. Center for the History of Medicine: OnView. Retrieved October 4, 2015, from http://collections.countway.harvard.edu/onview/items/show/6277

Ursyn, A. (1997). Computer art graphics integration of art and science. *Learning and Instruction, the Journal of the European Association for Research on Learning and Instruction, 7*(1), 65-87.

Ursyn, A. (1997). Computer Art Graphics Integration of Art and Science. *Learning and Instruction. The Journal of the European Association for Research on Learning and Instruction*, *7*(1), 65–87. doi:10.1016/S0959-4752(96)00011-4

Ursyn, A. (2015). Cognitive Learning with Electronic Media and Social Networking. In A. Ursyn (Ed.), *Handbook of Research on Maximizing Cognitive Learning through Knowledge Visualization* (pp. 1–71). Hershey, PA: IGI Global. doi:10.4018/978-1-4666-8142-2.ch001

Ursyn, A., & Sung, R. (2007). Learning science with art.*Proceeding, SIGGRAPH '07 ACM/SIGGRAPH 2007 educators program.*

USDA. United States Department of Agriculture. (2013). *Food Safety and Inspection Service*. Retrieved March 1, 2015, from http://www.fsis.usda.gov/wps/portal/fsis/topics/food-safety-education/get-answers/food-safety-fact-sheets/egg-products-preparation/shell-eggs-from-farm-to-table/CT_Index

Uzwiak, A. (n.d.). *Vision*. Retrieved December 2, 2015, from http://www.rci.rutgers.edu/~uzwiak/AnatPhys/Sensory_Systems.html

V Factor | Natural History Museum. (2015). Retrieved August 27, 2015, from http://www.nhm.ac.uk/about-us/jobs-volunteering-internships/volunteering-interns-information/v-factor/index.html

Van De Bogart, W. (1970). Retrieved September 20, 2015, from http://www.earthportals.com/Portal_Messenger/synthesizermusic.html

Van Kleek, M. H. (1989). Hemispheric differences in global versus local processing of hierarchical visual stimuli by normal subjects: New data and a meta-analysis of previous studies. *Neuropsychologia*, *27*(9), 1165–1178. doi:10.1016/0028-3932(89)90099-7 PMID:2812299

Van Wagenen, W. P., & Herren, R. Y. (1940). Surgical division of commissural pathways in the corpus callosum relation to spread of an epileptic attack. *Archives of Neurology and Psychiatry*, *44*(4), 740–759. doi:10.1001/archneurpsyc.1940.02280100042004

Verhagen, P. W. (1994). Functions and Design of Video components in Multi-Media applications: a Review. In J. Schoenmaker & I. Stanchev (Eds.), *Principles and tools for instructional visualisation*. Enschede: Faculty of Educational Science and Technology, Anderson Consulting - ECC.

Veronikas, S., & Maushak, N. (2005). Effectiveness of Audio on Screen Captures in Software Application Instruction. *Journal of Educational Multimedia and Hypermedia*, *14*(2), 199–205.

Virtual Research Systems. (1998–2000). *Company Profile: Virtual Research Systems, Inc.* Retrieved September 4, 2015, from http://www.virtualresearch.com/company.html

Vogt, N. P., Cook, S. P., & Smith Muise, A. (2013). A New Resource for College Distance Education Astronomy Laboratory Exercises. *American Journal of Distance Education*, *27*(3), 189–200. doi:10.1080/08923647.2013.795365

Von Brücke, F. (1935). Über die Wirkung von Acetylcholine auf die Pilomotoren. *Klinische Wochenschrift*, *14*(1), 7–9. doi:10.1007/BF01778952

Vosniadou, S. (2007). Conceptual change and education. *Human Development*, *50*(1), 47–54. doi:10.1159/000097684

Wade, N. J. (2000). *A Natural History of Vision*. Cambridge, MA: MIT Press.

Wade, N. J., & Tgo, T. T. (2013). Early views on binocular rivalry. In S. M. Miller (Ed.), *The constitution of visual consciousness: Lessons from binocular rivalry, Edition: Advances in Consciousness Research* (Vol. 90, pp. 77–108). Philadelphia: John Benjamins Publishing Company. doi:10.1075/aicr.90.04wad

Waldman, D. (1992). *Collage, assemblage, and the found object*. New York, NY: Abrams.

Walker, P., & Powell, D. J. (1979). The sensitivity of binocular rivalry to changes in the nondominant stimulus. *Vision Research*, *19*(3), 247–249. doi:10.1016/0042-6989(79)90169-X PMID:442549

Wallace, D. F. (2004). *Everything and More: A Compact History of Infinity*. W. W. Norton & Company.

Wallach, H. (1935). Uber visuell wahrgenommene Bewegungsrichtung. *Perception*, *25*, 1319–1368.

Wandell, B. A. (n.d.). *Useful Numbers in Vision Science*. Retrieved September 4, 2015, from http://web.stanford.edu/group/vista/cgi-bin/wandell/useful-numbers-in-vision-science/

Wang, C. X., & Dwyer, F. M. (2006). Instructional effects of three concept mapping strategies in facilitating student achievement. *International Journal of Instructional Media*, *33*, 135–151.

Wang, P., Ma, T., Slipchenko, M. N., Liang, S., Hui, J., Shung, K. K., & Cheng, J.-X. et al. (2014). High-speed Intravascular Photoacoustic Imaging of Lipid-laden Atherosclerotic Plaque Enabled by a 2-kHz Barium Nitrite Raman Laser. *Scientific Reports*, *4*, 6889. doi:10.1038/srep06889 PMID:25366991

Wang, V. C. X. (Ed.). (2009). *Handbook of Research on E-Learning Applications for Career and Technical Education: Technologies for Vocational Training*. IGI Global. doi:10.4018/978-1-60566-739-3

Ward, M., Grinstein, G. G., & Keim, D. (2010). *Interactive data visualization: foundations, techniques, and applications*. Natick, MA: A K Peters Ltd.

Ward, A. J., Krause, J., & Sumpter, D. J. (2012). Quorum decision-making in foraging fish shoals. *PLoS ONE*, *7*(3), e32411. doi:10.1371/journal.pone.0032411 PMID:22412869

Warrick, D. R., Bundle, M. W., & Dial, K. P. (2002). Bird Maneuvering Flight: Blurred Bodies, Clear Heads1. *Integrative and Comparative Biology*, *42*(1), 141–148. doi:10.1093/icb/42.1.141 PMID:21708703

Weatherby, C. (2010). Study ranks salmon eggs as one of the three roes richest in omega-3s. *Vital Choice, Wild Seafood and Organics*. Retrieved March 1, 2015, from https://www.vitalchoice.com/shop/pc/articlesView.asp?id=948

Webber, A. L., & Wood, J. (2005). Amblyopia: Prevalence, Natural History, Functional Effects and Treatment. *Clinical & Experimental Optometry*, *88*(6), 365–375. doi:10.1111/j.1444-0938.2005.tb05102.x PMID:16329744

Weil, A. (1984). *Number Theory: An approach through history from Hammurapi to Legendre*. Boston: Birkhäuser.

Weissmann, D. H., & Banich, M. T. (1999). Global local inference modulated by communication between the hemispheres. *Journal of Experimental Psychology. General*, *128*(3), 283–308. doi:10.1037/0096-3445.128.3.283 PMID:10513397

Weissmann, D. H., & Banich, M. T. (2000). The cerebral hemispheres cooperate to perform complex but not simple tasks. *Neuropsychology*, *14*(1), 41–59. doi:10.1037/0894-4105.14.1.41 PMID:10674797

Wei, Z., Li, B., & Xu, C. (2009). Application of waste eggshell as low-cost solid catalyst for biodiesel production. *Bioresource Technology*, *100*(11), 2883–2885. doi:10.1016/j.biortech.2008.12.039 PMID:19201602

Wells, H. G. (1995). *The Time Machine*. Dover Publications.

Wescher, H. (1971). *Collage*. New York, NY: Abrams.

Wheatstone, C. (1838). Contributions to the physiology of vision – Part the first. On some remarkable, and hitherto unobserved, phenomena of binocular vision. *Philosophical Transactions of the Royal Society of London. Series B, Biological Sciences*, *128*, 371–394. Retrieved from https://www.stereoscopy.com/library/wheatstone-paper1838.html

When. (2015). *Beach*. Retrieved August 13, 2015, from http://us.when.com/wiki/Beach?s_chn=1&s_pt=aolsem&type=content&v_t=content

Wiesel, T. N., & Hubel, D. H. (1963). Effects of visual deprivation on morphology and physiology of cell in the cat's lateral geniculate body. *Journal of Neurophysiology*, *26*(6), 978–993. PMID:14084170

Wilde, O. (2001). *The Canterville Ghost*. Dover Publications.

Wilson, E. O. (2014). Life on Earth. Palo Alto, CA: iBooks.

Wittmann, B. (2013). Outlining species: Drawing as a research technique in contemporary biology. *Science in Context*, *26*(2), 363–391. doi:10.1017/S0269889713000094

Wittrock, M. C. (1989). Generative processes of comprehension. *Educational Psychologist*, *24*(4), 345–376. doi:10.1207/s15326985ep2404_2

Wolfram, E. (1975). *History of collage: An anthology of collage, assemblage and event structures*. New York, NY: Macmillan.

Wolman, D. (2012, March14). The split brain: A tale of two halves. *Nature*, *483*(7389), 260–263. doi:10.1038/483260a PMID:22422242

Wood, D., & Fels, J. (2008). *The natures of maps: Cartographic constructions of the natural world*. Chicago: University of Chicago Press.

Wyatt, M. (1834). *Algae Danmonienses*. Torquay: Cockrem.

Wyeld, T. G. (2015). Re-Visualising Giotto's 14th-Century Assisi Fresco "Exorcism of the Demons at Arezzo". In Handbook of Research on Maximizing Cognitive Learning through Knowledge Visualization (pp. 374-396). Hershey, PA: IGI Global.

Yap, C., Ebert, C., & Lyons, J. (2003). *Assessing students' perceptions of the engineering profession*. Paper presented as the South Carolina educators for the practical use of research annual conference, Columbia, SC.

Yeager, J. (2015). *12 Eggscellent Things You Can Do with Eggshells*. Retrieved February 23, 2015, from http://www.goodhousekeeping.com/home/green-living/reuse-eggshells-460809

yiddishwit. com. (2015). Retrieved March 3, 2015, from http://www.yiddishwit.com/gallery/eggs-smart.html

Yoo, S., Kokoszka, J., Zou, P., & Hsieh, J. S. (2009). Utilization of calcium carbonate particles from eggshell waste as coating pigments for ink-jet printing paper. *Bioresource Technology*, *100*(24), 6416–6421. doi:10.1016/j.biortech.2009.06.112 PMID:19665373

Zeki, S. (1993). Vision of the brain. Wiley-Blackwell.

Zeki, S. (2009). Splendors and miseries of the brain: Love, creativity, and the quest for human happiness. Wiley-Blackwell.

Zeki, S. (2011). *The Mona Lisa in 30 seconds*. Blog. Retrieved December 12, 2011, from http://profzeki.blogspot.com/

Zeki, S. (1998). Art and the Brain. *Journal of Conscious Studies: Controversies in Science and the Humanities*, *6*(6/7), 76–96.

Zeki, S. (2001). Artistic creativity and the brain. *Science*, *293*(5527), 51–52. doi:10.1126/science.1062331 PMID:11441167

Zheng, R. (2007). Cognitive functionality of multimedia in problem solving. In T. Kidd & H. Song (Eds.), *Handbook of Research on Instructional Systems and Technology* (pp. 230–246). Hershey, PA: Information Science. doi:10.4018/978-1-59904-865-9.ch017

Zheng, R. (2010). Effects of situated learning on students' knowledge gain: An individual differences perspective. *Journal of Educational Computing Research*, *43*(4), 463–483. doi:10.2190/EC.43.4.c

Zheng, R., & Cook, A. (2012). Solving complex problems: A convergent approach to cognitive load measurement. *British Journal of Educational Technology*, *43*(2), 233–246. doi:10.1111/j.1467-8535.2010.01169.x

Zheng, R., & Dahl, L. (2010). Using concept maps to enhance students' prior knowledge in complex learning. In H. Song & T. Kidd (Eds.), *Handbook of research on human performance and instructional technology* (pp. 163–181). Hershey, PA: IGI Global. doi:10.4018/978-1-60566-782-9.ch010

Zheng, R., McAlack, M., Wilmes, B., Kohler-Evans, P., & Williamson, J. (2009). Effects of multimedia on cognitive load, self-efficacy, and multiple rule-based problem solving. *British Journal of Educational Technology*, *40*(5), 790–803. doi:10.1111/j.1467-8535.2008.00859.x

Zheng, R., Miller, S., Snelbecker, G., & Cohen, I. (2006). Use of multimedia for problem-solving tasks. *Journal of Technology, Instruction. Cognition and Learning*, *3*(1-2), 135–143.

Zheng, R., Smith, D., Hill, J., Luptak, M., Hill, R., & Rupper, R. (in preparation). Exploring the roles of modality and crystalized knowledge in older adults' information processing. *University of Utah.*

Zheng, R., Udita, G., & Dewald, A. (2012). Does the format of pretraining matter? A study on the effects of different pretraining approaches on prior knowledge construction in an online environment. *International Journal of Cyber Behavior, Psychology and Learning*, *2*(2), 35–47. doi:10.4018/ijcbpl.2012040103

Zheng, R., Yang, W., Garcia, D., & McCadden, B. P. (2008). Effects of multimedia on schema induced analogical reasoning in science learning. *Journal of Computer Assisted Learning*, *24*(6), 474–482. doi:10.1111/j.1365-2729.2008.00282.x

Ziman, J. M. (1968). *Public knowledge*. Cambridge, UK: Cambridge University Press.

Zimmer, C. (2009, April 15). The Big Similarities & Quirky Differences Between Our Left and Right Brains. *Discover*. Retrieved October 1, 2015, from http://discovermagazine.com/2009/may/15-big-similarities-and-quirky-differences-between-our-left-and-right-brains

Zimmerman, B. J. (1998). Academic studying and the development of personal skill: A self-regulatory perspective. *Educational Psychology*, *33*(2-3), 73–86. doi:10.1080/00461520.1998.9653292

Zimmerman, B. J. (2001). Theories of self-regulated learning and academic achievement: An overview and analysis. In B. J. Zimmerman & D. H. Schunk (Eds.), *Self-regulated learning and academic achievement: Theoretical perspectives* (2nd ed.; pp. 1–38). Mahwah, NJ: Lawrence Erlbaum.

Zooniverse. (2015). Retrieved August 27, 2015, from https://www.zooniverse.org/#/

About the Contributors

Anna Ursyn, PhD, is a professor and Computer Graphics, Area Head at the School of Art and Design, University of Northern Colorado, USA. She combines programming with software and printmaking media, to unify computer generated and painted images, and mixed-media sculptures. Ursyn had over 30 single juried and invitational art shows, participated in over 100 juried and invitational fine art exhibitions, and published articles and artwork in books and journals. Research and pedagogy interests include integrated instruction in art, science, and computer art graphics. Since 1987, she serves as a Liaison, Organizing and Program Committee member of International IEEE Conferences on Information Visualization (iV) London, UK, and Computer Graphics, Imaging and Visualization Conferences (CGIV). From 1997 she serves as Chair of the Symposium and Digital Art Gallery D-ART iV. This is Anna's fifth book published with the IGI-Global. Website: Ursyn.com.

* * *

Mohammad Majid al-Rifaie is a researcher at Goldsmiths, University of London. His background is in computing and journalism and his artistic interests focuses on the interconnections between artificial intelligence, swarm intelligence, robotics and digital art. He has several publications in the field of swarm intelligence and biologically inspired algorithms (Stochastic Diffusion Search, Particle Swarm Optimisation, Genetic and Differential Evolution algorithms), analysing their performance and providing possible integration strategies. Many of Mohammad's projects on the aforementioned fields have been well received and sponsored by external entities (i.e. Luz by VIDA Fundación Telefónica competition of art and artificial intelligence 13th edition, Swarms Go Dutch by the American University of Paris and Mr Confused Robotic Household by Goldsmiths Annual Fund, University of London).

Jean Constant is past professor of Visual Communication, lecturer and researcher in mathematical visualization projects. He just completed a yearlong, 12 mathematical visualization program: 1 program a month, 1 image a day project assignment. He won a best use of mathematics in an artwork award in the 2015 Mathematics and Art conference at the Baltimore University, MD. He is also reviewer for the American Mathematical Society publications. Past and present affiliations include the Bridges Mathematics and Art organization, the Imaginary.org educational program of the MFO (Oberwolfach Research Institute for Mathematics), the D-Art, Information Visualization Society.

Robert C. Ehle, PhD is the Emeritus Professor of Music Theory and Composition, Electronic Music, and Acoustics, University of Northern Colorado, USA. He continues to serve on dissertation committees as an advisor to honors students. In Who's Who in American Music, Ehle is listed as a composer of electronic music. He received a Bachelor of Music degree from Eastman School of Music in Rochester, New York, and his Ph.D. from the University of North Texas in Denton, Texas in 1970. His interests pertain to occurrences and events associated with the experience of composing, playing, or listening to music. His research involved study on virtual music, experiments on the nature of pitch and psychoacoustics of resultant tones. He developed a bootstrap theory of sensory perception – the prenatal origins of musical emotion as the case for fetal imprinting. Among his compositions are A Sound Piece, A Space Symphony, The City is Beautiful and A Whole Earth Symphony. He received the Dallas Symphony Rockefeller Foundation Award in 1966. Ehle received a 2008-09 ASCAPLUS Award in the concert music division from the American Society of Composers, Authors and Publishers.

Michael Eisenberg received his doctorate in computer science at the Massachusetts Institute of Technology in 1991. In 1992 he joined the faculty of the University of Colorado, Boulder, where he is now Professor of Computer Science and a member of the Institute of Cognitive Science. His research interests focus on novel uses of technology in mathematics and science education, with a particular focus on designing creative, conceptually rich activities for children.

Donna Farland-Smith is an Associate Professor of Science Education in The School of Teaching & Learning at The Ohio State University. Her research focuses on students' perceptions and attitudes toward science and scientists as well as the characteristics of scientists that most positively affect the girl's perception of scientists. She has over a decade's experience in the classroom and previously taught science all grades K-12. Of particular interest to Dr. Farland-Smith is research that focuses on students' perceptions and attitudes toward science and scientists in both formal and informal settings. She was the director of the Side-by-Side with Scientists Camp at OSU-M, which served over 350 girls in the local community. Her research at the camp focused on middle school girls' attitudes and perceptions of scientists who are working 'side-by-side' with scientists as well as the characteristics of scientists that most positively affect the girl's perception of scientists.

Kevin D. Finson is a professor of science education and co-director of the Bradley University Center for STEM Education. He received his degrees from Kansas State University, including a B.S. in Science Education (1975), M.S. in Curriculum and Instruction with a focus on science (1980), and Ph.D. in Science Education (1985). He began his career teaching earth and physical sciences at the high school level, then moved into teaching earth, life, and physical sciences at middle school. His college teaching has included earth science and geology, physical science, energy curriculum, teaching pedagogy and methods, and instructional theory. He is very active in the Association for Science Teacher Education (ASTE). He has authored over seventy journal articles, five books, and several book chapters. Finson has been involved in research dealing with perceptions of scientists, drawings of scientists, conceptual change as revealed in drawings, and in visual data for over 25 years.

Maura C. Flannery is a Professor of Biology and Director of the Center for Teaching and Learning at St. John's University, New York. She holds bachelors and master's degrees in biology and a Ph.D. in science education. For 30 years she wrote an award-winning monthly column for *The American Biology*

Teacher resulting in two published collections: *Bitten by the Biology Bug* (selected for "Resources for Science Literacy" by the AAAS Project 2061) (1991) and *D'Arcy Thompson's Ice Cream and Other Essays from Biology Today* (2001). She has also published articles in such journals as *Leonardo, Perspectives in Biology and Medicine, Science Education, Plant Science Bulletin, Archives of Natural History,* and *Spontaneous Generations.* She is a Fellow the History and Philosophy Section of the AAAS, serves on the advisory board for the BioQUEST Curriculum Consortium, and was selected as a 2000-2001 Carnegie Scholar.

Gregory P. Garvey, MFA, MSVS, BS is the Chair of the Department of Visual and Performing Arts and also serves as Director of the Game Design and Development Program. In the digital games industry he has worked at Parker Brothers and Spinnaker Software. Previously at Quinnipiac University he was the Visiting Fellow in the Arts and also was an Associate Artist of the Digital Media Center for the Arts at Yale University. Prior to joining Quinnipiac University, he was Chair of the Department of Design Art at Concordia University in Montreal and was a member of the Board of Directors of the Montreal Design Institute. From 1983-85 he was a Fellow at the Center for Advanced Visual Studies at MIT.

Md. Fahimul Islam, was a Lecturer, Dept. of Computer Science & Technology, Faculty of Science & Technology, Atish Dipankar University of Science & Technology. He was the Coach for the ACM International Collegiate Programming Contest-Asia Dhaka Regional site for three consecutive years from 2010-2012. He is passionate about programming and system development. He completed his Bachelor in Computer Science & Engineering from Darul Ihsan University and Masters in Telecommunications Engineering from East West University. He has several publications in the Computerized Embedded System development, Robotics and Telecommunication field. He is currently working towards his Masters degree in Computer Science as a CUNY student in USA.

Hervé Lehning During the time of his studies (école normale supérieure, agrégation de mathématiques (France: PHD level) and a master in history), Hervé Lehning worked in the computing industry as an operator and then a computer analyst. It explains that his first works concerned the use of computers in mathematics, and cryptography. Then he taught mathematics and computer science in engineering schools (école d'ingénieurs de Tunis and école centrale de Paris), and in the CPGE system in Janson de Sailly (Paris). After having published several academic books with Masson, he joined *Tangente*, a French maths magazine and became its editor in chief. He has also been a contributor of *Pour La Science* (the French version of the *Scientific American*) and *La Recherche*. Hervé Lehning illustrates himself his books and articles with drawings and photographs, which leads him to exhibit his work. Website Hervé Lehning www.lehning.eu herve@lehning.eu

Dennis Summers In addition to degrees in the visual arts Summers also received a BA in chemistry, and has had a life-long interest in the sciences. Most of his artwork has been inspired by scientific concepts, and been about linking these ideas to ideas from other intellectual domains, and crafting a visual environment where this may be understood more or less intuitively depending on the background of the participant. The extensive preparatory research has been manifested as large scale multi-media installation art, book arts, conceptual art, interactive digital projects, and digitally created videos. He has also worked as a commercial digital media artist specializing in 3D animation since the mid-nineties, and was an associate professor for almost a decade. Some of his commercial work has included scientific and engineering visualization. For thirty years almost all of his artwork has been some form of collage. His artwork has been exhibited globally, and is in the collections of several major museums. He has had articles published in Leonardo, and regularly delivers presentations at the conferences of The Society for Literature, Science and the Arts (SLSA). His involvement with SLSA led to spending the last several years researching collage historically, aesthetically, and philosophically. This chapter contains excerpts from a much longer manuscript about collage.

Yiqing Wang is a faculty member teaching at Shanghai Normal University, P.R.China. She graduated with B.A. in English Education and received her M.A. in English language and literature in 2007, majoring in linguistics. She specializes in English education for non-English majors. She has published articles in *Shanghai Research on Education*, *Journal of Language and Literature Studies*, *Journal of Shanghai Normal University*. She has co-authored numerous books including *Special Training to Obtain High Scores for Reading Section, TEM8*, *A Set of Knack for College English Reading*, and *Guided Reading in English Literature*. She is also a First Prize winner for Excellent Teaching Award at Shanghai Normal University. She was a one-year visiting scholar at University of Missouri-Columbia.

Theodor Wyeld is currently a Lecturer in Digital Media at Flinders University, Australia. He is on a number of review committees and has published widely. Attracting more than 2 million in grants and venture capital, he is the inventor of 'thereitis.com'. The patented 'thereitis' technology is the culmination of 10 years of research in usability studies on how to better support the sorting of large collections of images. Between 2004 and 2009 he was a research associate with the Australasian Cooperative Research Centre for Interaction Design. He has an extensive background in architecture, planning and digital media production. He is completing a PhD in Cognitive Psychology at the Swinburne University of Technology, Australia.

Robert Zheng is a faculty member in the Department of Educational Psychology, University of Utah, USA. His research includes multimedia, cognition, visual learning, web-based design and implementation. He is the author of six edited books and has published more than fifty research papers, chapters and conference proceedings in the above areas. Some of his publications appear in high impact outlets including *British Journal of Educational Technology*, *Journal of Computer Assisted Learning*, *Educational Technology Research and Development*, and *Journal of Educational Computing Research*. He is also the founding editor of *International Journal of Cyber Behavior, Psychology and Learning*.

Index

E

F

G

H

I

K

L

M

O

P

R

S

T

U

V

W

CPSIA information can be obtained
at www.ICGtesting.com
Printed in the USA
BVOW10*0755140716
455433BV00001B/2/P

9 781522 504801